THE CARCASSES OF THE MAMMOTH AND RHINOCEROS FOUND IN THE FROZEN GROUND OF SIBERIA

BY

I. P. TOLMACHOFF

ARTICLE I
VOLUME XXIII

PHILADELPHIA:

THE AMERICAN PHILOSOPHICAL SOCIETY

104 South Fifth Street

1929

LANCASTER PRESS, INC.
LANCASTER, PA.

CONTENTS

PREFACE

THE present paper was primarily in the form of an address delivered, in 1921, by the writer before the Geological Department of the Tôhocu Imperial University of Sendaï, Japan. Never failing interest in this subject, perhaps even increased during the last few years, brought the writer to a decision to elaborate and complete the address referred to and publish it in a form of an article. The accomplishment of this task happened to be more difficult and took more time than the writer expected, and two more books on the same subject had appeared in meantime, before the writer was through with his work. These books are: "The Mammoth and Mammoth Hunting in Northeast Siberia" by Basset Digby, published in London, England, in 1926, and Pfizenmayer, E. W., "Mammutleichen und Urwaldmenschen in Nordost-Sibirien," published in Leipzig, Germany, also in 1926. The question was naturally aroused, if a new treatise on the same subject would be worth publication. After some consideration the writer has decided to finish and publish his work. If he was right, or wrong in doing so, it is to an eventual reader to decide.

The difficulties in accomplishing this paper were chiefly dependent upon the lack of special literature on this subject. Although the writer had the opportunity to use in Pittsburgh the Library of the Carnegie Museum and the Carnegie Public Library, in New York libraries of the American Museum of Natural History and of the American Geographical Society, in Washington, D. C. libraries of Congress and of the U. S. National Museum, in Chicago the Public Library, the John Crerar Library, and the Chicago University Library, he still lacked a great deal of important data, because in all these libraries Russian publications have been very incomplete and fragmentary. This paper, in the form in which it is now published, could be written only owing to the kind help of the Russian friends of the writer. Among them R. Th. Gekker and A. I. Tolmachoff delivered a number of different publications and data not easily available otherwise. The writer feels especially obliged to R. Th. Gekker, Curator of the Geological Museum of the Russian Academy of Sciences in St. Petersburg, who did a great deal of work in the libraries and archives of that city and supplied the writer with a number of quotations from different Russian publications which the writer lacked and which were impossible to secure from Russia. Through R. Th. Gekker the writer has received also from A. A. Byelinizki-Birula, Director of the Zoological Museum of the Russian Academy of Sciences, very important data on the fossil Siberian rhinoceros belonging to the collection of that museum. E. E. Ahnert delivered valuable information on the localities of the mammoth in the Russian Far East and in Northern Manchuria; also the literature on the subject, referring to these regions of Eastern Asia. These few lines are a weak expression of the gratitude of the writer to these gentlemen.

In spite of all this assistance the writer has been still unable to procure a number of publications which he would have liked to consult for his work. To get them, mostly from Russia, required so much time that the accomplishment of the paper would have been postponed indefinitely.   Very often the writer was forced to make reference to some works only through the other authors who had had the chance to use these works before.   Such a reference is always marked accordingly in the present paper.   Especially important in this direction was Howorth's book "The Mammoth and the Flood," in which an amazing amount of literary data has been brought together by that author.   In all these cases the reference to volumes, pages, etc., of the original work belongs to the auxiliary author. The writer feels necessary to emphasize that, because he had a chance to compare some quotations with an original work and discovered a few mistakes, which usually were not real errors, but dependent only upon the difference of edition of quoted works.

INTRODUCTION

In no other country of the world are the remnants of the mammoth and, to a lesser extent, of the diluvial rhinoceros and of other fossil mammals of the same geological age so familiar to everybody, as they are in Siberia. In no other country, except Siberia, have these remnants such an economic importance. Since time out of mind fossil ivory was used by natives, in their simple housekeeping, for very different purposes as a hard homogeneous material which could be worked about so easily as wood and in many cases could replace metals very conveniently. It is quite possible that the mammoth was hunted by the primitive man armed with spears and arrows made of ivory. Fossil ivory since the dawn of civilization has been used also for artistic purposes, for the making of small decorative objects often of great perfection and beauty, also for sculpture.[1] Good pictures of Yakutish ivory work are given by Pfizenmayer in his book. Bones have been used for different purposes as well. Of the ribs of the mammoth, for example, Yakuts used to make spoons.[2] Fossil bones were also much used in Russia for making animal charcoal. In the stores of bone-burning factories it was possible to fish out good specimens of bones of extinct animals, of course, without any reference to a locality.[3] Thin plates cut from the horns of the fossil rhinoceros are elastic in the highest degree. They were, therefore, much used by Yucaguirs of Northeastern Siberia to line their bows, and very eagerly sought for.[4] Spoons, forks, pipes, etc., are made of these horns as well.[5] Meat of the mammoth used to be not only devoured by dogs and wild animals, but also utilized by natives in their fox traps, chiefly because of its peculiar strong smell.[6]

The mammoth has been very common in the Siberian folklore. As a child the present writer was told that the mammoth had been such a large animal that Noah could not take it in his ark during the Deluge. Another story told was that the mammoth had been saved by Noah in the ark, but perished later, after it left the ark. Soaked soil could not bear such a ponderous animal and it sank into the underground where it has been found. Other stories suggest that the sunken mammoth is still alive and dwells in the underground. Many of these stories were published by travelers through Siberia of the seventeenth and eighteenth centuries. Mixed with the legends were real and correct data on the mammoth founded on direct observations. Explanations of the extinction of the mammoth, of preservation of its frozen carcasses, etc., told to travelers by ignorant local people, often corresponded to those given later by European students. For example, the Russian inhabitants of Western Siberia told Ysbrand Ides that the mammoth was very similar to the recent elephant, but its tusks were stronger and not so straight. The animal lived in Siberia before Noah's Flood, when the climate was warmer. The drowned floating bodies

[1] Howorth, H. H., "The Mammoth and the Flood," p. 50. Pfizenmayer, E. W., "Mammutleichen," p. 258, plate facing p. 280.

[2] Bunge, A., "Die Lena-Expedition," S. 51.

[3] Tscherski, I. D., "Beschreibung der Sammlung," S. 396.

[4] Erman, Ad., "Reise um die Erde," II, S. 263. Spasski, G., "Zoological Discoveries in Northeastern Siberia," p. 352.

[5] Pfizenmayer, E. W., "Mammutleichen," S. 244.

[6] Kutomanov, G. N., "Rapport sur une mission," p. 380.

vii

were, after the Flood, forced into subterranean cavities, became frozen and preserved in this way, because after the Flood climate became colder. Y. Ides points out that the former warm climate is not a necessary condition to explain this event, because the bodies could be brought by Flood from the southern countries for many hundreds of miles.[1] In other cases he was told by Siberian natives that the discovery of carcasses of the mammoth on high shores was the result of their accidental appearance to the surface from the underground where the animal used to dwell beyond the influence of air and light which were supposed to be deadly to it.

Discovery of bones of *Elephas* and *Rhinoceros* became a great puzzle to scientists and laymen alike in Europe, as everybody compared them to animals of warm climate and could not imagine that these bones were remnants of animals found on the very place where they once lived. In his book "The Mammoth and the Flood" Howorth carefully brought together the different explanations of these discoveries that prevailed at that time in Europe, having now, of course, only an historical interest.

A still greater puzzle were frozen carcasses of the mammoth which had been known in Siberia and China centuries before mammoth bones were found in Europe; but European scientists learned of them only after the conquering of Siberia by Russians. The explanations of these discoveries by Siberian natives and Chinese philosophers, reflected in local folklore, proved to their credit that they considered these remnants belonged to animals which lived in the same region, although in quite unusual conditions.

When Siberian localities became known to Europe, and in Europe itself were found more remnants of the mammoth than were possible to explain by referring to the elephants of Hannibal run astray a.s.o., the Noachian Deluge was used to explain the means by which anything might be transported. Some scientists accepted the theory that the Deluge had transported floating bodies of dead animals, as did Pallas in his description of the discovery of the *Rhinoceros*. Others believed that the animals were driven before the approaching Deluge into inhospitable country and perished there, or were trapped and killed by the Flood. In Pallas' time the frozen ground of Siberia was an established fact, and he simply accepted the theory that the same conditions prevailed there during the Deluge as well. "C'est pourquoi l'animal transportée des pays méridionaux à l'époque du déluge pouvait se conserver," said he.[2]

Although a great naturalist and a good and punctilious observer, Pallas did not try to imagine the method of transportation, for thousands of miles, of carcasses, or the mechanism of their enclosure in frozen ground. Deluge was something mysterious and miraculous, as well as everything connected with it, and there was no room for an exact scientific inquiry.

When Cuvier showed that the mammoth and fossil rhinoceros were specifically different from recent tropical animals, and that both were well protected against cold with a fur, especially heavy in the case of the mammoth, the theory of their tropical origin was replaced by the quite logical conclusion that both had been natives of the country where their carcasses were found. But as their domicile in the Arctic appeared improbable, it was suggested that their home had been in Central Siberia, and from there their remnants and whole carcasses were floated down by the great Siberian rivers, Ob, Yenisei, Lena, etc.,[3]

[1] Ides, Y. E., "Three Years Travels," p. 26.
[2] Pallas, P. S., "Voyages du Professeur Pallas, V.," p. 215.
[3] Howorth, H. H., "The Mammoth and the Flood," p. 60.

and in this way distributed over the whole of Siberia. This theory met with strong criticism. Cuvier [1] emphasized that the bones found in the Far North exhibited no marks of detrition. Hedenström [2] also suggested that carcasses carried by rivers must have been destroyed long before their arrival in their present localities in the North. It was also shown that, conforming to this theory, remnants must be more numerous in Southern Siberia than in Northern one, which is contrary to the fact. By the end of the eighteenth century it was also well established that the northern race of the mammoth had been distinguished by its smaller size from that found in Southern Siberia. The "floating" theory was supported by Lyell and, in somewhat limited form, by Middendorff and Bunge in its application to some special cases. It could be not applied to the carcasses of mammoth undoubtedly buried on the very spot where they met their death, discovery of which has proved beyond doubt that the mammoth was living in those regions of Northern Siberia. As in the tundra ground along with these carcasses were discovered remnants of trees and bushes which in the Age of the mammoth had been grown within the area of recent tundra, it was suggested that the mammoth used to live in Arctic Siberia under milder climatic conditions than the present ones. The change of climate for the worse, i.e. for the present condition, was the cause of the extinction of the mammoth. In such a general form these suggestions may be still found in textbooks, but they could not be reconciled with all the observed facts referring to the Mammoth-localities, when considered in detail. A desperate attempt to bring about such a reconciliation was made by Howorth, but he was forced to resurrect the theory of cataclysms and take recourse to the Flood, a fervent, although solitary, advocate of which he remained till his death, in 1923. According to Howorth, the mammoth used to live in the same areas of Northern and Southern Siberia where its carcasses have been found; these areas enjoyed at that time much milder climate. Then occurred the Flood, and the mammoths perished by drowning, their carcasses becoming buried in silt. "Immediately afterwards the same ground became frozen, and the same climate became Arctic, and this not gradually and in accordance with some slowly continuous astronomical or cosmical changes, but suddenly and per saltem." [3] According to Tscherski,[4] Howorth's ideas about the sudden extermination of the mammoth were also temporarily accepted by Lapparent. In later editions of his book Lapparent attributes the extinction of the mammoth only to a gradual increase in the coldness of the North Siberian climate, connected with the decrease of the supply of food.[5]

The discovery of frozen carcasses of the mammoth within the circumpolar region has been explained in a singular way by Gardner, the author of the theory on Central Sun, according to which our Earth is represented as something similar to a nutshell with the kernel removed. The shell is opened on both poles with apertures of some size. In the centre of such an empty ball is located an inner sun giving light and warmth to the inner side of the shell and provoking there a life more or less similar, or more or less different from this one on the surface of the globe. The mammoth is still living on the inner side of the earth shell. Sometimes it happens to approach the opening on the North Pole and

[1] Cuvier, G., "Recherches sur les ossements fossiles," I, p. 202.
[2] Hedenström, M., "Otrivki o Sibiri," p. 122.
[3] Howorth, H. H., "The Mammoth and the Flood," p. 96.
[4] Tscherski, I. D., "Beschreibung der Sammlung," S. 463, a footnote.
[5] Lapparent, A., "Traité de Géologie," 5 ed., p. 1686.

carelessly walks out to the surface of the earth.   Immediately it is killed by Arctic conditions, its carcass becomes frozen and in such a way remains preserved for generations in surprisingly fresh conditions.[1]

The study of the mammoth made great advancement during the present century, chiefly owing to a few successful expeditions commissioned to the Northeastern Siberia by the Russian Academy of Sciences.   The greatest of them is the establishment of the fact that the mammoth used to live in climatic conditions closely corresponding to the present ones.   In the recent literature the so-called mammoth-question is usually considered from this point of view.

[1] Gardner, M. B., "A Journey to the Earth's Interior," p. 47.

THE CARCASSES OF THE MAMMOTH AND RHINOCEROS FOUND IN THE FROZEN GROUND OF SIBERIA

By I. P. TOLMACHOFF

IVORY INDUSTRY IN NORTHERN SIBERIA

FOSSIL ivory was called in Siberia "Mamontova Kost,"[1] meaning "Bone of the mammoth," the name which is still used. Not so common is another picturesque name, connected with ivory localities on islands and on bottom of sea. "Ribya Kost," which means "Fish Bone."[2] Fossil ivory has been exported from Siberia and European Russia since very ancient times. For the last two centuries this trade has been carried on quite regularly, giving the local population a very decent income. Being important economically this industry has contributed very much to the accumulation of information about the mammoth. Owing to ivory the mammoth became known a very long time ago. Ivory hunters had discovered all the localities of frozen carcasses of this animal which were later examined by scientists, and some of which found their way into different museums. The statistics of the trade are also very interesting, as they give a fairly good idea of the number of mammoths which were discovered in different times. Some details and figures concerning the ivory industry of Siberia would be, therefore, not superfluous in the present paper.

The first mention of the mammoth is found in Chinese ceremonial books of the fourth century B.C.[3] It was certainly connected with fossil ivory brought from Siberia to China at that remote time. There are later records of fossil ivory exported from Russia to Southern Europe and Central Asia in the tenth century.[4] This ivory was, probably, not from Siberia, but used to be found on the Wolga River near the location of the present town of Simbirsk.[5] Of Siberian ivory was made, presumably, the throne of the Great Mogol Khan Kuyuk, which shows that in the thirteenth century fossil ivory was known in Mongolia in large amounts.[6] There is no exact data as to the export of Siberian ivory to China, but, probably, since the old days this commerce was carried on for centuries in a very regular way. Concerning this trade in the seventeenth and eighteenth centuries Stralenberg speaks of it as of a very common thing. Says he in his work: "A great many of these teeth which are white are carried for sale to China."[7] Ysbrand Ides also mentions

---

[1] Howorth, H. H., "The Mammoth in Siberia," p. 413.
[2] Howorth, H. H., "The Mammoth and the Flood," p. 29.
[3] Howorth, H. H., "The Mammoth and the Flood," p. 78.
[4] Howorth, H. H., "The Mammoth and the Flood," p. 80.
[5] Pallas, P. S., "Voyages du Professeur Pallas," I, p. 214.
[6] Howorth, H. H., "The Mammoth and the Flood," p. 79.
[7] Howorth, H. H., "The Mammoth and the Flood," p. 52.

mammoth teeth as an important item of trade carried on by Russian dwellers in a northern town on the Yenisei River.[1]

In Western Europe the first fossil ivory became known in 1611 when a mammoth tusk was brought to London by one Jonas Logan, who had bought it from Samoyeds on the Pechora River.[2] In European Russia, although fossil ivory had been abundantly found for a very long time, it became a regular commodity only after the conquest of Siberia by Ermak in 1582. How important fossil ivory had been considered at that time is very conspicuously proved by the desire of Russian Czars to monopolize this trade.[3] The ivory industry in Siberia has developed on a very considerable scale since the middle of eighteenth century, after the discovery of the first island in the group of New Siberian Islands, christened later the Bolshoi Lyakhov Island, in 1712, by the cossack Vagin,[4] and its exploration by the Siberian trader Lyakhov, in 1770. After this followed the discovery of other islands of the same group, by Lyakhov and other Siberian cossacks and trappers.[5] Probably, these islands were known and exploited, as ivory mines, much earlier. At least Avril, who traveled in Russia in 1685, learned from a Russian whom he calls Mushim Pushkun, the Voevoda of Smolensk and former Intendant of the Government of Siberia "that at the mouth of the Lena there was a spacious island very well peopled, and which is no less considerable for hunting the Behemot, an amphibious animal, whose teeth are in great esteem."[6] Mushim Pushkun, probably, was referring to the mammoth, but not to the walrus, as it would be possible to suggest, because, according to Nordenskiöld,[7] the walrus is not found between Khatanga and Chaun Bay. The walrus has been not exterminated here recently, as the same fact was already stated by Erman.[8] The New Siberian Islands used to be visited yearly in summer by ivory hunters, who were going to the islands in spring and returning in the fall, crossing the straits both ways on sledges with dog teams. They did not make any excavation or digging to any extent, but were just looking for ivory in the cliffs along the seashore, on rivers and creeks, on lakes, or collecting ivory in shallow places in sea near the islands. If such a hunter happened to come across a tusk protruding from the cliff, but still firmly fixed in the ground by the other end, he put a mark on it, the claim of his possession never being disputed or ignored by other hunters, and he could come to the same spot the next year, or two years later.

Statistics available on the ivory industry of Siberia are very incomplete, often covering widely separated periods. Nevertheless they give a very good idea about the immense number of mammoths discovered and still buried in the frozen ground of Siberia. Thus, North-Siberian cossack Sannikov brought, in 1809, from the New Siberian Islands 250 poods (9000 lbs.) of ivory, which corresponds approximately to the amount of ivory from 80–100 animals. Another ivory collector returned, in 1821, from the same islands with a double amount of ivory, 500 poods (18000 lbs.). After Stschukin, about 1000 poods (36000 lbs.) of ivory used to be sold at Yakutsk every year during the first half of nineteenth century, but twice within the period of time between 1825 and 1831 this amount reached 2000

---

[1] Ides, Y. F., "Three Years Travels," p. 107.
[2] Howorth, H. H., "The Mammoth and the Flood," p. 48.
[3] Cuvier, H., "Recherches sur les ossements fossiles, I," p. 142.
[4] Toll, Ed., "A Sketch of the Geology of New Siberian Islands," p. 2.
[5] Toll, Ed., "A Sketch of the Geology of New Siberian Islands," p. 2.
[6] Howorth, H. H., "The Mammoth and the Flood," p. 49.
[7] Nordenskiöld, A. E., "Die Umsegelung Asiens und Europa, I," S. 405.
[8] Erman, Ad., "Reise um die Erde, II," S. 264.

poods (72000 lbs.).   Besides this trade at Yakutsk, there were from 80 to 100 poods (2880–3600 lbs.) sold at Turukhansk, 75–100 poods (2700–3600 lbs.) at Obdorsk, and some found its way also to Tobolsk.[1]   Middendorff[2] supposing that the ivory of at least 100 mammoths was delivered yearly to market, estimated the number of animals which had been discovered for two hundred years before his time at 20000, a figure which has been considered much too low by Nordenskiöld.[3]   Argentov, in 1857, and Klutrov, in 1856, speak of great boats on the Lena River laden with mammoth ivory.[4]

In 1872, 1630 very fine mammoth-tusks were brought to England; and in 1873, 1140 were brought, weighing from 140 to 160 lbs. each.[5]   Not all of the mammoth ivory coming to England is good, perhaps a half being rotten; specimens, however, are found as perfect and in as fine conditions, as if recently killed.[6]   Digby who had a chance to see plenty of tusks at Yakutsk and on the Lena River, says that "A great deal of the stuff (tusks) is fit only for burning, to make India ink, and is not worth the heavy cost of transport abroad for that,"[7] but also that "Two or three that I examined were as modern elephant tusks. They must have come straight out of clean ice."[8]

In the period of time from 1887 to 1893 was sold annually at Yakutsk 1100–1750 poods (39960–63000 lbs.) of ivory for the price of 24–37 rubles a pood (33–51 cents a pound); from 1894 to 1897 was sold there in different years 1460–1750 poods (52560–62700 lbs.) for 29–35 rubles a pood (40–48 cents a pound).   The yearly yield of fossil ivory has gradually decreased during the last decades, although in 1910 there were sold at Yakutsk 1900 poods (68400 lbs.), and in 1913 (the latest available figures) 1600 poods (57600 lbs.) for the average price of 53 rubles a pood (73 cents a pound).   In the same year was delivered to Yakutsk from the embouchure of the Lena River 1300 poods (46800 lbs.) of ivory.   The price of ivory on the northern shore of Siberia at the same time was 40–75 rubles a pood, the difference depended upon the quality of ivory, thus closely corresponding to the price at Yakutsk; while in European markets (in London) it was at least ten times higher.   An average yearly figure for ivory sold at Yakutsk is estimated by local statisticians at 1500 poods (54000 lbs.).   Supposing that every animal could deliver an average of 8 poods (288 lbs.) of ivory; they estimate that every year 187 mammoths must be found to supply the market with the given amount of ivory and that during the two and a half centuries, since the Russian occupation of the country at least 46750 animals must have been discovered.[9]

As a pair of tusks only in rare cases weighs 8 poods, and as the tusks from New Siberian Islands as well as from the northern shore of mainland, which make the bulk of those sold at Yakutsk, are never over 3 poods of weight each, the average weight given above cannot

[1] Howorth, H. H., "The Mammoth and the Flood," p. 51–52.
[2] Middendorff, A. Th., Sibirische Reise, IV," I, S. 278–279.
[3] Nordenskiöld, A. E., "Die Umseglung Asiens und Europa," I, S. 365.
[4] Middendorff, A. Th., "Sibirische Reise, IV," I, S. 278–279.
[5] Howorth, H. H., "The Mammoth and the Flood," p. 52.
[6] "Encyclopedia Britannica, XI.," edition, XV, p. 92.
[7] Digby, B., "The Mammoth," p. 154.
[8] Digby, B., "The Mammoth," p. 176.
[9] The figures for the last thirty years have been brought together in Yakutskaya Okraina, August 12, 1912, a newspaper published at Yakutsk.   These data are reprinted and somewhat supplemented in a Russian article by W. M. Zenzinov, "Sketches on the Trade in the North of the Territory of Yakutsk, Moscow," 1916, pp. 70–71.   In his article "With an Exile in Arctic Siberia" (National Geographic Magazine, XLVI, 1924) the same author, speaking of fossil ivory (p. 718), says: "In the past century the yield has been estimated at from 20 to 30 tons."   According to Pfizenmayer (Mammutleichen, S. 256) the northern merchant Sannicov estimated, in 1908, the yearly yield as 2000 poods, or 72000 lbs.

be taken as such. Accordingly the number of animals delivering yearly fossil ivory to market must be higher and not lower than 250 specimens.

The second figure, 46750 animals, appears to be exaggerated, because the export of ivory in the seventeenth and eighteenth centuries was certainly smaller than during the last hundred years, and the same average figure of animals found every year should not be taken for the whole period of 250 years. But the average number of animals discovered every year, 187, has been found, as notes above, too low for the last hundred or hundred and fifty years (since the starting of the ivory industry on the New Siberian Islands). Not all the discovered mammoths had delivered a marketable ivory, therefore not all were included in the number of yearly discovered animals. In ancient days ivory was used much more for domestic purposes than now, and a smaller export did not mean discovery of fewer animals. Taking all that into consideration, the above figure cannot be considered exaggerated, but rather a small one.

Some data in Pfizenmayer's book could even bring one to the conclusion that the above figures are too small, but there is in the book a *lapsus calami* which needs a correction. According to Pfizenmayer, Bunge's expedition to the New Siberian Islands collected during three summers, in 1882–1884, in the islands Lyakhov, Kotelni, and Fadyev about 2500 first grade tusks.[1] It would correspond to 1250 animals, or 417 as an average for a year, a figure exceeding that given above twice and even much more, if one would take into consideration that along with "Erstklassige Mammutstosszähne" must also have been found poorly preserved ones, that the figures given by the present writer refer to the ivory industry of the whole Northeastern Siberia, not to the New Siberian Islands alone, and that the expedition referred to was in the islands only for the summer of 1886, as in 1882–1884 Bunge was engaged in another expedition to the delta of Lena. As matter of fact the figure 2500 refers to all bones and tusks collected by both expeditions referred to,[2] among which the bones of the mammoth were present only in a small proportion. On the next page of his book [3] Pfizenmayer gives statistics in part repeating those brought together by the writer in this paper, in part closely corresponding to them, but does not try to bring these data in reconciliation with his figure of 2500 tusks collected by Bunge's expedition.

All these figures show how common are the remnants of mammoth in the frozen ground of Siberia, and how common the animal must have been in its time. From the technical point of view it is of interest to mention that Siberian localities have been considered "inexhaustible as a coalfield and in future, perhaps, the only source of animal ivory." [4]

[1] Pfizenmayer, E. W., "Mammutleichen," S. 255.
[2] Tscherski, I. D., "Beschreibung der Sammlung," S. 2.
[3] Pfizenmayer, E. W., "Mammutleichen," S. 256.
[4] "Encyclopedia Britannica, XI.," Edition XV, p. 92.

HUNTING FOR THE MAMMOTH

BEING so familiar in Russia the remnants of the mammoth very early attracted the attention of Russian scientists and students who had come to that country from abroad. Peter the Great, with whose name are connected many innovations in the intellectual life of Russia, was also interested in these remnants and issued, for example, an order to find out to which animal belonged the mammoth horns.[1] Another decree ordered that the bones of mammoth must be delivered to the Kunstkamera, the name of the first scientific Museum in Russia, which has been developed now into a number of museums of the Russian Academy of Sciences.

Since that time the study of the mammoth has become a tradition of the Russian Academy of Sciences which, during the two hundred years of its existence, has sent scientific expeditions to different parts of Siberia to examine the localities of carcasses of mammoths, the discovery of which from time to time was reported to the Academy. In these undertakings the Academy was invariably supported by the Russian Government, all expenses of the expeditions being always paid by the state treasury.

To promote discoveries of this kind the Academy promised a money premium to every discoverer of skeletons or carcasses of large fossil animals, if such a discovery would be immediately reported to St. Petersburg. A special announcement worked out by Baer, Brandt, and Middendorff referring to the discoveries of this kind and to the promised premium was sent, in 1860, by the Academy to the Governor of the Archangel Government, to the General-Governor of Eastern Siberia and to the General-Governor of Western Siberia. In 1880, the same leaflet was reprinted and distributed, through Government officials, travelers, traders, etc., among the population of Northern Siberia. The text of this announcement translated into English is as follows:

"Bones of gigantic prediluvial animals, as of the mammoth and others, happen to be found in tundras of Siberia, on the shores of streams, rivers, and lakes, as well as on the sea coast where, during a flood, or tide, bluffs are underwashed, and landslides originate. Among all these bones are utilized only tusks, called also horns or moustaches, which traders used to purchase. Besides that, so-called bird talons (in reality they are the horns of a prediluvial rhinoceros) are used in construction of bows. The other bones of these prediluvial animals do not find any utilization and decay on the tundra where they happen to be found here and there. They are absolutely valueless and are not worth a mention.

"Sometimes it happens also that out of frozen ground appears a complete skeleton of a prediluvial beast, bone by bone, properly arranged, not disconnected bones only. Such skeletons while within the ground and in complete order, even without any visible horns, are very much needed by scientists. If such skeletons should be found anywhere, the Academy would send a scientist to examine them on the very spot. The Academy therefore promises everybody, a native, hunter, trader, or official alike, that it will pay a premium of a hundred rubles to the first one who finds such a complete skeleton, and, having marked the spot, at once reports the matter to the next, his chief. The latter one will immediately

---

[1] "The Pacific Russian Scientific Investigations, Geology," by A. Kryshtofovich, p. 41.

report to the Governor of his province.   In the report it is necessary to explain the locality as clearly as possible, to give the distances from some known town, or village, or hut, to tell of which river is the stream, on the shore of which has been found the skeleton, the tributary.

"Do not report any groundless gossip.

"If after the examination by the scientist commissioned by the Academy, the skeleton is recognized to be very good, the Academy will pay to the discoverer of the skeleton, besides the hundred rubles already paid, fifty rubles more.   The discoverer is also at liberty to sell the tusks to whom he wishes.   If, as it has happened now and then, the prediluvial beast should appear complete with meat and hide, it would be a great opportunity.   It would be necessary to hurry the report.   If the commissioner of the Academy should find even a little of the meat and hide not decayed, the Academy will pay three hundred rubles to the discoverer of the prediluvial beast, who at once reported the matter.

"Besides that the commissioner would report to the Academy, if anybody should get a honorary reward for his work and zeal in the interests of the cause."

The premium has not met with as much success as had been expected.   It took a long time to bring the news of the premium to the knowledge of the people in the Far North of Siberia.   Besides that, the natives, the first discoverers of frozen carcasses of ancient animals, were usually not very enthusiastic to hunt for the promised 300 rubles, or even for 1000 rubles when the premium was increased to this amount by the Siberian General-Governor Anuchin.[1]   They thought the premium could not recompense them for all the troubles connected with the arrival of an expedition and with the travel of government officials, as they had the chance to learn by sad experience during the Adams' expedition to the delta of Lena River, in 1806, which had given the local population much trouble. A good example of this behavior is the history of the well-preserved carcass of the mammoth found by natives, in 1857, at the mouth of Lena River, on the Mostak Island, but not reported at all to authorities or to the Academy of Sciences either.   When Ispravnik (a chief of police) of Verkhoyansk district happened to learn of this discovery and asked for details, he was told that the carcass had been destroyed and carried away with water, after which he, naturally, did no further questioning.[2]   As matter of fact the remnants of this mammoth were found by Bunge twenty-five years later, on the same spot.   Even Bunge was not told about it immediately after his arrival at the Lena Delta, although he asked particularly for mammoth carcasses.   He learned of this mammoth, perhaps, only for the reason that it was impossible to keep any longer a secret, the mammoth was located only twenty-five miles southwest of the meteorological station of the expedition.   In the same way Maydell during his hunting for the mammoth more than once came across an unwillingness to say anything of carcasses of mammoth, which attitude, in his opinion, "had been the result of the Adams' expedition, which is still remembered resentfully by all the Yakuts; so that, whenever possible, they conceal all finds, fearing to be forced to work and provide haulage."[3]   Remnants of a mammoth found on the Kolyma River, of which Maydell learned during this journey, had been known to natives for some three years, but had been kept secret by them, as they were fearing to be compelled to dig out the bones

[1] Toll, Ed., "Die fossile Eislager," S. 81.
[2] Bunge, A., "Die Lena-Expedition," S. 51.
[3] Maydell's letter to Dr. L. Schrenck, of February 19, 1869, reprinted in Digby, B., "The Mammoth," p. 83.

and transport them.   They had also forbidden the Yakut, who was in Maydell's service to acquaint him with the find.[1]

The natives were also not very anxious to look for the frozen carcasses, because in their superstition they believe that such a discovery sometimes could have bad consequences for the discoverer.   Adams, while on his expedition, was told, for example, about a Tungus who died, with all his family, after he had had the bad luck to look at a frozen mammoth.[2] Even in recent times the native who came across the Beresovca mammoth was much afraid of his discovery.   Pfizenmayer tells us in his book that during the excavation work on the River Beresovca the expedition used to be visited by natives.   While on his last visit a Lamut, who in meantime became a good friend of the party, in saying good-by, expressed his hope and sincere wish that Pfizenmayer would be saved of any wretchedness on account of the carcass of the mammoth.[3]   Even local Russian populations are not less, or only a little less, superstitious as compared with the natives.   When, seven years later, Pfizenmayer again visited Northern Siberia he was told of the misfortunes affecting almost everybody who had been in any way connected with the discovery and excavation of the Beresovca mammoth.   The cossack Yavlovski, who had reported this mammoth, became insane and perished as the result of drunkenness.   Ispravnik (chief of police) Horn who had visited the locality and reported the matter to the Academy of Sciences, died shortly after, only two days before the day he was to receive a cross of honor bestowed on him by the Czar for his service in this case.   The untimely death of the leader of the expedition, Herz, who passed away two years after the expedition, was attributed also to the mammoth. The good health of Pfizenmayer gave the people no trouble, as they believed his sad fate was only postponed for a while.[4]

At the same time, the natives are hunting very eagerly for ivory, which, perhaps, proves that the discovery of carcasses of the mammoth is not such a common thing, as of its skeleton, skull, and isolated bones, which provoke no respect from the local people.

News of the discovery of frozen carcasses used to reach the outside world only when a rumor became known to local traders, priests, or Government officials, the people who could expect to earn something from every expedition, much more from a successful one, without danger of losing anything in any case.   It was only the good will of a Government officer which caused him to forward farther a report on a carcass found by natives or hunters, as he must be more or less sure that the matter was worthy of attention, and that he would not be reprimanded by his chief, the Governor of his province, through whom was carried on all the correspondence with St. Petersburg.   Therefore, they were acting differently.   Some of them, probably afraid of possible troubles, used to try to conceal discoveries and urged the natives not to report them, even ordering the latter whipped[5] when they were not willing to keep the matter secret.   Other more ambitious officials were sometimes too eager to report immediately a new-found carcass without having checked the discovery.   Only a few of them could be given credit for a preliminary investigation of the locality where a frozen animal had been reported found by natives.   As

[1] Maydell's letter to Dr. L. Schrenck, of April 17, 1870, reprinted in Digby, B., "The Mammoth," p. 97.
[2] Howorth, H. H., "The Mammoth and the Flood," p. 83.
[3] Pfizenmayer, E. W., "Mammutleichen," S. 164.
[4] Pfizenmayer, E. W., "Mammutleichen," S. 221.
[5] Toll, Ed., "The fossil Glaciers of New Siberian Islands," Russian Edition published by the Russian Geographical Society in 1897, p. 123.

matter of fact, in most cases no carcass was found at all by the expeditions commissioned by the Academy of Sciences to the reported localities, although they had a good chance to accomplish a great deal of scientific work and contributed much to the knowledge of the mammoth.   At the same time almost every expedition to Northern Siberia used to come across the remnants of mammoth which often had been known to local people for years, but were never reported to the Academy, although some of them, perhaps, had been discovered in very good conditions.   So far as the writer is able to recollect, the premium referred to, during a hundred years of hunting for mammoth, was paid only once, for the discovery of the famous mammoth from the Beresovca River.   Cossack Yavlovski who had reported this discovery to the Academy, received in premium a thousand rubles which on the same evening he gambled away.   He was also rewarded with a silver medal.[1]

The rumor of the discovery of the Beresovca mammoth, its successful transportation to St. Petersburg, pecuniary and other rewards given for the discovery and for the work accomplished later, spread very quickly over Northern Siberia, often in an exaggerated form, and gave an impetus to the hunting for mammoths and to a rather undesirable marketing of found or supposed finds of carcasses.   Having been afraid that the speculation could go too far and in some cases contribute to the destruction of found carcasses, the Academy, in 1910, worked out a new law proclaiming as national property all the carcasses of the Pleistocene animals found in the frozen ground of Siberia, as well as in other parts of Russia.[2]   Once found they have to be reported at once to authorities and delivered to Russian scientific institutes.   The discoverer must be awarded with the premium in amount 300–500 rubles plus the market price of tusks, if they should be present.   After some alterations the law was worked out in the following form in which it was submitted, in 1914, to the Gosudarstvennaya Duma, by which, however, it was not passed, because its regular work has been interrupted since the World War and Russian Revolution.

1. The complete carcasses, or their parts, as well as all the remnants of the mammoth, extinct rhinoceros, and other extinct animals are a national property.

2. Everybody who happens to find the above mentioned remnants is obliged to report to the local executives, and through them, or directly, to the Imperial Academy of Sciences, Imperial Universities, Geological Committee of the Department of Commerce and Industry, Catherine II School of Mines in St. Petersburg and at Ecaterinoslav, or any other Government Institution in possession of geological and zoological museums or collections.

3. Local executives must guard the remnants of extinct animals before the arrival of people commissioned to dig them out.

4. Everybody who discovers such remnants and reports them, will be remunerated according to regulations which formerly were published by the Senate.

5. Systematic excavation of the remnants of large vertebrates on the State lands is permitted to all Government Institutions mentioned above in p. 2, also to Russian Natural Science Societies.   All collections become a property of organizations which were paying for the expenses of excavation.   Private individuals may be granted the privilege of carrying on the same work only in exceptional cases, every time with special permission by the Minister of Education and with the consent of Departments interested in this matter.

[1] Pfizenmayer, E. W., "Mammutleichen," S. 188.
[2] Meeting of the Physico-Mathematic Section of the Russian Academy of Sciences, March 3, 1910: Bull. Acad. Sc., IV, p. 587, St. Petersburg, 1910.

6. Excavation of remnants of extinct animals on private lands may be carried on only with the permission of the proprietors of these lands. In cases of exceptional scientific importance when such a permission would riot be granted, institutions and societies mentioned above in p. 2 are at liberty to request an expropriation of localities of remnants referred to on basis of article 575 and later civil laws (Russian Code, Vol. X, part 1, edition 1900 and 1912).

7. An export of the remnants of animals mentioned in p. 1 is granted only by permission of the Minister of Education and with the consent of Departments interested in this matter.

At very last time Academy of Sciences published and distributed a new announcement. It has on top a picture of a mammoth wandering on the snow covered shore of a river or lake. Its content is as follows.

On the discovery of fossil animals.

While collecting mammoth bone, you can come across a carcass, skeleton, or skull of some animal unfamiliar to you, or not living more in your region, as, for example, of a mammoth, rhinoceros, wild ox, wild horse, wild sheep, washed out somewhere in a ravine, near rivers and streams, also on sea shore, or appeared within a landslide. Such a discovery has a great scientific interest. Therefore the Yakutsk Commission of the Academy of Sciences announce that every prompt report of such a discovery will be rewarded. A reporter will receive, according to scientific value of found remnants, *a cash reward up to 500 rubles and different fabrics, tobacco and food supplies up to the value of 200 rubles.*

The expeditions at different times sent by the Academy to Northern Siberia were hunting for the mammoth. The wooly rhinoceros (*Rhinoceros tichorhinus* Fisch.) found in the same conditions, used to attract less attention and was reported only in rare cases, perhaps, because its remnants have only a small commercial value in comparison with the mammoth. Besides, carcasses of rhinoceros are found more rarely. Carcasses of other animals, as of the musk, ox, horse, etc., must be very common in the frozen ground of Siberia, but local people usually do not pay any attention to them, and stories of their discoveries used to be told only occasionally to the scientists who happened to visit the particular spot where such an animal had been found.

A few preliminary remarks are necessary concerning the frozen ground of Siberia, in which are found frozen carcasses. It is necessary to distinguish the frozen ground, i.e. sand, clay, etc., transformed, with the water frozen within, into a peculiar rock, and pure ice found in frozen ground in masses of very different dimensions, from thin layers and small lumps to the accumulations composing large mounds. Frozen ground, inclusive ice, is dependent, as to its origin, upon climatic conditions: the average annual temperature below the freezing point, and dry climate connected with scarcity of snow in winter. The origin of ground ice, besides that, is dependent upon some special conditions, the discussion of which would be beyond limits of the present article.

MAP OF LOCALITIES

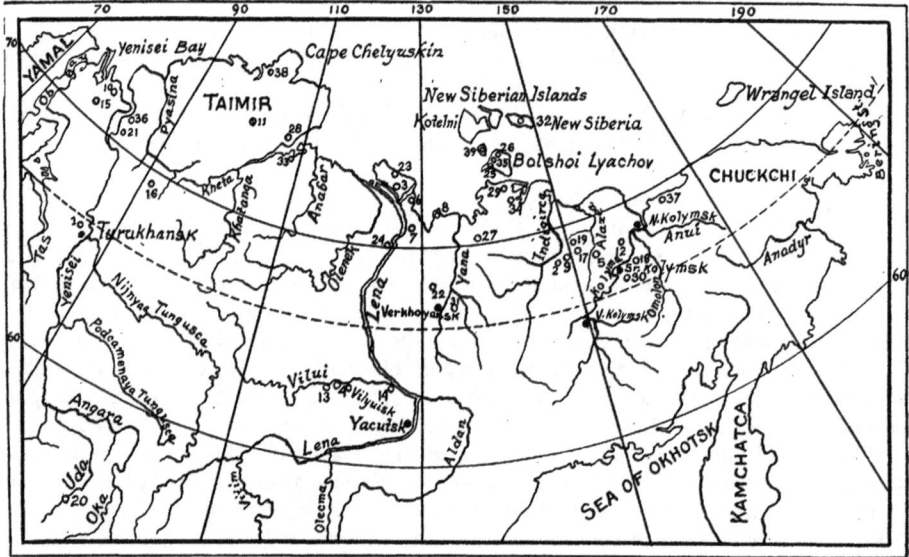

LIST OF LOCALITIES OF FROZEN CARCASSES GIVEN ON THE MAP

1. Ysbrand Ydes, 1692 . . Mammoth
2. Messerschmidt, 1724 . . . Mammoth
3. Adams, 1780 . . . . Mammoth
4. Pallas, 1771 . . . . Rhinoceros
5. Sarychev, 1787 . . . . Mammoth
6. Adams, 1799 . . . . Mammoth
7. Adams, 1797 . . . . Mammoth
8. Potapov, 1800 . . . . Mammoth
9. Rozhin, 1839 . . . . Mammoth
10. Mochulsky, 1839 . . . . . Mammoth
11. Middendorff, 1843 . . . . . Mammoth
12. Khitrovo, 1854 . . . . Mammoth
13. Stubendorff, 1858 . . . . . Rhinoceros

14. Kolesov, 1863 . . . . Mammoth?
15. Schmidt, 1864 . . . . Mammoth
16. Schmidt, 1866 . . . . Mammoth
17. Maydell, 1867 . . . . Mammoth
18. Maydell, 1870 . . . . Mammoth
19. Maydell, 1870 . . . . Mammoth
20. Tscherski, 1875 . . . . Rhinoceros
21. Nordenskiöld, 1876 . . . . Mammoth
22. Gorokhov, 1877 . . . . . . Rhinoceros
23. Bunge, 1857 . . . . Mammoth
24. Bunge, 1879 . . . . Mammoth
25. Bunge, 1866 . . . . Mammoth
26. Toll, 1860 . . . . Mammoth

27. Toll, 1863 . . . . Mammoth
28. Burimovich, 1899 . . . . . . Mammoth
29. Toll, 1891 . . . . Mammoth
30. Herz, 1900 . . . . Mammoth
31. Pfizenmayer, 1901 . . . . . Rhinoceros
32. Brusnev, 1903 . . . . Mammoth
33. Tolmachoff, 1905 . . . . . . Mammoth
34. Vollosovich, 1907 . . . . . Mammoth
35. Stenbok-Fermor, 1906 . . Mammoth
36. Kootomanov, 1909 . . . . Mammoth
37. Soloviev, 1910 . . . . Mammoth
38. Transehe, 1915 . . . . Mammoth
39. Andrews, 1923 . . . . Mammoth

RECORDED DISCOVERIES OF THE CARCASSES OF THE MAMMOTH AND RHINOCEROS

TRAVELERS of the seventeenth and eighteenth centuries, when dealing with Siberia, give a lot of information on the carcasses of mammoths found in the frozen ground of Siberia, including the descriptions of a few particular discoveries which are recorded in the following. Even speaking of different legends on the mammoth among the natives and Russians in Siberia they give much information referring not to the legendary animal, but to real carcasses of it which from time to time used to be found in different parts of Siberia.

Witsen during his stay in Russia, in 1686, brought together a lot of information on the mammoth, which refers to the carcass of the animal. He says that a mammoth was of a dark-brown color and emitted a great stench. Its tail was like a horse, and its feet short.[1]

The first direct mention of a carcass of a mammoth discovered in the frozen ground of Siberia belongs to Ysbrand Ides who, in 1692, was sent by Peter the Great as an envoy to China (1).[2] On his way through Siberia he had along with him a man who used to travel annually for the collection of fossil ivory, and who told him he had once found a head of a mammoth in a piece of frozen earth which had tumbled down.[3] The soft parts of the head were putrefied, but the bones were still colored with blood. He found also a frozen foot of the girth of a man, which with the assistance of his companions he cut off and took to Turukhansk. With some difficulty he also broke out the teeth which, he said, were placed before the mouth like those of an elephant. The mention of Turukhansk brings Nordenskiöld to the conclusion that the mammoth under consideration was found somewhere on the lower Yenisei, or anyhow not very far from this river.[4] To this mammoth, probably, Pfizenmayer refers,[5] while speaking of the carcass of a mammoth found by cossacks, in 1692, on the River Yenisei. Ydes was already familiar with the fact that these elephants used to be found on high banks of rivers, imbedded in the frozen ground from which they were washed out during the spring flood. Remnants of the mammoth (tongues and legs, as mentioned by Ides) happened to be found particularly often on the shores of the rivers Yenisei, Turukhan, Mangamzea, Lena and near Yakutsk, to as far as the Frozen Sea.

J. B. Müller, one of the Swedish prisoners of war in Siberia, in his memoir on the customs of Ostyaks, published in 1720, among other stories on the mammoth speaks of the bloody bones of this animal and of clotted blood within the cavities of its bones.[6]

Laurence Lange, in his narrative of a journey to China, speaking of the mammoth, reports that several people assured him that they had seen the bodies of this animal with flesh and blood still remaining.[7]

[1] Howorth, H. H., "The Mammoth and the Flood," pp. 74 and 80.
[2] Figure in brackets in this, as in other similar cases, refers to the number of locality on the map.
[3] Ides, Y. E., "Three Years Travels," p. 25.
[4] Nordenskiöld, A. E., "Die Umseglung Asiens und Europa," I, S. 365.
[5] Pfizenmayer, E. W., "Mammutleichen," S. 23.
[6] Howorth, H. H., "The Mammoth and the Flood," pp. 75 and 80.
[7] Howorth, H. H., "The Mammoth and the Flood," pp. 74 and 80.

Tatischev, the Chief of the Altaï Mining District at the beginning of the eighteenth century, speaks also of the mammoth bones still colored with blood, which used to be found by natives.[1]

Dr. D. G. Messerschmidt who had been sent by Peter the Great to Siberia on a special mission, to study its natural history, brought back a short report on remnants of the carcass of a mammoth found on the Indigirca River along with the skeleton of another mammoth (2). For some reason he found it necessary to get a solemn confirmation of this discovery written by an eye-witness in the following form: [2]

"Whereas Mr. Messerschmidt entreated me to let him know where the head of the mammoth with its teeth and other parts were found; as I was an eye-witness to the digging it up I thought proper to give him this short account thereof in writing. That head was found by a certain Russian soldier, Vasili Erlov, on the eastern bank of the river Indigirca, not far from the rivulet Volocovoi Ruchei. After it was discovered, I, being at leisure, was present and eye-witness to the digging up of this skeleton or bones; and further likewise on the other bank of the same river, which bank is named Sztanoiyar, I saw a piece of skin putrefied, appearing out of the side of a sand-hill, which was pretty large, thick-set, and brown, somewhat resembling goat's hair, which skin I could not take for that of a goat, but of a Behamoth; inasmuch as I could not appropriate it to any animal that I knew. This I certify by this Latin testimonial for the present and even hold it my duty to give a more circumstantial verbal account whenever Her Imperial Majesty shall be graciously pleased to lay Her royal commands on me."

Dated at Irkutsk, Feb. 10, 1724. (Signed) Michael Wolochowich.

Khariton Laptev during his cruising, in 1739, along the northern coast of Siberia, east of the Lena River, had the opportunity to hear much about the discovery of mammoth corpses, as well as to observe their remnants in the ground. "On the banks of several rivers on the tundra whole mammoths with their tusks are dug out with thick hides on them. Their hair and bodies are, however, rotten, while the bones, except the tusks, are also decaying," says he in a rather general way. In a short description of some fossil heads given by him we recognize not the mammoth, but a rhinoceros.[3]

During his expedition to the delta of Lena River, in 1806, Adams was told by a Tungus that an animal similar to that examined by Adams, covered with hair, had been discovered a number of years before at the mouth of Lena River (3). The Tungus who had been unlucky enough to look at the animal had died immediately, with all his family. The recorded discovery, probably, took place sometime during the second half of the eighteenth century.[4]

It happened to be the carcass of a rhinoceros (*Rhinoceros tichorhinus* Fisch.) found, in 1771, by Yakuts on Vilui River, about 25 miles above the small town Vilyuisk,[5] which was not only discovered or mentioned by some traveler, but for the first time a part of it, although not the whole (a head, two legs, and a piece of hide), was delivered to St. Petersburg and deposited at the Museum of the Academy. The importance of this discovery was admirably expressed by Cuvier, who said: "Il est heureux du moins que les parties

[1] Howorth, H. H., "The Mammoth and the Flood," pp. 76 and 80.
[2] Howorth, H. H., "The Mammoth and the Flood," p. 81.
[3] Middendorff, A. Th., "Sibirische Reise, IV," I, S. 277.
[4] Howorth, H. H., "The Mammoth and the Flood," p. 83; Middendorff, A. Th., "Sibirische Reise, IV," I, S. 276.
[5] Pallas, P. S., "Voyages du Professeur Pallas, V," pp. 215-218.

les plus essentielles de ce monuments d'un genre et d'une date si extraordinaire, soient désormais à l'abri de la destruction." [1] A forefoot and the upper part of a hind leg later were burned through careless drying.[2] The animal was found on the low shore of the river, partly embedded in sand, and had been preserved in frozen sandy ground on the high banks of the river (4). The head and legs were chopped off by aborigines and sent to Yakutsk and Irkutsk. The head and two legs (anterior and posterior one) were delivered to Irkutsk where Pallas at that time happened to be, and he immediately brought together all available information concerning the locality and described the remnants.[3] According to Pfizenmayer these precious remnants were presented to the Empress Cathrine II for the Kunstkamera by the Archbishop of Tobolsk.[4]

In 1787 a carcass of a mammoth was found on the Alazea River. The discovery was reported to the Captain Sarychev of the Russian Navy who at that time traveled from Sredne-Kolymsk to Yakutsk. It was described as a skeleton of a great animal of which only one half was visible, washed out of the sand bank of the Alazea River (5). It was about the size of an elephant, was found in an upright position, still retained its skin, and, in some places, its hair. A recent heavy fall of snow, combined with the necessity of making a long detour prevented Sarychev from visiting the locality, and from allowing his companion, Dr. Merck, to go over there, although the latter was very anxious to investigate the locality.[5]

The first carcass of a mammoth, the remnants of which have found their way into museums, had been found, in 1799, in the delta of the Lena River, at the latitude of 72° and 130° east of Greenwich, near the Cape Bycov (6) by a Tungus named Shumakhov who had noticed at first only an indeterminable, but queer looking mass within the frozen ground on a cliff. After that he used to visit the place every year and observe it, as more and more of the animal appeared out of the ground. On the fifth year the cliff underthawed, and the carcass slipped down to the sandy shore where it could be well examined. At that time Tungus chopped off its tusks and bartered them for a value of fifty rubles. About the same time the carcass had been seen by a local trader, by name Boltunov, who described [6] it and prepared a rough, schematical drawing of the animal, which, as an original or a copy, was sent by Adams to Blumenbach at Göttingen and later was reproduced by Baer.[7]

In 1806 to Yakutsk happened to come the zoologist, M. F. Adams, a member of the Russian Academy of Sciences, who at that time traveled through Siberia with Count Golovin, Russian Ambassador to Pekin. Adams learned about the discovery of the carcass and immediately left for the Lena delta. In the meantime the carcass suffered very much. The trunk which Boltunov had well described, was no longer there, nor was the short (about 10 inches long) tail. Of two ears, each over ten inches long, was found only one. One eye was found still keeping its color, destroyed later in the process of drying out. Other soft parts, with the exception of the skin on the head, on a foot, and on the side on

[1] Cuvier, G., "Recherches sur les ossements fossiles, II," I, p. 88.
[2] Middendorff, A. Th., "Sibirische Reise, IV," I, S. 272.
[3] Pallas, P. S., "De reliquiis animalium exoticorum. After Tscherski, I. D., Beschreibung der Sammlung," S. 3.
[4] Pfizenmayer, E. W., "Mammutleichen," S. 243.
[5] Middendorff, A. Th., "Sibirische Reise, IV," I, S. 277.
[6] Published by Severgin (in Russian) in Tekhnologicheski Journal, 111, 4, p. 162, St. Petersburg, 1806. After Baer, K. E., Neue auffindung eines vollständigen Mammuths, S. 278.
[7] Baer, K. E., "Fortsetzung der Berichte über die Expedition," plate.

which the animal had lain, had been destroyed completely by wild animals and dogs which local Yakuts had fed on Mammoth meat during the shortage of dog food. Shumakhov described the animal to Adams as very well fed and fat. It was a male with a long mane. The part of the hide preserved in the ground was covered with thick hair. Adams secured a portion of this hide which was so heavy that ten men with difficulty dragged it to the bank. He also collected about a pood (36 lbs.) of long hair, which lay scattered about the ground round about.

This carcass of the mammoth undoubtedly was the best one ever found. If a scientist could have arrived at the locality during the first four years when the Tungus discoverer used to watch patiently his animal still frozen in the ground within the cliff, he could have examined the body just in the condition in which it had been buried, without any more recent damage.

The tusks of the animal had been cut in pieces and sold in Moscow, a long time before it was examined by Adams. After the Adams inquiry, they were found to be ten feet long and weighed 360 pounds. To complete the skeleton, Adams had purchased at Yakutsk two tusks and fixed them later to the skeleton. According to Pfizenmayer [1] they had been restored (when and where (?)) from fragments (three in the right tusk, two in the left one), the interstices between filled up with some mastic so perfect that this nature of the tusks has been never noticed before, and Brandt even emphasized that the tusks were of a single piece, although not belonging to the specimen. [2] Pohlig, who had the chance to examine remnants of the mammoth at the St. Petersburg Zoological Museum, also failed to notice the composite character of the tusks, but showed that both of them had been taken from different and smaller specimens than the Adams mammoth. [3] In spite of that, the position of the tusks, which Adams fixed without much reason in such a way that the ends were widely separated and turned over to the right and left sides respectively, became commonly known and generally recognized. It happened, perhaps, because of the reproduction of the picture of the Adams mammoth by Cuvier who sanctioned the restoration, not having been able to discover all its defects. [4] As it was proved about a hundred years later, the tusks of the mammoth were approaching each other at their distal ends. [5] With the exception of tusks and a forefoot the skeleton was nearly complete. Bones at Yakutsk were cleaned by Adams from meat and ligaments by boiling.

Besides those possessed by the Zoological Museum of the Russian Academy of Sciences, a few pieces of skin and some hair, through the Russian Ambassador Golovin, found their way into the Natural Science Museum at Stuttgart, Germany, [6] and some also were sent to Sir Joseph Bancs and deposited by him at the Royal College of Surgeons in London. [7] A piece of the hide happened to be at the Zoological Museum in Berlin, where its hair was examined and described by Möbius. [8] Middendorff also saw hair of this mammoth at the University Museum in Moscow. [9]

[1] Pfizenmayer, E., "Morphologie von Elephas primigenius Blum.," S. 540. Also Mammutleichen, S. 240.
[2] Brandt, J. F., "Mittheilungen über die Gestalt," S. 96, footnote.
[3] Pohlig, H., "Monographie über die fossilen Elephanten," S. 323 and 388. After Pfizenmayer, E. W., "Mammutleichen," S. 241.
[4] Cuvier, G., "Recherches sur les ossements fossiles," 1, pl. xl.
[5] Pfizenmayer, E., "Morphologie von Elephas primigenius Blum.," S. 531.
[6] Brandt, J. F., "Einige Wörte über die Haardecke des Mammuths," S. 348.
[7] Middendorff, A. Th., "Sibirische Reise," IV, 1, S. 278.
[8] Möbius, K., "Die Beharung des Mammuts. After Pfizenmayer, E., Morphologie von Elephas primigenius Blum.," S. 527.
[9] Middendorff, A. Th., "Sibirische Reise, IV," I, S. 278.

The description of the locality as given by Adams [1] was not quite exact and provoked the false idea that the mammoth had been buried in ice. Toll proved later that this mammoth, as many others, was found in frozen ground, and the ice mentioned by Adams had been within ice cliffs near by the locality and underlain the mammoth bearing layer of frozen ground.[2] In spite of that, Pfizenmayer again speaks of this mammoth as of one found "within a gigantic piece of fossil coast-ice." [3] However, in another place he suggests that the animal had plunged into a crevasse and had been buried there within silt quickly frozen under a low temperature.[4] Adams also examined the nearby shore hills in one of which his mammoth had been found. They were covered on the top with tundra from which protruded the pieces of buried wood and plenty of tusks of mammoth in a surprisingly good state of preservation.

According to the Russian mining engineer Zlobin who visited the place in 1830, with a companion of Adams, trader Belcov, and later told Middendorff [5] of his observations, the mammoth had been found in a secondary location, as the carcass had slipped down the hill 35 feet high.[6]

The last time the locality was examined was during the Lena expedition, 1882–1884, by A. A. Bunge, who gave a detailed geographical and geological description of the Bycov Peninsula called Tumus or Tumul Peninsula by Yakuts.[7] According to Bunge the peninsula is a part of the delta of the Lena River, the carcass of the Adams mammoth was therefore buried within old river deposits and had been brought to its burial place by the Lena River, as were of course, the remnants of other Post-Pliocene mammals found here. Toll denies the delta origin of the Bycov Peninsula and considers it a part of the northern shore of Siberia, having the same geological structure a long distance east of the mouth of the Lena River.[8] As the shore referred to is composed mostly of silt and has been originated by means of the work of rivers, the question on the Bycov Peninsula, as a part of the delta of the Lena River, arouses only an academic interest.

Describing the locality of his mammoth, Adams speaks also of another carcass of a mammoth found two years before his own discovery, on the banks of the Lena River, a long way from the sea (7). This locality was not visited by him or by anyone else, and the carcass referred to has been lost to science.[9]

Tilesius,[10] while on his way to Kamchatka in 1805, was told by one Potapov, a Russian seaman, that a short time before, i.e. at the very beginning of the nineteenth century, the latter had seen on the shores of the Polar Sea, a mammoth with skin (8). Potapov presented Tilesius with a bunch of the hair from this carcass, which Tilesius in turn sent on to Blumenbach.

Shortly after the Adams expedition, in 1809, a Russian Government official, Hedenström by name, visited and later described the New Siberian Islands.[11] He was not a

[1] Adams, M., "Relation abrégée d'un voyage à la mer glaciale." The writer consulted the Russian translation of this article published in the *Sibirski Vestnik*, 1820, X, p. 307, St. Petersburg, 1920.
[2] Toll, Ed., "Die fossilen Eislager," S. 9.
[3] Pfizenmayer, E. W., "Mammutleichen," S. 24.
[4] Pfizenmayer, E. W., "Mammutleichen," S. 132.
[5] Middendorff, A. Th., "Sibirische Reise, IV," I, S. 294.
[6] Brandt, J. F., "Mittheilungen über die Gestalt," S. 103.
[7] Bunge, A., "Die Lena Expedition," S. 40–46.
[8] Toll, Ed., "Die fossilen Eislager," S. 14.
[9] Howorth, H. H., "The Mammoth and the Flood," p. 83.
[10] Howorth, H. H., "The Mammoth and the Flood," p. 83.
[11] Hedenström, M., "Travel to the Ice Sea," Hedenström, M., "Otrivki o Sibiri," p. 129.

scientist and was not looking for the carcasses of a mammoth, but could not pass by abundant fossil bones preserved just so well as the bones of recent animals could be.   In a rather humorous way he tells, how he collected a large sack of bones of a mammoth, still containing marrow, with the intention of bringing them back and utilizing their fat for some pharmaceutic purpose, or for perfumery.   During his travel home the bones by chance happened to be brought into a house and put too closely to a fire with the result that the fat flowed out, and Hedenström was deprived the opportunity to prepare a sensational "Pommade à Mammouth" to use his expression.   This sad story shows very well, how fresh the remnants of these animals used to be.   He was surprised that the marrow "in spite of its old age" did not emit a putrid scent.[1]

Maydell during his travels in Northeastern Siberia was told that nearly thirty years before, i.e. about in 1839, an ivory hunter by the name of Rozhin had found a carcass of a mammoth on the Shangin River, a tributary of river Indigirka, about a hundred miles above its embouchure (9).   The mammoth had been found in an upright position with its head and forefeet, all covered with hair, protruding from the bluff.   The remnant, i.e. the largest part of the carcass, had been still preserved within ground.   Nothing else was known later about this specimen, probably, one of the best, which has been therefore lost to science.[2]

In 1839 a partially destroyed frozen carcass of a mammoth had been found by Samoyeds on Tas River, as it was reported by them, washed out of the bank of the river.   Speaking of the animal they told, among other things, about a black tongue of the length of a month-old reindeer calf, which could have been nothing else than a trunk, afterwards destroyed completely.   They also spoke of the flapping ears of the animal.   At that time a Russian entomologist Mochulsky happened to be at Tobolsk and to learn about this discovery. He undertook the necessary steps to save for science what was possible to save.   A merchant of Berezof, Trofimof, visited the locality and brought to Obdorsk the parts of the skeleton, some hair, probably, from the mane of the animal, and a few pounds of flesh.   From Obdorsk these remnants have found their way into the Museum of the University of Moscow.[3]   Soft parts were collected and delivered in a shapeless heap.   They were microscopically examined and described by Glebov [4] who found within them tissues, hide, fat, and brain.   The remnants of the dried brain were sent on from Moscow to the St. Petersburg Academy of Sciences.   The locality was given by Trofimov on the left side of the Yenisei River, about fifty miles from its mouth, not far from the river itself.   According to Schmidt, who had also the chance to visit these regions, the specimen under consideration had been found near Zimovie Krestovskoye close to the Arctic Sea (10).

During his travel in Northern Siberia, in 1843, Middendorff chanced to discover the remnants of a half-grown mammoth, which he found on the Taimir River, near the mouth of the latter, at the latitude of about 75° (11).   The skeleton was fairly complete and intact, but the bones of it were rotten, softened, and covered with black, fatty soil over an inch thick.   As an organic substance was found in this soil later, it probably had originated from the soft parts of the carcass.   The animal laid on the left side in a layer of sand and

[1] Hedenström, M., "Otrivki o Sibiri," p. 121.

[2] Maydell, G., "Reisen und Forschungen im Jakutischen Gebiet," I, S. 426, Anm. 77.

[3] Middendorff, A. Th., "Sibirische Reise, IV," I, S. 272, Anmerkung.

[4] Gleboff, Recherches microscopiques sur les parties molles du mammouth: after Middendorff, A. Th., " Sibirische Reise, ᵀV," I, S. 272, Anmerkung.

clay, 5 to 7 feet below the surface, on a bank of the river, which was 42 feet high and composed of coarse sand, with boulders. In the opinion of Middendorff the mammoth in this particular case had not been buried on the very spot of his death, but its carcass had been brought by the river to its present location from more southern parts of Siberia.[1]

In 1848, the carcass of a mammoth was reported found on the Indigirca River by Benkendorff, a member of a Russian topographical expedition to Northern Siberia. The carcass, as it was told, was detached from the frozen banks of the river, and was careering about in the flood, when Benkendorff came across, secured it with a rope, examined the body, also the content of its stomach, but lost it when a sudden rush of water carried it away.[2] No attention is paid to this "discovery" in papers on the Mammoth of Russian scientists, as for example, by Baer, Brandt, Schrenck, Schmidt. Howorth quite correctly considers it a fiction written just for a boys' book.[3] The article was reprinted by Middendorff,[4] but he only "wished not to deprive pleasure to his readers." Such an expedition never took place to this part of Siberia. The first steamer arrived to the Lena River only with the Nordenskiöld expedition in the "Vega," in 1881. However, the mentioning of this article is necessary, because, apparently, sometimes it happened to be considered not quite from the point of view advocated by the writer.[5]

In 1854 the Museum of Geographical Society at Irkutsk, Siberia, got a foot of a mammoth from the Archbishop Nil, covered with hair. It was an only remnant of the carcass of the mammoth which had been found, presumably, in very good condition a few years before on the Kolyma River (12) by the local missionary Khitrovo who reported the matter to his chief and sent him a foot. The Archbishop, on the advice of one Stschukin, who learned about this discovery, presented it to the Museum of the Geographical Society at Irkutsk. Here the foot was seen by Schmidt who visited Irkutsk during his Mammoth expedition. The mammoth in this particular case had fallen down into the river from the underwashed shore cliff. Khitrovo reported also a putrid scent noticeable near the locality.[6]

In 1858, on the Vilui River, eighty miles above the town Vilyuisk, near the place known as Kentik (13), had been washed out a complete skeleton of *Rhinoceros tichorhinus* Fischer along with some soft parts, for example, six pectoral vertebrae of it were firmly connected by ligaments. These remnants were presented by Stubendorff, the Governor of the Territory of Yakutsk, to the Irkutsk Museum and then given over to the Russian Academy of Sciences. The Rhinoceros described by Pallas, perhaps, had been found at the same locality.[7]

In 1866 the Academy was told that about three years before Yakuts had found on Vilui River, near the mouth of the latter, the remnants of a large animal, mammoth or rhinoceros, covered with a skin, which they reported to the Yakutsk trader Kolesov (14). No investigation was made by Kolesov, or by the Academy either, which left all the business in the care of the East Siberian Branch of the Russian Geographical Society at Irkutsk.

[1] Middendorff, A. Th., "Sibirische Reise, I," Ss. 205–206; Bd. IV, I, Ss. 275, 284.
[2] Körber, Ph., "Kosmos für die Jugend."
[3] Howorth, H. H., "The Mammoth and the Flood," p. 90.
[4] Middendorff, A. Th., "Sibirische Reise," Bd. IV, Th. II, S. 1081.
[5] Sucachev, V. N., "Examination of Plant Remnants," p. 2.
[6] Brandt, J. F., "Zur Lebensgeschichte des Mammuth," Anhang, pp. 117–118. Also: Brandt, J. F., "Einige Wörte zur Ergänzung," p. 362.
[7] Tscherski, I. D., "Beschreibung der Sammlung," S. 31.

So far as is known, nothing was done by this scientific body either, and the carcass, if there was one, was destroyed and lost.[1]

In the same year, the Academy got word of the remnants of a mammoth found in 1864 by a Yurak (a native tribe of Northern Siberia) in the tundras between Taz and Yenisei Rivers, at the source of the River Gida (15), where he was looking for his reindeers and came across the tusk of a mammoth protruding out of the ground.[2] After some digging he discovered the head of a mammoth. He broke, or cut off a tusk, took a piece of skin, and brought all to Dudinka, a small Russian settlement on the lower Yenisei River, where it provoked a sensation unfavorable to the better preservation of such discoveries, as some people visited the locality and tried to dig out something more, destroying what had not been already destroyed. The rumor of the new found mammoth spread over Siberia and reached one Gulyaev, who at that time happened to be at Barnaul, a small town in the Altaï Region, but who had some personal connections with the far northern inhabitants. He was interested in science, and immediately appreciated the importance of this discovery: so he reported it to a member of the Academy, Dr. Baer, who brought the matter to the attention of the Academy which was thus notified of the discovery just two years after the mammoth had been found.[3] The Academy immediately decided to send over an expedition in charge of Fr. Schmidt, afterwards a member of the Academy. For this expedition, 4800 rubles [4] were assigned by the Russian State Treasury. As the particular spot where the mammoth had been found was not quite certain, Fr. Schmidt was advised to watch in the tundras for the smell which could be originated from the rotten remnants of a mammoth.[5]

Schmidt's expedition, so far as the carcass of the mammoth was concerned, was a complete failure, a few isolated and broken bones, pieces of skin and plenty of hair being all that was brought to St. Petersburg, but Schmidt collected very important data on the locality itself. For the first time a geological section of the Post-Pliocene strata was established for Northern Siberia and the relations between the mammoth-bearing strata and other ones, especially the deposits of Arctic transgression were determined. Concerning the locality of his mammoth, Fr. Schmidt came to the conclusion that the animal had been buried on the very spot where it had died, or had been moved only a very little.[6]

While in the North Fr. Schmidt learned of another skeleton of the mammoth found in the Avamskaya Tundra and eventually secured a number of bones and a quantity of the hair from this specimen (16). The lot consisted of foot-long hairs, probably from the mane, and short wooly hairs, two inches long.[7]

In 1867, a Tungus, by the name of Phoca, came across a foot of a mammoth protruding more than two feet from the ground, in the tundras between the rivers Alazea and Indigirca in Northeastern Siberia (17). There was neither meat nor skin preserved, but only liga-

---

[1] Meeting of the Physico-Mathematic Section of the Russian Academy Sc., March 22, 1866: Mem. Ac. Sc., IX, p. 166, St. Petersburg, 1866.
[2] Meeting of the Physico-Mathematic Section of the Russian Academy Sc., January 11, 1866: Mem. Acad. Sc., IX, p. 81, St. Petersburg, 1866.
[3] Baer, K. F., "Neue Auffindung eines vollständigen Mammuths," p. 230. Baer, K. E., "Fortsetzung der Berichte über die Expedition," p. 513.
[4] Meeting of the Physico-Mathematic Section of the Russian Academy Sc., February 8, 1866: Mem. Acad. Sc., IX, p. 87, St. Petersburg, 1866.
[5] A personal communication of late Fr. Schmidt, to the writer.
[6] Schmidt, Fr., "Vorläufige Mittheilung." Schmidt, Fr., "Resultate der Mammuthexpedition."
[7] Howorth, H. H., "The Mammoth and the Flood," p. 87.

ments were found, besides the bones. The next year the same native visited the locality again, was unable to find the foot of the previous year, but discovered a small part of another one.

At that time happened G. Maydell to be traveling in Northeastern Siberia. The Academy after it had learned, from the Ispravnik (chief of police) of the Verkhoyansk District, of this discovery, asked Maydell to examine the locality discovered by Tungus Phoca, and assigned for this purpose 1500 rubles. The locality was fixed by Maydell on the small river Kovshechya (Zuskendunu in Tungusic, Khomos-Urakh in Yacutish) which enters the Arctic Sea about 40 miles west of Alazea River. The Kovshechya River is composed of two branches: the eastern one, by name Ulakhan-Khomos-Uryakh, and the western one known as Alshygy-Khomos-Uryakh, on which the mammoth was found. A reliable man, sent to the spot by Maydell, found only a foot, a piece of skin, and a skull frozen in the ground on the bottom of the valley. The man thought that the carcass of the mammoth had been washed out of the cliff, fallen down into the stream, and gradually been destroyed by water.[1]

The Kovshechya River had been mentioned by Wrangel's expedition, under the name Vshivaya or Pila, as a river which washed down off its shores many mammoth bones.[2] All this part of Northern Siberia has been known since ancient times to be extremely rich in remnants of large fossil animals. Here has been found, for example, in 1787, the mammoth mentioned above.

It was hardly surprising, therefore, that, in 1870, Maydell learned of another mammoth found on the right side of the Kolyma River (18), between Nijne-Kolymsk and Sredne-Kolymsk towns, as well as of the third one discovered at a small creek, Shadran (19), 25 miles west of the first locality on river Kovshechnaya. At the second locality there were found only bones of the animal, piled together. At the third place Maydell found a foot and a part of another one, also a layer composed of the hair of a mammoth mixed with earth. The remnants were found here in a narrow edge-like divide between two runs, and were also washed out by water. In Maydell's opinion in this case the carcass of a mammoth had fallen into the creek, gradually decayed, and at last been destroyed by the stream.[3]

In 1875, Tscherski, commissioned by the Siberian Branch of the Russian Geographical Society, happened to explore, in Southern Siberia at the latitude about 54° 25′ and about 98° 35′ E., a cave located about 40 miles south of Town Nijne-Udinsk, therefore known in Russian literature as the Nijne-Udinsk Cave. In frozen ground within the cave, among the remnants of various animals, he found a piece of hide covered with hair, which he identified *Rhinoceros tichorhinus* Fischer (20). Although this remnant could be not compared with the carcasses formerly found in Northern Siberia, it certainly is worthy of mention on account of its geographical position, so far distant from the northern localities.[4]

In 1876, Nordenskiöld collected a few bones and pieces of the hide of a mammoth shown to him by natives on the Yenisei River, at the latitude 71° 28′, at the mouth of the

[1] Maydell, G., "Reisen und Forschungen im Jakutischen Gebiet."
[2] Wrangel, F., "Narrative of an Expedition," p. 220.
[3] Schrenck, L., "Bericht über neuerdings im Norden Sibiriens angeblich zum Vorschein gekommene Mammuthe," Ss. 147–173. Maydell's letters published in this article have been reprinted by Digby in his book. Descriptions of the localities, with some comments, is given also by Toll in "Die fossilen Eislager," pp. 18–25.
[4] Bull. East Siberian Branch, Russian Geogr. Soc., Irkutsk, VI, 5 and 6, p. 211; VII, 2 and 3, p. 78; X, 1 and 2, p. 28. After Tscherski, I. D., "Beschreibung der Sammlung," S. 12.

Mesenkin River (21).   The hide was 20–25 mm. thick and appeared to be naturally tanned. Presumably these remnants had been washed out of a tundra bank by the Mesenkin River.   In the neighborhood was found also a skull of the musk-sheep.[1]

In 1877, the Academy got a rather vague report on a carcass of a mammoth, presumably, discovered in the Kuznetzki Alatau Mountains in Southern Siberia.   Zoologist Polyacov immediately left St. Petersburg for Altaï in charge of an expedition for which the Academy had assigned a thousand rubles.[2]   The locality was said to be in the valley of a small creek, Nicolca, a tributary of the Kundat River, which flows into Kiya River, at the gold placer Zolotoi Bugorok.   What was considered the skin of a mammoth appeared to be mountain-leather, a mineral aggregate, which had been found immediately underlying the goldbearing sand layer, about 15 feet below the surface.[3]   The chief of the local police, through whose hands had passed all the news of the mammoth discovery, and who had officially checked the report, felt that he was responsible for Polyacov's expedition and scolded the discoverer, a Siberian peasant.   The chief especially reproached and treated him as a liar for the statement that the latter ate "the skin," which consideration had, probably, been for the chief a decisive argument as to the reality of the mammoth find and a sufficient reason for reporting the discovery to the Academy.   The man obstinately affirmed he truly had eaten the supposed skin, but added: "seasoned with butter, what is not possible to eat."

In 1877, a carcass of a *Rhinoceros tichorhinus* Fischer perfectly preserved, with skin and hair, had been found by Yacut Gorokhov at the latitude of about 68° 30′ on the Khalbugaï Creek which flows from the right side into river Bytentaï, a left tributary of the Yana River (22).   With the assistance of his son, Gorokhov chopped off a foot and the head of the animal, but left the body, which was destroyed the next year by spring water.   The head was sent to Irkutsk Museum of the East Siberian Branch of the Russian Geographical Society, where it was identified and described by Tscherski.[4]   The further history of this remarkable relic is certainly worth recording.   In 1879 the great Irkutsk fire destroyed the Museum along with all collections and the library.   The head of the rhinoceros escaped the same fate only because of the fact that a few months before it had been sent to the anthropological exhibition in Moscow.   After the exhibition the head was transferred to St. Petersburg, to the Academy to which it was presented by the Irkutsk Branch of the Russian Geographical Society.[5]   At the present time this specimen is exhibited at the Zoological Museum of the Academy, just in the same condition, as was brought to Irkutsk from Northern Siberia, i.e. dried, or mummified.   The history of this rhinoceros is not quite correctly given by Pfizenmayer.[6]   According to him the carcass was discovered in 1879.   Gorokhov sent the head to some merchant at Irkutsk, and this one presented it to the Academy.   By L. Schrenck, differing with Tscherski, this rhinoceros was identified

[1] Nordenskiöld, A. E., "Die Umseglung Asiens und Europas," I, S. 371.

[2] Meeting of the Physico-Mathematic Section of the Russian Academy Sc., May 3, 1877: Mem. Academy Sc., XXX, p. 50, St. Petersburg, 1877.

[3] Meeting of the Physico-Mathematic Section of the Russian Academy Sc., September 27, 1877: Mem. Acad. Sc., XXX, p. 81, St. Petersburg, 1877.

[4] Bull. East Siberian Branch, Russian Geographical Society, Irkutsk, IX, 5 and 6; X, 1 and 2. After Tscherski, I. D., "Beschreibung der Sammlung," S. 12.

[5] Meetings of the Physico-Mathematic Section of the Russian Academy Sc., August 28 and September 25, 1879: Mem. Acad. Sc., XXXV, pp. 111 and 116, St. Petersburg, 1879.

[6] Pfizenmayer, E. W., "Mammutleichen," S. 243.

as *Rhinoceros merckii* Jaeg.;[1] but Tscherski later verified his first identification.[2] Schrenck also suggested that this rhinoceros, like the recent one, had been deprived of fur. As Pfizenmayer has shown, the hair was destroyed, probably, through careless transportation of the head for more than two thousand miles.[3] In 1885, the locality was examined by Toll who gave its detailed description. According to Toll the carcass was washed out owing to very high level of water in the year of discovery, when the high shore of the river was underwashed, and the carcass slipped to the lower shore. In Toll's opinion the carcass was buried in the old river channel where it had been brought by the water. In other words the locality belonged to the higher, older terrace of the river.[4]

During the Lena expedition, in 1882–1884, commissioned by the Russian Academy of Sciences for the meteorological work in Northern Siberia to the mouth of the Lena River, where a special meteorological station was built up for this purpose,[5] a member of the expedition, Dr. Bunge, did a great deal of exploration and travel in the delta and paid much attention to the localities of carcasses of mammoth. He visited and closely examined the locality of the Adams mammoth. He also investigated the locality of the mammoth found, in 1857, by natives at Island Mostakh (23), but not reported to authorities and destroyed as much by natural causes as by man.[6] When this mammoth had been found its head, bearing tusks, which had appeared out of the ground first, was chopped off and sold to the local trader, Shakhurdin. The skin, according to natives, was about two inches thick, and so well preserved that it could be used to make dog harnesses. The fat was a little yellowish on the surface, but snow white deeper. It was used by natives to lubricate small local boats known as vyetca. The flesh, pink on the surface, was bright red deeper. The natives did not try to eat any of it themselves.[7] During a number of years and with the assistance of dogs and wild animals the carcass, probably one of the best ever found, was destroyed completely, and Bunge after excavating (in the meantime the carcass has been covered with sand), could collect only fragments of bones bearing the traces of axes, plenty of hair, remnants of food from the stomach of the animal, excrement, remnants of fat and of ligaments. According to Bunge, it was a young animal. The examination of the excrement of this mammoth, made by Famintzin, proved the presence of a vegetable matter within, but gave no particularly important results.[8]

At the same time Bunge learned of another mammoth found, in 1879, on Moloda Creek, a left tributary of the Lena River, above the settlement Sictakh, more than 400 miles up the river from the meteorological station (24). Here also, from a sandy bluff, at first appeared the head of the animal, the tusks of which were chopped off immediately. Bunge was unable to visit the locality, and this mammoth has been lost to science. It had never been reported to authorities.

In 1885–1886 the Academy commissioned Bunge and Toll to go to the New

---

[1] Schrenck, L., "Der erste Fund einer Leiche von *Rhinoceros Merkii* Jaeg." After Tscherski, I. D., "Beschreibung der Sammlung," S. 12.

[2] Tscherski, I. D., "Beschreibung der Sammlung," S. 13.

[3] Pfizenmayer, E. W., "Mammutleichen," S. 245.

[4] Toll, Ed., "Die fossilen Eislager," S. 36.

[5] It was the Russian share in a great scientific enterprise undertaken at that time by different countries in Arctic regions.

[6] Bunge, A., "Die Lena Expedition," S. 52–96 (Nachrichten über Mammuthcadaver im Unteren Lena-Gebiet).

[7] Bunge, A., "Die Lena Expedition," S. 51.

[8] "Meeting of Physico-Mathematic Section of the Russian Academy Sc., January 21, 1886: Mem. Acad. Sc., LII," p. 173, St. Petersburg, 1886.

Siberian Islands. The expedition examined also the adjacent part of Arctic Siberia.[1] During their trips each of the travelers happened to come across remnants of mammoth. In no case were they complete, but the observations made on the spot contributed a great deal to the natural history of the mammoth.

Pieces of skin and plenty of hair of a mammoth were found by Bunge protruding out of a frozen bluff on the coast of Bolshoi Lyakhov Island (25). Destroyed and incomplete bones of the same specimen were found below the spot. The marrow within the bones was chalk-like, but fresh enough to be immediately devoured by dogs. Presumably it was a complete carcass of a mammoth a few years before the arrival of the expedition.[2] Bunge's guide told him that a few years before, he had found at the bottom of the same bluff a complete carcass of a musk ox which he was able to describe so well that it was no trouble for Bunge to identify the animal. Carcasses of different animals, skeletons and isolated bones used to fall out of the bluff during the whole warm season. After a while bodies and skeletons used to be destroyed by warmth, streams of water running from the bluff, and by the waves of sea. Temporarily, bones can be buried again, plunging in the soft ground at the bottom of the bluff, ground originating from the mud streams running from the face of the bluff, or being covered with mud. Owing to the presence of a frozen ground below the thawed surface, large bones, such as the tusks of a mammoth, cannot plunge deeply, but small bones are usually buried completely. Later they can be washed out by waves and easily collected, at low tide, which usually is provoked by a favorable wind, when the shallow sea around the New Siberian Islands dries out to a great distance. A great amount of ivory used to be collected in this way by ivory-hunters. Traveling along the shore Bunge noticed also an odor of decomposition in the thawed ground, which is in his opinion, probably, peculiar for all the earthy deposits of the New Siberian Islands.

Another locality was visited by Toll on the northern shore of the Bolshoi Lyakhov Island (26). A mammoth had been discovered there by the hunter Boyarski, in 1860. Boyarski accompanied Toll to the very spot and only for this reason could the latter examine the locality, as no remnants of animal were present. The part of the bluff containing the carcass of the mammoth had been completely destroyed about 1863. The mammoth had been found by Boyarski in an upright position, frozen within the clay and sand pockets located between two ice masses composing here the cliff, and with its posterior part protruding out of the bluff.[3]

Toll, in 1886, also examined a locality of a mammoth on the mainland, at the latitude of 70° 20′, on the Boryurakh Creek, a right tributary of the Chendon River which enters the Arctic Sea about a hundred miles east of river Yana (27). Only fragments of bones, a few of soft parts, and hair were found in this locality examined by the expedition 23 years after the mammoth, probably a head only, had been discovered by Tungus Sleptzov who carelessly had chopped off the tusks and destroyed the specimen. In Toll's opinion, in this case incomplete remnants of a mammoth had been buried on the ice of flood ice, i.e. aufeis, when during the flood season they were covered with silt and later frozen within.[4]

Besides remnants of the mammoth Bunge-Toll's expedition collected also a number

[1] Beiträge zur Kenntniss des Russischen Reiches, III Folge, Bd. III, after Toll, Ed., "Die fossilen Eislager," S. 49.
[2] Toll, Ed., "Die fossile Eislager," S. 50.
[3] Toll, Ed., "Die fossilen Eislager," S. 53.
[4] Toll, Ed., "Die fossilen Eislager," S. 40.

of bones of other Post-Tertiary mammals which were identified and described in detail by Tscherski.[1]

In 1889 a discovery of a complete mammoth was reported to the Academy by the General-Governor of Eastern Siberia, as found somewhere near river Anabar in Turukhansk district of Northern Siberia.[2] Information following gave the locality on the Balakhna River near Khatanga Bay (28) at the latitude 73°. The Academy commissioned Toll to investigate this locality, but owing to the condition of his health he could not enter this enterprise.[3] A few years later, in 1893, he happened to be in these regions, but did not mention this locality. Probably, it was a rumor of no importance. This mammoth sometimes has been mentioned as the Burimovich mammoth.[4] Burimovich was the Ispravnik (chief of police) of Turukhansk district, who first has delivered the news of this discovery.

In 1891, Tscherski offered a new plan of mammoth-hunting according to which a scientist must stay in the Far North of Siberia for a couple of years and in this way to have an opportunity of checking immediately all reported discoveries of carcasses of mammoth.[5] For a scientist it would be a voluntary exile which Tscherski elected for himself, his wife, and their son of eleven years of age, when, in 1891, he left St. Petersburg for Northeastern Siberia with the intention of staying there for four years. He expected to do a regular geological work and at the same time to listen to all rumors referring to mammoth-localities. Very unfortunately his health was broken at that time and the next year he died during the boat travel down the Kolyma River.[6]

Tscherski had no chance to learn of any new mammoths, or to discover one himself either, although just at this time, in 1891, a mammoth has been discovered in Northern Siberia, on the Sanga-Yurakh River, about 250 miles east of the settlement Ust-Yansk (29), and the discovery reported to the Academy by a local trader Sannikov. Word about it was sent over to Tscherski,[7] but he could not get it. This locality was examined, in 1893, by Toll who was commissioned by the Academy to accomplish, so far as it was possible, and as time and money permitted, the work which had been started by the late Tscherski. Toll found only destroyed bones, pieces of hairy skin, and much of hair, all deposited within the alluvium of the Sanga-Yurakh River. The carcass of the animal in this case had been washed out by the river many years before, and the locality had no special interest even from a geological point of view.[8]

Probably the most important discovery of a mammoth was made, in 1900, in Northeastern Siberia, about 200 miles northeast of the small town Sredne-Kolymsk, on the river Beresovca, the right tributary of the Kolyma River (30). As usual the discoverer was a native, Lamut S. Tarabukin. In August, 1900 while hunting for a reindeer he came across a tusk of a mammoth weighing about 166 English pounds. Looking for another tusk he

[1] Tscherski, I. D., "Beschreibung der Sammlung."

[2] "Meetings of Physico-Mathematic Section of the Russian Academy Sc., April 25 and September 5, 1889: Mem. Acad. Sc., LXXI," pp. 79 and 127, St. Petersburg, 1890.

[3] Toll, Ed., "Eine Reise nach den Neusibirischen Inseln," S. 132.

[4] Digby, B., "The Mammoth," map.

[5] Tscherski, I. D., "Beschreibung der Sammlung," S. 454.

[6] On the last days of the Siberian Traveler I. D. Tscherski.

[7] "Meeting of the Physico-Mathematic Section of the Russian Academy Sc., January 15, 1892: Mem. Acad. Sc., XLIX," p. 54, St. Petersburg, 1892.

[8] Toll, Ed., "Eine Reise nach den Neusibirischen Inseln."

discovered a well preserved head of a mammoth bearing only one tusk of much smaller dimensions than the first one, about 63 lbs. as it was found later.

As Lamuts believe that the excavation of a mammoth produces sickness, Tarabukin was rather afraïd of his discovery, did not touch the carcass, but returned immediately to his camp and told two other Lamuts about the mammoth. The next day they visited the locality, chopped off the tusk, but did not touch the carcass. Examining the locality, the Lamuts came to conclusion that the head of the animal had appeared out of the ground during the previous season, i.e. in 1899. The tusks were later sold to a Russian cossack, Yavlovski, who learned about the mammoth on that occasion from the Lamuts, and persuaded them to show him the locality. After the discovery had been checked, Yavlovski received from the Lamuts their claim to the mammoth, reported the matter to local authorities and, through their assistance, to the Academy. The carcass he covered for a while with sand and stones. The news of this mammoth arrived to St. Petersburg in April, 1901. It was immediately resolved to send an expedition composed of three people: the leader, a zoologist, O. Herz, a taxidermist, E. Pfizenmayer, and a geologist, D. Selivanov. 16300 rubles were assigned to the Academy from the State treasury for this expedition. Later this sum was increased by a few thousand rubles, a part of which was given by the Grand Duke Constantine, the President of the Academy at that time, who returned to the Academy his salary of the President to cover some extra expenses of the expedition.

The expedition left St. Petersburg on May 3/16, 1901, in June arrived in Yakutsk and immediately left for Sredne-Kolymsk. During the summer the country is practically impassable, and usually nobody tries to cover the distance, about 1500 miles, between Yakutsk and Sredne-Kolymsk except in winter, when horse and reindeer sledges are used. During the summer the journey can be made only on horseback, using pack horses for carrying baggage. The expedition took more than three months to cover these 1500 miles. The drawbacks and difficulties of such a trip could be appreciated only by one who himself had the misfortune to travel through the same region and under the same conditions. The geologist of the expedition, a young strong man, but lacking sufficient training, was completely broken down and stopped all work about at the end of the journey, when less than a hundred miles separated him from the mammoth. A lively description of this journey has been given by Pfizenmayer in his book,[1] often quoted by the writer.

The work of excavation was carried on with great energy and skill and accomplished in a month, between September 11/24 and October 11/24. Soft parts were treated in the usual way, but great part brought to St. Petersburg frozen and only later prepared for a permanent preservation. Thanks to the Russian winter it was also possible to bring to St. Petersburg two large pieces of ground ice from the locality and have time, before the warm season, for their examination by the writer.

The mammoth was found in the best imaginable condition and comparatively little spoiled by wild animals. It has been exhibited in the Zoological Museum of the Academy as a stuffed animal with the skeleton exhibited nearby separately. The pose given to the specimen corresponds to that in which the animal was found, as if trying with its last strength to go out of some trap into which it had happened to fall. Perhaps the animal had broken through into a crevice, as thought Herz,[2] or plunged into soft ground, as sug-

---

[1] Pfizenmayer, E. W., "Mammutleichen."
[2] Herz, O. F., "Frozen Mammoth in Siberia," p. 617.

gested by the writer,[1] while on its pasture-ground, and died of injuries received (the pelvis, a forefoot and a few ribs were found broken, as well as the indication of a strong hemorrhage) and also of suffocation in mud. The death by suffocation is proved by the erected male genital, a condition inexplicable in any other way. However, the carcass was found, not on the very spot where the animal had perished, but within the landslide which, along with the carcass, slid down from the upper border of the high terrace of river Beresovca,[2] these slides caused by the thawing of rock ice underlying the tundra. The flesh was so fresh and appealing that dogs devoured every piece thrown to them. Such investigations as those on the histology of stomach tissues were accomplished later with great ease. Blood, collected in great masses, owing to hemorrhage, was found to be in such a good state of preservation that it could be examined about as easily as the blood of recent animals.[3] According to Pfizenmayer it was even possible to establish the relationship of blood of the mammoth and the Indian elephant.[4] Concerning the preservation of blood it is necessary to mention that Neuville and Gautrelet, who examined the blood of the mammoth from the Bolshoï Lyakhov Island in the Museum of Paris, in a nearly similar state of preservation, do not confirm the conclusions of Russian students as to the extremely unaltered character of the blood.[5]

It is beyond the limits of the present paper to speak of all the scientific work done on the remnants of this mammoth.[6] As to the shape of the animal, in the Beresovca mammoth have been discovered a number of new characters. The tail of the mammoth was found to be much shorter than that of the Indian elephant, but much thicker in its basal part. Connected with the tail the mammoth had a peculiar cover of the anus in the form of a fold of the skin. Differing from other elephants the feet of the mammoth had only four toes each. The spiral-like tusks were not turned towards the outside, but had their ends directed inwards and downwards. The animal probably had no mane, as usually suggested in descriptions of the mammoth.[7] It is also necessary to mention that the Beresovca mammoth has been identified by Hay on the basis of description by Zalensky,[8] as a new species, *Elephas beresovkius* sp. n.[9]

In a rather unusual way a few remnants of this mammoth have found their way into the U. S. National Museum which purchased them, in 1922, from Pfizenmayer. Everything collected during the Beresovca expedition was the property of the Academy. Pfizenmayer had no right to keep in his hands the specimens referred to, much less to sell them.[1]

During his travel to the Beresovca River, in 1901, Pfizenmayer discovered near the small town Verkhoyansk, in the bed of the Khoptolog Run (31), a skull and other bones of a destroyed skeleton of a *Rhinoceros tichorhinus* Fischer, which still preserved a few remnants of ligaments and other soft parts.[11]

[1] Tolmatschow, I. P., "Bodeneis vom Fluss Beresovka," S. 444.
[2] Pfizenmayer, E. W., "Mammutleichen," S. 128.
[3] Bialinitzki-Birula, T. A., "Observations histologiques," p. 10; Zaleskii, W. W., "Etude microscopique," p. 33.
[4] Pfizenmayer, E. W., "Mammutleichen," S. 165.
[5] Neuville, H. et, J. Gautrelet, "Observations faites sur le sang du Mammouth," p. 108.
[6] Different articles written on this mammoth, or in connection with it, were published by the Russian Academy of Sciences, in Russian, in a set under general title, "Resultats scientifiques de l'expédition organisée par l'Académie Impériale des Sciences pour la fouille du Mammouth, trouvé sur la rivière Bèrèzowka en 1901."
[7] Pfizenmayer, E., "Beitrag zur Morphologie von *Elephas primigenius* Blum.," S. 527.
[8] Zalenskii, W. W., "Osteological and Odontological Researches."
[9] Hay, O. P., "Observations on Some Extinct Elephants," p. 4.
[10] Report of the U. S. National Museum for the year, ending June 30, 1922, p. 80.
[11] Pfizenmayer, E. W., "Mammutleichen," S. 85.

In 1903, Engineer Brusnev, a member of the Russian Arctic Expedition, during his travel on Island Novaya Sibir came across the remnants of a mammoth (32). After two days of work he realized that no carcass was present, as he found only some odoriferous badly putrefied flesh, among other remnants part of a decayed trunk, a broken tusk and plenty of hair mixed up with clay.[1]

The writer, during his Khatanga expedition, in 1905, found on the southeastern coast of the Khatanga Bay, at the latitude about 73° 15', bones of the pelvis and of a hind foot of a mammoth protruding out of frozen bluff in a more or less upright position (33). Although the presence of, at least, a complete head, or a skull could be suggested here, no excavations were made, as the expedition was short of time, had few people and no tools. As a special expedition could arrive over there only in the next year, or even two years after the visit referred to, and the moment was very unfavorable for finding necessary funds, no arrangement was made later in St. Petersburg, and no expedition was sent to dig out these remnants which, therefore, have been lost for science.

A rich locality of fossil ivory discovered by Lyakhov, in 1750, "between rivers Khatanga and Anabar" probably had been found on the southeastern shore of the Khatanga Bay.[2]

In 1907, the Governor of the Territory of Yakutsk reported to the Academy a new mammoth found by a Lamut, V. Dyacov (34), on the shore of the river Sanga-Yurakh (the same river where a mammoth-locality was examined by Toll, in 1893), about 200 miles northeast of a small Russian settlement Kosachye on river Yana.[3] In February, 1908, the Academy sent over an expedition in charge of geologist C. A. Vollosovich along with the taxidermist E. Pfizenmayer, for which purpose was assigned by the State treasury a sum of 16928 rubles.

From the time of the first report of the discovery, the locality was guarded, by the order of the Governor. In spite of that, the carcass was found in rather poor condition, many parts missing, and all scattered around, although in Vollosovich's opinion it was found just on the very spot where the animal had found its end, trapped in a mud stream after hopelessly having tried to free itself from the treacherous catch.[4] In Pfizenmayer's opinion this locality was secondary. Primarily, the carcass had been frozen in ground on the slope of hills bordering the valley on the right side, near the locality. The carcass was uncovered by spring water, gradually washed out, and brought down into the run bed, where it was found and examined by the expedition.[5] According to Vollosovich much of the carcass had been destroyed by wild animals immediately after the death of the mammoth and before it was protected by a cover of mud. In Pfizenmayer's opinion ice foxes used to feast upon the carcass after it was uncovered.

Of special interest in this case was the discovery of remnants of a trunk which at that time was not known exactly, in the mammoth. Worthy of mention also is the comparatively small size of this mammoth, although it was a full grown animal. Pfizenmayer suggested that it must have been a female,[6] which is supported by Nasonov,[7] but even for

[1] Brusnev, M., Report of the Leader of an Expedition to New Siberian Islands, p. 192.
[2] "Account of Russian Sea Travels," p. 168. Wrangel, F., "Narrative of an Expedition," p. 460.
[3] Meeting of the Physico-Mathematic Section of the Russian Academy Sc., January 8, 1908: Bull. Acad. Sc., II, p. 339, St. Petersburg, 1908.
[4] Vollosovich, C. A., "On the digging out of the Sanga-Yurakh Mammoth, in 1908," p. 453.
[5] Pfizenmayer, E. W., "Mammutleichen," S. 225.
[6] Pfizenmayer, E. W., "Mammutleichen," S. 227.
[7] Nasonov, N. V., "On the Remnants of the Carcass of the Mammoth from the Sanga-Yurakh River," p. 1320.

a female it was an undersized individual. Vollosovich considered such a decrease in the size of the mammoth as an indication of the beginning of the extinction of the race.[1]

According to Digby a lock of the hair of this mammoth can be seen in an exhibition case at the British Museum in London.[2]

While in the North, Vollosovich heard a rumor of another well preserved mammoth found, in 1906, by a trader, A. Gorokhov, on the Bolshoï Lyakhov Island, at the source of the Eterikan Creek (35). Here in the valley of a small nearly dry run was discovered the skull of the animal, frozen in the ground and still covered here and there with a hairy hide. A trunk, "a tube about seven feet long," as it was described by Gorokhov, also covered with skin was found as well, but for some reason he chopped it off, and broke out a tusk as well. As the presence in the ground of other remains of the mammoth appeared very probable, Vollosovich asked the party of ivory hunters, who were ready to leave for this island, to find this locality again (Gorokhov had died in the meantime) and collect the best preserved parts of the mammoth. To finance this undertaking Vollosovich borrowed money from local people, expecting that all the expenses would be covered later by the Academy.

In 1908 the hunters found the locality and remnants of the mammoth, still buried within the frozen ground of the run, and started the excavation. A new party continued the work in 1909 and finished it in 1910, but was unable, because of shortage of dogs, to bring everything to the continent and left behind a part of remnants. Digby is certainly right in saying: "The problem of hunting ivory in the New Siberians is less the difficulty of finding tusks than the difficulty of getting them away."[3] During the summer of 1910 the mammoth was preserved for some time in the frozen ground on the lower Lena, then with the last steamer sent over to Yakutsk and in December forwarded to St. Petersburg, where it was kept for a few years in a refrigerator. In the meantime a piece of hide covered with hair was sent to Paris.[4]

According to the report of the collectors, the carcass was found lying on the left side which, still frozen in ground, was therefore better preserved. The upper part of the carcass, probably, had been destroyed shortly after the death of the mammoth by wild animals. From this mammoth were delivered to St. Petersburg: a skull with the left tusk, upper lip and the left eye; the most important parts of the skeleton; pieces of skin from the head and back, with the left ear; skin from the hip, with the tail, also from different parts of body; penis and a few lumps of putrefied meat; four feet of which the left hind one had been preserved completely down from the knee, the other ones only in their lower parts. The hair of this mammoth has been distinguished by the great variety in color on different parts of body, as well as by the length of hairs, which has been explained by Vollosovich as a result of the seasonal change of hair, and as an indication that the animal had perished late in the summer. The remnants of the food from the stomach were not well collected, perhaps even mixed up with plant-fragments brought to the place with water later. Anyhow, they are similar to those found within the stomach of the Beresovca mammoth and mostly consist of grass and of a little moss. The well preserved feet of this mammoth have very peculiar hoofs, such as are found now in the cattle dwelling on the wet ground.

[1] Vollosovich, C. A., "On the Digging out of the Sanga-Yurakh Mammoth in 1908," p. 456.
[2] Digby, B., "The Mammoth," p. 212.
[3] Digby, B., "The Mammoth," p. 151.
[4] Vollosovitch, C. A., "Le mammouth de l'ile Bolchoï Lakhovsky," p. 310.

It was a good adaptation for marshy pasturages, but made mud streams, originating from the thawing ground, more dangerous for the mammoth than for its contemporaries. In Vollosovich's opinion this mammoth, like that of Sankha-Yurakh, had found its end in a stream of mud.[1]

The further fate of this mammoth is worth relating. For a long time Vollosovich could not get from the Academy the reimbursement of money spent for this supplementary and successful expedition. Being unable to get out of trouble he asked a friend of his, Count Stenbok-Fermor, for assistance. The latter immediately paid the whole sum of money, but, in 1914, presented the mammoth to the Jardin des Plants in Paris.[2] The reason for such a generous gift was the hope of being decorated with the Légion d'Honneur and, in the capacity of a possessor of this decoration, of having at his funeral a military band playing.[3] For some peculiar reason the gentleman was as much interested in this band as in the decoration itself. According to Digby the mounting of the skeleton of this mammoth is nearing completion at the Paris Museum.[4] Among the papers published in France on this mammoth it is necessary to mention this one by Depéret and Mayet [5] who have made a new subspecies *Elephas primigenius sibiricus* D. & M. If the mammoth of Siberia has to be considered specifically different from the Blumenbach's species, this name, conforming to the rules of priority, must be replaced by that offered by Hay a year before— *Elephas beresovkius* Hay.[6]

In 1909, the Academy learned of a mammoth found by a Samoyed in the tundras east of the Yenisei River not far from the small settlement Golchikha, who sold his find to a Russian trader.[7] The locality was then examined by the local trader, Byegichev, who located it at the sources of Creeks Kazachya and Poperechnaya, about 20 miles northeast of river Yenisei (36). He reported to the Academy[8] that the mammoth primarily had, probably, been in very good condition, but later the carcass was greatly damaged by wild animals and natives. After this disappointing report the locality was left by the Academy without any further attention.

In 1912, a new discovery from the same locality was reported to the Academy by a local merchant, Kucherencov, who described the mammoth as well preserved. In 1913, the locality was examined by Kutomanov commissioned by the Academy.[9] He found the carcass completely destroyed and could collect only the skull without tusks, isolated bones of the skeleton, pieces of hide, hair a.s.o. He was told that shortly after the discovery of the mammoth a piece of flesh had been sent to the local museum at Yeniseisk. The flesh was fresh and fat. He was unable to find it later at the Yeniseisk Museum. Kutomanov could positively establish the fact that this mammoth had been found, in 1908, by an Yurakh who sold it to a Russian trader. As matter of fact the latter did not know what

[1] Vollosovitch, C. A., "Le Mammouth de l'ile Bolchoï Lakhovsky," p. 325.
[2] "La Nature, 42 Année, I Sém., No. 2128 (Mars 7, 1914)," p. 240.
[3] A personal communication to the writer by the late Vollosovich, like a number of other details given in the above history of this mammoth, which Vollosovich's article partly lacks.
[4] Digby, D., "The Mammoth," p. 212.
[5] Depéret, Ch. et L. Mayet, "Monographie des Eléphants pliocènes."
[6] Hay, O. P., "Observations on Some Extinct Elephants."
[7] Meeting of the Physico-Mathematic Section of the Russian Academy Sc., April 29, 1909: Bull. Acad. Sc., III, p. 809, St. Petersburg, 1909.
[8] Meeting of the Physico-Mathematic Section of the Russian Academy Sc., May 26, 1910: Bull. Acad. Sc., IV, p. 1158, St. Petersburg, 1910.
[9] Kutomanov, G. N., "Rapport sur une mission," p. 377.

to do with his purchase and after his death the mammoth became again *res nullius*. The rediscovery of the mammoth, in 1912, was provoked only by the more enterprising nature of Kucherencov who at that time became interested in this business. Although the report by Kutomanov did not definitely state that his mammoth and that reported to the Academy, in 1909, are the same, and although there are some differences in the nationalities of discoverers mentioned in both cases, as well as in the names of purchasers of the mammoth, there can be little doubt about it. Some uncertainty in this relation could be perhaps explained by the fact that not everything concerning this discovery has been reflected in the publications of the Academy, and some correspondence between the scientists connected with this body and local people remained private. As a matter of fact the Academy, in organizing the expedition of Kutomanov, was hunting for the same mammoth to which no attention had been paid two years before.

The poor condition in which the mammoth was finally found put an end to all the speculation and to the many groundless hopes aroused in connection with its discovery. At the same time it automatically finished all the claims and, perhaps, saved the Academy possible trouble. To avoid this in the future the Academy worked out a special law, as had been told above, protecting fossil remnants from possible speculation.

In 1911, the Academy got news of a skull of a mammoth found by natives on the shore of the Arctic Ocean on the east side of the Cape Maly Baranov.[1] As usual the tusks were broken off immediately and sold to a local trader, Soloviev, who visited the locality and through local officials reported the matter to the Academy (37). He saw only the skull, but was unable to dig deeper. As all the correspondence on this subject stopped after the first letters received by the Academy from Yacutsk, and official reports, it is suggested that after closer examination the locality was not found worthy of attention. It might, perhaps, be of some interest to mention that an official who was much impressed by the small size of tusks, only about 25 lbs., expressed an opinion that the mammoth, probably, used to change periodically its tusks like the reindeer does horns. Thus, the mammoth under consideration had young tusks not yet grown adult size.

In 1915, during the spring and summer excursions (May–July) of the Hydrographic Expedition of the Arctic Ocean, under Capt. B. A. Vilkitzki, a frozen carcass of a mammoth, with tusks eight feet long, was discovered in the Haffner Fiord, on the northern cape of the entrance into the fiord, at the northern latitude of about 76° 30′ and longitude 116° 15′ East (38). The coasts here are frozen earthen banks, 30–50 feet in height, covered with hilly tundra. No attempt at excavation was made by the party which had discovered the locality.[2]

A very fine skull of a Siberian mammoth, probably female, was purchased, in 1923, by the British Museum from ivory merchants.[3] It was probably found in the New Siberian Islands (39). The skull still contains the remnants of ligaments. The ivory of the tusks, both of which were present in their natural position, making this specimen especially interesting, "is in an extraordinary fresh condition."

In the summer of 1926 information was given by a visitor to the Zoological Museum

[1] Meetings of the Physico-Mathematic Section of the Russian Academy Sc., January 19, 1911, and March 9, 1911: Bull. Acad. Sc., V, pp. 272, 282 and 480, 487, St. Petersburg, 1911.

[2] Transehe, N. A., "The Siberian Sea Road," p. 391. For most of the details the writer is obliged to personal communication by Transehe.

[3] Andrews, C. W., "Note on the Skull and Mandible of Siberian Mammoth."

of the Academy about the remnants of the carcass of the mammoth found, in 1922, by gold prospectors in Transbaikalia, in frozen ground on the Kara River, the left tributary of the river Shilka. The Academy immediately sent to the locality the geologist, R. Ph. Gecker, who found a skull and fragments of tusks buried seven meters below the surface, within the frozen drift deposited by the Kara River.[1]

[1] Information Bulletin published by the Russian Academy of Sciences, No. 8, October 1, 1926, p. 2; (No. 11, November 20, 1926), p. 11.

ORGANIZATION OF THE EXPEDITIONS AFTER THE MAMMOTH AND RHINOCEROS

IN THE preceding description the writer has brought together thirty-nine discoveries known to him of carcasses of mammoth and rhinoceros found in the frozen ground of Siberia during a period of time covering more than 225 years. In this number are not included, although mentioned above, Benkendorff's mammoth, 1846, the report on which was undoubtedly a fiction; Polyacov's mammoth, 1877, when no mammoth was found at all; and Gecker's mammoth, 1922. In the latter case only parts of a skeleton were found. All recorded cases are of very different values, so far as the preservation of carcasses is concerned. A few of them refer to more or less complete carcasses, most of them only to parts. In some cases only bones were found, with a few remnants of soft parts, or hair, pieces of skin, or of meat alone. From a purely theoretical point of view, the preservation of a complete carcass, or of a few ligaments on bones, or of a piece of hide, is exactly the same phenomenon, dependent upon the same special conditions, which has to be explained in the same way. For this reason discoveries of a more or less complete carcass of a mammoth, or of isolated and small remnants of soft parts have been treated alike by the writer.

The number of all discoveries is certainly very small. In a country where the ivory of, at least, 250 animals is collected yearly, the greatest part of it out of frozen ground, the number would be increased hundreds or thousands of times, if it were possible to register all the cases in which soft parts were found along with bones. The abundance of remnants of these animals is shown by the fact that near the cliffs in which carcasses are found, one usually perceives a putrid smell, although no rotten remnants may be seen. In Pfizenmayer's opinion too, the carcasses must appear oftener than they are reported.[1]

The number of possible discoveries cannot be correctly appreciated by the number of reported cases. It is quite certain that only a small part of such discoveries used to be reported. Superstition, dread of troubles connected with the arrival of an expedition and with participation in its work (which, for the local population, often used to be compulsory), the meager chance of getting a premium, etc., usually led the discoverer to content himself with picking up only the tusks of a mammoth, leaving the carcass undisturbed, if he had found one. It is also more than certain that the remnants of mammoth or rhinoceros discovered, even taking into account those which were known to local population, but not reported to officials, might be only a part of all possible discoveries of this kind on the shores of numberless creeks, rivers, and lakes of Northern Siberia. Immense areas of that country are so sparsely populated that, according to available Russian statistics, in many regions there, every individual has to his or her account "over" a hundred square versts (about 44 square miles) of land. As the settled population is concentrated along the rivers, and even nomads are not distributed uniformly, waste areas are practically deserts, only occasionally visited. During summer, the most favorable time for the discovery of frozen carcasses, all journeys of any length are practically stopped, except the travel by boat on rivers and lakes, or along the sea shore. All occasional summer trips from

---

[1] Pfizenmayer, E. W., "Mammutleichen," S. 149.

41

temporary dwellings are necessarily very short. All long wanderings of the nomads, dependent upon their reindeer, used to be made during the spring and fall, and usually followed well established routes. Winter travel between dwellings also varies very little in different years. Besides, the winter season when the country is covered with snow for nine months, is especially unfavorable for such discoveries. Even professional ivory hunters used to work in rather limited area, visited year after year. Thus the chance is very small of coming across a frozen animal which has just appeared out of the ground, and many remnants of this kind must be destroyed by putrefication, wild animals, and flood water before they are discovered by anyone. All these facts suggest that the chance of discovering a good specimen of a frozen mammoth or rhinoceros is still present, and could be increased by a rational organization of scientific expeditions to the Northern Siberia. So far the history of most of the expeditions which the Academy used to send to the mammoth-localities every time a rumor of a discovery reached St. Petersburg, has been a series of bitter disappointments for the Academy and for the scientists commissioned by the Academy, who, after long and hard travel over thousands of miles, arrived at the places only to dig out a few bones and poor remnants of soft parts. It was usually not an absolute waste of time, money, and work only because the commissioned scientists used the opportunity to make far reaching researches into unknown or very little known areas of Northern Siberia. In this way they contributed much to the knowledge of the mammoth, although they had rather poor luck in completing the museum material relating to this animal.

The reasons for such ill luck vary a great deal. It was customary to attribute the misfortune to the belated arrival of an expedition to the reported locality. As a matter of fact, this was a true cause only in the case of the Adams mammoth which had been discovered, probably, in perfect condition and was decaying for seven years. In other cases the discoverer reported remnants which had been washed out from a primary locality a long time before, destroyed, while uncovered and buried again. It was also unreasonable to speak of delay when a locality was examined by an expedition twenty or more years after the discovery of a carcass by local people. In other cases a carcass had been more or less destroyed immediately after the death of the animal, and only poor remnants were left. These were reported by discoverers incapable of appreciating all the different conditions and who, in most cases, could not prove the reality and value of a discovery without excavating, at great expense and with danger of spoiling a locality. It distinctly shows that the expeditions sent after mammoths and dependent exclusively upon the data delivered by local people might only in rare cases be expected to be successful, as the discovery of a carcass even in such a case would be always a matter of chance.

As has been mentioned above, in 1891, Tscherski presented a new plan for hunting the mammoth, according to which a scientist must stay for a while in the mammoth-country watching for the possible discovery of a new carcass. This plan appeared as a result of the experience of practically all the expeditions to Northern Siberia, everyone of which chanced to come across the remnants of a mammoth-carcass which had never been reported to Russian officials and remained unknown to anybody except a few natives. A good example of the importance of a scientist being present on the spot was Maydell, who, in a short time was able to examine three localities of mammoths the report of anyone of which would have been sufficient to send a special and expensive expedition from St. Petersburg. Having recognized the localities as worthless, Maydell saved for the Academy a great deal of trouble and money. The rationality of the plan was criticized by Toll,[1] chiefly in connection with social and general conditions of Northern Siberia; but Tscherski's plan was certainly sound, as well as his idea of doing the mammoth-hunting in connection with broad geological and geographical investigations of the country. It could give good results, however, only if a scientist were able to cover with his trips a large area of the country under investigation, and come in contact with as many natives, especially ivory hunters, as possible. Certainly, an expedition must be directed to the most promising localities, as, for example, the New Siberian Islands and the Arctic shores of the Territory of Yacutsk, areas regularly visited by ivory hunters with whom is always possible to make some arrangement concerning skeletons and carcasses of mammoths, rhinoceros, etc. to the mutual interest of science and of the ivory hunter himself. The latter usually is interested only in tusks of a mammoth. Breaking them off out of the ground, he usually does not pay much attention to what has been concealed in the ground behind the tusks. It could be a skull of a mammoth, a skeleton, or even a carcass. He never has any means of making an excavation to decide this question, and only if he found a carcass in a land slide or some soft parts protruding out of the ground, would he report the matter to a trader or a Government official.. A scientific expedition sent after the mammoth has to make mammoth-hunting just as interesting commercially to an ivory hunter, as is the collecting of tusks. But even if he were interested in the excavation of a mammoth or a rhinoceros and supplied with all the necessary means of making excavations, an ivory hunter could not become a substitute for a scientific expedition, for which there would still be much to do. In spite of the fact that many scientists have examined mammoth-localities, the conditions in which the carcasses used to be found, the geology of the localities, etc., there is still much uncertainty in regard to many questions connected with the mammoth and the conditions of its localities, as well as with the geology of mammoth bearing strata. It was partly dependent upon the fact that in most cases scientific expeditions were dealing only with natural outcrops and had very little chance to make large excavations, deep pits, or drillings to get the materials which would replace speculations with firmly established facts. Scientific expeditions sent after the mammoth must, therefore, be familiar with the score of all problems connected with the mammoth and mammoth-localities, and well supplied with all the necessary instruments and machinery for detailed investigation of localities. They will certainly have a greater chance than the former expeditions had of discovering the carcasses of a mammoth or rhinoceros and, even in case of failure in this particular direction, would be able to make a number of important observations on the occurrence of the mammoth, and to make general investigations of this little-known country.

Lately Tscherski's plan has been again advocated by Pfizenmayer. In his opinion a scientist must establish himself at Verkhoyansk, which he selected as the most central point in Northeastern Siberia, and from there organize expeditions to reported localities of mammoth-carcasses.[2]

[1] Toll, Ed., "The fossill Glaciers of New Siberian Islands." Russian edition published by the Russian Geographical Society, p. 123. St. Petersburg, 1897.
[2] Pfizenmayer, E. W., "Mammutleichen," S. 321.

GEOGRAPHICAL DISTRIBUTION OF THE MAMMOTH

THE geographical distribution of the remnants of the mammoth, rhinoceros, and their contemporaries is very extensive. Pallas said [1] that in all Asiatic Russia, from the Don as far as the peninsula of Chukchis, there was not a river or a stream, especially of those flowing in the plains, on the banks, or in the bed of which there have not been found bones of elephants, or of other animals foreign to the present climate. This statement might be supplemented, as to the west the mammoth and its companions can be traced as far as the Pyrenees, and to the east, over the Bering Strait, into Alaska.

In Eastern Siberia the mammoth has been found as far north as the Taimir Peninsula, in the latitude 76° 47' [2] and on Bennet Island,[3] in the latitude 76° 38'. In the southern part of Eastern Siberia the mammoth was found in Transbaïkalia, i.e. about 27 degrees south of the most northern points of its distribution. A "mammoth" discovered in a number of places in Northern Manchuria does not belong to *Elephas primigenius* Blum., but to another fossil species.[4] The remnants are not distributed equally over this immense area, but increase in number towards the northern regions of the country. On the New Siberian Islands they used to be found, along with the remnants of other extinct Post-Tertiary mammals, in such an extreme abundance [5] that these islands might be called a real cemetery, or, because of the abundance of tusks, found there, an ivory mine. Digby suggests, however, that mammoth-bones are by no means scarce in Southern Siberia either.[6] These differences in distribution may be attributed only in part to the more favorable conditions for preservation in Arctic regions of Siberia, as compared with its more southern sections. The accumulation of fossil remnants within the river deposits in the Far North might be partly explained by the drifting of complete carcasses and bones by ancient rivers. It might also be connected with seasonal migrations of the mammoth. But both of these agents would have had only a limited extension, because the mammoth of the New Siberian Islands and of the Arctic shore of the mainland belonged to a special race distinguishable from the South Siberian variety by its smaller size, most plainly shown by the smaller size of its tusks. According to Hedenström, tusks are smaller and lighter in weight the further one advances towards the North, so that it is a rare occurrence on the islands to find a tusk of more than three poods in weight, whereas on the continent they are said to weigh as much as twelve poods.[7] Hedenström's data, referring to the northern race of the mammoth, were checked later by Middendorff.[8] The small size of the mammoth found on the Sanga-Yurakh River has been emphasized by Nasonov and Vollosovich.[9]

The frozen carcasses of the mammoth have, up to the present time, been found exclusively in the northern part of Eastern Siberia, the most western localities among them

[1] Howorth, H. H., "The Mammoth and the Flood," p. 54.
[2] Wrangel, F., "Narrative of an Expedition," p. 436.
[3] Toll, E. V., "Short Report for the Period June 7 to November 8, 1902," p. 158.
[4] Tolmatchew, V. I., "Remains of a Mammoth found in Manchuria," p. 5.
[5] Hedenström, M., "Otrivki o Sibiri," p. 122.
[6] Digby, B., "The Mammoth," p. 52.
[7] Hedenström, M., "Otrivki o Sibiri," p. 122.
[8] Middendorff, A. Th., "Sibirische Reise, IV," I, S. 278.
[9] Cmp. above, p. 37.

44

being on the Yenisei River, or only a few miles west of it (Schmidt's mammoth—16). The Beresovca mammoth is usually considered as the easternmost frozen carcass,[1] but soft parts of this animal have also been found in the frozen ground of Alaska, although not in such good condition as those in Siberia.[2]

In Western Siberia the remnants of the mammoth are known from the extreme North to the shore of Lake Aral,[3] in about 45° latitude.   They belong, here, to the typical *Elephas primigenius* Blum., as the writer had an opportunity of verifying.[4]   Frozen carcasses of the mammoth have not yet been found in Western Siberia, and all known remnants are represented by more or less complete parts of the skeleton.

In Northern Europe the remnants of the mammoth are known east of the White Sea, where in the basin of the Pechora River they are just as numerous as in corresponding parts of Western Siberia, being represented, however, only by bones.   West of the White Sea and of Lakes Onega and Ladoga, remnants of the mammoth are rare.   According to Lyell, Sweden and Scandinavia in general, probably, even lacked the mammoth.   The rare specimens of mammoth-bones found there were, in his opinion, brought there by ice or otherwise.[5]

Such a distribution of the mammoth could not be governed only by chance, but must depend upon some natural cause which might be, perhaps, connected with the Scandinavian ice sheet which the mammoth tried, so far as was possible, to avoid.   As the remnants of the mammoth have been found within the glaciated part of Europe, the animal must have wandered great distances following the retreat or advance of the Scandinavian glacier. The waste plains of Northeastern Siberia were never covered with the ice-sheet of a glacier, and mammoths, as well as their contemporaries could wander unmolested over their pasturages, perhaps, migrating according to seasons.   Long existence under fixed physico-geographical conditions allowed the uninterrupted progress of evolution, and resulted in the development of a new race of Siberian mammoth somewhat different from the typical European *Elephas primigenius* Blum.   This was suggested a long time ago by Howorth [6] and also mentioned by Russian students; but the distinguishing characters of the Siberian form have been formulated only during the last few years [7] almost simultaneously, although independently, by Hay who described, in 1922, a new species, *Elephas beresovkius* Hay, and by Depéret and Mayet who established, in 1923, a new variety or subspecies, *Elephas primigenius sibiricus* D. & M.   The North Siberian mammoth originated from the South Siberian or European form in just the same way, as the latter had originated from *Elephas trogontherii* Pohlig and *Elephas antiquus* Falc., i.e. through the further decrease in the size of the dental plates and in the thickness of the layers of enamel.

Of the thirty-nine recorded discoveries thirty-four refer to the mammoth and five to the rhinoceros.   This relation can be explained not only by the fact that the rhinoceros used to receive less attention from the ivory hunters, but probably also by the greater rarity of the former; and it must also be dependent upon the original habits of both animals. It was often noticed that remnants of the mammoth are frequently found together in great

[1] Digby, B., "The Mammoth," p. 139.
[2] Digby, B., "The Mammoth," p. 142.
[3] Cuvier, G., "Recherches sur les ossements fossiles, I," p. 151.
[4] Tolmachoff, I., "In Berg's Lake Aral," p. 521.
[5] Howorth, H. H., "The Mammoth and the Flood," p. 101.
[6] Howorth, H. H., "The Mammoth and the Flood," p. 56.
[7] Cmp. above, p. 38.

masses, according to Matiushkin "forming immense local accumulations which become both richer and more extensive the further one advances to the north," [1] and that those of the rhinoceros are found separately.   It makes plausible the suggestion that the habits of these extinct animals closely corresponded with those of recent elephants, which usually wander in herds, and of recent rhinoceroses which prefer solitude.

[1] Wrangel, F., "Narrative of an Expedition," p. 179.

VEGETATION AND CLIMATE OF ARCTIC SIBERIA DURING THE AGE OF THE MAMMOTH

ASSOCIATED with bones and carcasses of the mammoth and other animals different plant remnants used to be found in the same horizon. The first Russian colonists in the Northern Siberia discovered in the tundra, far from the present forest, remnants of trees buried in the ground, which they used to call Adamovchina and to distinguish from Noevchina, the latter name being applied to drift wood carried out into the ocean by Siberian rivers.[1] On the island of New Siberia Hedenström discovered immense accumulations of buried trees and, referring to them, called the bluffs on the shore "The Wooden Hills."[2] Middendorff, Schmidt, Toll, Vollosovich a.o. had opportunities of collecting, within the tundra ground, leaves, roots and fine branches of plants like *Alnus fruticosa* and *Betula alba*, which are not to be found there now, but grow in more southern latitudes. These facts brought local people and scientists alike to the conclusion that it had not been very long since trees used to grow within the recent tundra region much farther north than they do now. Not all of the observations were found to prove this theory, however. For example, plants of "The Wooden Hills" have been found to be of Miocene age. Many of the trees found in the tundra ground did not grow there, but were brought by rivers, or sea currents, deposited in the tundra, and are now found far from the shore, due to the uplift of these areas. In spite of that, there still remains a number of facts which undoubtedly argue for a more northern limit of forests in Arctic Siberia during the time shortly preceding the present one. The next quite natural conclusion was that the climate of Arctic Siberia at that time was milder than it is now. Howorth[3] was even ready to attribute to Northern Siberia, during the Age of the mammoth, a climate corresponding to the recent one of Lithuania. This was certainly a great exaggeration not corresponding with the known facts. No one of scientists who were familiar with the recent and subfossil flora of Northern Siberia was going so far. The shifting of the forest limits could be measured only through a few degrees of latitude, and subfossil forest flora found in the ground of the recent tundra is represented by Arctic and Subarctic flora, not by that of more moderate regions. Considering these facts, any theory as to a milder climate in that time should, in the opinion of the writer, be accepted only with great reservations. The advancement of the tundra towards the South may be dependent not only upon the change of climate for the worse, but upon other physico-geographical conditions as well. Northern Siberia, in spite of its severity of climate, has the northern forest limit in all the world, going towards the North beyond the parallel of 72° and in protected places, as in the Khatanga valley, jutting out towards the North about twenty minutes more. From here in both (western and eastern) directions the forests retreat southwards. In Northeastern Siberia, on the Chukchi Peninsula, only poor shrubs are known, and these in protected places, near the Polar Circle, i.e. more than six degrees south as compared with the valley of the Khatanga River. At the same time, the average yearly temperature of the Chukchi

[1] Adams, "Travel: Sibirski Vestnik, 1820, part X," p. 324.
[2] Hedenström, M., "Otrivki o Sibiri," p. 128.
[3] Howorth, H. H., "The Mammoth and the Flood," p. 561.

Peninsula is higher than that in the North of Central Siberia. The forest line in Northern Siberia follows, roughly speaking, the Arctic coast, but nowhere approaches the ocean. We can imagine that, if the northern shore of the Chukchi Peninsula were to increase, for two–three degrees of latitude, it would be accompanied by an advancement of forests towards the North and their encroachment upon the tundra. When *Alnus fruticosa* was growing on the New Siberian Islands they were connected with the continent which at that time thus had protruded about four degrees farther north as compared with the recent shore line of the mainland. The retreat of the forests might have been caused by the separation of the New Siberian Islands, although the climate, generally speaking, can have suffered very little change, if any.

Some data referring to the flora of the Age of the mammoth have also been received through the examination of remnants of undigested food found in the mouth and stomach of the mammoths and rhinoceroses. The first investigations of this kind refer to the Siberian fossil rhinoceros and had as material a very small amount of vegetable matter found on the teeth of the animal and examined by a number of observers. Brandt found bits of coniferous wood and remains of a seed. Meyer found the seed of an *Ephedra*. Mercklin distinguished the wood of a willow. Schmalhausen found remains of monocotyledons and dicotyledons, and recognized traces of graminaceous plant, and of an ericaceous one, the latter probably *Vaccinium Vitis Idaea*. Among the remains of coniferae were those of a *Picea* (*?obovata*), of an *Abies* (*?sibirica*), of a *Larix* (*?sibirica*). There were also found the remains of a *Betula*, of a *Salix*, and of an *Ephedra*. All these plants are still growing in Siberia.[1] Tscherski also came to the same conclusion after his work on the same subject.[2]

Although, according to Wright, "The stomachs of some of the mammoths have been found containing leaves of trees whose present habitat is hundreds of miles south of the locality where the animal perished,"[3] as matter of fact, the first detailed examination of undigested food from the stomach of a mammoth was made only after the discovery of the Beresovca mammoth. Previously, Famintzin[4] had examined the excrement of a mammoth, brought by Bunge from the Lena River, but the investigation proved only the presence of vegetable matter without any particular result. In the remnant of food found in the stomach of the Beresovca mammoth Sucachev[5] identified: *Hypnum fluitans* (Dill) L., *Aulacomnium turgidum* (Wahlnb.) Schwaegr., *Alopecurus alpinus* Sm., *Beckmannia cruciformis* (L.) Host., *Agropyrum cristatum* (L.) Bess., *Hordeum violaceum* Boiss. & Huet., *Carex lagopina* Wahlenb., *Ranunculus acris* L., *Oxytropis sordida* (Willd.) Trautv. All of these species are typical representatives of a meadow flora of Northern Siberia at the present day. Leaves and branches of bushes were not found, although they had been not lacking on the shores of Beresovca. In summer the mammoth was a grass-eater who, like the recent reindeer, preferred this food to any other. It certainly had no difficulty in picking up even the lowest grass of the tundra with its trunk, and, probably, never tried to "Graze close to the tundra like oxen" which, according to Howorth,[6] would make its

---

[1] Howorth, H. H., "The Mammoth in Siberia," p. 557.

[2] Bull. East Siberian Branch of the Russian Geographical Society, VII, Nos. 4–5, Irkutsk, 1876. After Tscherski, I. D., "Beschreibung der Sammlung," S. 453 and 458.

[3] Wright, G. F., "Asiatic Russia, II," p. 579.

[4] Meeting of Physico-Mathematic Section of the Russian Academy Sc., January 21, 1886: Mem. Acad. Sc., LII, p. 173, St. Petersburg, 1886.

[5] Sucachev, V. N., "Examination of Plant Remnants," p. 15.

[6] Howorth, H. H., "The Mammoth and the Flood," p. 59.

existence in the tundra impossible. This selective taste of the animal does not permit possible conclusions to be drawn from these data without any reservations, but, so far as the examined plants are concerned, we can join Osborn's statement that the climate at that time was not milder, nor more frigid than that prevailing now in this part of Siberia.[1] On the strength of his new investigations, A. I. Tolmachoff also emphasizes a close similarity of the present flora of Northern Eurasia with that of the Age of the mammoth.[2] Plants in the stomach of the mammoth from the Bolshoï Lyakhov Island were poorly preserved and not as well collected as in that from Beresovca; but they were also represented by grasses and, perhaps, some moss, and again corresponded with the flora of the recent tundra. Both mammoths referred to had perished during late summer, or early fall, as has been shown by the remnants of their food examined. During winter time, the food of the mammoth would have been composed of leaves, small branches, and bark of trees, probably reindeer moss often growing on trees, etc. The remnants of plants described from the teeth of the Siberian fossil rhinoceros might correspond to the winter diet of these animals. To find this food, the mammoth had to leave the tundras and migrate for a few hundred miles towards the South, to the forests, as reindeers do at the present time.

Thus, if the retreat of the forests in Northern Siberia may be considered an established fact, we do not find that the mammoth enjoyed a milder climate, or was in need of it for its existence. So far as food is considered, it suffered, probably, no privations, because in nearly all cases of carcasses of mammoth, discovered, they belonged to well fed and often fat animals, of robust health.[3]

Not only the examination of flora supports the theory of climatic conditions in the Age of the mammoth similar to the present ones of the Northern Siberia, but also the discovery of frozen carcasses of the mammoth and rhinoceros, the origin of which we cannot understand as the result of any other conditions than those of an Arctic or Subarctic climate. The presence of frozen ground, for example, appears to be quite indispensable.

We could certainly easily imagine the mammoth living comfortably in a much milder climate, as suggested by Howorth, but we can just so well imagine the evolution of the mammoth being the result of its adaptation to gradually changing climatic conditions. As suggested by Tscherski,[4] climatic conditions in Northern Siberia were changing for the worse, very slowly and gradually; and the mammoth, living in the same area for a long time, could have easily and without having suffered any harm adapted to new conditions. However, such a process of adaptation must have been accomplished in very ancient times, as the mammoth was undoubtedly already well adapted to the surrounding conditions of severe Siberian climate, probably no less than is the recent reindeer. Referring to a doubt of the possibility of so large an animal finding enough food in the tundra, the writer likes to remember the surprise of Nordenskiöld when the latter found that reindeers, killed during his expedition in Spitzbergen, in October, 1872, were so fat that their necks were not sharply separated from the heads. Nordenskiöld's question, "How this animal can collect such a mass of fat in Spitzbergen where the vegetation is so scanty and the summer so short"[5] may be answered only by the statement that we cannot always understand the limits of adaptation of wild animals to surrounding conditions.

[1] Osborn, H. F., "Age of Mammals," p. 420.
[2] Personal communication to the writer.
[3] Howorth, H. H., "The Mammoth and the Flood," p. 178.
[4] Tscherski, I. D., "Beschreibung der Sammlung," S. 475.
[5] Nordenskiöld, A. E., "The Arctic Voyages," p. 200.

It cannot be denied that the idea of the Siberian mammoth as a northern animal undoubtedly has gained ground during the last few years. It is shown, for example, by restorations in which the mammoth is almost always pictured in a winter environment, walking over ice and snow through a stunted Arctic forest. Depéret and Mayet who described the mammoth of the Bolshoï Lyakhov Island, attributed to it not only an adaptation to Arctic conditions, but even an Arctic origin. Its distribution to the South they connect with the increased coldness of climate southwards, following the advance of the ice of the Glacial age. "Nous sommes amenés par ce raisonment à admettre pour le Mammouth Sibérien une origine et une centre de dispersion tout differents de de ceux du Mammouth normal, et à le considerer comme un *rameau spécial* indépendent, d'origine nordique (Asie septentrionale), dont les représentants se sont avancés plus ou moins loin vers le Sud, à la faveur du grand refroidissement final du Quaternaire." [1]

[1] Depéret, Ch. et L. Mayet, "Monographe des Eléphants pliocènes," p. 190.

GEOLOGY OF THE MAMMOTH AND RHINOCEROS LOCALITIES

EVERYWHERE carcasses of the mammoth and rhinoceros were found, they had been buried within the frozen ground of tundra near its upper surface and usually on comparatively elevated points, on the top of bluffs, etc. This has long been known and, according to Wrangel, "The best mammoth bones are found at a certain depth below the surface" and "more in elevations situated near higher hills than along the low coast, or on the flat tundra." [1] Often mammoth localities are on the highest points of the tundra. The occurrence of the mammoth at high levels was also noticed in Alaska on the cliffs in the Kotzebue Sound, which in their features closely correspond with the cliffs on the shores of the New Siberian Islands, or on the Arctic coast of Northeast Siberia.[2] Bones and tusks of the mammoth were also often found protruding from the ground on a high tundra. Excavation often disclosed the remnants of a complete animal which had been buried there. Carcasses and isolated bones also used to be found on the bottom of valleys, or on tide-flats, as near the New Siberian Islands, having been washed out of cliffs or rolled down in frozen masses by underwashing of the cliffs by spring floods. Examples of this mode of occurrence are the rhinoceros found, in 1877, on the Khalbugaï Creek, and the Adams mammoth which slipped down to the beach after the cliff had been underwashed.

Mammoth-bearing drift deposits sometimes have a thickness tens of feet, sometimes they are spread out in comparatively thin layers. In some localities, as in the one of Schmidt's mammoth have been discovered, underneath these deposits, the sediments of the last Arctic transgression. In Northeastern Siberia they are usually underlain by layers of rock ice, and very often, in this case, are reduced to a thickness of only two or three feet. An inaccurate expression by Adams created the idea that his mammoth had been frozen within ice. But after the detailed consideration of this matter by Toll there is no more doubt that this mammoth like others had been frozen within the driftground underlain by rock ice. Theoretically, it is possible to imagine carcasses enclosed within ice, but as matter of fact, neither mammoth nor rhinoceros was ever found in such conditions, as Howorth has already emphasized.[3]

The uppermost position of mammoth-bearing deposits, covering sediments of the Arctic transgression, corresponds exactly with the systematic position of the Siberian mammoth as the youngest member of the group of fossil elephants. In European Russia where the mammoth-bearing strata often are found together with moraines, or are partly composed of glacial material, the true mammoth belongs to the upper Glacial stage, and the European mammoth must have been a contemporary of the Siberian one, or perhaps the latter was the successor of the European one, but not vice versa. Elephant bones discovered in European Russia within the older morainique material were identified as *Elephas trogontherii* Pohlig.[4] The position of the mammoth within the youngest part of the mammoth-bearing horizon was also emphasized by Tscherski.[5]

1 Wrangel, F., "Narrative of an Expedition," p. 275.
2 Howorth, H. H., "The Mammoth and the Flood," p. 266.
3 Howorth, H. H., "The Sudden Extinction of the Mammoth," p. 313.
4 Tolmatschow, I. P., "Fouilles de l'*Elephas trogontherii* Pohl.," p. 259.
5 Tscherski, I. D., "Beschreibung der Sammlung," S. 40, footnote.

The strata in which are buried the bones and carcasses of mammoths, rhinoceroses, and other extinct mammals and remnants of plants, are represented by sandy, clayish, or loamy sediments of different thickness. Marine shells or marine mammals have never been discovered in them, and these sediments may be only of fresh water or terrestrial origin. The writer, having examined the geological specimens brought from the locality of the Beresovca mammoth, could realize that the earth strata in which the carcass was found had, to a great extent, taken their origin from the drift brought by rain and snow water from the neighboring hills, surmounting the river terrace on which the mammoth was found.[1] Where similar orographic conditions are present, tundra ground could easily originate in this way. But all open tundras of Arctic Siberia usually lack these conditions. In a few cases, also within the tundra ground, lake sediments were discovered. In most cases tundra deposits are formed in connection with the work of rivers which carry to the sea great amounts of silt which are deposited in deltas, or within estuaries and bays, and distributed along the shore. The greatest part of the Arctic shores of Eurasia is undoubtedly composed of materials delivered from the mainland by rivers. Owing to recent changes of sea level, in some places, for example, in Lena Delta, river deposits have been found 200 and more feet above the sea level. The close connection of tundra deposits with river drift on the Yenisei River was mentioned by Nordenskiöld.[2] He emphasizes also the fact that shells, when they are found in the tundra sand, all belong to living types of the Arctic sea.

The shores of Arctic Eurasia have not only been uplifted, but partly have been submerged as well, with the result that the New Siberian Islands, for example, which not a very long time ago were parts of the continent, have now been not only separated from it, but even partly destroyed by the victorious sea-waves. Mammoth-bearing strata of the New Siberian Islands originally were undoubtedly dependent upon the silt brought from the mainland by Siberian rivers, and connected with the corresponding sediments on the shore of the continent. Toll considered it possible to reconstruct the former channels of rivers between the New Siberian Islands and the mainland,[3] although it was rather against his suggestion as to the origin of the rock ice of the New Siberian Islands, which in his opinion, is a remnant of the Glacial ice sheet. Also, on the mainland the mammoth was not always found in recent river valleys, or within deltas, but, just as on the islands, in the sediments deposited by former rivers the channels of which were obliterated later. Certainly some remnants of the mammoth were found outside of any river valleys, as, for example, on Kotelny Island where they were discovered by Toll [4] about 1000 feet above sea level, and where animals had perished during their wanderings over divides.

In spite of the work of a number of keen students who had the chance to visit and examine mammoth localities, there is no unanimous opinion as to the composition of Post-Tertiary strata of North Siberia, the origin of different horizons, their relations to each other, and consequently, the stratigraphic position of the mammoth horizon. A lot of confusion is also caused by the presence of rock ice among the Post-Tertiary strata of Northern Siberia and by the difference of opinion as to its origin and its stratigraphic importance. In Toll's opinion, rock ice, as already has been mentioned

[1] Tolmatschow, I. P., "Bodeneis vom Fluss Beresovka," S. 448.
[2] Nordenskiöld, A. E., "The Arctic Voyages," p. 331.
[3] Toll, Ed., "Die fossilen Eislager," S. 79.
[4] Toll, Ed., "Die fossilen Eislager," S. 62.

above, is a remnant of the Glacial ice sheet, and it underlies the mammoth-bearing strata. As in Western Siberia these strata cover the sediments of Arctic transgression, the latter one and rock ice may be correlated with each other; therefore, the mammoth-bearing deposits (called also the mammoth-horizon or tundra-horizon) appear to be the uppermost horizon among the Post-Tertiary strata, as has been stated above.[1] Vollosovich, who had an opportunity to examine the same localities, as Toll had done, has distinguished two horizons of rock ice, which he has called Lower and Upper ice, both separated by the loam horizon which he has also called the mammoth-horizon, because, in his opinion the mammoth and rhinoceros have been limited to this horizon. To this horizon exclusively he also attributed the remnants of *Alnus fruticosa, Betula alba,* etc. The origin of the Lower ice he connects with the Ice Age of the Northern Hemisphere, although he is not so decisive to its glacial nature, as was Toll. The Upper ice must, therefore, belong to the second Glacial period, although it may be not a remnant of a former glacier either, and the mammoth-horizon belongs to the Interglacial period. He has also given the following scheme of the Post-Tertiary history of the New Siberian Islands, starting from the bottom.[2]

1. Lower rock ice of the Bolshoï Lyakhov Island corresponding with the greatest glaciation of the North.

2. Loam sediments containing remnants of poor meadow and shrubbery flora.

3. Loam deposits with *Alnus fruticosa, Betula alba,* and grasses. Mammoth and rhinoceros the most important mammals.

4. Upper ice. Dying out of gigantic mammals.

5. Loam deposits with *Betula nana* and *Salix* (different species). Many sporophytes in the meadow flora. Horse as the most important mammal. Beginning of the Arctic transgression.

6. Loam deposits with rare *Betula nana* and common *Salix polaris.* Musk ox and deers. Arctic transgression, with *Yoldia arctica.* Separation of the New Siberian Islands from the mainland.

7. Emergence of islands of New Siberia and Thadeevsky. Retreating of the sea and tendency towards the connection of islands with the continent. The recent tundra flora. Dominant mammal, reindeer.

This scheme was from the beginning complicated a little by the discovery on the island of New Siberia of rock ice covered and underlain with sediments of Arctic transgression.[3] Considering rock ice as a horizon of independent stratigraphical position, as, indeed, Toll and Vollosovich did, it would be necessary to speak of two Arctic transgressions of different age, which are not included in the above scheme.

According to Vollosovich's scheme the mammoth belongs to an older horizon than has been accepted in this paper; a horizon which is not only older than the Arctic transgression, but even precedes the second Glacial period. Owing to the difference of opinion as to the stratigraphic importance of rock ice and especially to its correlation with different glaciations, it is more convenient to pay attention only to the relations of the mammoth horizon to the Arctic transgression, a real and infallible measure stick of Post-Tertiary stratigraphy of Arctic Eurasia, which in Western Siberia was found below the mammoth-bearing strata. Vollosovich did not try to reconcile his scheme with this firmly established

[1] Toll, Ed., "Die fossilen Eislager," S. 76.
[2] Pavlova, M., "Description of Fossil Mammals," p. 36.
[3] Pavlova, M., "Description of Fossil Mammals," p. 38.

fact. Neither did he explain the difference of opinion concerning *Alnus fruticosa* which in the New Siberian Islands had been discovered first by Toll in the ground of the upper recent tundra, where the latter located, of course, the mammoth-horizon. Vollosovich did not correct, either, statements which he must consider erroneous in the observations by Toll. The rich ivory localities in some parts of Kotelny Island, ivory collected on the surface of tundra, Vollosovich is inclined to consider originated from the older tundra uncovered or only slightly covered there with new sediments.[1]

Such a change of the stratigraphic position of the mammoth is not supported paleontologically. As was stated above, the Siberian mammoth, by its specific characters, belongs to the youngest generation in its family. In the position suggested by the above scheme, it would have had to approach older elephants, nearer to *Elephas trogontherii* Pohlig, or even be replaced by the latter form.

Unhappily for the scheme referred to, it has a very insufficient foundation, as it is only an interpretation of the observations by Vollosovich on an ice cliff on the shore of Bolshoï Lyakhov Island. The ice outcrops here twice, in the upper part of the cliff in the form of a nearly vertical wall, and in the lower, more regular slope of the shore, near the sea. Both outcrops are separated from each other by irregular accumulations of drift which, in Vollosovich's opinion correspond to an intermediate layer, to his mammoth-horizon. However, in the opinion of other explorers, who had the opportunity of observing the same or similar ice cliffs in different parts of Arctic Siberia, these drift accumulations had originated from the streams of mud running down, caused by the thawing of the ice cliff, are only deposited on the slope and do not separate Upper and Lower ice layers which are connected below the accumulation of drift. The difference in profile of both parts of the outcrop is explained by the fact that the lower part of the ice, when not underwashed by sea waves, in spring remains covered with snow for a long time after the upper part had been exposed to direct sun rays, being also protected in summer with deposited silt. As the result of such a condition the upper part is thawing much more quickly and retreating farther from the shore than the lower one. As matter of fact, the same ice cliff on Bolshoï Lyakhov Island was examined by Toll who discovered, within the silt deposits on the slope, remnants of *Alnus fruticosa*, but did not hesitate to consider them washed out from the ground of upper recent tundra and brought down by mud streams. He disregarded the idea of the stratigraphic independence of loam deposits on the slope and of the different Lower and Upper ice layers. According to Vollosovich, the deposits of his mammoth-horizon originated from mud streams as the result of thawing of rock ice, but it was the old Lower ice and old tundra on its surface, which delivered these streams of mud. All plant remnants found by Vollosovich were buried in a secondary location, and he never had a chance to examine the old tundra ground in its primary condition, but always in the form of such mud stream deposits. The mammoth horizon, after Vollosovich, is sometimes covered with lake deposits, and sometimes underlain with them, as, for example, at the locality of the mammoth on the river Sanga-Yurakh, where rock ice was absent.

Vollosovich's suggestion brings to attention a very important question, but his data are not of the sort to be taken without reservations. Besides the difference of opinion between Vollosovich and Toll, mentioned above, as to the geology of Bolshoï Lyakhov Island, there is another between Vollosovich and Pfizenmayer, concerning the locality on the Sanga-Yurakh River, a secondary one in the opinion of the latter.[2] Vollosovich's

[1] Vollosovitch, C. A., "Le mammouth de l'île Bolchoï Lakhovsky," p. 315.
[2] Cmp. above, p. 36.

observations must first be checked on the very spot before changing the stratigraphic position of the mammoth-horizon, so far established as the highest one among the Post-Tertiary strata in Northern Siberia and elsewhere.[1]

The old tundra ground must have originated, according to Vollosovich, from the silt brought down from the surmounting Tertiary hills, i.e. in the same way as have the earth strata on the terrace of Beresovca River, in rather exceptional conditions, which cannot exist elsewhere, as has been explained above. The origin of deposits of the upper recent tundra is not considered by him at all, but, probably, he was ready to explain it in the same way and meet, therefore, the same objections which are made to the universal application of this kind of explanation.

Great confusion also exists concerning rock ice. The writer cannot consider this problem just now, but would like to emphasize the fact that Toll's suggestions on the glacial nature of rock ice must be completely abandoned. Rock ice is a product of recent climatic conditions of Northern Siberia and would originate whenever these conditions prevailed. Its origin, therefore, was not confined to any particular geological moment, and rock ice must not be considered as a well-defined horizon. Among many theories trying to explain its origin no one could be recognized as being fully satisfactory in all cases, although every one of them is good for some particular case. In some instances, rock ice may be even younger than the strata within or below which it happens to be found, being in this case in the nature of a dyke, even of an intrusive one.

[1] Depéret, Ch. et L. Mayet, Monographie des Eléphants pliocènes.

CONDITIONS OF PRESERVATION OF FROZEN CARCASSES OF THE MAMMOTH AND RHINOCEROS

The most difficult part of the mammoth question is to find out the ways in which a carcass could be quickly buried and saved from decaying. It appears quite mysterious, if one tries to look for possible explanation in the familiar conditions of a moderate climate; but can be easily understood, if one takes into consideration the climate of recent Arctic Siberia which corresponds to the climate of the Age of mammoth, as has been shown above.

During his travel in Siberia, Middendorff,[1] on the shore of the Sea of Okhotsk, came across the carcass of a whale which had been buried within the drift accumulated by waves and preserved so well that a few weeks later its fat was found good enough to be used for food. An animal protected in this way from wolves and foxes could stay till the next winter, be frozen and have a chance to remain intact for a long time. Middendorff suggested that the carcass of a mammoth like that found by him on the Taimir River could be brought down a river to its mouth, covered here with silt, frozen, and preserved in the frozen ground, for an indefinite length of time. Middendorff's theory certainly could be considered valid for the localities in deltas and mouths of rivers, or on a sea shore. Bunge enlarged on Middendorff's theory by suggesting that an already frozen carcass of a mammoth could drift and be buried in the same way.[2] During his Vega travel Nordenskiöld chanced to discover in the ground, on the sea shore of the Chukchi Peninsula, remnants of a whale which had been buried and preserved in just the way explained by Middendorff. It was a skeleton of *Balaena Mysticelus* still partially covered with skin and with deep red, almost fresh, flesh adhering to those parts of it which were frozen in the ground. According to Chukchis, no whale had stranded there in the memory of man, therefore the animal must have been buried many scores of years before. Nordenskiöld, describing this discovery, as an example of protection against putrefaction of flesh of gigantic sea animals by means of preservation in the frozen soil of Siberia, refers to it, as "a parallel to the mammoth mummies, though from a considerably more recent period."[3]

Localities in which a mammoth obviously was found buried on the very spot where it had died, could not be explained according to Middendorff and Bunge; but a very satisfactory explanation, in the opinion of the writer, was offered a long time ago by J. E. Brandt and then enlarged upon and completed by other scientists. Brandt was very much impressed by the fact that remnants of the mammoth, carcasses and skeletons alike, sometimes were found in poses which indicated that the animals had perished standing upright, as though they had bogged. In the case of the skeleton of a mammoth found in such a pose near Moscow, Russia, Brandt suggested that the animal must have sunk into soft mud.[4] Concerning conditions of preservation he says: "Wurde das Moskauer Government damals einen ewig gefrorenen Boden besessen haben und noch bis auf heute besitzen

[1] Middendorff, A. Th., "Sibirische Reise, I," I, S. 236.
[2] Bunge, A., "Die Lena Expedition," S. 46.
[3] Nordenskiöld, A. E., "Die Umseglung Asiens und Europas, I," S. 476.
[4] Howorth, H. H., "The Mammoth and the Flood," p. 158.

ähnlich wie der Norden Sibiriens, so würde das fragliche Mammuth wohl als ganzes Cadaver zum Vorschein gekommen sein." Such accidental plunging into soft ground Brandt considered as one of the most important reasons of the death of the mammoth. If it happened in the fall, the carcass could be frozen shortly after and thus preserved for a long time. A very important supplement to this theory was made by Al. Brandt who suggested that the mammoth could be trapped in streams of mud having originated through landslides.[1]

The present writer in his description of the geology of the locality of the mammoth on river Beresovca suggested that the animal had been trapped in soft ground when pasturing on the river terrace. As a matter of fact, the swamps and bogs of a moderate climate, with their treacherous pits, in Northern Siberia, owing to the permanently frozen ground, could exist only in quite exceptional conditions, as those observed by Pfizenmayer in the Yana Region.[2] As Pfizenmayer speaks also of a permanently frozen ground there, thawing, in the summer, only 50–70 centimeters from the surface, it is rather difficult to understand from his description the possibility of existence of the swamps referred to. Al. Brandt's allusion to mud streams, mentioned above, certainly, has, therefore, a special importance for Northern Siberia, where such streams used to originate through the melting of frozen ground and of rock ice which is always covered with loam layers, as well as having loam masses included within the ice itself. Mud originated in this way is very soft and at the same time extremely sticky. A few inches of it are practically impassable for a man, a foot or a little more was, probably, sufficient to stop a mammoth. During his first travel to Northern Siberia Vollosovich happened to be trapped in such a stream which he tried to cross. After some unsuccessful attempts he was released only with assistance of his guides. The next morning he examined the treacherous spot and found his tracks firmly frozen under a new layer of mud.[3] A mammoth once trapped within the mud must have succumbed after a short, desperate, but unsuccessful struggle, during which the Beresovca mammoth, for example, had broken its pelvis and other bones. Recent animals, horses and cows, once trapped in mud, very quickly give up any resistance and remain immovable, waiting for their fate, even though uninjured. Once trapped in a moving mud stream the body of a mammoth made a kind of a dam against which the mud piled up until it could overflow the body and finally suffocate the animal. As long as the latter was alive and could move a little, it was protected against the attacks of wild animals. The mud cover used to give some protection later as well. If the accident happened in the fall, the covering of the carcass and its freezing could go on hand in hand, and in a short time the carcass would be completely frozen. In Vollosovich's opinion, the mammoth collected by him on the Sanga-Yurakh River had perished and had been buried in this way.[4] He arrived at the same conclusion as to the mammoth from Bolshoï Lyakhov Island. The death from suffocation of the latter specimen was proved in the same way as in the case of the Beresovca mammoth, through the erection of its penis.[5] The death from asphyxia was proved also for *Rhinoceros tichorhinus* Fischer from Vilui, an examination of the head of which revealed

---

[1] Brandt, Al., "Kurze Bemerkung über aufrecht stehende Mammuthleichen." After Fr. Schmidt, "Vorläufige Mitteilung," S. 97.

[2] Pfizenmayer, E. W., "Mammutleichen," S. 94.

[3] Personal communication to the writer by the late Vollosovich.

[4] Vollosovich, C. A., "On the Digging out of the Sanga-Yurakh Mammoth, in 1908," p. 453.

[5] Vollossovitch, C. A., "Le mammouth de l'ile Bolchoï Lakhovsky," p. 337.

that the blood-vessels and the fine capillaries were filled up with brown coagulated blood, which in many places still preserved its red color.[1] The death from suffocation was suggested for the specimen of rhinoceros from the Khalbugaï Creek as well.[2] It is certainly not possible to find out, in either case, if asphyxia was the result of entrapment within mud, or of drowning.

In the sixties of the last century the Russian Academy of Sciences was hunting for mammoths in Western and Eastern Siberia. On each expedition, Schmidt and Maydell, respectively, came across only poor remnants of mammoth, although Maydell had a chance to visit three different localities. All hopes of discovering a complete new carcass were apparently frustrated. It brought Schrenck to conclusion that, although remnants of mammoth with soft parts are very common in Northern Siberia, complete carcasses are extremely rare and required the death of an animal and preservation of its carcass under quite exceptional conditions. As the only complete carcass of a mammoth known at that time, was that of the Adams mammoth, found, as it was erroneously suggested, enclosed within ice, Schrenck describes the probable conditions in which mammoths (and rhinoceroses) perished, and in which their carcasses were preserved, in the following way: mammoths in their wandering happened to break through large accumulations of snow in narrow valleys, canyons, or under cliffs. Once plunged down, they were unable to get out, and gradually sank deeper and deeper and became well protected against the warmth of the next summer, as well as against wild animals, and preserved complete for a number of years to come.[3] As has been mentioned above, the Adams mammoth was not found within ice, but in frozen ground. The head of *Rhinoceros tichorhinus* Fischer from the Khalbugaï Creek, whose lack of any soil particles brought Schrenck to the conclusion that this animal had also been found within ice, was washed twice or three times before its arrival at St. Petersburg, and this animal had also been found in frozen ground.[4] With the exception of these two specimens no one has been known whose death and preservation could be explained in the way described by Schrenck. Besides, the physico-geographical conditions required by the theory have been nowhere known in Siberia, and it is rather difficult to imagine them having existed there in former periods. Howorth correctly emphasized this fact.[5]

In spite of that, Schrenck's theory has found some adherents. Bunge [6] thought that in this way could have originated the carcasses found in deltas to which they had been drifted, frozen, from the upper part of streams. Nehring [7] also considered such accidents quite possible, and Kayser is ready to recognize Nehring's explanation as well.[8] In recent times Schrenck's theory has been accepted without any reservation by Digby. According to his book "The cold-stored mammoths and wooly rhinos that have survived in flesh-and-blood . . . were just a very few which happened to fall into a deep, steep-sided crevasse, filled with snow, on the eve of, or during, a blizzard, which filled in the hole behind them, when they themselves did not fill it in by their struggles." [9] On the basis

[1] Howorth, H. H., "The Mammoth and the Flood," p. 184.
[2] Howorth, H. H., "The Mammoth and the Flood," p. 185.
[3] Schrenck, L., "Bericht über neuerdings in Norden Sibiriens angeblich zum Vorschein gekommene Mammuthe," S. 173.
[4] Toll, Ed., "Die fossilen Eislager," S. 39 and 48.
[5] Howorth, H. H., "The Mammoth and the Flood," p. 95.
[6] Bunge, A., "Die Lena Expedition," S. 46.
[7] Toll, Ed., "Die fossilen Eislager," S. 81.
[8] Kayser, Em., "Formations-Kunde," S. 525.
[9] Digby, B., "The Mammoth," pp. 55 and 138.

of the same theory Digby explains why the frozen rhinoceros is less often found than the mammoth. "Rhinoceros being built like the bows of a ship could drive a tunnel in snow to the open end of the gully. Therefore it is seldom found cold-storaged. . . . Other contemporaries of mammoth were not heavy enough to go snowdrift, or managed to tunnel out." [1]

Although carcasses of extinct animals have not been found enclosed within ice, it is of some interest to mention that an animal could meet death and become enclosed within ice in the so-called aufeis,[2] which has a wide distribution in Northeastern Siberia and sometimes covers surfaces of many square miles with a layer of ice often of a great thickness, which can easily last over summer, or even for an indefinite number of years. It was considered impossible by Howorth [3] who, probably, was not very familiar with this phenomenon and while having said that "the ice in the rivers is completely melted during the summer," he referred to the usual winter ice cover of rivers in moderate climate. Under certain conditions, for example, after heavy snow, the aufeis would be passable only with difficulty, and such a heavy animal as the mammoth could be easily trapped within it, and doomed to destruction. The carcass could be covered with water which, at a temperature many degrees below zero, would be quickly transformed into ice, and preserved for an indefinite time. The aufeis might be covered with drift and transformed into rock ice. In number of years, the rock ice might be perhaps destroyed through a gradual deepening of the particular valley, and the carcass would be found buried in frozen ground on the terrace of a river.[4]

To explain the way in which carcasses were enclosed into ice, as well as the origin of ice itself other theories were also offered, which are certainly only of historical interest now. For example, according to Gümbel,[5] the carcasses of extinct animals found in Northern Siberia were brought over there from Southern Siberia enclosed within the ice, presumably, of ancient glaciers. Heer [6] also found possible such a transportation of carcasses enclosed within ice. The destruction of an animal and its preservation within glacier ice could take place in just the same way, as it sometimes happens to unhappy glacier climbers of our days. According to James Geikie, during the Glacial epoch great snow drifts accumulated and became consolidated. Over this ice, mosses and lichens crept until a tundra was formed over solid ice, a condition to be noted in some places now. Later this ice may melt away in places leaving the tundra apparently firm. In such traps as these many of the great animals might be caught and perish.[7] Trapped in this way a mammoth, or other animal, could become frozen very quickly. Maydell mentioned such traps in the tundra over the crevices in rock ice,[8] and Herz explained the death of the Beresovca mammoth by its plunging into an ice crevice.[9] But even in such a case a mammoth would be buried in frozen ground; and only under quite exceptional conditions be found entombed within ice, when it had been, for example, covered with snow during a blizzard.

[1] Digby, B., "The Mammoth," p. 58.
[2] The name along with its German spelling has been introduced into American literature by Leffingwell: U. S. G. S. Prof. Paper 109, p. 158.
[3] Howorth, H. H., "The Mammoth and the Flood," p. 95.
[4] Toll, Ed., "Die fossilen Eislager," S. 39.
[5] Tscherski, I. D., "Beschreibung der Sammlung," S. 49, footnote.
[6] Tscherski, I. D., "Beschreibung der Sammlung," S. 463, footnote.
[7] Wright, G. F., "Asiatic Russia, II," p. 580.
[8] Toll, Ed., "Die fossilen Eislager," S. 22.
[9] Herz, O. F., "Frozen Mammoth in Siberia," p. 617.

In spite of the similarity to a real ice-box, of the frozen ground of Northern Siberia, and of the possibility for the carcasses enclosed within to be preserved almost indefinitely, there are still some details in the matter which need an explanation. Between the moment of the death of a mammoth and that of its transformation into a frozen carcass and burial within the natural refrigerator must have passed some time during which the carcass surely suffered some decay, although current opinion attributes to the meat of a mammoth an almost absolute freshness. Howorth, for example, compares it with "the flesh recently taken out of an Esquimaux cache or a Yakut subterranean meat-save." [1] As matter of fact such freshness is a legend. The only proof of it is bright red color of flesh and white or yellowish of fat, and the fact that the flesh used to be devoured with avidity by dogs and wild animals. But the same meat was absolutely unpalatable for an adventurous scientist. All stories published in newspapers of this country of a dinner in St. Petersburg where the meat of the Beresovca mammoth was served, are a hundred per cent invention. [2] All travelers also used to say that the carcasses of the mammoth as a rule had an intolerable putrid smell. As in no case a scientist had a chance to examine mammoth flesh immediately after the animal had been discovered, but usually a year, two, or more later, it appeared correct to attribute these conditions to putrefaction which took place after the uncovering of a carcass. But a strong smell is peculiar to the mammoth localities and to the ground within which remnants are buried, even when they are concealed within and, presumably, still firmly frozen. No process of decay is possible under temperatures below the freezing point, and in the case of the mammoth, rhinoceros, etc., it did not take place; because if it had, after many thousand years of decaying even though it were a gradual process, no soft parts would have been preserved. The smell in the ground may, therefore, be the result of the putrefaction started immediately after the death of an animal, before it became permanently frozen, and may be called fossil as well as a carcass itself. The putrid smell of a mammoth is different from that of other putrefied flesh, but more penetrating and very appealing to wild animals. For this reason the flesh of the mammoth is often used by natives as bait in their fox traps. An examination of the flesh and fat of the mammoth from Beresovca River has also shown that they suffered a deeply penetrating chemical alteration as a result of the very slow decay which was going on, probably, in an airtight medium. [3] It would be possible to refer these alterations to the time immediately following the death of an animal when it was, for example, covered with drift, but not yet definitely frozen, or was for a while in water, etc. Like common decay these chemical processes had to be suspended so soon as ground and the carcass became firmly frozen.

Decay of organic matter in Arctic climate, so far as results are concerned, is going on differently than putrefaction in moderate climate, and much more slowly. A good illustration of that has been given by the Adams mammoth which was examined four years after it had slipped down to the shore and had, all this time, remained uncovered. Its carcass had suffered very much, but chiefly through wild animals and dogs. A few soft parts happened to remain intact, were gradually dried out and in this form brought to St. Petersburg. The Beresovca mammoth also remained uncovered and not well protected for two summers, before the arrival of an expedition. In Arctic regions the highest summer

[1] Howorth, H. H., "The Mammoth and the Flood," p. 93.
[2] Gardner, M. B., "A Journey to the Earth's Interior," p. 44.
[3] Bialinitzki-Birula, F. A., "Observations histologiques," p. 19.

temperature of the air is still lower than in moderate ones.   The warm time of a year is much shorter and is practically confined to the nightless period of summer.   During other summer time warm days are alternating with cold nights when the temperature often goes down towards freezing.   The process of putrefaction goes on very slowly in Arctic regions also, because of the great purity of air and, perhaps, also due to the absence of insects. In Green Harbor, on Spitzbergen Islands, Nordenskiöld, in 1868, saw on the shore twenty-four white-whales killed by fishing vessels.   Although the carcasses were "exposed day and night to the direct action of the sun's rays, there was no sign of putrefaction, and the entomologist of the expedition could not capture a single fly or other flesh-loving insect upon them." [1]   More recently air in Novaya Zemlya was found practically germless.[2]   In Northeastern Siberia putrefaction is also delayed because of the very dry climate which, so far as the amount of precipitation is concerned, might be compared with the steppes and deserts of Middle Asiatic territories east of the Caspian Sea.   In such conditions sometimes a carcass of an animal could stay over the warm season without becoming very much decayed. Much depends upon the time when an animal died.   If it happened during early spring or winter, and the carcass were not exposed to direct action of the sunshine, during the spring it would thaw in the day time and freeze again at night, drying out all the time, days and nights alike, without undergoing putrefaction, as it would happen in a warm region; putrefaction is handicapped by the sterilization produced by the night freezing. In this way a carcass might be mummified, completely or partly, before the warmest time of the year and better resist decaying processes in the summer.   Fall offers the same or perhaps better conditions, because a carcass would be partly mummified before winter, making the whole period favorable to preservation; longer.

How important all these processes are, and how they delay putrefaction, has been proved by everyday experience and often profitably utilized.   Traveling in the mountains of Southern Siberia the writer, for example, realized that the supply of meat could be kept in fresh, or in quite palatable condition, if the meat every night were taken out of bags or other containers and hung up to be freely affected by the night breeze and decreased temperature.   North Siberian natives utilize the early spring time to prepare dried meat. They cut fresh meat taken for drying into pieces or strips and hang up in the shade, never under the sun rays, in places with good circulation of air, affected by the breeze.   The meat remains frozen the greatest part of the day, but thaws a little about noon when it becomes affected by warm dry air.   The days grow longer, and the meat every day thaws for a longer time, but every day it also becomes dryer and dryer.   In about two or three months the meat is ready and may be used raw, boiled, or grilled in every form being quite a palatable product, suitable for preservation for an indefinite time.

The writer also chanced to find, in 1905, near Lake Yesei, in Northern Siberia, at the latitude about 68° 30′ and 102° East of Greenwich, a body of Tungus-shaman transformed into a mummy by the climatic conditions referred to.   The body was found well dried, in a so-called "hanging-tomb," a strong wooden coffin fixed on a wooden structure three or four feet above ground, where it had been buried more than fifty years before. Brought to St. Petersburg and placed in a museum this body has remained without any noticeable decay, although in entirely different climatic conditions.

[1] Nordenskiöld, A. E., "The Arctic Voyages," p. 137.
[2] Science, LXIX, No. 1789, Suppl., p. XIV.

The recent climate of Northern Siberia is such that we easily can imagine conditions under which a carcass of an animal died on the land would last over summer, without complete decay, become frozen during fall and winter and, if covered with silt, landslides, etc., be preserved within the frozen ground for an indefinite number of years. It is, therefore, unnecessary to look for different physico-geographical conditions, when the recent Siberian ones so readily explain the origin and preservation of the carcasses of the mammoth. As we have seen before, such conditions were prevailing at the Age of the mammoth in Siberia. The burial of some mammoths was considered above in the case of the animals trapped in mud.   As has been shown by the plants found in its stomach, it happened during late summer or early fall, when the middle part of summer was already over, nights had started to lengthen, and mud streams were at their greatest size, being easy traps and giving good conditions for a comparatively quick burial.   In this case wild animals were contributing more to the destruction of the carcass than decay.   The mammoth examined by Vollosovich on the Sanga-Yurakh River had been spoiled by ice-foxes immediately after its death and before it was buried.   But even in the case of a mammoth trapped within mud, only the upper surface of the carcass was available for a feast, and this became smaller and smaller due to the continuing flow of mud over it.   Usually a carcass could only be spoiled by devourers and only occasionally completely destroyed.   Animals perished in winter, since they were frozen, were more or less protected against the attack of wild animals by the hardness of their frozen carcass, which increased with the lowering of temperature.   They were also covered with snow drifted over the obstacle.

In recent Siberia we also meet physico-geographical conditions under which a carcass will undergo very slow decay, probably much slower than on the land, and at the same time will have nearly perfect protection against different carnivores.   These conditions must have existed in the Age of the mammoth, and recent observations can be applied to the mammoth.   An animal could plunge through the ice of a river or lake, or simply drown, and its carcass, under the climatic conditions of Siberia, and due to the presence of frozen ground, could be frozen to the bottom and eventually covered with silt.   It could also be moved by the stream along the bottom, or later, on acquiring buoyancy, drift down the river and be buried again on the bottom or shore of a river, lake or sea.   Owing to the low temperature of water in North Siberian rivers the decaying of a drowned carcass must go on very slowly in it.   If a carcass happened to be covered with silt, it must have been at that time in fairly fresh condition.   After such a covering, the process of decomposition must be retarded still more.   To all these phenomena, details of which are dependent upon the nature of North Siberian rivers, surprisingly little attention was paid until recently. Occasionally, and in very unusual and tragic conditions, the writer was able to realize all the importance of the events referred to for the question under consideration.

In 1920, during the Russian civil war, thousands of people were killed at Nicolaevsk, a small town on the Amur River, and hundreds of bodies thrown into the river, under the ice in winter, overboard from a tug in the spring. The killing stopped only in June when the destroyed town was abandoned by the population.   In July of the same year the writer happened to be at Nicolaevsk and for seventeen days stayed on board of a steamer occasionally anchored near the very spot where the killing had been going on in winter and spring.   Every day we could observe floating bodies which had left the bottom, where they had lain for a few months, and were drifting down the river.   Some of them later were found on the shores, others, probably, brought to the sea, or sunk again.   The number of bodies which left the bottom during seventeen days was estimated at about 150.   It was, therefore, a regular phenomenon not an occasional event.   In some way the bodies were fixed to the bottom during a few months, presumably frozen there, because every time, as the body was observed coming to the surface at the first moment, it gave an impression of jumping out of the water, as if it had been freed only due to some excess of buoyancy.   The cold water into which the bodies has been thrown preserved them for a number of months very little decomposed, if changed at all; only in the middle of summer, when the water became warmed through, did decomposition start.   The preserving properties of cold water are well known in Northern Siberia.   Matyushkin tells us that in hunting for reindeers, crossing rivers on their way back from the North in the middle of August, "The deer which have been killed are sunk in the river, the ice-cold water of which preserves them for several days, till there is time to prepare them for winter use." [1]   The floating bodies on the Amur River were all badly decomposed, but hardly more than, in usual conditions, a body drowned for a week or two would be.   If they were frozen in the ground and discovered thousands of years later, they certainly would be considered very well preserved.   All these events were observed in Southern Siberia which enjoys a comparatively moderate climate.   In Northern Siberia the bodies could remain on the bottom all through the summer, be buried within the drift accumulated around an obstacle, and preserved frozen for an indefinite number of years to come.   After thousands of centuries the river could deepen its channel and abandon the former one.   In this case the carcass buried in the way referred to, would be found on the river terrace, or, speaking more broadly, within old river deposits, like the Adams mammoth, Khalbugaï rhinoceros, etc. Such happenings certainly could not be considered as every day events, but, neither were they unusual or exceptional.   A distribution of the skeletons of fossil animals in old river channels has been established not once.   Such is, for example, a very known locality of *Iguanodons* at Bernissart, Belgium.   One of the most interesting localities of this kind was discovered more than a quarter of a century ago in Northern Russia, on the shore of the river Syeverhaya Dvina, where a number of perfectly preserved skeletons of Permian reptiles and amphibians were found within the bed of an ancient river, filled up completely with drift.   All the skeletons were found enclosed in peculiar concretions in most cases reproducing fairly well the general shape of an animal.   They were oriented at the locality all in the same way, undoubtedly according to the direction of a flow, and there could be no doubt about their transportation by a river.   Animals had probably drowned and were preserved in the cold water from very quick decomposition, as well as from attack by wild animals.   In the writer's opinion it was even possible that they were buried as carcasses, and that the origin of the concretions was connected with a gradual decaying of soft parts.   In the case of the mammoth, we also have examples of such a concretion *in statu nascendi*.   This was the Middendorff mammoth found on the Taimir River, soft parts of which were already replaced by mineral material containing some organic substance. If this mammoth were protected against destruction, its skeleton in due time could be found enclosed within a concretion more or less similar to those from the river Syevernaya Dvina. We can suggest a still closer analogy between the conditions of preservation of mammoth carcasses within the river drift and those of the Permian fauna in the Syevernaya Dvina

[1] Wrangell, F., "Narrative of an Expedition," p. 186.

Territory, because in both cases climatic conditions, presumably, were somewhat similar. Permian animals were found there within the region of the *Glossopteris* flora, a product of a rather cold climate. Besides that, Amalitzki, to whose credit belongs the discovery of this fauna, was ready to attribute a glacial morainic origin to some parts of the Permian strata outcropping in the Syevernaya Dvina Region. This conclusion is certainly apt to bring a number of far-reaching suggestions concerning this Permian fauna, but it would be beyond the limits of the present paper to dwell longer on them.

As follows from the review just accomplished, among the many theories offered and supported by different scientists in trying to explain the conditions of death, burial, and preservation of carcasses of the mammoth and rhinoceros, some are absolutely improbable, others probable, but not supported by facts, and two not only can stand criticism, but are supported by real observations. These two explanations accepted by the writer are: the mammoth had perished and had been buried in mud streams caused by the thawing of frozen ground and rock ice; or, the mammoth had drowned in rivers or lakes, especially during winter or early spring, been frozen to the bottom, and buried in drift on the very spot, or drifted down stream, often as a frozen carcass, and buried somewhere in the lower part of the rivers within their deltas, or embouchure sediments. The examples of the first case are the mammoths found on the Beresovca River, on Sanga-Yurakh River, and on the Bolshoï Lyakhov Island. The examples of the second case are the Taimir mammoth of Middendorff, the Lena mammoth of Adams, rhinoceroses found on the Vilui and Khalbugaï Rivers, and perhaps, at least partly, carcasses found on the New Siberian Islands. Because of all the conditions already considered, one would not expect, in the first case, to find a carcass perfectly preserved and unmolested by wild animals. But such localities give a better idea about the surroundings of the mammoth, its habits, and its geological position. In the second case one might expect to find the carcass of a mammoth in perfect condition, so far as its preservation was concerned. But geological and other data which are collected in such localities could not be compared with those achieved in the first case, and such localities have to be considered as secondary ones.

EXTINCTION OF THE MAMMOTH

THE writer has already said [1] a few words on the extinction of the mammoth and given a short review of different explanations of this phenomenon, all of which have now only an historic interest and can only be called fantastic, comparatively recent theories of Howorth and Gardner included. In a rather curious way these theories repeat, in somewhat modernized form, the tales of Siberian natives, reflected in Siberian folklore, and old Chinese traditions. Unfortunately we are unable to replace them by new ones which could harmonize with all accumulated data and stand criticism from different quarters, but must be·satisfied with more or less probable suggestions. It seems to the writer that most of the students who had a chance to work on the mammoth-question came to the same sad conclusion. The problem is extremely difficult. We must explain the extinction of an animal which was living in great numbers, apparently very prosperously, over a large area, in variable physico-geographical conditions to which it was well adapted, and which died out in a very short time, geologically speaking. The difficulty of problem, perhaps, is well illustrated by the·fact that Arctic scientific travelers from whom it would be more natural to expect a solution, are very cautious in their speculations. If they sometimes used to touch on the question of the extinction of the mammoth and its contemporaries, they did not consider it for the whole mammoth-country, but were referring only to some limited area. In the same way, Russian scientists familiar with the problem did not try to find its general solution. Brandt and Schrenck, for example, used to consider an accidental extermination of an individual mammoth and did not approach the problem of the extinction of species. Middendorff and Schmidt dealt with the living conditions of the mammoth-habitat in the extreme North of Siberia. Toll and Vollosovich tried to understand the extinction of the mammoth and other Post-Pliocene mammals, but took into consideration only the territory of the New Siberian Islands, which does not clear up the whole question Tscherski and Pavlova, who identified and described the fossil mammals from the New Siberian Islands and Arctic part of the mainland, also considered the problem of extinction in connection with the change of climate, but were not more successful than their predecessors. Pavlova·has recognized that the problem still waits for new·observations on the spot and for new data, to be solved more or less satisfactorily, a hope which still remains frustrated. A number of different universal theories explaining the extinction of the mammoth were offered usually by scientists who were not familiar enough with the mammoth-question in detail and could not appreciate all the present and former conditions of the regions in which the mammoth had lived. It is also necessary to mention that the extinction was considered by many scientists to be a result of an extermination which, in the opinion of the writer, is mostly incorrect. Extermination destroys an individual, extinction—a species. The former works in one generation, the latter goes on through many generations. Extermination is a result of some exterior agents, usually is local and could not result in an extinction of a species over a large area, except in catastrophes like those·which were previously advocated.

[1] Cmp. Introduction.

The recent explanations of the extinction of the mammoth may be placed in the three following categories: (1) Destruction by man; (2) defects of organization having resulted in poor adaptation to surrounding conditions; (3) change of physico-geographical conditions.

Extermination by man has been eagerly advocated recently by Digby [1] who follows L. Laloy in this.[2] It went on by hunting and direct killing, or by pursuing and pushing the mammoths into inhospitable parts of the country, where they were destroyed by the unfavorable physico-geographical conditions. Such ideas appeared more than once before, but were always discarded as unreasonable.[3] Human society in Post-Pliocene was too scarce to be dangerous to huge flocks of mammoths distributed over large areas. Northern Siberia was, probably, no more populous than now. Accordingly, Reid [4] accepted the extermination of the mammoth by man only in Europe. In Siberia, in his opinion, the mammoth was gradually killed by the increase of cold and want of food. According to Pfizenmayer, extermination of the mammoth by man took place in Middle Siberia where it had come from the North because of the change of climate.[5] The primitive man was armed very poorly and, probably, was more inclined to avoid the big brute than to chase it. The ancient natives of Siberia had plenty of other game more available than the mammoth, like their recent African brothers who, according to Digby,[6] "surrounded by mealie gardens, fowls and great herds of antelope, left the great-tusked trampling brutes alone." To African natives the elephant was, probably, more dangerous than they to it; and, perhaps, the elephant used to kill as many natives as they kill elephants.[7] The extermination of the recent elephant in Africa started only since the arrival of the white man with his rifle and a great outside market for ivory. In a picturesque way, Digby describes how the prehistoric man caught the mammoth [8] in booby traps exactly in the same way as do African natives nowadays, although the making of a large pit in the frozen Siberian ground offered almost insurmountable difficulties, and natural clefts suitable for this purpose were practically absent there. Extermination of the American bison, which Digby compares with that of the mammoth was also accomplished by white intruders. American Indians lived for centuries along with these animals, had plenty of food and skins, but were in no danger of being deprived of this storage in the years to come. As matter of fact, primitive man always was and has been very wise as to the utilization of natural resources. He never used to kill more animals than he needed for his household and knew exactly his needs. Only the intervention of cultured colonists who opened a market for game, skins, and furs, could make a native forget his wise economy. Unmolested, or only seldom attacked by man, the mammoth had no serious enemies among his carnivorous contemporaries either.[9]

Concerning the defective adaptation to surrounding conditions, it is necessary to say that in no one case was it possible to discover in the frozen carcasses of the mammoth any bad effects of conditions under which the animal used to live. The animals were always well fed and fat, sometimes too fat, as the Adams mammoth, the belly of which,

---

[1] Digby, B., "The Mammoth," p. 33.
[2] Laloy, L., "Le régime alimentaire du Mammouth: L'Anthropologie, XVII," p. 234, Paris, 1906.
[3] Howorth, H. H., "The Mammoth and the Flood," p. 172.
[4] Reid, Cl., "The Sudden Extinction of the Mammoth," p. 44.
[5] Pfizenmayer, E. W., "Mammutleichen," S. 148.
[6] Digby, B., "The Mammoth," p. 65.
[7] Neuville, H., "On the Extinction of the Mammoth," p. 328, footnote.
[8] Digby, B., "The Mammoth," p. 65.
[9] Howorth, H. H., "The Mammoth and the Flood," p. 172.

according to its native discoverer, was below its knees.[1] The animal must have had a comfortable living, which, under the conditions of the Arctic climate, was possible only if it were well adapted to those conditions.   It must have been well protected against cold and been able to find plenty of food.   Vollosovich pointed out the undersized dimensions of his Sanga-Yurakh mammoth as a prophecy of the extinction of the race.[2] We certainly must recognize the fact that a regular decrease in size affecting a species is a danger-signal, but we must not consider the small-sized races of a species as groups becoming extinct. In many cases a decrease in size may be an adaptation to some special conditions, not dangerous, but favorable for the race.   We do not know whether this was the case with the mammoth and its northern race, but might suggest so, on the ground of analogy to the recent reindeer which might be compared with the mammoth, so far as its geographical distribution, character of living, and feeding habits are concerned.   The reindeer living in the forests of Northern Siberia belongs to so-called *Tungusian* or *Lamutian* race which is higher, heavier, and stronger than different races bred along the Arctic coast of Europa and Asia.   Only this race can be used for riding even by a heavy man, but we have no ground for considering the northern race as a dying one.   It is just as well adapted to the Arctic conditions, and the writer was told by Siberian natives that in the fall, when the tundra is covered with the first snow, it finds its food more easily than does the southern one. Still smaller is the Spitzbergen race which lives very well in those desolate Arctic islands.[3]

In recent times, the large curved tusks of the mammoth attracted the attention of some scientists, as the thing which had given the mammoth less resistance as compared with Indian and African elephants.   It was suggested that formerly the mammoth used to live in the forest, where it had ample opportunity to rub its tusks upon trees and prevent them from growing beyond limits.[4] Then it migrated to the tundra, and its tusks started to grow more than necessary and brought about its extinction.   In a similar way a squirrel which was fed on a soft food could be killed by the unlimited growth of its incisors.   As a matter of fact, the mammoth never completely abandoned forests for the tundra, but stayed in forests or wandered in the tundras alike.   Besides that and quite contrary to the theory, the tusks of the more southern and therefore forest-loving variety were larger than those of the tundra dweller.   Curiously enough, according to Howorth, "The arboreal nature of the food of the mammoth is again proved by the inordinate length of its tusks as contrasted with the short tusks of the grass-eating Indian elephant."[5]

After the histological examination of the skin of the Siberian mammoth, Neuville came to the conclusion that the animal, in spite of its heavy fur, had, in its skin, a very poor protection against cold.[6] At the same time it was unable to leave the country, which on account of cold had become very unhospitable for it, because the structure of its feet did not allow the mammoth quick locomotion.   In a special article the writer tried to show that the histological structure of the skin of the mammoth, as described by Neuville, was a good adaptation and protection against a low temperature,[7] not to mention the fur itself which exactly corresponded to the fur of other Arctic animals.[8] The structure of the

[1] "Sibirski Vestnik, 1920, part X," p. 320.
[2] Vollosovich, C. A., "On the Digging out of the Sanga-Yurakh Mammoth, in 1908," p. 456.
[3] Nordenskiöld, A. E., "The Arctic Voyages," p. 86.
[4] Neuville, H., "On the Extinction of the Mammoth," p. 333, footnote.
[5] Howorth, H. H., "The Mammoth and the Flood," p. 69.
[6] Neuville, H., "On the Extinction of the Mammoth," p. 336.
[7] Tolmachoff, I. P., "Note on the Extinction of the Mammoth," p. 68.
[8] Pfizenmayer, E. W., "Mammutleichen," S. 229.

feet was also a very good adaptation to the soft ground of the mammoth pasturages.[1] As to its walking abilities, they also were, probably, not so bad. Toll found bones of the mammoth on the central plateau of Kotelny Island, about a thousand feet above sea level, and remarked that the mammoth must have been a good walker.[2]

All theories as to a defective organization of the mammoth which presumably brought it to extinction, meet a very strong objection in the fact that the mammoth was not the only animal which had lived and died out in Northern Siberia since the Post-Pliocene. *Rhinoceros tichorhinus* Fischer was its typical companion, as well as a number of other animals abundantly found in different parts of Siberia, among other places also on the New Siberian Islands. The effects of organization destructive for one species could not be responsible for the extinction of another one, but, as a matter of fact, all these animals died out more or less simultaneously and probably from the same cause. It is easy to understand therefore that the explanation of extinction of this fauna through the action of changed physico-geographical conditions acquired more and more supporters.

It was mentioned above that during the Post-Pliocene, forests in Siberia, probably, penetrated farther towards the North than now, and the climate of Northern Siberia was somewhat milder than the present one. The influence of this change, not a very great one, anyway, must not be overestimated for the simple reason that it could affect only the most northern limit of the distribution of Post-Pliocene mammals. Change of climate could only go on very gradually and slowly, and animals affected by it had plenty of time to become adapted to new conditions, or to migrate southwards and to find conditions corresponding to their former habitat. Concerning the mammoth we saw above that it, probably, lived happily through all changes of climate and did not suffer at all from the severe climate of its last days. It also had enough food all the time. Direct influence of changed climate could not, therefore, be considered responsible for the extinction of the mammoth, and we have to examine it in connection with other events. In the opinion of Toll, the dying out of the mammoth and its contemporaries on the New Siberian Islands could be caused by the separation of the islands from the mainland,[3] this in connection with the change of climate for the worse and the decrease of food, lack of which could not be filled through migration, produced very unfavorable conditions of living, especially of feeding the very abundant mammalian fauna of islands and resulted in its extermination. According to Vollosovich, extinction of the mammoth and rhinoceros on the New Siberian Islands had been already accomplished before the separation of islands from the mainland, thus could have been affected only by the change of climate connected with the decrease of food.[4] If the separation of the islands from the continent and each other were complete, it could also bring their animal population to painless extinction by the close interbreeding so fatal to small separated communities. But this interbreeding could not have been so close as to become destructive. Besides, the islands for about eight or nine months remain firmly connected with the mainland, as well as together, by frozen sea. A seasonal migration of reindeers to the islands and back to the continent is going on now regularly every year, and Toll met a flock of them, about thirty animals, so far north as the Bennet Island.[5] In

[1] Vollossovitch, C. A., "Le mammouth de l'île Bolchoï Lakhovsky," p. 336.
[2] Toll, Ed., "Die fossilen Eislager," p. 62.
[3] Toll, Ed., "Sketch of the Geology of New Siberian Islands," p. 15.
[4] Pavlova, M., "Description of Fossil Mammals," p. 36.
[5] Toll, Ed., "Short Report for the Period June 7 to November 8, 1902," p. 158.

the same way certainly the migration of the mammoth and its contemporaries went on. It is also probable that the abundance of the former fauna on the New Siberian Islands was not as great as it appears to be, owing to the great profusion of fossil remnants; because a large part of them has been found there in a secondary locality, having been brought from the adjacent parts of the continent by ancient rivers.

To these migrations between islands and mainland, as well as between different islands perhaps, may be applied the speculations of Wright referring to the plunging, through ice, of animals during the early fall and late spring migrations.[1] Such accidents were quite possible, as they happen now, but they had nothing to do with the extinction of the mammoth. According to Bell, change of climate for the worse was contributing to the extinction of the mammoth chiefly in connection with its accustomed migrations. In his words: "As the climate gradually became more and more severe, and the summers shorter and shorter, the inertia of this migratory spirit continued, and large herds of mammoth from time to time were caught in the fearful blizzards, so common now during the early autumn in Northern Siberia, and perished from cold and hunger."[2] Such accidents could also take place, but they did not help much towards the general extinction of the mammoth and, as a matter of fact, were independent of the change of climate referred to. In the North they could happen, even if the climate were much milder than the present one. With increasing coldness of climate the distance of migrations must become shorter. Besides, the change of climate was not so great as to have a noticeable influence upon the duration of warm season.

Thus, no one of three possible lines of explanation of extinction of the mammoth can stand criticism and give a satisfactory solution of problem which, in the opinion of the writer, must be considered from a quite different point of view, namely, as an example of a very well-known phenomenon of extinction, in different geological periods, of species, genera, families and even of faunas. In all these cases a group of animals was replaced by another one, when physico-geographical conditions did not become destructive for the former, which was sufficiently proved by the survival of the isolated representatives of the first group. In this way mammals in Tertiary came into possession of the position which during the Mesozoic was the indisputable property of reptiles. Lower Paleozoic seas used to swarm with trilobites as did the Mesozoic ones with ammonites, and both these groups died out without any special reason. Explanation given by different paleontologists of the extinction of these groups do not satisfy us any more than those referring to the extinction of the mammoth. For example, the extinction of trilobites has often been explained by the appearance in the Lower Paleozoic of fishes which fed on trilobites. The latter could be exterminated in this way, but such an extinction must have a character of a momentary catastrophe when suffering animals would be destroyed in no time, geologically speaking. We know that the extinction of trilobites was going on through a number of geological periods, and they were extinguished like a lamp which gradually has less and less oil. This oil in the case of an organism is its vital force or, more exactly, its ability to reproduce. Replacement of reptiles by mammals is a still more mysterious phenomenon, as we lack the evidence of any direct struggle between the two groups. We may say that mammals have taken in nature the place already abandoned by reptiles, or

[1] Wright, G. F., "Asiatic Russia, II," p. 581.
[2] Wright, G. F., "Asiatic Russia, II," p. 581.

which the latter were ready to vacate.   An appeal to changing climatic conditions, as the cause of extinction in the cases referred to, is usually unsuccessful as well.   In most cases we do not know exactly all the conditions of corresponding periods and must suggest something; and in any case those conditions could exterminate a race only in an accidental, catastrophical way.   If the change of conditions were going on slowly and gradually, as, of course, would be most natural, organisms would have plenty of time for adaptation to new conditions, or for migration.

It is very important to notice that in all cases extinction was accompanied by the peculiar development of different morphological characters of a given group.   It was usually a dernier cri in the development of some structures which an animal or group of animals was striving to develop and perfect during all its life, and which often became developed *ad absurdum*.   These structures are well known to paleontologists and have often been considered as cases of over-specialization,[1] being connected with extinction, or as a prophecy of an approaching extinction; measures unconsciously and tentatively taken by organisms to avoid an extinction, "as if heroic efforts were being made to maintain the race."[2]   Examples could be given in hundreds among different groups of organisms. We can consider these aberrant structures as inevitable companions of extinction, perhaps, as causes of it.   We can suggest that in their struggle to accomplish some peculiar structures, organisms sometimes can exhaust their vital forces and be doomed to destruction.   This exhaustion affects the reproductive abilities of an organism and causes a species or group of species to undergo gradual painless extinction without any direct influence of physico-geographical conditions, appearance of new enemies, etc.   In connection with this suggestion we must remember the well-known biological fact that lower groups of organisms are much more prolific than the higher ones.   The same fact applies to different forms of the same class.   The age of puberty appears later among higher organisms than among the lower ones, which decreases the reproductive ability of the former.   Highly specialized forms are often sterile.   As specialization is a result of adaptation of an individual to given conditions, we arrive to a rather paradoxal conclusion that the great achievement of an individual may become destructive for the species.   In such an over-specialization, accompanied by decrease in reproduction which might not become a complete sterility, we must look for a general cause of extinction.   Extermination through natural enemies or change of surrounding conditions might be accompanied by extinction only in rare, rather exceptional and usually catastrophical cases.   It was, perhaps, just the reason of the complete failure of many attempts to explain extinction in the light of extermination.

Returning to the mammoth, we find in this animal a few characters of extreme specialization, as, for example, the structure of its molars.   The type of the structure had started in *Elephas antiquus* Falc.   The intermediate form, *Elephas trogontherii* Pohlig, had more numerous dental plates, more closely arranged and covered with thinner enamel. *Elephas primigenius* Blum. is a further step in development in the same direction, especially well expressed in the Siberian variety with its numerous dental plates and a very thin enamel covering.   Another over-specialization is in the size and in the form of tusks of the mammoth.   The form was such that a mammoth could use its tusks as a recent elephant uses them, probably only while a young animal.   For the adult individual they were nearly

[1] Gregory, W. K., "Two Views of the Origin of Man," p. 601.
[2] Schuchert, Charles, "Historical Geology," p. 210.

useless. The mammoth also had a more specialized foot than any other elephant, having only four toes as compared with the five of other elephants.[1] Its tail had a form which could be only explained as a special adaptation against cold. For the same purpose was developed a separated cover of the anus, in the form of a skin fold,[2] and the fur of the animal. No doubt, individually the mammoth was not weaker, or more poorly adapted, to its living conditions than its ancestors, but, being the last member in a particular line of evolution, it was thus doomed to extinction, by causes which were not external, but concealed in the character of the species itself, in its decreased ability of reproduction. Probably the similar consideration might be applied to the fossil rhinoceros which was very different from its recent tropical relatives, having been, like the mammoth, wonderfully adapted to the severe climate of Siberia.

The fate of the mammoth and rhinoceros was shared by a number of their contemporaries. The same considerations might be applied to all of them, but, as matter of fact, we know less about them than about the mammoth and rhinoceros who overshadowed their less imposing companions. We do not know even, how exactly they might be called contemporaries. The only geologist who tried to find a proper place in a geological succession for different Pleistocene mammals of the New Siberian Islands, where this fauna has been known better than anywhere in Siberia, was Vollosovich. But, as has been shown above, his geological scheme may not be accepted without checking, and his distribution of fossil mammals must also be revised before it can be depended upon. Besides that, Bunge, Toll, and Vollosovich did not distinguish between primary and secondary localities of fossil mammals in the New Siberian Islands. Our knowledge of different fossil mammals there is very unequal. We have, for example, little doubt that the musk ox was an Arctic animal, like its recent representatives, and that it used to live and die out along with the mammoth and rhinoceros. But we know, for example, very little about the tiger the remnants of which were found in the New Siberian Islands. Was it also an animal well adapted to Arctic conditions, or did it lack such an adaptation, making the change of climate referred to above fatal for it? Did it formerly live in the New Siberian Islands, or were the few bones found brought over there by rivers?

The writer does not pretend that his explanation of the extinction of the mammoth is anything more than a suggestion which appears to him more or less plausible. He would like only to emphasize once more that the extinction of species is seldom dependent upon the same causes as an extermination of individuals belonging to this species. In many cases and, probably, in most cases the cause of extinction may be entirely different from the cause of destruction of an individual. A race might be not weakened at all, might even become stronger than before and be doomed to destruction because of high specialization which affects the ability of reproduction and brings species, apparently vigorous, to extinction. High specialization in some particular line or lines, perhaps all characters which tend to bring an individual to a high perfection, at the same time may be fatal for the corresponding species.

[1] Pfizenmayer, E. W., "Mammutleichen," S. 153 and 239.
[2] Pfizenmayer, E. W., "Mammutleichen," the picture facing p. 161.

BIBLIOGRAPHY ON THE MAMMOTH AND FOSSIL RHINOCEROS OF SIBERIA [1]

Account of Russian Sea Travels from the Siberian Rivers to the Arctic Sea: Sibirski Vestnik, 1822, XVII, pp. 39–48, 117–128, 185–196; XVIII, pp. 305–314, 379–398; XIX, pp. 167–180 (in Russian).

ADAMS, M.—Relation d'un voyage à la mer glaciale et decouverte des restes d'un mamouth: Journal du Nord, XXXII, p. 633, St. Pétersbourg, 1807.

ANDREWS, C. W.—Note on the Skull and Mandible of Siberian Mammoth exhibited in the British Museum (Natural History): Ann. Mag. Nat. Hist., 9 Ser., XII, pp. 322–325, London, 1923.

BAER, K. E.—Neue Auffindung eines vollständigen Mammuths, mit der Haut und den Weichttheilen, in Eisboden Sibiriens, in der Nähe der Bucht des Tas: Bull. Acad. Sc., X, pp. 230–296, St. Petersburg, 1866.

BAER, K. E.—Fortsetzung der Berichte über die Expedition zur Aufsuchung des angekündigten Mammuths: Bull. Acad. Sc., X, pp. 513–534, St. Petersburg, 1866.

BIALINITZKI-BIRULA, F. A.—Observations histologiques et microchimiques sur les tissus du Mammouth: Resultats scientifiques de l'expédition organisée par l'Académie Imperiale des Sciences pour la fouillée du Mammouth trouvé sur la rivière Bérézowka en 1901, II, pp. 10–14, St. Pétersbourg, 1909 (in Russian).

BOLTUNOV.—Description of the Lena Mammoth: Technologuicheski Journal, III, p. 162, St. Petersburg, 1806 (in Russian).

BRANDT, AL.—Kurze Bemerkung über aufrecht stehende Mammuthleichen: Bull. Soc. Nat. Moscou, III, p. 97, Moscow, 1867.

BRANDT, J. F.—Mittheilungen über die Gestalt und Unterscheidungsmerkmale des Mammuth oder Mamont (*Elephas primigenius*): Bull. Acad. Sc., X, pp. 93–111, St. Petersburg, 1866.

BRANDT, J. F.—Zur Lebensgeschichte des Mammuth: Bull. Acad. Sc., X, pp. 111–118, St. Petersburg, 1866.

BRANDT, J. F.—Einige Wörte zur Ergänzung meiner Mittheilungen über die Naturgeschichte des Mammuths: Bull. Acad. Sc., X, pp. 361–364, St. Petersburg, 1866.

BRANDT, J. F.—Einige Wörte über die Haardecke des Mammuth in Bezug auf gefällige schriftliche Mittheilung des Hrn. Professor O. Fraas über die im Stuttgarten Königl. Naturalienkabinet aufbewahrten Haut- und Haarreste des fraglichen Thieres: Bull. Acad. Sc., XV, pp. 347–351, St. Petersburg, 1871.

BRUSNEV, M.—Report of the Leader of the Baron Toll Relief Expedition to the New Siberian Islands: Bull. Acad. Sc., XX, Phys. Math. Section, pp. 161–194, St. Petersburg, 1904 (in Russian).

BUNGE, A.—Die Lena Expedition 1881–1884: Beobachtungen der Russischen Polarstation an der Lenamündung, I, Anhang, S. 1–96, St. Petersburg, 1885.

BUNGE, A. UND E. TOLL—Expedition nach den Neusibirischen Inseln und dem Jana-Lande: Beiträge zur Kenntniss des Russischen Reiches, III Folge, III, S. 1–363, 6 Karten, St. Petersburg, 1887.

CUVIER, G.—Recherches sur les ossements fossiles, où l'on retablit les charactères des plusieurs animaux dont les révolutions du globe ont détruit les espèces. Troisième édition, I and II, Paris, 1825.

DEPÉRET, CH. ET L. MAYET—Monographie des eléphants pliocènes de L'Europe et de l'Afrique du Nord: Annales de l'Université de Lyon, n.s., Iéfasc. 43, pp. 89–224, Lyon et Paris, 1923.

DIGBY, BASSET—The Mammoth and Mammoth-Hunting in Northeast Siberia, pp. 1–224, with photographs and a map, London, 1926.

Encyclopedia Britannica, XI Edition, Cambridge, England, 1911.

ERMAN, AD.—Reise um die Erde durch Nord-Asien und die beide Oceane in den Jahren 1828, 1829, und 1830, Abtheilung I, Historischer Bericht, I–III, Berlin, 1833–1848.

FAMINTZIN, A. S.—On the Excrement of the Mammoth: Mem. Acad. Sc., LII, p. 173, St. Petersburg, 1886 (in Russian).

GARDNER, MARSHALL B.—A Journey to the Earth's Interior or Have the Poles Really Been Discovered, pp. 1–69, Aurora, Illinois, 1913.

Germless Island in the Polar Seas: Science, LXIX, No. 1789, Suppl., p. XIV, 1929.

GLEBOV—Recherches microscopiques sur les parties molles du Mammouth: Bull. Soc. Nat. Moscou, XIX, p. 108, Moscou, 1846.

GREGORY, W. K.—Two Views of the Origin of Man: Science, LXV, No. 1695, p. 601, New York, 1927.

HAY, OLIVER P.—Observations on Some Extinct Elephants, pp. 1–19, Washington, D. C., 1922.

HEDENSTRÖM, M.—Travel to the Ice Sea and to the Islands Situated East of the Lena Embouchure: Sibirski Vestnik, 1822, XVII, pp. 27–38, 99–116, 171–184; XVIII, pp. 245–258, 291–304, 359–378; XIX, pp. 1–18, 85–106 (in Russian).

HEDENSTRÖM, M.—Otrivki o Sibiri, St. Petersburg, 1830 (in Russian).

HERZ, O. F.—Frozen Mammoth in Siberia: Ann. Rep. Smithsonian Institution for the Year ending June 30, 1903, pp. 611–625, Washington, 1904.

HOWORTH, HENRY H.—The Mammoth in Siberia: Geol. Magazine, VII, pp. 408–414, 491–501, 550–561, London, 1880.

HOWORTH, HENRY H.—The Sudden Extinction of the Mammoth: Geol. Magazine, VIII, pp. 309–315, 569–572, London, 1881.

HOWORTH, H. H.—The Mammoth and the Flood, pp. I–XXXII, 1–464, London, 1887.

IDES, YSBRAND E.—Three Years Travels from Moscow over Land to China, London, 1706 (English edition).

KAYSER, EMANUEL—Lehrbuch der geologischen Formationskunde, 6 and 7 Auflage, I and II, Stuttgart, 1923 and 1924.

KÖRBER, PH.—Kosmos für die Jugend. Blick in die Schöpfung der Welt und in die Kulturgeschichte der Menschheit vom Anfang bis zum Gegenwart: Nürnberg, 1862.

[1] Only the papers quoted in this article are mentioned in this list.

KUTOMANOV, G. N.—Rapport sur une mission à l'embouchure du Eniseij pour les fouillée du cadavre d'un mammouth: Bull. Acad. Sc., VIII, pp. 377–388, St. Petersburg, 1914 (in Russian).

LALOY, L.—Le régime alimentaire du Mammouth: L'Anthropologie, XVII, p. 234, Paris, 1906.

LANG, HERBERT—Problems and Facts about Frozen Siberian Mammoths (*Elephas primigenius*) and their Ivory: Zoologica, IV, pp. 25–53, New York, 1925.

LAPPARENT, A.—Traité de géologie, 5-ième édition, Paris, 1906.

LEFFINGWELL, ERNEST DE K.—The Canning River Region, Northern Alaska: U. S. Geol. Survey Prof. Paper 109, pp. 1–251, Washington, 1919.

MAYDELL, GERHARD—Reisen und Forschungen im Jakutischen Gebiet Ost-Sibiriens in den Jahren 1861–1871, I and II, St. Petersburg, 1893 and 1896.

MIDDENDORFF, A. TH.—Reise in den äussersten Norden und Osten Sibiriens während der Jahre 1843–44 mit allerhöchsten Genehmigung auf Veranlassung der Akademie der Wissenschaften zu St. Petersburg, ausgefuhrt und in Verbindung mit vielen Gelehrten herausgegeben, I–IV, St. Petersburg, 1848–1885.

MÖBIUS, K.—Die Beharung des Mammut und der lebenden Elephanten, vergleichend untersucht: Sitzungberichte Berliner Acad. Wiss., Berlin, 1892.

NEUVILLE, H.—De l'extinction du Mammouth: L'Anthropologie, XXIX, pp. 193–212, Paris, 1918–1919.

NEUVILLE, H.—On the Extinction of the Mammoth: Ann. Report, Smithsonian Inst., 1919, pp. 327–338, pls. I–III, Washington, 1921.

NEUVILLE, H. ET J. GAUTRELET—Observations faites sur le sang du Mammouth offert au Musée par le Compte Stenbock-Fermor: Bull. Mus. Nat. Hist. Nat., XX, pp. 106–109, pl. 1, Paris, 1914.

NORDENSKIÖLD, A. E.—The Arctic Voyages of Adolf Erick Nordenskiöld, 1858–1879, pp. I–XIV, 1–447, with illustrations and maps, London, 1879.

NORDENSKIÖLD, A. E.—Die Umsegelung Asiens und Europas auf der Vega, I and II, Leipzig, 1882.

OSBORN, HENRY FAIRFIELD—The Age of Mammals in Europe, Asia and North America, illustrated, pp. I–XVII, 1–635, New York, 1910.

Pacific (The)—Russian Scientific Investigations: Acad. Sc. U. S. S. R., pp. 1–190, with maps and portraits, Leningrad, 1926.

PALLAS, P. S.—Voyages du Professeur Pallas dans plusieurs provinces de l'empire de Russie et dans l'Asie septentrionale, I–VIII et Atlas, Paris, 1794.

PALLAS, P. S.—De reliquiis excticorum per Asiam borealem repertis complementum: Novi Commentarii Acad. Sc., XVII, St. Petersburg, 1772.

PAVLOVA, MARY—Description of Fossil Mammals Collected by the Russian Arctic Expedition, 1900–1903: Mem. Acad. Sc., Physico-Math. Section, XXI, No. 1, St. Petersburg, 1906 (in Russian).

PFIZENMAYER, E. W.—Beitrag zur Morphologie von *Elephas primigenius* und Erklärung meines Rekonstruktionversuchs: Verhandl. Min. Ges., II Ser., Bd. XLIII, Ss. 521–542, 1 Taf., 5 Fig., St. Petersburg, 1905.

PFIZENMAYER, E.—A Contribution to the Morphology of the Mammoth, *Elephas primigenius* Blumenbach; with an Explanation of my Attempt at a Restoration: Ann. Rep. Smithsonian Institution for the year ending June 30, 1906, pp. 321–333, Washington, 1907.

PFIZENMAYER, E. W.—Mammutleichen und Urwaldmenschen in Nordost-Sibirien, S. 1–341, mit 118 Abbildungen und 3 Karten, Leipzig, 1926.

POHLIG, H.—Dentition und Craniologie des *Elephas antiquus* Falc. mit Beiträgen über *Elephas primigenius* Blum. und *Elephas meridionalis* Nesti: Nova Acta Academiae Caesareae Leop. Carol., LIII, No. 1; LVII, No. 5, Ss. 1–472, Taf. I–XVII, textfig. Halle, 1888 (1889) to 1891 (1892).

REID, CL.—The Sudden Extinction of the Mammoth: Geol. Magazine, IX, pp. 43–44, London, 1882.

SCHMIDT, FR.—Vorläufige Mittheilung über die wissenschaftliche Resultate der Expedition zur Aufsuchung eines angekündigten Mammuthkadavers: Bull. Acad. Sc., XIII, Ss. 97–130, St. Petersburg, 1869.

SCHMIDT, FR.—Wissenschaftliche Resultate der zur Aufsuchung eines angekündigten Mammuthkadavers von der Kaiserlichen Akademie der Wissenschaften an den unteren Jenissei ausgesandten Expedition: Mem. Acad. Sc., VII Ser., XVIII, No. 1, Ss. 1–169, 5 Taf., Eine Karte, Textfiguren, St. Petersburg, 1872.

SCHRENCK, L.—Bericht über neuerdingst im Norden Sibiriens angeblich zum Vorschein gekommene Mammuthe, nach briefliche Mittheilung des Herrn Gerhard v. Maydell nebst Bemerkungen über den Modus der Erhaltung und die vermeintliche Häufigkeit ganzer Mammuthleichen: Bull. Acad. Sc., XVI, Ss. 147–173, St. Petersburg, 1871.

SCHUCHERT, CHARLES—Historical Geology, New York, 1924.

SPASSKI, G.—Zoological Discoveries in Northeastern Siberia: Sibirski Vestnik, 1822, XVIII, pp. 349–354 (in Russian).

SUKACHEV, V. N.—Examination of Plant Remnants Found Within the Food of the Mammoth Discovered on the Beresovca River, Territory of Yakutsk: Resultats scientifiques de l'expédition organisée par l'Academie Imperiale des Sciences pour la fouilée du Mammouth, trouvé sur la rivière Bérézovka en 1901, III, pp. 1–18, pls. I–IV, 2 figs., Petrograd, 1914 (in Russian).

TILESIUS VON TILENAU, W. G.—De sceleto mammonteo Sibirico anno 1797 effosso: Mem. Acad. Sc., V ser., V, St. Petersburg, 1812.

TOLL, ED.—Mittheilung über eine Reise nach den Neusibirischen Inseln und längst der Eismeereküste ausgeführt im Jahre 1893: Petermann's Mitth., XL, Ss. 131–139, 155–159, Gotha, 1894.

TOLL, ED.—Die fossile Eislager und ihre Beziehung zu den Mammuthleichen: Mem. Acad. Sc., VII Ser., XLII, No. 13, St. Petersburg, 1895.

TOLL, ED.—A Sketch of the Geology of the New Siberian Islands and the most important Problems of the Exploration of Arctic Regions: Mem. Acad. Sc., VIII Ser., Phys. Math. Sect., IX, 1, pp. 1–20, 2 maps, St. Petersburg, 1899 (in Russian).

TOLL, E. V.—A Short Report for the Period of June 7 to November 8, 1902: Bull. Acad. Sc., Phys. Math. Sect., XX, pp. 158–160, 2 plates, St. Petersburg, 1904 (in Russian and German).

TOLMATSCHOW, I. P.—Fouilles dans le gouvernement de Nijni-Novgorod à la recherche des restes d'un exemplaire de l'*Elephas trogontherii* Pohlig: Bull. Acad. Sc., XVIII, Phys. Math. Sect., pp. 251–262, St. Petersburg, 1903 (in Russian).

TOLMATSCHOW, I. P.—Bodeneis vom Fluss Beresovka: Verh. Min. Ges., II Ser., XL, Ss. 415–452, Taf. V–VIII, 4 Fig., St. Petersburg, 1903.

TOLMACHOFF, I. P.—On a Mammoth from Lake Aral, in Berg's Lake of Aral: Bull. Turkestan Branch, Russian Geographical Society, V, p. 521, St. Petersburg, 1908 (in Russian).

TOLMACHOFF, I. P.—Note on the Extinction of the Mammoth in Siberia: Am. Journ. Sc., XIV, pp. 66–69, New Haven, 1927.

TOLMATCHEW, V. I.—Remains of a Mammoth found in Manchuria: Review of the Manchuria Research Society, No. 6, pp. 1–5, Harbin, 1926 (in Russian).

TRANSEHE, N. A.—The Siberian Sea Road. The Work of the Russian Hydrographical Expedition to the Arctic 1910–1915: Geogr. Review, XV, pp. 367–398, figs. 1–33, a map, New York, 1925.

TSCHERSKI, I. D.—Beschreibung der Sammlung posttertiärer Säugethiere (Die wissenschaftliche Resultate der von der Kaiserlichen Akademie der Wissenschaften zur Erforschung des Janalandes und der Neusibirischen Inseln in den Jahren 1885 und 1886 ausgesandten Expedition): Mem. Acad. Sc., VII Ser., XL, 1, S. I–V, 1–511, 6 Taf., 2 Figs., St. Petersburg, 1890.

TSCHERSKI, I. D.—On the Last Days of the Siberian Traveler I. D. Tscherski: Mem. Acad. Sc., LXXII, pp. 1–7, St. Petersburg, 1893 (in Russian).

VOLLOSOVICH, C. A.—On the Digging Out of the Sanga-Yurakh Mammoth, in 1908: Bull. Acad. Sc., III, pp. 437–458, St. Petersburg, 1909 (in Russian).

VOLLOSSOVITCH, C. A.—Le mammouth de l'île Bolchoï Lakhovsky (Iles de la Nouvelle Sibérie): Verh. Min. Ges., II Ser., L, pp. 305–338, pls. XII–XVI, 1 fig., St. Petersburg, 1915 (in Russian).

WRANGEL, FERDINAND—Narrative of an Expedition to the Polar Sea in the years 1820, 1821, 1822 and 1823, Second Edition by Edw. Sabine, pp. I–XIX, 1–525, 1 map, London, 1844.

WRIGHT, G. F.—Asiatic Russia, with Maps and Illustrations, I and II, New York, 1902.

Yakutskaya Okraina, a Newspaper, Yakutsk, 1912 (in Russian).

ZALENSKII, W. W.—Osteological and Odontological Researches on the Mammoth (*Elephas primigenius* Blum.) and Elephants (*El. indicus* L. and *El. africanus* Blum.): Resultats scientifiques de l'expédition organisée par l'Académie Impériale des Sciences pour la fouille du mammouth, trouvé sur la rivière Bérézowka en 1901, I, pp. 1–124, pls. 1–24, St. Petersburg, 1903 (in Russian).

ZALENSKY, W.—Ueber die Haupteresultate der Erforschung des im Jahre 1901 am Ufer der Beresowka entdeckten männlichen Mammutcadavers: C. R. d. Séances du Sixième Congrès Internationale de Zoologie, pp. 67–86, Berne, 1904.

ZALENSKIJ, W. W.—Études microscopiques de quelques organes du mammouth: Resultats scientifiques de l'expédition organisée par l'Académie Impériale des Sciences pour la fouille du mammouth, trouvé sur la rivière Bérézowka en 1901, II, St. Petersburg, 1909 (in Russian).

ZENZINOV, V. M.—On the Trade in the North of the Territory of Yakutsk, Moscow, 1916 (in Russian).

ZENZINOV, W. M.—With an Exile in Arctic Siberia. The Narrative of a Russian who was compelled to turn Polar Explorer for two Years: Nat. Geogr. Mag., XLVI, pp. 695–718, Washington, 1924.

INDEX

TRANSACTIONS

OF THE

AMERICAN PHILOSOPHICAL SOCIETY

HELD AT PHILADELPHIA

FOR PROMOTING USEFUL KNOWLEDGE

———

NEW SERIES—VOLUME XXIII

PART II

———

Fish Skulls: A Study of the Evolution
of Natural Mechanisms

WILLIAM K. GREGORY

———

PHILADELPHIA:

THE AMERICAN PHILOSOPHICAL SOCIETY

104 SOUTH FIFTH STREET

1933

FISH SKULLS: A STUDY OF THE EVOLUTION OF NATURAL MECHANISMS

BY

WILLIAM K. GREGORY

(Read December 2, 1932)

ARTICLE II
VOLUME XXIII

PHILADELPHIA:

THE AMERICAN PHILOSOPHICAL SOCIETY

104 South Fifth Street

1933

CONTENTS

CONTENTS

TO THE MEMORY

OF

BASHFORD DEAN

PREFACE

THE vast and scattered literature of ichthyology contains hundreds of figures and descriptions of the skulls of teleost fishes, both recent and fossil. Monographs of outstanding value such as those of Allis, Starks, Ridewood, Jungersen, Kishinouye and many others have been devoted to the anatomy and osteology of particular types or groups of teleosts, while every systematist has used skull characters in his definitions of the swarming orders, suborders, families, genera and species. Nevertheless it has seemed worth while to bring into existence the present collection of drawings of teleost skulls and to attempt a new review of the field as a whole, with special reference to problems of evolution.

The specimens studied are for the most part in the American Museum of Natural History, New York, but not a few were kindly placed at my disposal in the British Museum (Natural History), through the courtesy of Director Tate Regan and Mr. J. R. Norman. For many oceanic and deep-sea forms I am indebted to the generosity of Dr. William Beebe, Director of the Department of Tropical Research of the New York Zoological Society, who, during our long voyage on the *Arcturus* and in his laboratory at Bermuda, gave me every facility for the study of his collections.

During the years 1926–1928 Mrs. Louise Nash made, under the author's direction and for the present work, a considerable number of drawings of teleost skulls representing many of the orders and suborders. In these semi-diagrammatic, largely free-hand drawings the artist has, it seems, successfully seized the more salient characteristics; but precision in measurements is not claimed for them. In 1929, 1930, 1931 and 1932, Mrs. Helen Ziska contributed to the series a still larger number of carefully measured drawings.

The number of illustrations prepared for the present paper is doubtless inadequate to set forth the protean modifications of the fish skull. A thousand illustrations would still permit only a sparse selection of the principal types. But rather than defer publication indefinitely, I have thought it more useful to bring together and publish the somewhat scant material now in hand.

The aphorism "Analysis must precede synthesis" was long since adopted as the official motto of American ichthyology and has gradually been accepted as binding by ichthyologists the world over. As a result, however, analysis has so far outrun synthesis that recently even the great monographs have been concerned almost exclusively with the routine discrimination of families, genera, species and subspecies, and with the construction of ingenious artificial keys. These of course are indispensable, since they enable one to sit down quickly and write more or less correct labels for large numbers of new specimens, without having to bother at all about the real relationships of any of the fishes in hand. Two or three recent papers, however, which are of broader scope, raise the hope that more ichthyologists may become actively interested in the relationships as well as in the differences between fishes, and that the unfortunate and unnecessary separation of taxonomy, from both phylogeny and the study of nature's mechanisms, may be completely abolished in this country and abroad.

WILLIAM K. GREGORY

PUBLISHED BY PERMISSION OF THE TRUSTEES OF THE AMERICAN MUSEUM OF NATURAL HISTORY

AMERICAN MUSEUM OF NATURAL HISTORY New York, June 12th, 1932

FISH SKULLS: A STUDY OF THE EVOLUTION OF NATURAL MECHANISMS

THE BEGINNINGS OF THE FISH SKULL

THE typical fish skull, or syncranium (Fig. 1), notwithstanding the intricacy of its details, is generally recognized to be composed of two sharply contrasting divisions, which may be called the neurocranium, or braincase, and the branchiocranium. The main parts of the neurocranium are: first, a series of inner, or endosteal, elements that surround and protect the olfactory, optic and otic capsules and the anterior end of the notochord; second, a series of superficial ectosteal, or derm bones. These derm bones were originally similar to ganoid scales in microscopic structure but have long since lost their enamel-like surface. Some of them have become pitted, tunneled, or inflated by the lateral line organs which pass over or through them. They are also modified in various ways by the deeper layers of the skin which now often covers them. The adult teleost endocranium may also be considered as a complex of four intergrading parts surrounding the orbits; these may be named the ethmovomer block, the interorbital bridge, the cranial vault and the keel bone or parasphenoid.[1]

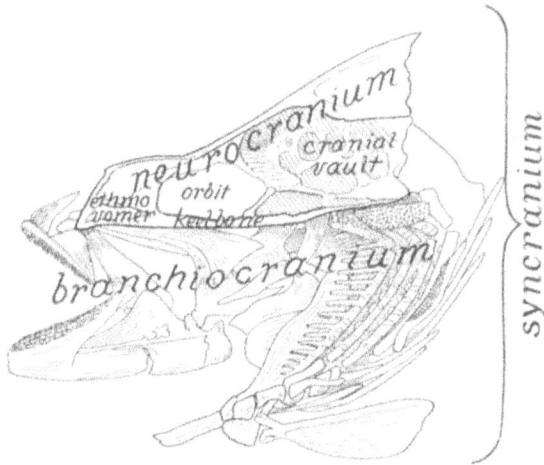

FIG. 1. The syncranium and its parts. Diagram showing sagittal section of the neurocranium and inner aspect of right half of branchiocranium in a typical teleost fish (*Roccus lineatus*).

The branchiocranium includes the mandibular region (comprising the oromandibular arch and attached derm bones), the hyal region (the hyoid arch and opercular series) and the branchial arches with their attached dermal plates.

The modern fish skull is subdivided into a large number of separate bones, each one

[1] For a discussion of the functions of these four parts, see p. 434.

rising from its own center of ossification. There is reason to believe, however, that the subdivision of the skull into separate bones has been conditioned chiefly by the necessities of growth and nutrition and that originally the endocranium was a continuum and the dermocranium consisted of a shell of ectosteal tissue, covering the chief functional regions or organs. Even now after the separate bones have enjoyed many millions of years of individuality, they are primarily regional subdivisions of functionally organic groups or tracts as well as organs in themselves.

In nearly all the hosts of typical fishes the syncranium is concerned with the pursuit and capture of living prey, the exceptions being few and peculiar forms such as the parrot-fishes and the like, which have given up this freely competitive roving life and become highly specialized for living either on aquatic vegetation or on sessile animals.

What then is the phylogenetic history of this peculiar and unique arrangement of parts? There is strong evidence for inferring that even before the Ordovician period the prevertebrates were bilaterally symmetrical, forwardly-moving animals in which the paired olfactory, orbital and otic capsules were arranged in an invariable antero-posterior order and in this fixed relation to the expanded oropharynx. The brain and cephalic sense organs were protected by the connective tissue and tough skin of the head and roof of the orobranchial chamber. The nervous system controlled the simple locomotor organs; these consisted chiefly of bilateral rows of zig-zag muscle segments tapering to the rear and causing undulations of the body as a whole. By means of its sense organs, nervous system and locomotor apparatus the primitive chordate was able to move toward suitable sources of energy, toward its mates and away from danger. The further inference is also highly probable that long before the typical fish stage was reached, the vertebrates already subsisted on living, moving prey, even if that prey was of quite small size.

The ancient ostracoderms, or pre-fishes, are first known from a single plate found in rocks of Middle Ordovician (Harding) age near Canyon City, Colorado. The class is represented in the Upper Silurian of Spitzbergen, Oesel, Norway, Scotland and North America by many genera belonging to several orders and to a considerable series of families. In the succeeding Devonian period the ostracoderms gradually declined and for the most part became extinct. The profound researches of Stensiö (1927) and Kiær (1924) have left no reasonable doubt however, that one or another of the ostracoderms gave rise to the modern class of cyclostomes, including the lampreys and hags, thus confirming the earlier views of Cope and others.

The true or gnathostome fishes are not known until the Devonian period and even up to the present time there are no known forms which definitely connect them with the ostracoderms. Nevertheless there is a common fundamental plan possessed by cyclostomes, ostracoderms and true fishes, especially in the arrangement of the parts of the brain and spinal nerves and general anatomy, and as investigations multiply there is less and less reason for doubting that the Agnatha (including ostracoderms and cyclostomes) and the Gnathostomata (or true fishes and higher vertebrates) at least have a common ancestral source, which would assuredly be nearer in most features to the agnathous than to the gnathostome ground-plan.

The earliest known gnathostomes, represented by the Palaeozoic sharks and ganoids and their modern descendants, are already so far advanced in their adaptations for swift swimming and predaceous habits that we necessarily seek for more primitive conditions in earlier horizons.

The agnathous ostracoderms of the Silurian period, taken as a class, reveal to us what those primitive characters were, even though they were perhaps not the characters that we expected. For as a class the ostracoderms are so inferior to the gnathostomes in their locomotor apparatus that they have even been assumed to be a specialized bottom-living group with no claim to be considered in the line of ascent to the gnathostomes. That was partly because it was further assumed that the continuous "headshield" must always be the result of the fusion of small polygonal plates. But Stensiö's intensive researches have revealed that the primitive ostracoderm shield was supported by a continuous endoskeleton without sutures, which was covered by a bony membrane.

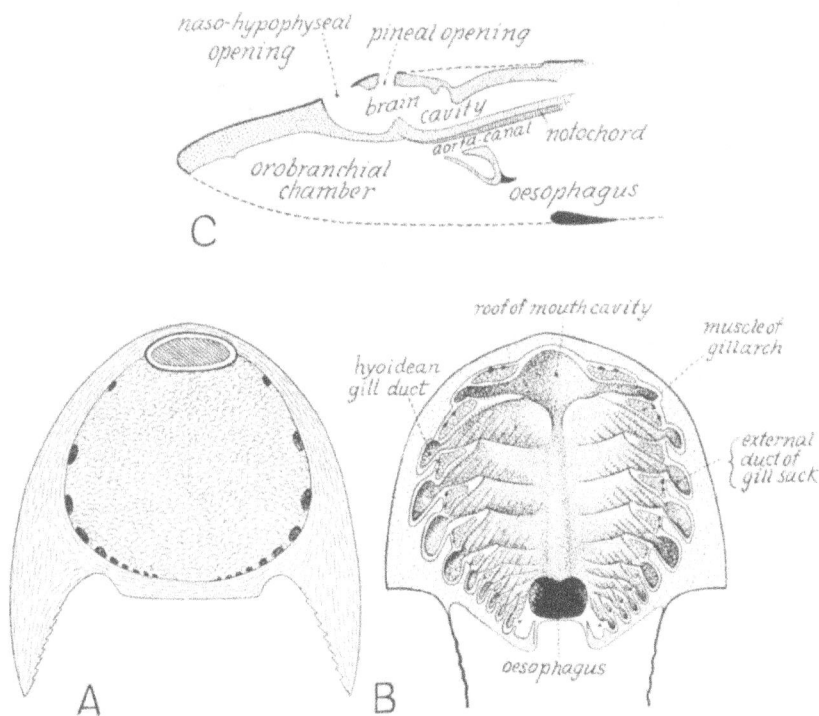

FIG. 2. Relations of the orobranchial chamber to the neurocranium in ostracoderms and cyclostomes.

A. Tentative restoration by Stensiö of the ventral aspect of the head of a cephalaspid, showing probable position of mouth and gill openings.

B. Roof of orobranchial chamber of cephalaspid, with gill-tubes, gill-clefts and interbranchial partitions, the latter extending downward from floor of neurocranium. Restoration by Stensiö.

C. Schematic median sagittal section of cephalaspid head-shield showing relations of neurocranium and orobranchial chamber. Exoskeletal bone in thick lines; perichondral bone layers in fine lines. After Stensiö.

DIFFERENTIATION OF THE NEUROCRANIUM AND THE BRANCHIOCRANIUM

The capacious orobranchial chamber of ostracoderms (Fig. 2) lay immediately beneath the brain and the cranial nerves and vessels; but there was no sharp separation between the neurocranium and the visceral arches as there is in existing fishes. The floor of the brain was also the roof of the orobranchial chamber and this continuous mesodermal tissue extended downward between the gill-chambers, moulding itself around them and preserving, so to speak, a natural cast both of the partitions between the gill pouches and of the tubes for the nerves and blood vessels.

Stensiö has demonstrated that many of the peculiar features of the central nervous system of the modern lamprey (Fig. 3A) were present in the ostracoderms, and for this and other reasons modern cyclostome structures may now be interpreted in the light of evidence from ostracoderm anatomy and *vice versa*. But whereas in the ostracoderms the orobranchial or visceral skeleton was merely the undifferentiated supporting tissue surrounding the gill-chambers, which was continuous above with the endoskeletal floor of the braincase, in adult lampreys the branchial skeleton (Fig. 3B) has become a somewhat irregularly fenestrated basket surrounded by thin muscle bands. This branchial basket differs from that of the shark in the lack of definite joints or articulations and in the somewhat irregular form of the cartilaginous tracts surrounding the branchial chambers. In embryo sharks the contrast between the neurocranium and the visceral arches is very sharp even from the first appearance of the gill-bars, but in the pre-gnathostome stage, as we have seen, this contrast was far less conspicuous.

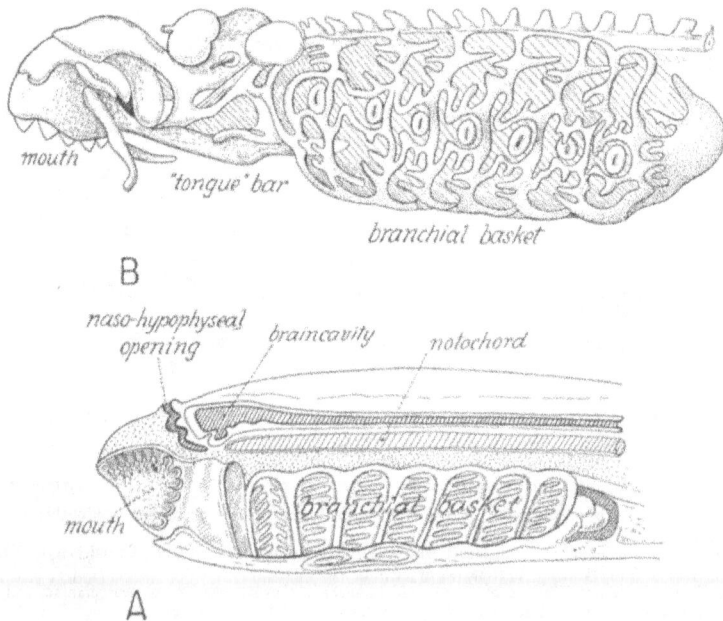

Fig. 3. A. Diagram showing relations of brain to orobranchial chamber in larval lamprey. · Based on figures of Parker and of Goodrich.

B. Cartilaginous syncranium of adult lamprey. After W. K. Parker.

According to Stockard the so-called tongue of cyclostomes represents the mandibular or Meckelian bars of gnathostomes, since it has similar relations to the visceral pockets and clefts.

In the earliest chordates the orobranchial cavity was doubtless already surrounded with sensory projections and the "head" was only the synthesis of paired sense organs and their cranial nerves, which were all designed either to direct the mouth toward the food or to suck the food into the mouth. Hence the building up of a syncranium of the teleost type was only a further development of an integration that was already well begun in the ostracoderms.

SERIAL HOMOLOGY OF THE JAWS WITH THE GILL ARCHES

In some of the anaspid ostracoderms the mouth cavity was protected externally by thin dermal plates, but if cartilaginous jaws were present they were not sufficiently calcified to be preserved in the fossils and hence the ostracoderms are often referred to as Agnatha, or jawless chordates. However this may be, Kiær (1924) has shown that in certain of these anaspids in some features the arrangement of the dermal plates on the outer surface of the orobranchial chamber is suggestive of the dermal jaws of true fishes. At this stage the gill-openings were small and circular, as in existing cyclostomes.

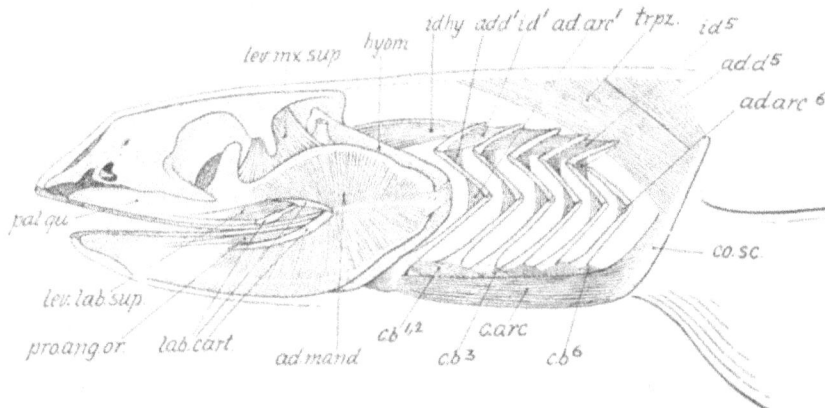

FIG. 4. Skull and visceral arches of *Chlamydoselachus*, with the deep muscles of the branchiocranium.

Based chiefly on the plates and descriptions of Allis (1923a), supplemented by a partial dissection of a specimen collected by Dr. Bashford Dean.

The deep muscles fall into two main groups: extensors of the oral and branchial arches, running anteroposteriorly, and flexors, running vertically.

Abbreviations (Allis):

ad. arc. = musculi adductores arcuales 1–6.
ad. d. = musculi adductores dorsales 1–5.
ad. mand. = musculus adductor mandibulæ.
carc = musculus coraco-arcualis.
cb = musculi coracobranchiales 1–6.
co sc = coracoscapular arch.
hyom = hyomandibular.
id = musculi interdorsales 1–5.

id hy = interarcualis between hyal and first branchial arch.
lab cart = labial cartilages.
lev lab sup = musculus levator labii superioris.
lev mx sup = musculus levator maxillæ superioris.
pal qu = palatoquadrate.
pro ang or = musculus protractor anguli oris.
trpz = musculus trapezius.

A comparison of the orobranchial region of an ostracoderm (Fig. 2A) with that of a typical gnathostome (Fig. 4) reveals, however, a striking difference. In the ostracoderms the mouth is much greater than the gill-openings and widely removed from them. In the primitive gnathostome, on the other hand, the mouth is in series with the gill-slits and only somewhat larger.   Here also the ossified membranes around the mouth have become enlarged and folded back over the hyoid arch, while the bony membrane on the latter has likewise become enlarged and folded over laterally to the remaining gill-arches.   In the ostracoderms there is no evidence that either the mouth or the minute gill-openings were supported by *jointed* arches like those of sharks.   But, as already noted, there was a continuous supporting tissue in the septa between the gill chambers.

It is hardly necessary to review in detail the evidence for the now commonly accepted conclusion that the primary jaws, or oromandibular arches of the gnathostome vertebrates are serially homologous with the branchial arches.   (See Goodrich, 1930, pp. 396–423.) Even in the adult shark (Fig. 4) the topographic correspondences of the jaws themselves to the branchial arches and of the adductor mandibulæ muscles and their nerves to the flexor muscles of the branchial arches and their nerves, reinforce the embryological evidence as recently set forth by Sewertzoff (1927).   Sewertzoff (1927, Taf. 29, fig. 10) has shown that in the embryo shark the labial cartilages which lie on either side of the oropharynx also have vestigial pouches suggesting those of the gill-arches.   These pre-oral cartilages reach a high degree of functional elaboration in the chimæroids, where they somewhat suggest small jaws lateral to the main jaws (Sewertzoff, 1927, Taf. 31).   According to Sewertzoff, the mandibular arch of gnathostomes represents the third of the series; this, together with the hyoid arch and five to seven normal gill-arches of sharks, would make nine to eleven in all.   The view of Ayers (1921) that the jaws of gnathostomes have arisen from rod-like structures in the velum of *Amphioxus* does not appear very tenable in the light of more direct evidence for the view summarized above.   The palæontological evidence, illustrated especially in the acanthodian and cladoselachian sharks (Fig. 5), reinforces the embryological and morphological evidence in favor of the strict serial homology of the oromandibular arch with the hyoid and branchial arches.

Very obscure and complex is the problem of the serial homologies of the dorsal segments of the mandibular, hyoid and branchial arches.   Allis (1915, 1923b, 1925b) has maintained that the trabeculæ of the embryonic chondrocranium represent "premandibular arches which have swung upward to fuse with the membranous brain case," a view originally proposed by Huxley (Goodrich, 1930, p. 238).   Also that the polar cartilages which connect the trabeculæ with the parachordals represent the dorsal elements (pharyngomandibulars) of the mandibular arches, of which the palatoquadrates represent the epimandibulars while the Meckel's cartilages represent ceratomandibulars.   But both Sewertzoff (1928, p. 202) and Goodrich (1930, p. 238) agree that the trabecular and polar cartilages are part of the axial skeleton or neurocranium.

The Hyomandibular Problem

As to the hyomandibular, Allis (1915, 1923b, 1925b) has shown that there are two radically different types: (a) the selachian type representing an *epihyal*, which is ventral to the vena capitis lateralis and (b) the teleostome type, which is dorsal to that vein and dorsal to the adductor hyomandibularis muscle (see Goodrich, 1930, Fig. 446).   That the selachian hyomandibular is an epihyal was also held by Luther (1909) who found in *Stegostoma tigrinum* that the hyomandibular had every appearance of being in series with the epibranchials, and that it also carried a small dorsal cartilage corresponding exactly to the pharyngobranchials (see Sewertzoff, 1927, pp. 447, 521).

Woskobojnikoff (1914, quoted by Sewertzoff, 1927, p. 479) held that the hyomandibular of the sharks does not belong originally to the hyoid arch at all, but represents a dorsal segment of the mandibular arch, hence a pharyngo-mandibular; but Sewertzoff (1927) shows that in its relations to the muscles and cranial nerves the hyomandibular of sharks corresponds to an epihyal, as held by Allis.

As to the hyomandibular of teleostomes, Allis (1918) regarded it as a complex of two parts; the anterior derived from the anterior branchial-ray bar of the hyal arch, and the posterior derived from the dorsal extra-branchial of the hyal arch. This conclusion rested in part on the fact that in the larval *Polypterus* there was a small and independent bit of cartilage posterior to the articular head of the hyomandibular, and that the so-called "accessory hyomandibular" of the adult *Polypterus* (see p. 111 below) was developed in relation to this piece. Edgeworth (1926), however, found that in an earlier stage of development the "small and independent bit of cartilage" was not an originally separate cartilage but was formed quite late in development and is due to a separation of a posterior process of the head of the hyomandibular from the main part; also that the osseous "accessory hyomandibular was a covering bone" (? of derm bone origin) and different in nature from the hyomandibular itself. After comparing the development of the hyomandibular of *Acipenser*, *Lepidosteus* and *Amia* with that of *Mustelus*, Edgeworth concludes (1) that the hyomandibular of teleostomes is a single structure and not the result of fusion of two skeletal structures; (2) that in spite of the fact that in teleostomes the head of the hyomandibular lies above the vena capitis lateralis, while in sharks it lies below it, there is no constancy of position of the head of the hyomandibular in relation to the auditory capsule in either sharks or teleosts and that the evidence indicates that the hyomandibular of teleostomes is fully homologous with that of sharks. Goodrich (1930, p. 419) concludes that "on the whole, for the usually accepted view that the hyomandibular is homologous in all these fishes, there is good evidence not only from embryology but also from palæontology. . . . That the articular head [of the hyomandibular] is an 'otic process' is doubtful; but it is not impossible that an articulation, originally ventral in Selachians, may have moved up to a new position by passing over the bridge forming the outer wall of the jugular canal into which the vein and nerve have sunk in Teleostomes (Stensiö, 1921)." Sewertzoff also from his embryological studies on sharks and teleostomes accepts the homology of the hyomandibular in the two groups. Accordingly, as regards the disputed question of the homologies of the hyomandibular in sharks and teleostomes, I adopt provisionally the view that the hyomandibular of the teleostomes is truly homologous with that of the Selachii (Edgeworth) but that it has shifted dorsally, passing over a groove and bridge containing the vena capitis lateralis and part of the facial nerve (Stensiö, 1921).

The symplectic of teleostomes appears to represent only the lower part of the hyomandibular, which in *Polypterus* is not yet separated off from the main part of the element.

With regard to the branchial arches, Allis pointed out in several important papers (1915, 1923b, 1925b) that in the gnathostome fishes as the mouth and branchial arches enlarged, two distinctly different forms of branchial arch arose: first, the Sigma-shaped arch of the Selachii in which the pharyngobranchials project postero-mesially, second, the V-shaped arch of the Teleostomi in which the pharyngobranchials project antero-mesially. Allis's theory (1925b) in brief is that the branchial clefts became prolonged dorso-anteriorly, causing the reduction of the posterior projections of the pharyngobranchials and the prolongation of the processes on the anterior sides of the pharyngobranchials; these newly prolonged processes then became segmented off and articulated with the epibranchials in the same way that the entire branchial arch originally became segmented. Thus the Σ-shaped curve would be changed into a $>$-shaped one.

ORIGIN OF THE OROBRANCHIAL APPARATUS OF GNATHOSTOMES

In the cephalaspid ostracoderms (cf. Fig. 2B) as described by Stensiö (1927) the first six gills of the ten gill sacs extend directly laterad, having small round external openings as in the lampreys. In the latter the branchial basket is a delicate stiffening skeleton of cartilage, without joints. In the primitive shark (Figs. 5, 6) as described by Allis (1923a) the jaws and gill slits are directed obliquely backward, the stout branchial skeleton is elaborately segmented and jointed; the jaws and branchial arches are surrounded externally by band-like constrictor muscles (Cs 1, 2, 3, etc.) and there are adductor and interbranchial muscles on the arches themselves (Fig. 4).

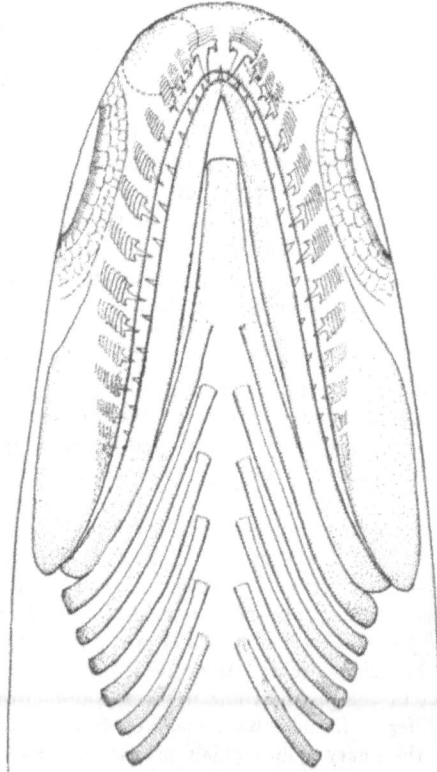

FIG. 5.   Underside of the skull of a Devonian shark, *Cladoselache fyleri.*   After Dean.

Based on several well-preserved specimens and showing the mandible in series with the lower bars of the branchial arches; the median pieces are not shown.

According to the hypothesis here put forward, the primitive chordates were comparatively sluggish, bottom-living forms with a large oropharynx, feeding on small organisms by ciliary ingestion. Progressive selection for larger prey and for free-swimming, predaceous habits led to the enlargement of the oromandibular and branchial arches and of their

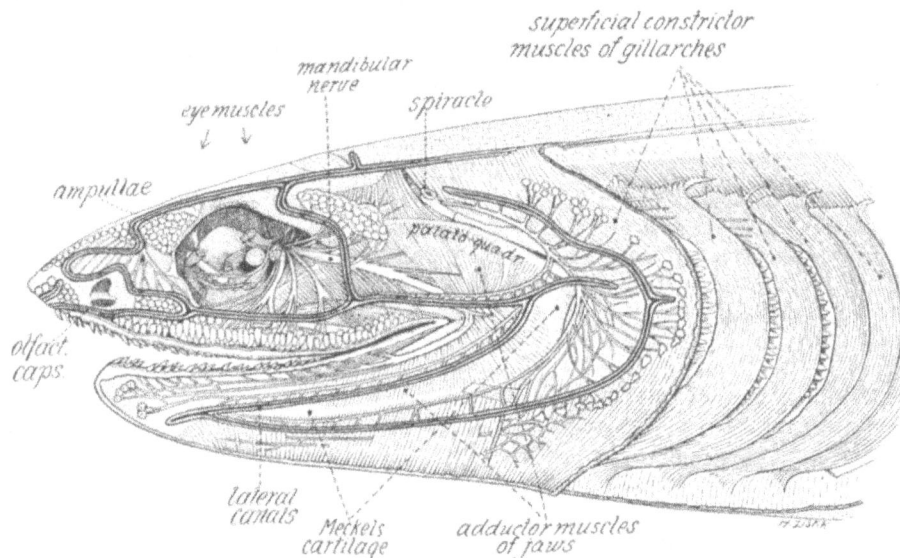

FIG. 6. Head of *Chlamydoselachus anguineus*. Redrawn and slightly simplified by Mrs. Helen Ziska after the color plate in Allis, 1923, Pl. IV.

constrictor, adductor and interbranchial muscles. The rhythmic contraction of these muscle bands in connection with both respiration and deglutition would tend to bend and fold the enlarging orobranchial arches and to break them up into joints. In this connection Allis has long recognized that the subdivision of the mandibular and branchial arches was conditioned by the development of biting jaws (1925a, p. 75). Sewertzoff (1927, p. 520), in comparing the embryo shark with the larval cyclostome, notes that the flexure of the orobranchial arches has been conditioned by the activity of these muscles. Goodrich (1930, p. 441) also remarks that "The segmentation of the arches is perhaps secondary; it is probably related to the development of special branchial muscles and allows the walls of the pharynx to be expanded and contracted for breathing and eating purposes." Further enlargement beneath the zone of constrictor muscles, together with the advantages of a narrower head in rapid swimming, would lead directly to the obliquity of the jaws and arches as seen from below (Fig. 5) and to their flexures in the lateral view. The enlargement and folding up of all these arches is evidenced by the sharp turning of the cranial nerves which supply the constrictor muscle bands, as shown (Fig. 6) in the dissections of *Chlamydoselachus* figured by Allis (1923a).

2

ORIGIN OF THE OPERCULAR SERIES

The beginnings of the opercular fold may be seen in the shark *Chlamydoselachus*, where there is a prolongation of the skin on the outer border of the gill-arches. In *Chimæra* there is a true opercular flap borne on the back of the hyoid arch and supported by cartilaginous streaks which have the appearance of being serially homologous with the extrabranchial cartilages of the branchial arches.

In the most primitive known actinopterans the oromandibular arch grows backward over the cheek and under the circumorbital bones; likewise the hyoid arch grows backward and its covering dermal fold gives rise to the opercular, subopercular, branchiostegals and interopercular, as can be seen in the larval stages of *Amia calva* and many other fishes. Tate Regan (1929, p. 313) has suggested that the interopercular appears to be the separated lower end of the subopercular. It is not found as such in the oldest actinopterans but may be represented by the first branchiostegal, since in the semionotids the latter element is becoming oblique in position and is manifestly equivalent also to the interopercular. The interopercular is always tied by ligament to the angular projection of the mandible. Below the interopercular there is a sharp crease separating the opercular from the branchiostegal series and permitting the former to move with the jaws, the latter with the branchial arches.

Ridewood (1904b) notes that the preopercular is associated with the "supratemporal" (= tabular, extrascapular, or scale-bone) and with the pterotic and the posttemporal in carrying a branch of the sensory canal system. Thus these elements, along with the circumorbital bones, the dermosphenotics and some others, have relatively constant relations with the "lateral-line" system of the head (cf. Goodrich, 1909, pp. 220–222).

As all ichthyologists know, the relations of the numerous sliding bony plates that cover the gill-chamber are also, on the whole, remarkably constant, at least in their broader features, in all the hosts of the teleosts, except when by degenerative specialization one or more of the standard elements may be reduced or even disappear. It is also evident that there is often a close correlation between the particular form and details of all these parts and the modifications of the jaws and gill apparatus that are connected with different types of food and feeding and of respiration. Moreover there is an equally close correlation between the head-form and body-form, the latter in turn being connected with the method of locomotion (see p. 431).

SUMMARY: FOUR CHIEF STAGES IN THE ORIGIN OF THE JAWS, BRANCHIAL ARCHES AND OPERCULAR ELEMENTS

The preceding tentative review of the chief stages in the origin and early evolution of the visceral skeleton may be summarized as follows:

I. Development of supporting tissue beneath brain-chamber and surrounding gill-pouches. This tissue was moulded around the septa between the gill-tubes and gill-chambers. It also surrounded many of the blood-vessels and nerves.

II. Origin of localized visceral arch skeleton (= oromandibular plus branchial arches) by *progressive fenestration* of the originally continuous supporting tissue mentioned above. Stages perfectly preserved in Stensiö's ostracoderms and in modern *Petromyzon*. At the beginning of this stage the organisms were slow-moving, *partly* bottom-living forms with flattened throats and domed heads, feeding perhaps by ciliary ingestion.

III. Selection for larger prey and more active swift-swimming led to:
(1) Need for improved respiration: hence

 (a) Great enlargement and folding of gill filaments (compare cyclostomes with sharks);

 (b) Strengthening of zonal and oblique muscles around gill region;

 (c) Bending of irregular gill-basket into gill-arches;

 (d) Formation of joints in gill-arches in adjustment to muscle pulls;

(2) Enlargement of oromandibular arch and its muscles with change of function from respiration to ingestion of struggling prey;

(3) Sharp flexure of oral and branchial arches backward, so that they overlap each other.

 IV. Continuance of (3) above, resulting in opercular flaps.

In the development of modern fish probably much of this evolutionary history has been greatly foreshortened, so that the elements of the visceral arches are emphasized in very early stages.

Synopsis of Ten Stages Leading to the Typical Percomorph Skull

The structural stages of evolution, several of which almost certainly do not lie in a direct phyletic sequence, are as follows:

1. Undiscovered and hypothetical prechordate stage, possibly resembling in fundamental characters the "*Tornaria*" larva of *Balanoglossus*.   Pre-Cambrian.

2. Undiscovered early protochordate stage, presumably like the early stage of *Amphioxus* in basic features; namely, with bilateral symmetry and a sequence of enterocoelic pouches giving rise to contractile locomotor sacs; head incipient.   Cambrian.

3. Primitive ostracoderm (Fig. 2), with typically chordate bilateral and cephalocaudal differentiation; head comprising capsules for olfactory, optic and static organs in antero-posterior sequence; "agnathous," i.e., jaw cartilages, if present, not different in character from the other branchial arches.   Ordovician.

4. Primitive undiscovered osteichthyan gnathostome with orobranchial arches shark-like but covered with ganoid plates and provided with true teeth (common ancestral stock of Crossopterygii and Actinopterygii).   Silurian.

5. Primitive palæoniscoid actinopterygian (e.g., *Cheirolepis*) (Fig. 12*A*); predaceous skull with large backwardly-inclined maxillary and suspensorium; eyes far forward; "asymmetric" arrangement of cheek plates; median and paired gular plates; parasphenoid not extending behind infundibulum.   Devonian.

6. Primitive protospondyl (*Acentrophorus*) (Fig. 21), with relatively small mouth; eyes dominant with "concentric" arrangement of circumorbital and postorbital bones, essentially as in larval teleosts of the present time.   Permian.

7. Primitive amioid (*Eugnathus*) (Fig. 27), with larger predaceous mouth; ganoine surface retained; median gular plate present; mandible retaining "splenial" (coronoid) elements.   Triassic.

8. Primitive teleostean (*Leptolepis*) (Fig. 30); ganoine surface reduced; "postorbitals" absent; median gular plate reduced or absent; maxilla forming greater part of oral border; two supramaxillaries.   Jurassic.

9. Progressive deep-bodied clupeoid (*Ctenothrissa*) (Fig. 42); transitional between numbers 8 and 10; two supramaxillaries.   Cretaceous.

10. Typical percomorph (Fig. 114); outer bones sunk beneath surface of skin and often more or less pitted along lateral line tracts; endosteal and ectosteal bones of braincase and mandible tending to lose separate identity; premaxillæ protrusile, excluding maxillæ from oral border; maxillæ serving as levers for depression of alveolar bar of premaxilla; one supramaxilla (except in the most primitive berycoids); mandible without "splenial" (coronoid) elements; typically six branchiostegals, of which four are attached to the epi- and cerato-hyal (Hubbs); orbitosphenoid absent, basisphenoid much reduced; supraoccipital forming a large median keel and in contact with frontal; occipital condyles tripartite. Upper Cretaceous.

Stages 5–10 of this outline of the evolution of the teleost skull are in conformity with those given by Smith Woodward in 1895, but have here been independently described and checked against the material.

Thus a typical percomorph syncranium as a whole appears rather widely different from that of a primitive palæoniscoid ganoid, yet, as we have seen, the number of important morphological differences between them is much less than the number of fundamental agreements.

Diagrams showing a tentative summary of the phylogeny of the principal groups of fishes from the ostracoderms to the most highly specialized teleosts are given in Plate II.

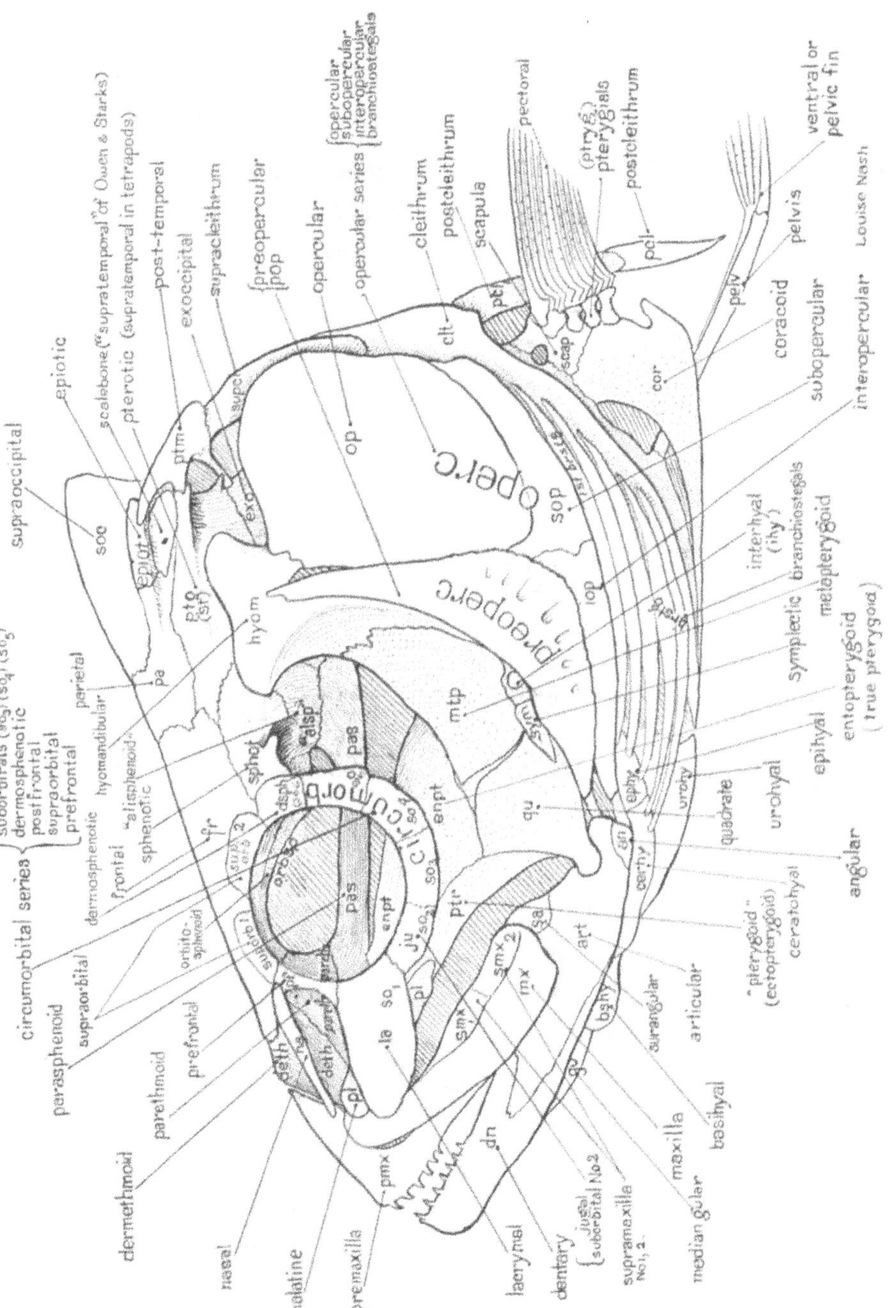

FIG. 7. The syncranium. A composite diagram based upon the study of skulls of about fifty species of teleosts, illustrating the positions of the bones and the names and abbreviations adopted in this work. Drawn by Mrs. Louise Nash, under the author's direction.

CLASSIFICATION AND NOMENCLATURE OF SKULL PARTS

THE relations of the parts of the syncranium may be summarized according to both their functions and their origins in the following scheme:

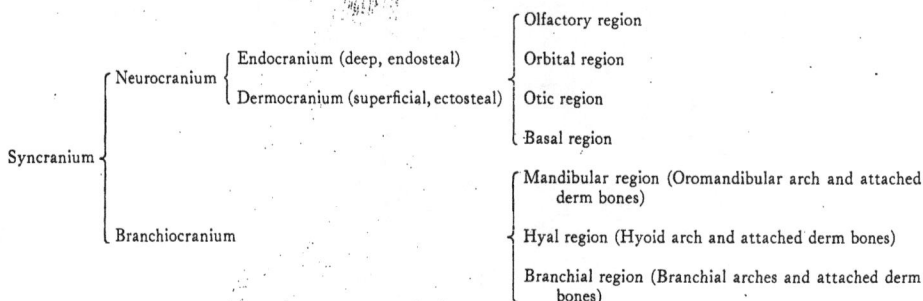

```
                        ┌ Endocranium (deep, endosteal)          ┌ Olfactory region
           ┌ Neurocranium ┤                                      │ Orbital region
           │            └ Dermocranium (superficial, ectosteal) ┤ Otic region
Syncranium ┤                                                     └ Basal region
           │                                                     ┌ Mandibular region (Oromandibular arch and attached
           │                                                     │   derm bones)
           └ Branchiocranium                                     ┤ Hyal region (Hyoid arch and attached derm bones)
                                                                 │ Branchial region (Branchial arches and attached derm
                                                                 └   bones)
```

A fuller subdivision of these parts is given below in Table I.

TABLE I

A CLASSIFICATION OF THE BONES OF THE TYPICAL FISH SKULL (SYNCRANIUM)

I. NEUROCRANIUM

A ENDOCRANIAL (deep)		*A'* DERMOCRANIAL (superficial)

(1) Olfactory Region

Ethmoid (originally preformed in cartilage)	covered dorsally by	Rostrals and Postrostrals of ganoids, some or all of which give rise to "Mesethmoid" [1] (= dermal mesethmoid)
	covered dorsolaterally by	Nasal (possibly derived from dorsal part of antorbital of palæoniscids, bearing nasal branch of supraorbital canal)
	covered ventrolaterally by	Adnasal (? derived from ventral part of antorbital of palæoniscids, bearing preorbital branch of suborbital canal)
	covered ventrally by	"Vomer" (= prevomer of tetrapods)
		Septomaxilla (in floor of nasal capsule)
Parethmoid (= lateral ethmoids, wrongly called "prefrontals")	covered dorsolaterally by	True Prefrontal (part of circumorbital series) sometimes fused with parethmoid

[1] Quotation marks used when implied homologies with tetrapod bones are incomplete, doubtful or incorrect.

(2) Orbital Region

Sclerotic bones

Circumorbitals: (*a*) suborbitals, including: "lacrymal" (so 1) or preorbital, so 2 (= "jugal"), so 3 (= true postorbital), so 4, so 5, so 6 (dermosphenotic), bearing the suborbital branch of the lateral-line canals; (*b*) supraorbitals (= pre- and postfrontals of tetrapods); (*c*) "prefrontal" (surface)

88

TABLE I—*Continued*

Orbitosphenoid

"Alisphenoid" (pterosphenoid, De Beer, 1926) } covered dorsally by   Frontal

> "Postorbitals": cheek plates covering the adductor mandibulæ muscles laterally in semionotids, *Lepidosteus*, primitive amioids, etc.; lost in teleosts (sometimes wrongly called suborbitals); not equivalent to true postorbital of tetrapods, which is derived from the inner or circumorbital row.

(3) Otic Region

Sphenotic	covered dorsolaterally by	Dermosphenotic ("so 6")
Pterotic	covered laterodorsally by	"Pterotic" (often fused with true pterotics; bearing part of the supratemporal canal = supratemporal of tetrapods; incorrectly called squamosal)

Parietal

Otic capsule {
- Proötic
- Opisthotic
- Exoccipital (often combined with opisthotic)
- Epiotic   often covered by and combined with   "Epiotic" (superficial)
- Supraoccipital (tectum synoticum)   often covered by and combined with   "Supraoccipital" (= dermosupraoccipital)

> "Scale bone" (tabular, extrascapular, "supratemporal" of Owen, Ridewood and Starks) covering posttemporal bone

Posttemporal (connecting pectoral girdle with skull)

(4) Basicranial Region

"Basisphenoid"

Basioccipital } covered ventrally by   Parasphenoid (= vomer of mammals)

II. Branchiocranium

B. Endognathal   B' Dermognathal

(1) Oromandibular Region

[? Labial cartilages of elasmobranchs]   replaced by   "Secondary upper jaw" (premaxilla, supramaxillæ 1, 2, maxilla)

Primary Upper Jaw (Palatoquadrate) comprising:[1]

(a) Autopalatine ("palatine," from anterior suprapterygoid)   covered ventrally by   Dentigerous palatine I (in primitive palæoniscids)

(b) Suprapterygoids (from 5 to 2, upper part of palatopterygoid of primitive palæoniscids, probably lost in teleosts)

(c) Metapterygoid (posterior end of palatoquadrate, connected with hyomandibular)

[1] See Watson, D. M. S., Proc. Zoöl. Soc. London, 1925, pp. 851–862.

TABLE I—*Continued*

(*d*) "Endopterygoid" or meso-pterygoid (= true pterygoid of tetrapods)	covered ventrally by	Dentigerous palatine II
(*e*) "Pterygoid" (= ectopterygoid of tetrapods)	originally covered ventrally by	Dentigerous part of ectopterygoid
Primary lower jaw (= Meckel's cartilage)		
Proximal part (Articular)	covered inferiorly by	Angular (usually reduced in teleosts)
	covered laterally by	Dermarticular (often fused with articular)
	covered dorsally by	Surangular (usually lost in teleosts)
	covered mesially by	Prearticular ("splenial")
		Sesamoid articular (a "tendon bone" marking insertion of part of adductor mandibulæ muscles (Ridewood, Starks))
Distal part (wanting in teleosts);	functionally replaced by	Coronoids ("splenial"; usually lost in teleosts)
		Dentary (the dominant dentigerous element in teleosts)
	covered by	Infradentaries 1, 2, 3 (anterior and posterior splenials of crossopterygians)

C ENDOHYAL		*C′* DERMOHYAL

(2) Hyoid Region

Hyomandibular	bearing posterolaterally	Opefcular
		Subopercular
	covered laterally by	Preopercular (bearing the hyomandibular branch of the lateral-line canals)
Symplectic (= distal segment of hyomandibular)		
Interhyal (= ? stylohyal of tetrapods)		Interopercular
Epihyal	bearing dorsally	Dentigerous plate
Ceratohyal	bearing ventrally	Branchiostegal rays (10 in Salmon, 34 in Tarpon, 6 in many Acanthopts.—Hubbs.)
Hypohyal		
Glossohyal (= basihyal of elasmobranchs, Ridewood)	covered dorsally by	Dentigerous plate
		Urohyal (a "tendon bone" arising in the septum between the longitudinal throat muscles)

D ENDOBRANCHIAL		*D′* DERMOBRANCHIAL

(3) Branchial Region

		The gill chamber is covered laterally by the Opercular, Subopercular and Interopercular and branchiostegal rays, all derived from the opercular fold
Pharyngobranchials 1, 2, 3, 4		"Upper Pharyngeals" (= dentigerous plates on fourth pharyngobranchials)
Epibranchials 1, 2, 3, 4		
Ceratobranchials 1, 2, 3, 4, 5		"Lower pharyngeal" (= dentigerous plate on fifth ceratobranchial)
Hypobranchials 1, 2, 3, 4	covered dorsally by	Dentigerous plates
Basibranchials 1, 2, 3	covered ventrally by	Gular plate of amioids, etc.

It was the aim of the earlier naturalists, including Cuvier and Owen, to identify in the fish skull those bony elements which appear to be homologous with corresponding bones in the skull of man. As noted by Williston (1914), in the adult human skull there are but twenty-eight bones (including the auditory ossicles), the primitive reptile has seventy-two separate bones in its skull, while in the oldest ganoid fishes there were more than one hundred and fifty. Consequently the supply of names from the human skull was soon used up and new names had to be invented for the remaining elements. Fish and man are very far apart, however, and in man the skull is so greatly distorted by the balloon-like expansion of the brain, by the reduction of the jaws and their firm union with the skull, that the equivalence of certain elements, such as the human squamosal, long remained in doubt.

Even in the times of Cuvier and of Owen comparison of the elements of the fish skull with the human type would have been still more difficult had it not been for the existence of living reptilian forms such as the crocodile, which serve in a measure to bridge the structural gap between the fish skull and that of man. Moreover since the days of the pioneers in skull morphology several generations of palæontologists have brought to light a whole series of fossil forms which enable us to follow step by step the transformation of the skull as a whole and the history of nearly every one of its elements. The present leaders in this line of work are Professor D. M. S. Watson of London and Professor Eric A:son Stensiö of Upsala. Watson has adopted a system of nomenclature of the elements of the fish skull which is based primarily on the names now commonly used for the corresponding elements in the skull of the oldest known amphibians and reptiles. Stensiö (1921, 1925) follows the same general method, but he has shown (e.g. 1921, p. 139) that certain of the bones present somewhat different combinations in different groups of Palæozoic fishes; it is also known that in the oldest fish skulls there are several elements, including the entire opercular series, which have been dropped out even in the earliest known tetrapods. In this country Dr. Starks (1926a) has demonstrated the difficulties in homologizing especially certain rather variable bones of the ethmoid region in different orders of teleosts; he has also described the various combinations and replacements of compound bones of endosteal or ectosteal origin, which further complicate the problem. Finally, despite all the work already done the gaps between known crossopterygians and palæoniscids, between known palæoniscids and holosteans, leave a few uncertainties as to the exact origin of several bones of the preorbital region of teleosts. Hence any system that may be used now must be more or less tentative and it must also be more or less of a compromise if it is to be practicable and serviceable.

For many years past I too have been interested in this question of the nomenclature of the skull elements of the lower vertebrates and have made repeated comparisons of skull patterns in the series from fish to man, especially endeavoring to equate the leading systems of nomenclature now in use in Europe and America (Gregory, 1917). In the American school of ichthyology as developed by Gill, Cope, Jordan, Evermann and Starks, the main object of the nomenclature of the fish skull was not so much to identify the fish-skull elements with those of man as to label the bones with appropriate names suggesting position in relation to certain landmarks. In selecting names for the elements of the skull I have tried to adopt those which are the most widely used, unless such names clearly imply incorrect homologies with the tetrapod skull. For instance, in common with Stensiö and Watson,

I reject the name *squamosal* for the bone which is lateral to the parietal, and adopt for this element the name *pterotic*, which is known to all ichthyologists. The superficial part of the pterotic is very probably equivalent to the true supratemporal of the earliest tetrapods, but as the name *supratemporal* has been applied to several different elements in this region of the fish skull, I choose to treat *supratemporal* as a synonym of pterotic.

In Table I the elements are classified according to their functional and topographic relations with the olfactory, optic, otic and encephalic organs, the primary and secondary jaws, hyoid and branchial regions. So far as practicable the distinction between endocranial (deep, endosteal) and dermocranial (superficial, ectosteal) bones has been indicated, but Starks has shown that in the cases of the ethmoid, sphenotic, pterotic, palatine, pterygoid and entopterygoid, we may be dealing either with an endocranial element or with a "derm bone" or with various combinations and replacements of inner and outer elements— all of which naturally make it impossible to avoid wrong implications of homology with corresponding elements of tetrapods whenever the same name covers different combinations in different groups.

The main steps in the development of the nomenclature adopted in this paper may be summarized as follows:

(1) The history of such skull bones as are retained in man have been followed backward to the oldest tetrapod stage.[1]

(2) "Williston's law"[2] of the progressive reduction in the number of skull bones in passing from lower to higher vertebrates has been critically examined and confirmed by the work of many authors, including Stensiö, Watson, Broom, Williston, Case, and von Huene.

(3) Such bones as do not survive into the mammalian stage but are known in the older reptilian and amphibian stages (including the supratemporal, intertemporal, postorbital, postfrontal, prevomers, ectethmoids, etc.) have been followed backward to the oldest tetrapod stage.

(4) The history of all bones which have been named first in the higher stages has been followed backward as far as possible.

(5) The history of those extra bones of the fish skull which are present in the older and tend to be lost in the later stages has been followed forward.

(6) The elements of the oldest tetrapod skulls have been compared and, so far as possible, equated with those of the oldest and most primitive fishes.

(7) (a) "Tetrapod" names have been used for certain bones when the homologies seem established beyond reasonable doubt (e.g. parietals, frontals); (b) "Fish" names have been retained when the homology with tetrapods is in doubt or when the appearance and function are extremely different in the two groups (e.g. the quadrate of fish is homologous with the incus of mammals, but to call it incus would be to introduce a new element of confusion in an already complicated nomenclature).

Elements of the Preorbital Region

In certain Devonian crossopterygians, in which the skull is probably on the whole much more primitive than that of teleosts, the dorsal and lateral surfaces of the rostrum

---

[1] For a summary of the main results in this field, see Gregory, 1927, The Palæomorphology of the Human Head: Ten Structural Stages from Fish to Man. Part I, Quart. Rev. Biol., II, No. 2, pp. 267–279; 1929, Part II, *Idem.*, IV, No. 2, pp. 233–247.

[2] This law is clearly formulated by S. W. Williston in his "Water Reptiles of the Past and Present," 1914, p. 3.

are covered by a system of numerous small plates. These included, according to Stensiö (1921, p. 134), the following series: (a) immediately above the premaxilla, a median and a lateral transverse pair of "rostrals"; (b) above (a) two median interrostrals one above the other; (c) above (b) two successive pairs of postrostrals, the more posterior pair articulating with the frontals; (d) on each side a longitudinal series of three "nasals" carrying the nasal branch of the supraorbital canal; (e) on each side, two successive antorbitals, between the nostril and the orbit. Thus there were some twenty plates above the rostrum. In the oldest actinopteran fishes the rostral system as described by Watson (1925) was less elaborate than that of the crossopterygian *Dictyonosteus* but still included one pair of rostrals immediately above the premaxillæ, two successive median postrostrals, one pair of large antorbitals lateral to the postrostrals.

In the teleosts this whole area is taken up by the so-called ethmoid, including the originally cartilaginous olfactory capsule, which is divided into an anterior or ethmoid region and paired lateral regions (variously called prefrontals, ectethmoids, lateral ethmoids or parethmoids).

Proethmoids are paired bones of uncertain origin covering the anterior corners of the lateral ethmoids in certain fishes; the "mesethmoid" is a median bone articulating in front with the premaxillæ and behind with the frontals. Besides these there are usually the paired nasals, sometimes paired dermal prefrontals and adnasals.

Thus the tracing of the homologies of the surface elements of the rostral and preorbital regions through the various families of chondrostean, lepidosteoid and amioid ganoids up into the lower teleosts is a difficult and hazardous undertaking due to various transformations, enlargements and shiftings in the different families. According to Watson (1925), the antorbital (or adnasal) of *Amia*, which bears a triradiate canal at the meeting-place of the suborbital and supraorbital canals, must be the homologue of the lacrymal of palæoniscids (the probable homologue of the tetrapod lacrymal), which bears a similar triradiate canal. But the adnasal or antorbital of *Amia* lies wholly in front of the circumorbital series and is closely associated with the anterior nares, the premaxilla and the nasal bones, while the lacrymal of palæoniscids lies well behind these elements and forms the lower anterior corner of the orbit, like the lower preorbital (our lacrymal) of *Amia*. Moreover, Watson figures a triradiate canal in *Coccocephalus* (Fig. 12C) in a bone at the lower anterior quarter of the orbit and this bone has every appearance of being homologous on the one hand with the lacrymal of *Cheirolepis* and on the other with the bone here called lacrymal in *Amia* (Fig. 28).

Repeated consideration of the literature of the preorbital and of specimens of early tetrapods, rhipidists, palæoniscoids, semionotids, eugnathids, *Lepidosteus*, *Amia*, etc., lead to the following conclusions:

(1) That the so-called mesethmoid (dermethmoid) of teleosts may be traced backward to one of the median rostrals (shown in *Amia*, *Lepidosteus*, *Eugnathus* and *Cheirolepis*) that originally lay immediately above the premaxillæ.

(2) That the anterior preorbital of teleosts is correctly homologized with Watson's lacrymal of *Cheirolepis* and of tetrapods.

(3) That the antorbital or adnasal of *Amia* (which is absent in teleosts) may be traced back to the antorbital of *Eugnathus* and of semionotids; this is postero-lateral to the nasals and to the nares. In *Perleideus* of the palæoniscoids the antorbital seems to have absorbed the lacrymal or driven it out. In *Cheirolepis* both are present.

(4) That the nasals of teleosts have come from the nasals of eugnathids and semionotids, which lie above the narial opening in front of the frontals.   In the palæoniscoids the nasals (of teleosts) may possibly be represented by the bones called by Watson postrostrals. Perhaps eventually the nasals may be traced to three sets of paired lateral "nasals" in the primitive rhipidists.   According to this interpretation of the facts the large size of the median ethmoid ("mesethmoid") and the small size of the nasals in typical teleosts reverse the conditions in their protospondyl predecessors; the teleosts have also lost most of the rostral mosaic, retaining only the median rostral in the form of a more or less enlarged "mesethmoid."

The mesethmoid of teleosts has been shown by Starks to vary enormously in the different groups.   "The scale-like, thin, superficial form of mesethmoid" (writes Starks, 1926a, p. 326), "that overlies the ethmoid cartilage and is doubtless of purely dermal origin, surely must be the most primitive, but it does not always co-ordinate with other primitive characters.   Some of the primitive fishes, as some of the plectospondyli, seem to have as complex and highly developed a mesethmoid as do the highly specialized and advanced spiny-rayed fishes.   The simplest form of it seems to appear first among the *Salmonidæ*, and especially among the *Argentinidæ*, though in the latter its exceedingly filmy condition possibly indicates some degree of degeneration.   On the other hand, the disk form of mesethmoid is found in more advanced families, as the *Atherinidæ* and *Poeciliidæ*, or such highly specialized forms as some of the *Labrididæ*.   In many of the higher forms, such as the last-mentioned family, its modification to the disk-like form seems to have been brought about by the development of the premaxillary processes, that lie over it and depress it."

In the higher or spiny-rayed fishes the mesethmoid is usually of dual origin, its surface being ectosteal and its interior endosteal (Starks, 1926a, p. 327).   The detailed changes of this element in the different groups have been described by Starks.

As Table I has to do chiefly with the normal elements of the typical fish skull, it does not include the following occasional elements of the ethmoid region, as described by Starks (1926a, pp. 332–335):

Pre-ethmoids.—Paired lateral ossifications in *Amia* and *Esox*, lying just above the vomer.   The palatines are attached or closely connected with them.   Somewhat similar elements in the Eventognathi (suckers and minnows) may be homologous with the pre-ethmoids, or may have arisen as epiphyses of the more or less ossified submaxillary rods (see below).

Submaxillaries.—"Sagemehl (1891) has applied this name to a pair of rods that connect the maxillaries with the pre-ethmoids in *Catostomus*.   They may be cartilaginous or partly ossified or wholly ossified.   In the carp-like fishes they are reduced to double concave disks of fibro-cartilage.   In most other fishes a thin, fibrous pad under each maxillary rests on the vomer and is apparently homologous with the submaxillary.   Occasionally the pads may ossify as described under *Sardinia* and other clupeoid fishes" (Starks, 1926a, p. 336). In other words, these elements appear to be like sesamoid or tendon bones of mammals.

Proethmoids of Esox.—Paired bones resting on the frontals behind and on the cartilage above the vomer in front.   Present in some other haplomous fishes and possibly combined with the nasals in *Tylosurus*.   Probably not equivalent with the mesethmoid (Starks).

Rhinosphenoid.—A median bone of unknown origin, forming a septum between the olfactory nerves as they issue from the orbitosphenoid.   Known only in two characinoid fishes (Starks).

REMARKS ON DR. LEIGHTON KESTEVEN'S NOMENCLATURE

In numerous papers Dr. Leighton Kesteven has conducted a painstaking and extensive investigation on the morphology of the teleost skull. In this he proposes many radical alterations in the long accepted homologizations of the elements of the fish skull with those of higher vertebrates, including man. For instance, he comes to the conclusion that the homologues of the human premaxilla and maxilla are to be sought not in the premaxilla and maxilla respectively of teleosts but in the vomer and palatine of teleosts; also that either the quadratojugal or the jugal of super-piscine vertebrates is represented by the so-called pterygoid of teleosts. Accordingly he changes the names of the fish premaxilla and maxilla to "premaxillary labial" and "maxillary labial" respectively; the vomer becomes the "premaxilla," the palatine, the "maxilla," while the pterygoid is called "quadratojugal."

I regret that I cannot accept these and similar conclusions of Dr. Leighton Kesteven's, first, upon broad grounds and secondly, in consideration of the merits of each case.

First, as to generalities. Our author's methods of investigating the problem of equating the elements of the teleost skull with those of higher vertebrates, including man, are, it is true, not essentially different from those of W. K. Parker, Huxley, Owen and most later morphologists, so that he is not to be criticized harshly for following in the well worn track. That time-honored method, against which I have protested for many years past, is essentially this: that if one is seeking to establish the homologies of a given element in a member of one class of vertebrates with a similar appearing or similarly situated element in a member of another class of vertebrates, one may do so by comparing *single pairs of forms chosen at will by the investigator, without regard to the magnitude of the phylogenetic gap between the members of each pair; further, that if in any such pair a given element in one is matched by a similar appearing element in the other, the two elements are assumed to be homologous without further inquiry as to whether the similar appearances may not be examples of analogy rather than of homology.*

This ancient method has immense apparent advantages over what I conceive to be the only reliable method of tracing homologies, in particular between members of different classes. First, it has always enabled morphologists (including embryologists) to write papers on the homologies and evolution of any particular organ without having to bother about the whole vast and complex background of the classification and evolution of the animals used in the comparison, without considering the palæontological history of the forms, without inquiry as to the evidence of changes in habits and correlative changes in structure, as affecting the position and appearance of the part or organ under investigation.

Secondly, as one can disregard the vast numbers of known living and fossil animals that bear the part under investigation, one can also conveniently limit his comparisons to almost any few forms that happen to be at hand.

Thus, hypothetical "phylogenies" of organs have been constructed as if they were entirely independent of the phylogenies of the animals that carried those organs.

For example, Leighton Kesteven (1926a, pp. 132–139) uses the conditions of the maxillæ in the eels for the identification of these bones in the fishes generally, thus ignoring the extreme specializations of the eels (see p. 202 below) which render them peculiarly unsuitable as a starting point for the establishment of homologies of their jaw elements with those of tetrapods and man.

Some of the other reasons why I cannot accept many of Dr. Leighton Kesteven's results are as follows:

By comparing chiefly a few modernized or highly specialized representatives of existing teleosts, reptiles, birds and mammals, Dr. Kesteven has persuaded himself that the so-called premaxilla and maxilla of the teleost are not respectively homologous with those of the higher vertebrates, including man. But he has not recognized that the premaxilla and maxilla of the teleost can be traced back with only relatively small breaks in the phylogenetic series to the premaxilla and maxilla of the palæoniscoid ganoids, such as *Cheirolepis* and its allies, which are generally and rightly regarded by all authorities as being the oldest and most primitive Devonian representatives of the entire actinopteran series. Now there seems to be no reasonable ground for suspecting that the premaxilla and the maxilla of these stem actinopterans are not respectively homologous with those of their contemporaries the very primitive crossopterygians *Osteolepis, Eusthenopteron*, etc. Assuredly the entire complex relations of the premaxillæ and maxillæ to the contiguous elements of the palate are strikingly similar in the basic Actinopteri and Crossopterygii and in this case the phylogenetic relationships are too close to justify an appeal to "convergence." If the homologies be admitted then the steps from the crossopterygians to the oldest amphibians, from these to the most primitive reptiles, thence up through ascending grades of the mammal-like reptiles to the lower mammals, Eocene primates, anthropoids and man, seem to be too closely graded to warrant hesitation in accepting the usual identification of the premaxillæ and maxillæ of teleosts as the upper outer jaw elements (Gregory, 1929).

Similarly, I can find no merit in Dr. Leighton Kesteven's suggestion that what were formerly labelled vomer (= prevomer) and palatine in the teleosts should now be called premaxilla and maxilla, because the very same evolutionary series mentioned above (which has been discovered by the cumulative researches of palæontologists) also tends strongly to confirm the view that the palatine and pterygoid of the teleosts were correctly identified by the older authors.

Similar considerations compel me to reject Dr. Kesteven's other identifications, e.g., the teleost parasphenoid with the paired pterygoids of higher vertebrates, including man.

ARE THE PREMAXILLÆ COMPOUND ELEMENTS?

In the course of his careful description of the cranial elements of the mail-cheeked fishes, Dr. Allis (1909, p. 24) comes to the conclusion that the ascending processes of the premaxillæ of these and other teleosts are primarily a pair of independent bones which have become fused with the premaxilla of lower teleosts. The latter, he thinks, have an articular process of the premaxilla but not an ascending process. These fused elements he compares with the median dermal ethmoid of *Amia* and with the paired "bone 2" of the ethmoid region of Huxley's description of *Esox*, which are the same as the "proethmoid" of Starks' description of that fish (1926a, p. 202).

After repeated consideration of Dr. Allis' discussion, I can only say that the evidence cited by him seems to be largely irrelevant to the point in question for the following reasons: first, not one of the teleosts cited by him shows any good evidence of the actual fusion of paired superficial ethmoidal elements with the premaxillæ, or even any evidence of the functional association of the proethmoids with the premaxillæ rather than with the ethmoid. (The supposed case of *Sphyræna* cited by Allis is discussed above.) The conditions in

Dallia pectoralis as figured by Starks (1926a, p. 204) are perhaps favorable to Dr. Allis' view, for in this form the proethmoids are large paired elements which extend over the tips of the ascending processes of the premaxillæ without, however, any tendency to unite with them. But this fish is assuredly very far off the line of ascent to the percomorphs. Secondly, none of the other teleosts cited by Allis can be considered to be closely related to the direct ancestors of the percoid and mail-cheeked fishes, while most of them are very widely removed from such relationship. Thirdly, the rostral cartilage now intervenes between the ascending processes of the premaxillæ and the dermal plate of the ethmoid. If the ascending processes of the premaxillæ are transferred portions of the ethmoid complex, how did the rostral cartilage come to lie between them and the remaining part of the ethmoid, especially if, as Allis rightly concludes (p. 28), "the rostral is quite certainly not a detached portion of the primordial cranium"?

Turning now to the more direct evidence, a personal review of the characters of the premaxillæ of representatives of most of the group mentioned in the present paper has convinced me that whenever the ascending processes are differentiated from the articular processes they are assuredly a part of the premaxillæ and not of the ethmoid complex. Even when the ascending and articular processes appear to be separated from each other at the base, as in the mail-cheeked fishes and in *Lophius*, a close examination with a pocket-lens will show that there is no suture between the ascending and articular portions of the premaxilla but that the apparent separation is due to (*a*) the different arrangement of the trabeculæ in the two parts, presumably in response to different stresses; and to (*b*) the occasional squeezing together of the ascending and articular processes so that the cleft between them becomes very narrow.

In *Fundulus*, which belongs to a group (Microcyprini) that stands more or less on the border between acanthopts and Haplomi, the premaxillæ have long ascending processes which have every appearance of being continuous below with the dentigerous part of the bone; posteriorly these tips overlap the mesethmoid (Starks, 1926a, p. 205).

Percopsis, which above all other living fishes has a claim to be regarded either as a direct descendant of the remote ancestors of the percomorphs or as a descendant of early isospondyls which were progressing in the direction of the percoids, lends no support to Dr. Allis' view that the ascending process was originally separate from the body of the premaxilla; for in this fish we see what is apparently an early stage in the differentiation of the ascending and articular processes from each other. Both processes are relatively short and the future cleft between them is represented only by a slight depression; the ascending process is indisputably continuous below with the articular process and with the dentigerous part of the bone, while the whole bone lies well in front of the ethmoid; the future rostral cartilage seems to be represented by the translucent material behind the ascending and articular processes, which has perhaps been secreted by the adjoining articular surfaces of the premaxillæ, maxillæ, vomer and ethmoids.

Finally, when we turn to the available palæontological evidence we find no confirmation of Dr. Allis' view. For in the very primitive Cretaceous berycoid *Hoplopteryx* as figured in detail by Smith Woodward (1902, Pl. III, Fig. 1), the ascending and articular processes of the premaxillæ are both plainly a part of that bone, while the mesethmoid is forked anteriorly much as in modern berycoids and percoids. In *Sardinoides* of the family Scopelidæ, which family Smith Woodward regards as standing near the line of ascent to the higher teleosts, the ascending process is plainly present but short and wide, while the ethmoid is undivided.   In brief, the ascending processes of the premaxillæ seem to have been developed *pari passu* with the rostral cartilage and with the ability to protrude the upper lip, and I can find no real evidence that they represent secondarily detached portions of the ethmoid complex.

THE EVOLUTION OF PARTICULAR SKULL TYPES

Lower Chordates

THE main object of this paper is the study of the origin and evolution of the teleost skull, but for the sake of completeness it seems desirable to insert brief notes on the skull structure of the lower chordates, chiefly as a key to the literature but also in order to put the teleost skull in its proper historical perspective.

Amphioxus was long supposed to be an ideally simple chordate and to furnish the key to the origin of the vertebrates (see Willey, "*Amphioxus* and the Origin of the Vertebrates"; Delage and Hérouard, "Les Procordés"). Even now Professor Howard Ayers (1921), a great authority on the minute anatomy of *Amphioxus*, tries to derive the fish skull and and jaws from this source. But in the light of the illuminating investigations of Kiaer and of Stensiö on the anatomy of the Silurian ostracoderms we may well suspect that the excessive simplification of the skull structures of *Amphioxus* is due, at least in part, to extreme degeneration and specialization acquired perhaps in connection with the habit of darting into the sand and drawing in microscopic food by ciliary ingestion. In order to establish his theme that *Amphioxus* is really primitive, with respect to the gnathostome chordates, Professor Ayers rejects the view that the jaws of fishes have been derived from gill-arches and sets up his hypothesis that they have been derived from parts of the velar skeleton of *Amphioxus*. This view, however, can hardly be taken seriously by those who appreciate the force of the evidence for the derivation of the primary vertebrate jaws from the branchial series. In brief, present evidence seems rather to favor the suggestion that *Amphioxus* is an extremely simplified descendant of some such early chordate as *Lasanius*, which has lost all of its exoskeleton and a good part of its endoskeleton as well. And the suggestive points of agreement of *Amphioxus* with the larvæ of the partly degraded cyclostomes (Delage and Hérouard, 1898, pp. 340–342) seem not averse to the possibility that *Amphioxus* represents a greatly degraded cyclostome, just as the latter is assuredly a degraded ostracoderm.

The skull of the cyclostomes has been intensively studied by many authors, including W. K. Parker (1883), Gaskell (1900, Pls. LVI, LVII), Goodrich (1909, 1930) and Stensiö (1927). Both Gaskell and Stensiö have shown the striking similarities of the larval lamprey head to that of the cephalaspid ostracoderms and in the light of much evidence it seems highly probable that the lamprey skull type has been derived from a cephalaspid-like type in the following way: (1) the bony exoskeleton has lost its bone cells and become membranous; (2) thorny epidermal teeth have developed around the sucker-like mouth; (3) a rasping apparatus has developed out of the so-called tongue, which is a specialized part of the branchial apparatus; (4) the rest of the branchial arches have been displaced backward; (5) the cartilages that support the sucker and its teeth have also been enlarged; (6) the originally continuous cartilaginous septa between the gill-pouches have become fenestrated, giving rise to the branchial basket; (7) a special hydraulic organ, described by T. E. Reynolds (1931) has been developed in the oral chamber to assist in the sucking action of the mouth.

3   99

In spite of its present specialization the lamprey skull retains the fundamental features of the ostracoderm skull.   Its otic capsule, for example, has but two semicircular canals, like that of the cephalaspid ostracoderms and unlike those of all gnathostomes.

The head of the Devonian antiarch *Bothriolepis*, which has been fully described by Patten (1912) bears a rounded bony cephalic shield, which articulates with a bony thoracic buckler.   The shield and buckler are composed of numerous separate bones, many of which appear to be homologous with those of the arthrodires (Hussakof, 1906, pp. 130–132). According to Patten, small bony pieces around the mouth function as premaxillæ, maxillæ and dentaries even though they may not be homologous with them.

In the Devonian *Macropetalichthys* as described by Stensiö (1925) we find a fish with a shark-like brain and a well developed endo- and dermo-cranium.   Some of the derm bones are homologous with those of the arthrodires, others are not easy to identify.   Olfactory, optic and otic capsules are surrounded by a continuous matrix.   Some of the dermal exoskeletal plates bear channels for sensory canals.   On the whole, *Macropetalichthys* seems to be related to the elasmobranchs but to have advanced in the direction of the ganoids without being ancestral to them.

The cranial anatomy of the arthrodires has recently been very effectively dealt with by Heintz (1931, '32), who has given an excellent restoration of the skull of *Dinichthys*.   The jaws of this fish, as described by Adams (1919) and by Heintz, were operated by a set of muscles which apparently are wholly different from the jaw muscles of typical fish.   The old view that the arthrodires were related to the dipnoans now seems quite untenable. They were rather a wholly extinct group, which nevertheless represented an abortive attempt of a primitive gnathostome stock to rise from the elasmobranch to the teleostome grade.   While their olfactory, optic and otic capsules were doubtless homologous with those of teleostomes, it seems probable that the individual bones of the dermocranium were independently evolved in the two groups.   The arthrodire jaws and jaw muscles also appear to have been independently evolved.

Sharks, Rays, Chimæroids

According to the evidence adduced by Stensiö (1925, pp. 160–164; 187–189) it appears that the cartilaginous condition of the skull in modern elasmobranchs is not improbably a result of degeneration, as in the better known cases of the cartilaginous skulls of sturgeons, spoonbills, *Ceratodus*, salmon, etc.   Thus even the exoskeleton of modern sharks is retrogressive and now represented only by the skin and shagreen armor.   In this connection it must be admitted that in the oldest acanthodian sharks the exoskeleton was well developed (Dean, 1907) and that, according to Reis (1896) and Jaekel (1927), the skulls of acanthodians were much more primitive than those of modern sharks, showing subdivisions of the palatoquadrate arch and of the Meckel's cartilage that correspond respectively with those of the branchial arches.   Dean however (1907) was skeptical as to the reality of these subdivisions, while Goodrich (1909, p. 190) figures the palatoquadrate and Meckel's cartilage of an acanthodian as severally undivided and essentially shark-like in form.

By the time of the Upper Devonian cladodont sharks (*Cladoselache*) the exoskeleton had already been weakened to a condition of delicate shagreen denticles and feeble teeth. On the other hand, the Devonian genus *Gemündina*, recently redescribed by Broili (1930) from excellent material, is a depressed form with wide pectorals, in which the exoskeleton of the head and body was strongly developed. Among the Permian pleuracanth sharks, which were highly specialized for swamp life, the endocranium was strongly built. Cope believed that in life it was subdivided into segments—whence the name Ichthyotomi; but examination of Cope's material shows that the supposed sutures are merely fractures of the fossilized mass, and that the skull of *Diacranodus* is essentially identical with that of modern elasmobranchs.

The skulls of modern sharks and rays are dealt with from a systematic viewpoint in the works of Gegenbaur (1872), Tate Regan (1906a), Garman (1913), while the embryology of the skull has engaged the attention of many authors from Balfour (1876–1878) to Sewertzoff (1916, 1917, 1927). The cranial anatomy of *Chlamydoselachus* is fully described in a beautiful monograph by Allis (1923a). Many morphological problems of the elasmobranch head are admirably discussed in the recent work of Goodrich (1930) and De Beer (1928). The outstanding differences between the elasmobranch endocranium and that of primitive teleostomous fishes will be summarized below (p. 115), but it may be said here that in general the shark skull, besides lacking bony centers and derm bones, reflects the greater development of the olfactory as compared with the optic parts of the skull; that it has not yet developed either a parasphenoid or opercular bones; and that its hyomandibular is attached below the vena capitis lateralis instead of above it (Goodrich). The palatoquadrate has typically two main points of attachment with the skull by means of palatobasal and otic processes, whereas in the typical teleosts the posterior contact is lost and the quadrate is suspended by the hyomandibular. The typical shark skull shows an early stage in differentiation of the trigemino-facialis chamber, a region of considerable morphological importance in the higher types (Allis, 1914).

The general form and details of the neurocranium of sharks and rays are naturally influenced by many factors both external and internal. The elasmobranch skull is usually wider and flatter than the teleost skull since the head as a whole is typically of the same proportions. For various reasons it seems likely that the primitive chordates were not swift-swimming, pelagic types but partly depressed, partly bottom-living forms (Gregory, 1928, pp. 389, 416), which at various times gave rise on the one hand to specialized bottom-living types such as the cephalaspids, antiarchs and the rays, and on the other to fusiform, partly free-swimming types, including the anaspid ostracoderms, macropetalichthyids, coccosteids and earliest acanthodian sharks. The curious Devonian elasmobranch *Gemündina*, as fully described by Broili (1930), perhaps stands midway in its habitus between primitive, partly benthonic elasmobranchs and secondarily free-swimming ones. In one direction a further broadening of its already broadened pectorals, without reduction of the trunk or flanks, would produce a torpedo-like type, while a narrowing of its pectorals would give rise to a *Cestracion*-like type. Further differentiation in opposite directions would lead to the flattened eagle-rays and skates on one side, and to the swift pelagic mackerel sharks on the other. But nowhere in the elasmobranchs do we find indications that the remote ancestral stock were compressed, deep-bodied forms like the primitive spiny-finned teleosts. Hence the elasmobranch occiput is never covered on top by extensions of the dorsal body- or fin-muscles, the interorbital space is wide and the rostrum more or less broad and depressed. This depression and widening, pronounced even in *Mustelus*, becomes emphasized in the bonnet-shark (*Sphyrna tiburo*) and excessive in the hammer-head (*Zygæna*), where extreme transverse growth of the front end of the head has pushed the eyes and olfactory capsules out to the edges of the suddenly projecting shelves on either side of the head.   This flattened face serves apparently as a bow rudder and makes possible very quick turns in pursuit of fish (Nichols and Murphy, 1916).

The adaptive radiation of the teeth and jaws in the elasmobranchs is very wide, so there are naturally marked differences in the neurocranium.   At one extreme we have the highly predaceous, almost snake-like jaws of *Chlamydoselachus anguineus*, in which the long slender jaws are produced backward; the hyomandibular is consequently elongate and inclined backward with a corresponding loss of the contact between the otic region and the palatoquadrate.   This is the specialized hyostylic type.   The neurocranium is cartilaginous and relatively delicate.

At the other extreme of the true sharks we have the massive short jaws and short thick neurocranium of the Port Jackson shark (*Cestracion* or *Heterodontus*).   The suspensorium remains hyostylic but the palatobasal contact is widened.   The massive whorls of flattened teeth borne by the heavy jaws seem well adapted for the crushing of mollusc shells and crustaceans.   Between these extremes stand the moderate to small-sized jaws of the squaloid sharks, beset with rows of small sharp-edged teeth adapted for biting.   The suspensorium is still hyostylic.

The rostrum of primitive sharks is prolonged as a bluntly-rounded prominence in front of the olfactory capsules, perhaps to improve the stream-lining of the head as a whole, perhaps to afford an expanded area for the ampullæ of Lorenzini and other sensory organs. Primitively the skeleton of the rostrum was probably continuous (as in the Permian pleuracanths), but in typical sharks it has become fenestrated and reduced to one median and two lateral bars.   The variations in the rostral skeleton have been used by Tate Regan (1906*a*) in classifying the families of elasmobranchs.   In the rays the anterointernal edge of the pectoral fins is tied to the side of the rostrum and in the manta-rays the rostrum caves in in front, while long projections of the pectoral fins, which have a sharp elbow and a spatulate base, serve perhaps as oral palps or as funnels to direct the food into the mouth when the fish is moving forward.

The skates and rays show a remarkable progressive development of various types of dental mills.   In the rhinobatids the individual teeth are still distinct but the crowns are flattened into more or less rhombic bases and the teeth form a rounded, indented whorl on the border of the upper and lower jaws.   In the higher rays individual teeth are replaced by multitudes of minute vertically-placed fused rods, grouped into polygonal and broad rectilinear blocks.   The rectilinear series forms an anteroposteriorly elongate, convex tract, or lower millstone, which opposes a similar plate on the broad palatoquadrate.   The latter is merely appressed against the wide floor of the neurocranium but is supported, apart from ligaments, only by the stout hyomandibular.   The ceratohyal is widely separated from the Meckel's cartilage, whereas in normal hyostylic elasmobranchs it is tied by ligament to that cartilage.   Hence the word *euhyostylic* was coined to describe the suspensorium of the jaw in the rays (Gregory, 1904, p. 58).

The jaws of all elasmobranchs differ from those of teleostomes in that there are no dentigerous dentary, coronoid nor splenial bones in the lower, and no dentigerous premaxillæ, maxillæ, vomerine, palatal nor pterygoids in the upper jaw.   The view that the labial cartilages of sharks represent the cores of the premaxillæ and maxillæ of teleostomes seems to have nothing in its favor but the general topographic correspondence of these two sets of elements in two widely separated groups.

In typical sharks, rays and chimæroids the dentition is integrated from masses of small units, clustered in groups around the borders of the Meckelian cartilage and on the primary upper jaw. In the chimæroids there are tritoral masses, moulded into incisive plates. Thus there is a complete absence of the dentigerous dentary, palatine and pterygoid bones which are so characteristic of the teleostomous fishes. Nor do the teeth ever attain the same degree of regional differentiation which one sees, for example, in the sheepshead (*Archosargus*) among teleosts. Moreover, in the more primitive elasmobranchs the individual teeth are still recognizably homologous with the shagreen, or placoid scales, whereas in the oldest teleostomes the serial homology of teeth with bony ganoid scales is evident only on comparison of the histological elements involved. Elasmobranchs also lack differentiated teeth in the pharynx or on the branchial arches, which are a conspicuous feature in many teleostomes.

The chimæroids have been shown by Dean (1906) to form a peculiarly specialized early side branch of the palæozoic elasmobranchs and not to be in any phylogenetic sense intermediate between the sharks and the teleostomes, notwithstanding the fact that they have paralleled the latter in developing an opercular flap which in the side view conceals all of the gill openings but one, and that they parallel the teleostomes in the dominance of the eyes over the olfactory organs. The large size of the eyes causes a marked constriction of the interorbital region, which contributes further to the list of teleostome-like characters.

The chimæroids are remarkable for their extreme specialization for durophagous diet, since they have strong nibbling or biting plates in the front of the jaws and cutting or crushing plates on the margins of the jaws and roof of the mouth, the patterns varying greatly in the different families of Mesozoic and modern chimæroids. In order to afford a firm support for this powerful dentition the upper jaw is completely fused with the skull. This arrangement has long been called "autostylic," but as that term more properly applies to the different conditions observed in the Dipnoi, I proposed (1904) to call the arrangement seen in the chimæroids *holostylic*. According to Dean (1906) the hyomandibular is represented by a small dorsal piece of the hyoid arch. It seems probable, however, that the hyomandibular has been fused with the posterior border of the quadrate and that the mode of suspension seen in the chimæroids has been derived from the primitive amphistylic condition, in which the palatoquadrate was attached to the otic region both by its own otic process and by the hyomandibular (see Goodrich, 1909, pp. 170, 171). According to this view the hyomandibular still functions as part of the suspensorium.

The labial cartilages of chimæroids obtain an unrivalled development. They and their muscles are well described by Sewertzoff (1927), who believes that they represent pre-oral segments of the visceral arch series. They appear to function as movable lips.

As a result of his studies on the embryology of chimæroids, Dean tried to derive the group from some Palæozoic relatives of the cestracionts. But Tate Regan (1910b, pp. 836–837) held that in the Holocephali (chimæroids) the hyoid arch is essentially similar to the succeeding branchial arches, that the pharyngohyal is well developed and the hyomandibular is not attached to the cranium, and that the cestracionts are true Euselachii, inasmuch as their "hyoid arch is modified in connection with the suspension of the jaws; the pharyngohyal is absent and the hyomandibular is articulated to the cranium." If, however, the hyomandibular has been fused with the quadrate and with the cranium, as held by Goodrich, the supposed difference between the chimæroids and the cestracionts would be considerably diminished.   Nevertheless, the many other profound differences between these groups seem to afford ample justification for Tate Regan's conclusion.   More recently Smith Woodward (1920) has pointed out that in the Palæozoic petalodont sharks of the genus *Climaxodus* the teeth exhibit "a restricted area of highly vascular dentine much resembling a tritor in the dental plate of an ordinary Chimæroid," also that "a relationship between some of the Palæzoic Cochliodonts and the early Mesozoic Chimæroids has already been remarked upon by P. de M. Grey Egerton and O. Jækel"; finally, "that the presence of an apparently Chimæroid character in Elasmobranch teeth which are noteworthy for their slow and scanty succession may therefore have some special significance."   In any event, it may be regarded as established that the chimæroids are modified Palæozoic sharks of some sort and in no sense intermediate between sharks and bony fishes.

DIPNOI

The oldest known dipnoan (*Dipterus*), of the Middle Devonian, was a contemporary of the oldest true ganoid (*Cheirolepis*) and crossopterygian (*Osteolepis*) and even at that remote epoch already exhibited the chief characteristics of dipnoan fishes, especially in the skull and dentition.   The latter consisted chiefly of masses of conical tubercles arranged in radiating rows, like the sticks of a fan, one pair of such clusters being on the roof of the mouth; the opposing fan-like clusters were located on the splenials or inner bones of the mandible.   There was also a pair of small patches on the vomerine region.   The lower border of the mouth formed a hard curved rim; the front upper teeth were inconspicuous or wanting; there were no teeth on the dentary bone and (except in *Scaumenacia*) none on the premaxillæ or maxillæ.   Each individual tooth of the fan-like clusters was composed of tissues that corresponded with the several layers of the ganoine-covered scales.

The histology of the scales and skull-bones, the form and arrangement of the fins, afford strong evidence of remote common ancestry with the contemporary crossopterygians, but the dipnoans were much more highly specialized in that their dentition, instead of being of the predaceous type, was adapted chiefly for crushing hard substances, perhaps the shells of molluscs or crustaceans.   Hence arose the necessity for a firm foundation for the dental plates on the roof of the mouth and for well braced points of articulation of the mandibles with the palatoquadrate.   These effects were secured by firmly attaching the palatoquadrate to the underside of the cranium.   The hyomandibular, becoming free from its suspensory function, eventually dwindled into a vestige, which in modern dipnoans has been the subject of many investigations (see Edgeworth, Goodrich, De Beer).

Meanwhile in the line that led to the modern *Neoceratodus* the endocranium became massive but at the same time much less osseus and more cartilaginous, while the jaw muscles extended dorsad, lifting the dermocranial roof away from the endocranium.   At the same time the roofing bones, like the scales, tended to sink beneath a membranous surface layer, to become thin and, in *Ceratodus*, more or less horny.

Even in the Devonian *Dipterus* the bones of the skull-root are only in part and with considerable difficulty homologizable with those of the contemporary crossopterygian. Watson and Day (1916), Watson and Gill (1923) and Goodrich (1930, p. 304) have tried to equate the *Dipterus* elements with those of the earliest tetrapods but here the gap is still greater.   For there are three rows of bones above the orbits to the midline in *Dipterus* (Fig. 8C) but only two in tetrapods (Fig. 8B).   From this and other circumstances there is a wide difference between the systems of Watson (1925, Fig. 2, p. 197) and of Goodrich (1930, p. 304, Fig. 311) in the identification of the parietals, frontals and other bones of the dipnoan skull-roof. More concretely, the paired bones in front of the intertemporals of *Dipterus* are called frontals by Watson, parietals by Goodrich, while the large median element above the orbit is equated with the paired tetrapod nasals by Watson, but with the paired frontals by Goodrich. Watson's proposed solution of the problem makes it easier to visualize or conceive the homologies of most of the *Dipterus* skull-roof bones with those of both tetrapods and crossopterygians; also Goodrich's "frontals" of *Dipterus* are anterior to the interfrontal while the frontals of *Eryops* are behind the interfrontal.

In *Dipterus* the skull-roof is broken up into a mosaic of many small bones (Pander, 1858); in the higher dipnoans certain of the paired bones became dominant, this finally culminating in the modern *Protopterus* and *Lepidosiren* in two greatly elongate, rod-like bones that run parallel with the mid-line on the top of the skull. In these two genera also the fan-shaped dental crests acquire sharp edges and the fish has apparently become carnivorous. This line of specialization was begun in the Carboniferous genus *Uronemus*.

In the light of all the evidence to the contrary, it is rather surprising to find Dr. Naef (1926) defending the theory that the dipnoans have given rise to the amphibians, and figuring *Uronemus* as a structural ancestor to a purely hypothetical form that gave rise to a suspiciously dipnoan-looking tetrapod! It is quite true that *Uronemus* and *Conchopoma* apparently lack the typical dipnoan fan-shaped dentition on the roof of the mouth and inner sides of the mandible, but Watson and Gill (1923, p. 214) give reasons for the belief that these genera are true dipnoans which have lost the typical dipnoan plates and have acquired a secondary dentition of patches of small tubercles on the parasphenoid and inner sides of the mandible. But the skull-roof and paired fins of *Conchopoma* are described by Watson and Gill as essentially similar to those of the existing dipnoan *Ceratodus*, while the cranial roof bones of *Uronemus* are arranged "very nearly as in *Ctenodus*." Thus these Carboniferous genera are definitely excluded from close relationships with the contemporary Amphibia, which have a very different type of skulls, dentitions and paired appendages.

In conclusion, the dipnoans, as a whole, have specialized toward an aberrant pattern of the skull-roof and fan-shaped dental clusters on the roof of the mouth and inner sides of the mandible, while the earliest tetrapods retained practically the complete carnivorous dentition and every bone of the skull-roof of the contemporary crossopterygians except the opercular series.

CROSSOPTERYGII

The skull patterns of the palæozoic crossopterygians are of great morphological importance because it has been shown, chiefly by Watson, that on the one hand they stand not far from the ancestral line to those of the land-living vertebrates, while on the other hand they are related to those of the oldest actinopterygian fishes. Hence they furnish the necessary bridge (Fig. 8) across which the nomenclature of the tetrapod skull has been transferred to the skull of the primitive fish. The skulls have been described in many monographs and papers by Pander, Traquair, Watson, Goodrich, Bryant, Stensiö and others.

Some of the salient characteristics of the more typical palæozoic crossopterygian skulls are as follows: they are highly predaceous types, typically with large jaws and strong teeth with labyrinthine bases; the teeth comprise rows of smaller caniniform teeth on the lateral

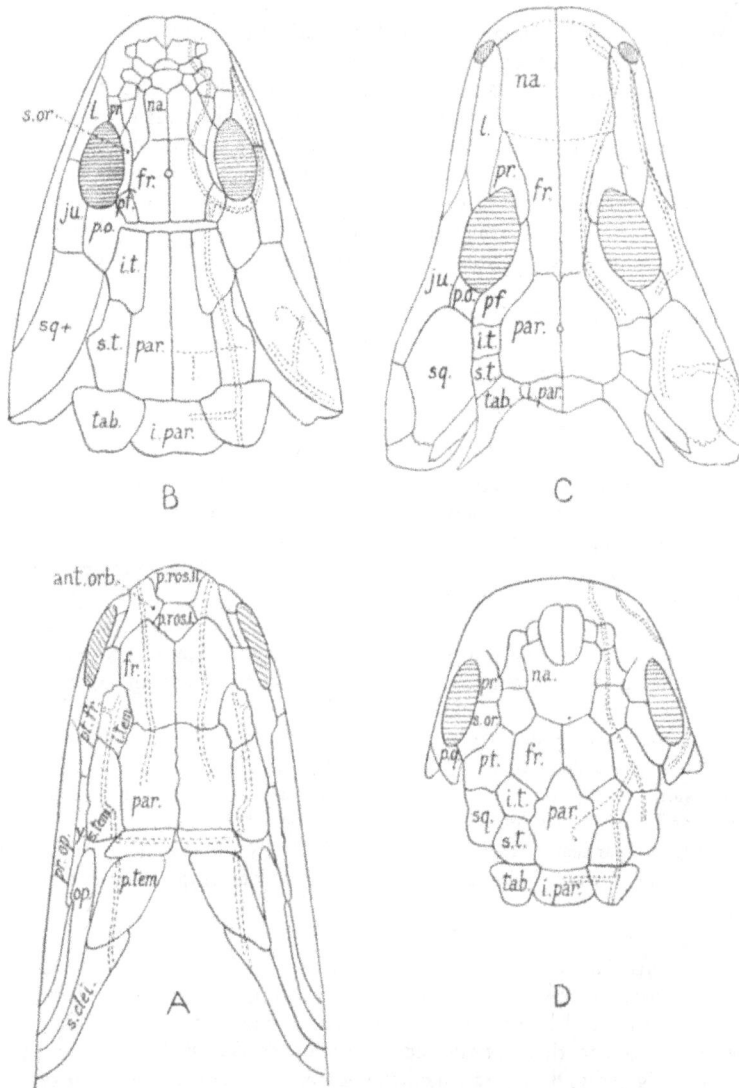

Fig. 8.   Comparison of skull patterns.
A. Primitive palæoniscoid, *Cheirolepis*.
B. Primitive crossopterygian, *Osteolepis*.
C. Primitive tetrapod, *Palæogyrinus*.
D. Primitive dipnoan, *Dipterus*.   All after Watson.

margins of the jaws and a few very large tusks on the dentigerous plates of the primary upper jaw and on the coronoid bones of the mandible; the large but feebly ossified hyomandibular is inclined backward and then sharply forward (Watson and Day, 1916, p. 16); the skull is rather low, with a blunt, rounded rostrum, eyes small and far forward; all the superficial bones, scales, etc., are covered in the most primitive genera with a shining armor of the "cosmoid" type, as described by Pander (1858) and Goodrich (1908); the nares, at least in some types, have an internal opening in the palate, as in primitive Stegocephalia.

The orbit is surrounded by a variable number of circumorbitals, homologous as a whole with those of primitive tetrapods. The cheeks are covered originally by two broad plates, apparently corresponding to the squamosal and quadratojugal of tetrapods and possibly in part with the "postorbitals" or cheek plates of semionotids (see p. 126 below). The preopercular is small (*Osteolepis*) (Goodrich, 1919), carrying the hyomandibular branch of the lateral-line canal, but sometimes covered by cheek plates. The interopercular and branchiostegals of higher fish are probably represented by so-called "lateral gulars." A row of infra-dentaries corresponds with the surangular, angular, postsplenial and splenial of the oldest amphibians (Watson, 1926).

The skull top presents the following conspicuous features: surface of rounded rostrum either more or less continuous (*Osteolepis*) or subdivided into numerous paired and unpaired plates (*Dictyonosteus*); a pineal foramen located between the short frontals (*Osteolepis*); parietals elongate. In the osteolepids the whole fore part of the skull could probably be turned upward, as a deep crease extends behind the upper jaws and cheek and behind the frontals, while the endocranium as a whole consists of two distinct parts, the rostro-orbital piece, including the vomer and short parasphenoid, being separated by an unossified space from the occipital region (Bryant, Watson). This is widely unlike the chondrocranium of primitive actinopterygians and its morphogenetic relations thereto are not well understood, although the facts themselves as to the osteology of the cranium of Palæozoic Crossopterygii of several genera are well known, through the labors of R. H. Traquair, W. L. Bryant, D. M. S. Watson and Eric A:son Stensiö.

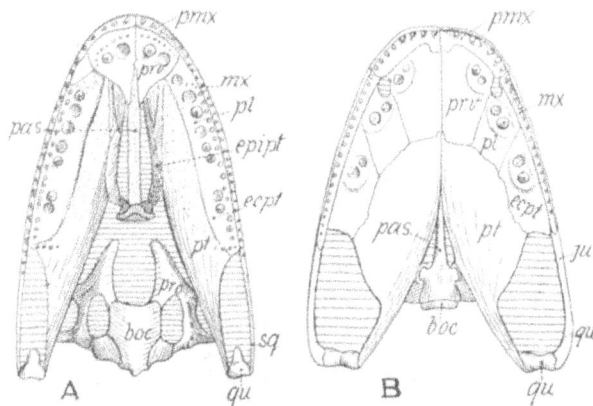

FIG. 9. Palatal view of skull of Devonian crossopterygian (A), and Lower Carboniferous amphibian (B).
A. *Eusthenopteron.* From Watson, mainly after Bryant.
B. *Baphetes kirkbyi.* After Watson.

The palate (Fig. 9) is characterized by the presence of two chief rows of teeth respectively on the outer or secondary and on the inner or primary upper jaw, and especially by the very short parasphenoid, which, unlike that of later fish, did not extend back under the occiput. According to Watson (1926, pp. 245–249), the massive palatoquadrate arch is ossified from about seven centers, the various regions giving rise to the palatal plate, true pterygoid (ectopterygoid), epipterygoid, three suprapterygoids (including the metapterygoid) and the quadrate. The arrangement of the numerous elements of the outer and inner upper jaws is strikingly similar in fundamental plan with that of the primitive Amphibia on the one hand and of primitive actinopterygian ganoids on the other (Watson, 1926, p. 198; 1925, p. 854; 1928, p. 62). There can therefore be little doubt that the premaxilla, maxilla, palatine, ectopterygoids, etc. of the crossopterygians are correctly equated with those of the earliest amphibians and earliest actinopterygian ganoids.

The hyomandibular in the rhipidistians *Eusthenopteron* (Bryant), *Rhizodopsis* (Watson and Day, 1916) and *Megalichthys* (Watson, 1926, p. 250) articulates above with the otic region of the neurocranium and at least in *Eusthenopteron* bears a process for the opercular. As noted by Bryant (p. 12) from excellent material, "it resembles very much its homolog in the Palæoniscidæ." In the Triassic coelacanth *Wimania*, as described by Stensiö (1921, p. 74), the main suspensor of the quadrate was the metapterygoid, which articulated with the side of the roof of the cranial vault. The hyomandibular (pharyngohyal) seems to have been cartilaginous, but the epihyal and ceratohyal were well ossified. Allis (1928, p. 207) infers that "*Wimania* is more primitive in several respects than the much earlier Rhipidistia," but Stensiö (1921, p. 135) concludes that ". . . as far as we can decide from the facts now known, the Coelacanthids ought to be taken as a highly specialized group, the ancestors of which are to be sought for among the primitive Rhipidistids."

In the coelacanth *Macropoma*, according to Watson (1921, p. 336), "there was a complete loss of the hyomandibular as a supporting element of the jaw. This loss is an exact parallel to that which has occurred in Tetrapods and Dipnoi." At the same time Watson adduces strong evidence that the coelacanths were derived from the osteolepids. Thus it seems that there is evidence that the coelacanths are much further from the actual ancestors of the teleosts than were the rhipidistids. Hence I doubt the propriety of assuming the conditions of the "pharyngohyal" of the coelacanth *Wimania* as structurally ancestral to different types of hyomandibular in Dipnoi, recent Chondrostei, certain palæoniscids and certain teleosts (Allis, 1928, p. 218). Edgeworth (1926) also adduced much embryological evidence for the old view that the hyomandibular of shark, ganoid and teleost is the same element throughout, a view which may be provisionally adopted in the present work.

POLYPTERINI

Polypterus is a heavily armored ganoid with a depressed reptilian-looking head not unlike that of *Ophiocephalus*. It is plainly predaceous in habit and its skull on the whole recalls that of the Upper Devonian Crossopterygian *Eusthenopteron*. However, according to Goodrich (1907, 1909, 1928), *Polypterus* does not belong with the Palæozoic crossopterygians at all, but represents a wholly different branch, allied rather with the primitive actinopterygian ganoids. Allis (1922) in a beautiful monograph on the cranial anatomy of *Polypterus* has compared the morphology of the skull with that of *Amia*, and in spite of the great gap between them has been able to establish the homologies of most of the elements.

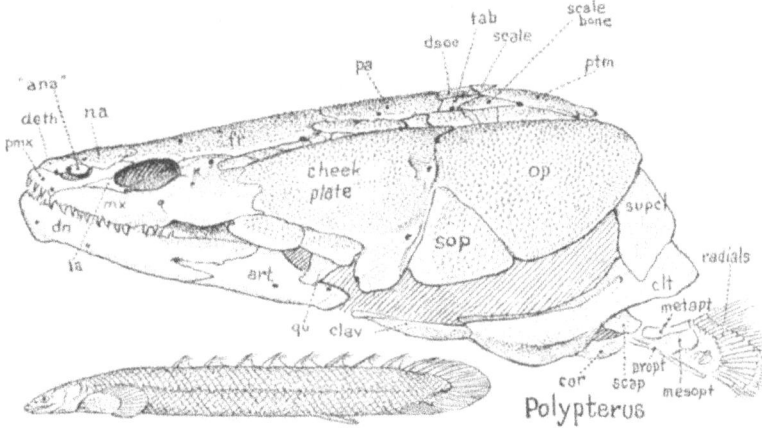

FIG. 10. *Polypterus* sp. Side view.

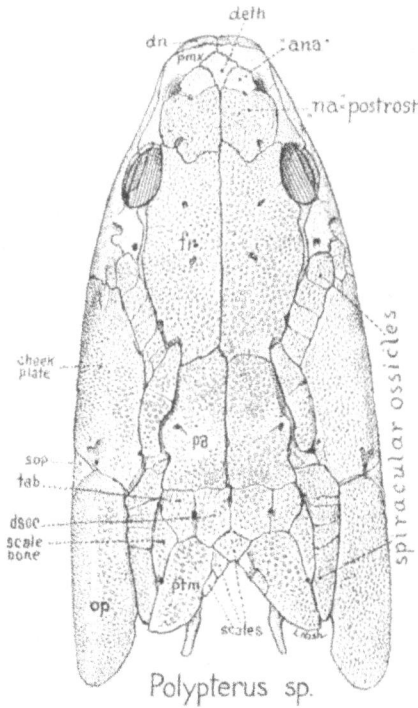

FIG. 11. *Polypterus* sp. Top view.

The presence of large paired gular plates and the general appearance of the skull are perhaps the most striking of the resemblances to the Palæozoic crossopterygians, but this is attributed by Goodrich (1928) to convergence, since he regards the paired gulars of *Polypterus*, which are derived from the median gular fold, as homologous not with the similar appearing plates of true Crossopterygii, but with "the two anterior lateral gulars, which are already becoming much enlarged in many Palæoniscids."

Goodrich has pointed out also that in all the more fundamental features of the scales, dermal plates, skull morphology and even in the paired fins and girdles, *Polypterus* appears to be allied with the Actinopterygii rather than with the true Crossopterygii, while especially strong evidence in the same direction is found in the morphology of the auditory labyrinth, pyloric cæca, urinogenital organs and brain.

Goodrich regards the Polypterini as being related more or less closely to the palæoniscids; Stensiö (1921, p. 135) concludes that his account of the morphology of *Dictyonosteus* and other coelacanthids "ought to show that neither the Coelacanthids nor the Rhipidistids are so closely related to the Polypterids as has been believed previously. The Polypterids represent rather, as Goodrich (1909, pp. 298–300) has pointed out, a type that is more closely allied to the Actinopterygii, although in my opinion they cannot be grouped with these either." In any case, *Polypterus* is of special interest since it is undoubtedly a specialized descendant of a very ancient line and because it shows what has happened to certain skull elements after divergence from the primitive palæozoic ganoids.

Among the archaic ganoidean characters of the *Polypterus* skull may be mentioned the following: (1) the retention of a heavy ganoid armor, histologically close to that of the oldest actinopterygian ganoids; (2) the retention of a spiracular cleft on the upper table of the skull; (3) the retention of large "parietal" bones (parieto-dermopterotics of Allis) on the upper surface of the skull, meeting each other in a long median suture; (4) the presence of a cheek-plate which is more or less homologous with the tetrapod squamosal and which is closely appressed to or united with the preopercular, bearing the hyomandibular branch of the lateral-line canal (Allis), as in palæoniscids (Goodrich); (5) the retention of paired prevomers; (6) the close relation of the ectopterygoids with the maxillæ; (7) the retention of the dentigerous premaxillæ and maxillae in their primitive position as an immovable rim bordering the dentigerous plates of the primary upper jaw; (8) the retention of large prearticular ("splenial") plates on the mandible; (9) the retention of a well ossified orbito-sphenoid part of the brain trough; (10) the generally primitive arrangement of the lateral-line canals on the skull.

Advancing specialization is indicated by: (1) the enlargement of the maxilla, which has apparently invaded the territory of the suborbital series, driving out or absorbing all but the lacrymal and establishing a wide contact with the enlarged cheek-plate; (2) the presence of a series of small plates, the spiracular ossicles of Allis (1922, p. 205), which may possibly have been derived by the backward growth and simultaneous fragmentation of the dermo-sphenotic; (3) the loss of the branchiostegals and lateral row of gulars; (4) the development of a cranial buckler covered by tough skin suggesting that of catfishes; (5) the union of the pterotics with the parietals; (6) the union of the 'alisphenoids' with the orbitosphenoids into a sphenethmoid.

According to Allis (1922, p. 248) the maxillary of *Polypterus* "has been formed by the fusion of two suborbital latero-sensory ossicles with a dental bone that quite certainly corresponds to the maxillary component of the superior maxillary bone of mammals." This type of maxillary is considered by Allis (1919, p. 386) to be not homologous either with the type of maxilla that is found in the Holostei or with the type that is characteristic of most Teleostei. The addition of suborbital elements bearing latero-sensory canals to the true maxilla makes the complex as a whole non-homologous with the maxilla of the Holostei and Teleostei. Nevertheless, the alveolar or dentigerous part of this bone appears to be fully homologous with that of other fishes. Indeed the palate of the palæoniscid *Elonichthys binneyi*, as described by Watson (1925, p. 854) would appear to give evidence that the premaxillæ, maxillæ and dentigerous palatines of that fish are respectively homologous with those of *Polypterus* on the one hand and those of *Amia* and the teleosts on the other. The hyomandibular of *Polypterus* bears a small "accessory hyomandibular" on its postero-superior border (Allis, 1922, p. 237, Pl. X), which is regarded by Allis as having arisen through the fusion of the bases of the posterior row of branchial rays of the hyal arch and as homologous with the posterior head of the hyomandibular of teleosts. Edgeworth (1926, p. 191), however, shows "that the hyomandibular of Teleostomi is a single structure and not the result of fusion of two skeletal structures. When, as in most Teleostomi, there is a foramen for the passage of the tr. hyomandibularis vii, this is due to the skeletal element —primarily situated only anterior to the nerve—spreading backwards. When, as in some Teleostomi, the hyomandibular has two articular heads, this is due to the gradual development of the primary single articular head into two heads, and is not due to the fusion of any extra cartilage."

As to the "accessory hyomandibular" of *Polypterus*, Edgeworth (p. 183) states that in a 75 mm. specimen of *Polypterus senegalensis* "the (osseus) 'accessory hyomandibular' fits on the back of the head of the hyosymplecticum [hyomandibular plus symplectic] and the lateral surface of its posterior process. . . . It is a covering bone for the dorsal posterior part of the hyosymplecticum." The head of the hyomandibular articulates with a cartilaginous socket which apparently represents the true endocranial part of the pterotic, while the dermo-pterotic has fused with the parietal.

The primordial cranium and visceral arches of a larval *Polypterus* have been described and figured by Budgett (1902, Pl. X); but a comparison with similar stages of *Heterodontus*, *Amia* and *Salmo*, figured by De Beer (1924a, 1927, 1928), and with the larval sturgeon figured by Sewertzoff, indicates that in general the chondrocranium and branchial arches in the larval stage of each foreshadow in most essentials the conditions in the respective adult, except where larval specializations interfere, and that neither *Polypterus* nor any of the others harks back to remote adult ancestors in its larval stages. Thus the lower end of its hyomandibular in the larval stage figured by Budgett does not project below the level of the attachment of the "stylohyoid" (interhyal) so as to give rise to a symplectic and thus it throws no additional light on the question whether a symplectic is absent in *Polypterus* or whether the attachment of the "stylohyal" has simply shifted down to the symplectic end of the hyomandibular.

PALÆONISCOIDEI

The basic pattern of the teleost skull was already established in the palæoniscids, which ranged from the Old Red Sandstone of Europe (Lower Devonian) into the Upper Jurassic. These relatively very primitive forms have long been well known through the researches of Traquair and others. More recently Stensiö (1921), D. M. S. Watson (1925, 1928) and

Brough (1931) have added greatly to the detailed morphology both of the braincase and of the surface bones of the skull. In the side view (Fig. 12A) of the skull of *Cheirolepis* by Watson (1925) we note that the surface bones may readily be grouped on a functional basis as follows: (1) rostro-nasal series; (2) bones of the skull-roof; (3) sclerotic plates; (4) circumorbital series; (5) oromandibular series; (6) preopercular plates; (7) opercular-branchiostegal series; (8) cleithral series, including the posttemporal and the dermal plates of the pectoral girdle.

This oldest and most primitive ganoid is a more or less predaceous fish with the large mouth and rather delicate teeth of a gulper. The arrangement of the dermal plates covering the jaws and opercular region depends upon the fact that the primary jaws are serially homologous with the hyoid and branchial arches and that the acts of ingestion, deglutition and respiration are structurally related. It is known from other specimens that the hyomandibular, which supports the jaws, is directed backward and this is reflected in the oblique position of the preopercular.

Some or all of the rostral and postrostral system of palæoniscids apparently gave rise by reduction in number or by coalescence to the "mesethmoid" of *Amia* and the teleosts. The antorbitals seem to be equivalent both to the nasals and adnasals of *Amia* and the teleosts (see p. 133 below).

The eyes are far forward, above the anterior end of the primary upper jaws and not as large as those of typical fishes. The sclerotic plates are not to be confused with the circumorbital plates of higher fishes. The circumorbital series as a whole is apparently homologous with those of the crossopterygians and oldest tetrapods but the determination of the homologues of the individual bones is uncertain, owing to the variability in the number of the plates in the different genera and families.

The preopercular is widely expanded on the upper part of the cheek above the maxilla and outside of the adductor mandibulæ muscles. In front of it, in *Elonichthys* and *Oxygnathus*, are two small cheek-plates which, as Watson notes, may be the source of the second postorbital row of plates of the Holostei (1928, pp. 58, 59).

Stensiö (1921, p. 139) has proposed the following hypothesis to account for the difficulty in homologizing the crossopterygian, palæoniscoid and teleost "squamosal" or cheek-plate elements.

"As we know, there is no cheek-plate in the Actinopterygians that can be directly compared with the squamosal of the Rhipidistids. Watson and Day (1916) have shown, as has been pointed out above, that in primitive forms among the Rhipidistids the squamosal consists of several independent elements, and I now think that this fact can guide us in judging of the conditions in the Actinopterygians. Thus one may venture the supposition that all the homologues of the elements forming the squamosal in the Rhipidistids have never been fused with one another into one bone-plate in the Actinopterygians, but that they were either distributed among the surrounding bones or with partial transformation had been fused into certain larger units which afterwards preserved their mutual independence or else were finally more and more reduced. In the Palæoniscids, Platysomids and Catopterids it is conceivable that the original squamosal elements were divided up between the maxillary and the preopercular and that they have even possibly provided material for the so-called postorbitals; in certain Protospondyli, such as the Semionotids and Eugnathids, they all probably provided material for the majority of the so-called postorbitals

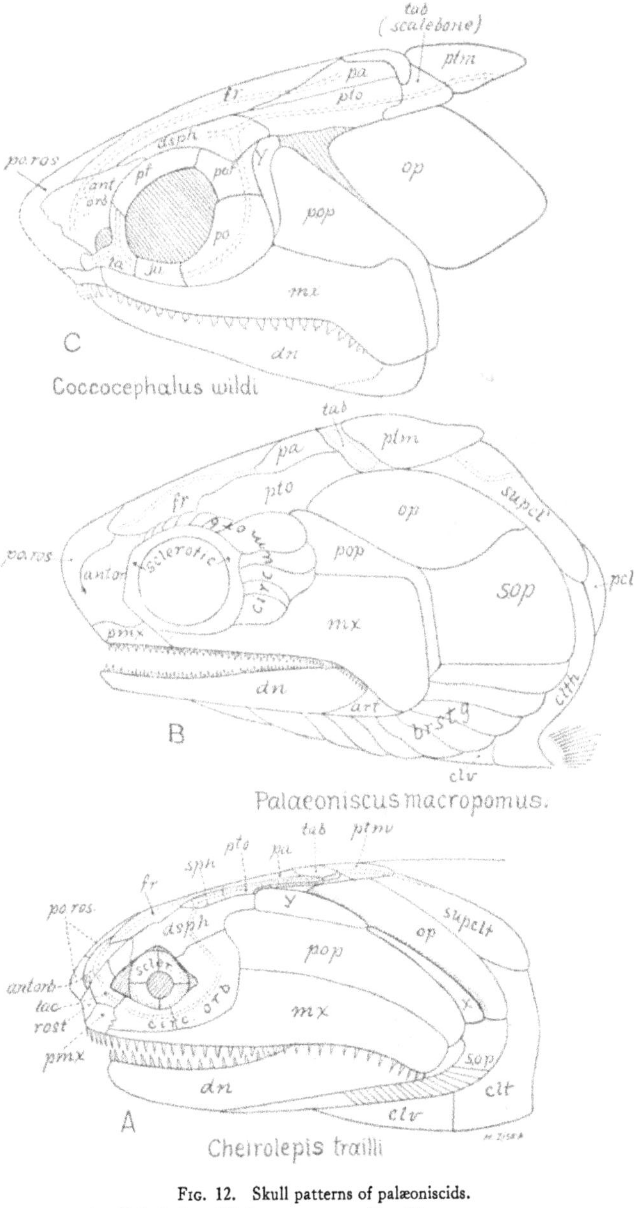

FIG. 12. Skull patterns of palæoniscids.

A. *Cheirolepis trailli*, Devonian age. After Watson.
B. *Palæoniscus macropomus*, Carboniferous age. After Traquair.
C. *Coccocephalus wildi*, Carboniferous age. After Watson.

In *Cheirolepis* the plate *Y* seems to correspond with the part of the "spiracular ossicles" of *Polypterus*. For the top view of this skull see Fig. 8 A, p. 106. The primitive paired rostrals and median postrostrals are well developed.

(suborbitals);—the ventral one, or some of them, can probably, as we have seen, be considered as corresponding to the quadratojugal; in the Teleosts the homologues of the
squamosal elements usually seem to be reduced.   Sometimes, however, it is possible that
parts of them may persist in the plates of the infraorbital chain, in the cases when these
plates are much developed and cover the cheek between the orbit and the preopercular
to any great extent."

The subopercular of the Palæoniscidæ was at first identified by Traquair (1877, p. 20)
as the interopercular, on account of the presence in *Rhabdolepis* and *Cosmoptychius* of a
small additional plate below the opercular; but in the later parts of the same monograph
(e.g. 1907, p. 94; 1909, p. 122) he labeled this element subopercular, which is the identification adopted by Watson.

The palate of *Elonichthys* and related genera, as described by Watson (1925, 1928),
closely resembles that of the primitive Crossopterygii in all important features.   The
fixed premaxillæ and maxillæ formed the margin for a row of tooth-bearing plates borne
by the primary upper jaw, which included ossifications named suprapterygoids (2–5),
metapterygoids, palatine (= anterior suprapterygoid), pterygoids (= meso- or internal
pterygoids of teleosts), ectopterygoids (= so-called pterygoids of teleosts).

The basipterygoid processes of the parasphenoid were immediately behind the orbits
on the anterior half of the skull and the parasphenoid, unlike that of later ganoids and
teleosts, did not extend back beneath the occiput (Watson, 1925).

The mandible of *Elonichthys* and its allies, as described by Watson, included an entire

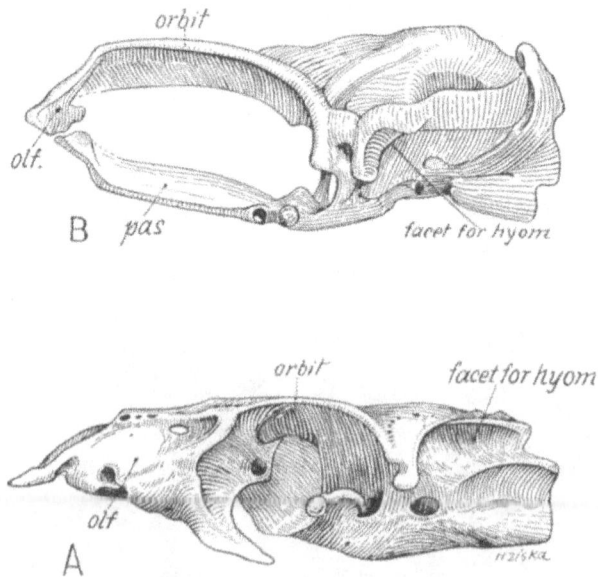

Fig. 13.   Neurocranium of (A) modern shark, *Chlamydoselachus* (after Allis) compared with (B) that of Carboniferous
palæoniscid (after Watson).
Note the relatively enormous orbit and very small olfactory capsule in (B).

Meckel's cartilage, the proximal end of which was ossified to form a large articular. A row of coronoid bones supported small denticles; the dentary was the principal jaw bone and supported the large teeth. The angular was of large size and seems to correspond with the dermarticular of typical fishes. The prearticular was a large flat plate. A conspicuous difference from the crossopterygian-tetrapod type was the absence of a separate row of infradentaries. The branchiostegal pieces—doubtless attached to the cerato- and epi-hyals—were wide and flat, the branchiostegal-opercular series not being interrupted by any obliquely placed interopercular. Thus the upper jaws and mandible of primitive palæoniscids as well as the surface pattern of the skull appear to be practically prototypal to that of higher ganoids and teleosts.

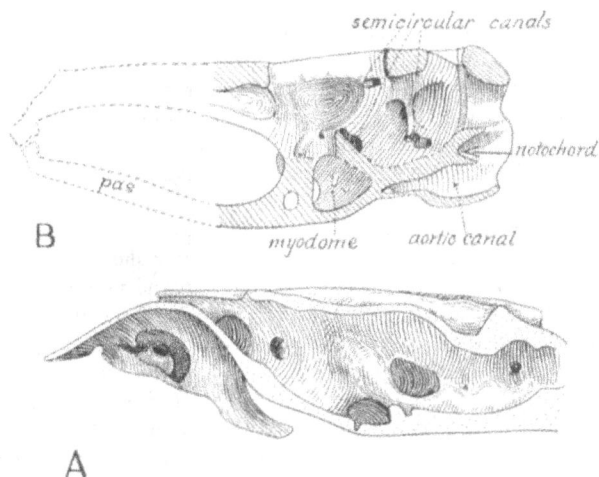

FIG. 14. Median sagittal section of neurocranium of (A) *Chlamydoselachus* (after Allis) and (B) palæoniscid (after Watson).

The neurocranium of various Palæozoic and later chondrosteans has been described by Stensiö in 1921 and 1925 and by Watson in 1925 and 1928. The neurocranium of the palæoniscids differs from the braincase of typical sharks, especially in the following points (Figs. 13, 14): (1) the nearly complete ossification of the braincase; (2) and the presence of separate ossific centres (at least in one genus) for the opisthotic, basioccipital and sphenotic; (3) the functional integration of the deep and surface ossifications; (4) the presence and importance of the parasphenoid or keel bone; (5) the reduction of the ventral part of the interorbital braincase to a thin septum; (6) the elevation of the interorbital brain trough above the parasphenoid; (7) the development of a myodome; (8) the presence of canals in the base of the cranium for the dorsal aorta and the efferent branchial arteries, as well as for the external and internal carotids.

Not improbably the parasphenoid bone may at first have been developed in order to stiffen the floor of the orbit, especially as the recti muscles extended backward in the manner described by Watson (1925, p. 849).

Many detailed differences seen between the palæoniscid and the shark in the interior

4

of the braincase are connected with the following contrasts in the brain and sense organs, among others pointed out by Watson (1925):

	Shark	Primitive actinopteran
Olfactory parts	Very large	Reduced
Optic parts	Moderate	Much enlarged
External rectus muscle of eye	Short	Produced backward (forming myodome)
Cerebellum	Moderate to large	Much enlarged and involuted

The olfactory or ethmoid region of the neurocranium of the palæoniscids figured by Watson is much less expanded anteroposteriorly than it is in typical teleosts. Processes homologous with the lateral ethmoids (parethmoids) of teleosts were present but not separately ossified from the general mass.

The occipital segment of the palæoniscoid cranium (Watson, 1928, Figs. 1–3; 1925, Fig. 12), consisting of the basioccipital and an ascending occipital plate, represents the neocranium or segmental portion of the cranium (Fürbringer, 1897; Gaupp, 1906) which is here shown in a very primitive stage. The possession of a neocranium is characteristic of the ganoids and teleosts, in contrast with the elasmobranchs, which have only a palæocranium, not extending behind the vagus nerve (Fürbringer, as quoted by Kindred, 1919, p. 23).

According to the views of Traquair, Smith Woodward and others, the Palæozoic palæoniscoids gave rise to the following groups: (1) the very short and deep-bodied Platysomidæ, which have a downwardly directed suspensorium and very small nibbling mouth; (2) the Trissolepidæ of the Permian; (3) the very peculiar Carboniferous genus *Phanero-rhynchus* of E. L. Gill (1923b), which is a palæoniscoid with a cartilaginous rostrum like that of *Acipenser* and a number of progressive semionotid-like characters in the fins (Watson, 1925); (4) the long-jawed Saurichthyidæ of the Triassic (cf. Stensiö, 1925); (5) the Jurassic Chondrosteidæ, which appear to be close to the lines leading to (6) the sturgeons and (7) the spoonbills; besides all these there were (8) the progressive Catopteridæ of the Triassic, which are normal fusiform fish with a fair-sized mouth. Neither the Catopteridæ nor any other known family of Chondrostei, however, appear to be directly ancestral to the typical holostean or protospondylous ganoids and later teleosts.

In the platysomids (as described chiefly by Traquair, 1879) the form of the head was clearly subordinate to the great depth of the body as a whole (Fig. 15) and in the extreme forms some of the skull plates are greatly lengthened vertically. This line of specialization finally culminates in extremely deep-bodied forms and the series as a whole has every appearance of being derived from such primitive fusiform types as *Cheirolepis* and *Palæoniscus*. In the Catopteridæ of the Triassic the skull pattern approaches that of the contemporary ancestors of the higher ganoids but only by way of convergence. Here again we have evidence that a short or moderate mouth and downwardly directed suspensorium have been derived from a large mouth and backwardly inclined suspensorium.

After a most painstaking description and analysis of the skull characters of the saurichthyids (Fig. 16), Stensiö concludes that they must be closely related to the palæoniscids, as maintained by Woodward since 1895 (p. vii); that while in some respects they are more primitive than the palæoniscids (e.g., in the retention of a quadratojugal, lost in the latter); on the whole they are considerably more specialized than these. Stensiö concludes (1925, p. 223) that "among the Chondrostei the saurichthyids are closely related to both palæoniscids and sturgeons (in the broad sense, including the acipenserids, polyodontids and chon-drosteids), but apparently mostly to the latter; that . . . they really seem to have evolved from the same ancestral form among the primitive actinopterygians as the sturgeons (in the broad sense); that this common ancestral form in its turn has not been a palæoniscid

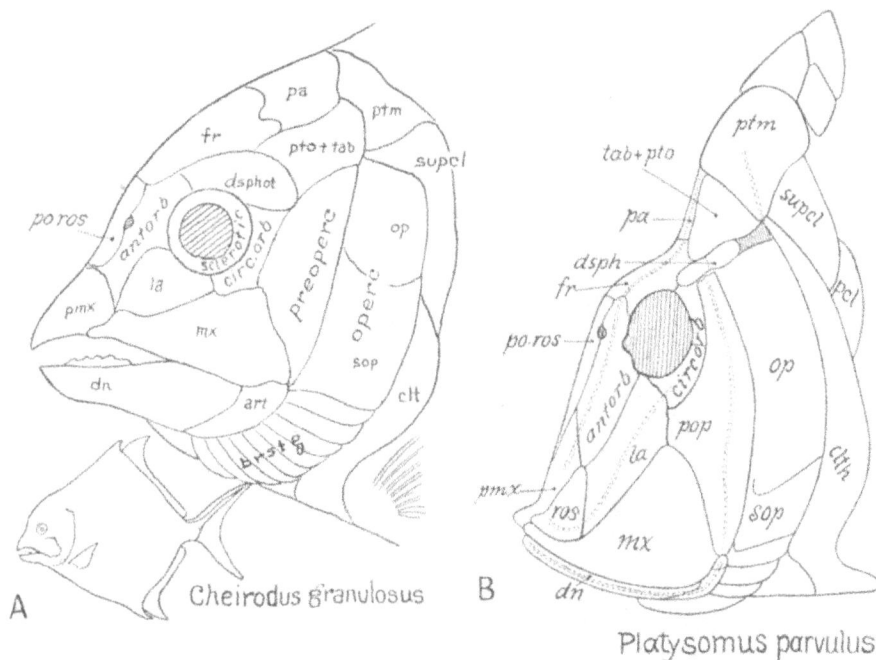

Fig. 15.   Skulls of deep-bodied derivatives of the Palæoniscidæ.

A. *Cheirodus granulosus*.   After Traquair.
B. *Platysomus parvulus*.   After Watson, but names of elements changed in some instances to conform to system herein adopted.

but must in certain respects . . . have been more primitive.   In reality it must have been closely related to some primitive, hitherto unknown type of actinopterygians from which the higher ganoids and teleosts also originated."   Stensiö also concludes that the saurich-thyids, like the sturgeons, palæoniscids, coelancanthids, dipnoans and arthrodires, form a degenerate series.   By this he means especially that in such series the adult endocranium is better ossified, less cartilaginous, in the earlier than in the later members of the series.

CHONDROSTEI (SPOONBILLS AND STURGEONS)

According to the views of Traquair (1877, p. 39) and most other palæoichthyologists the existing *Polyodon* (Fig. 17) represents a specialized and in some respects degraded de-rivative of the primitive chondrostean stock.   It perhaps owes its survival to the great development of the tactile snout.   The small eye remains above the front end of the upper jaws.   The primary jaws are very large, the long hyomandibular being directed obliquely backward and downward as in hyostylic sharks. The derm bones are reduced, almost vestigial. In spite of this specialization, however, the palæoniscoid heritage is still very evident, as noted by Traquair. The so-called opercular bone corresponds in position with the subopercular of the palæoniscoids and chondrosteids, and possibly the true opercular, already small in palæoniscids, has disappeared.

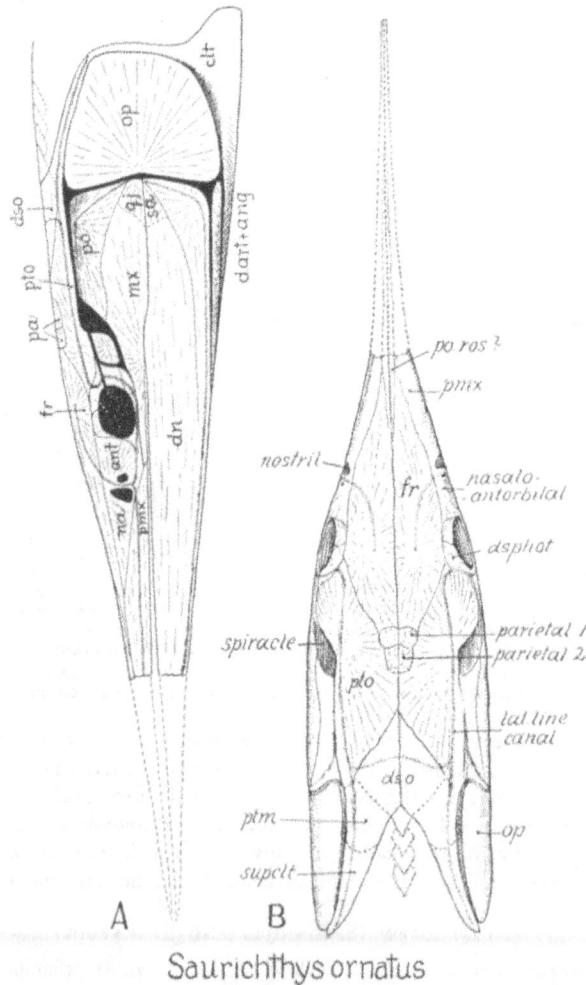

Saurichthys ornatus

Fig. 16. *Saurichthys ornatus*. After Stensiö.

A. Side view. Stensiö's nomenclature somewhat modified to accord with system adopted herein.

B. Top view. According to Stensiö, the bone marked *pto* represents a combination of the supratemporal (= pterotic), intertemporal and extrascapular (scale bone).

The sturgeon (Fig. 19) has specialized in the opposite direction from that of the primitive chondrosteans, as it has acquired an excessively small suctorial mouth which is withdrawn far behind the projecting rostrum. The enormous hyomandibular is directed

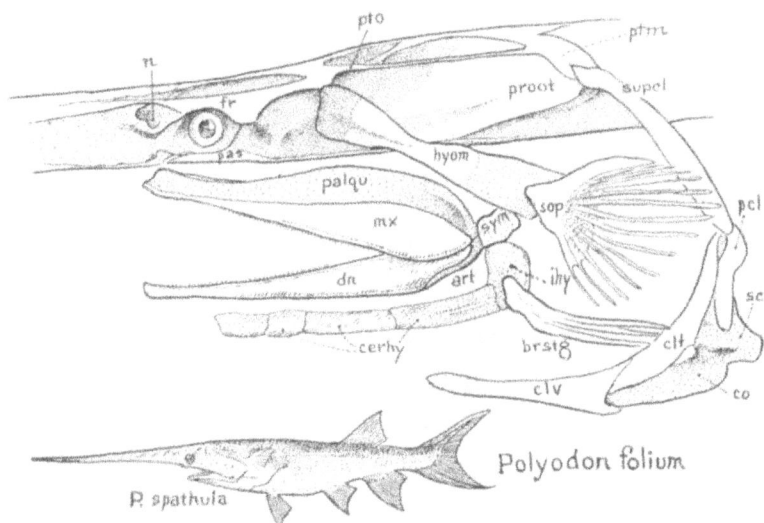

FIG. 17. Skull of *Polyodon folium.* Mainly after Traquair.

backward but the greatly enlarged symplectic reaches downward and forward. These elements are covered with broad and thick muscles, which alternately dilate and retract the hyomandibular, thus pumping water into the oropharynx. The eye remains relatively far forward above the anterior part of the upper jaw. The so-called opercular is free of the hyomandibular and probably corresponds to an enlarged subopercular. The whole snout and fore part of the braincase is warped downward above the capacious orobranchial cavity in order to bring the snout down parallel to the ground. The downward displacement of the rostrum probably increases the adverse leverages of the fore part of the skull upon the occipito-nuchal joint. Hence the extensive fusion of cervical segments with the occiput may be a means of compensating for this weakness. The rostral barbels are specialized tactile organs, more or less similar in function to those of siluroids, gonorhynchids, mullids, sciænids, gadids, etc.

The neurocranium of the sturgeon and spoonbill are largely cartilaginous but with more or less extensive centers of ossification. It has been assumed by Watson and Stensiö that this partly cartilaginous condition is due to retrogressive development (perhaps to the retention of early larval conditions in the adult). Sewertzoff, however, as a result of his embryological investigations (1928) challenges this view and concludes that the recent chondrosteans are much more nearly related to the elasmobranchs than was formerly suspected and that in many respects they are more primitive than the Palæozoic palæoniscids. He holds among other things that the numerous ossicles in the snout of the sturgeons are more primitive than the few rostral elements of the palæoniscids.

After a careful consideration of these opposing evidences and interpretations, I can only record my impression that the older view is by far the more probable, and that for many reasons, only a few of which may here be noticed.

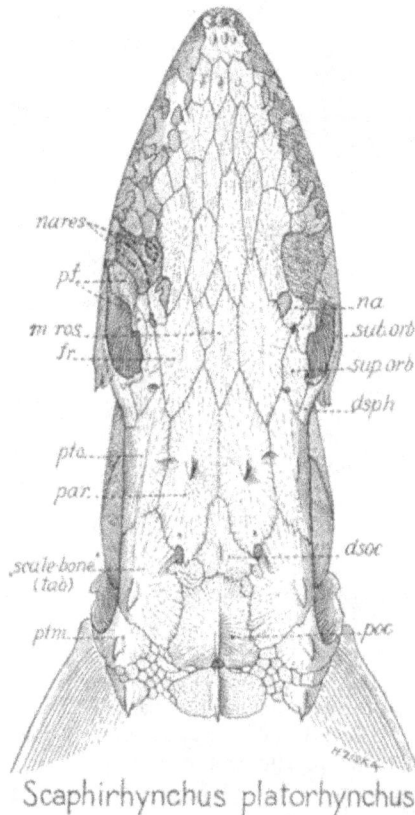

Scaphirhynchus platorhynchus

Fig. 18. *Scaphirhynchus platorhynchus.* Courtesy of Dr. E. W. Gudger.

Whatever may be said as to the sturgeon, it can hardly be doubted that the exoskeleton of the spoonbill (*Polyodon*) is in a highly retrogressive condition. In place of the fully formed ganoid scales of its palæozoic relatives it has a practically naked body with a few vestigial horny scales in the upper lobe of its heterocercal fin. Assuredly its so-called opercular plates have the appearance of being degenerate structures and the same is true of the thin derm-bones that overlie its palatoquadrate and Meckel's cartilage. In spite of its degenerations, however, the entire suspensorium is evidently of a modified primitive actinopteran type, differing from the elasmobranch especially in the presence of a large symplectic and of an "opercular" plate. The shoulder-girdle is that of an actinopteran, not that of a progressive shark. In view, therefore, of its degenerative specializations from an actinopteran starting-point, such peculiar resemblances to the sharks as the union of the opposite palatoquadrates beneath the braincase may well be regarded as examples of convergence.

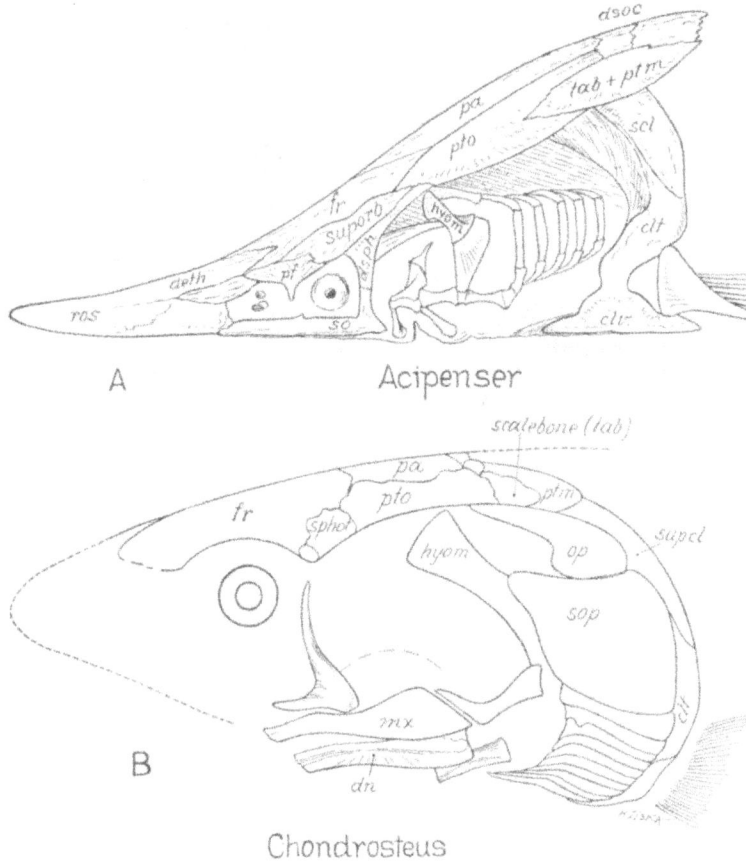

Fɪɢ. 19. Skulls of (A) modern sturgeon, and (B) its Jurassic ancestor, *Chondrosteus*. Both after A. S. Woodward.

The degenerative specialization of the sturgeon seems to me equally clear and convincing. The excessive concrescence of vertebral elements with the occiput, the presence of a valvula cerebelli, the extreme reduction of the mouth and jaws and their suctorial character, the development of rostral barbels, the loss of ganoine on the exoskeleton and the development of ridged bony plates on the skin, are a few of the many aberrant specializations of the sturgeons away from a typical actinopteran starting-point. The resemblances of the rostrum, mouth and lips of the sturgeon to those of *Acanthias*, noted by Sewertzoff (1928), are so superficial that they indicate with high probability that such resemblances between the modern chondrosteans and the elasmobranchs are largely convergent and I would even doubt the proposed homologization of the processi palatobasales laterales in the representatives of two such widely divergent classes of fishes.

In conclusion, it may be suspected that many of the embryonic characters of the sturgeon neurocranium and branchiocranium instead of being reminiscent of far-off pre-Palæozoic gnathostomes may rather be anticipatory of peculiar specializations of the

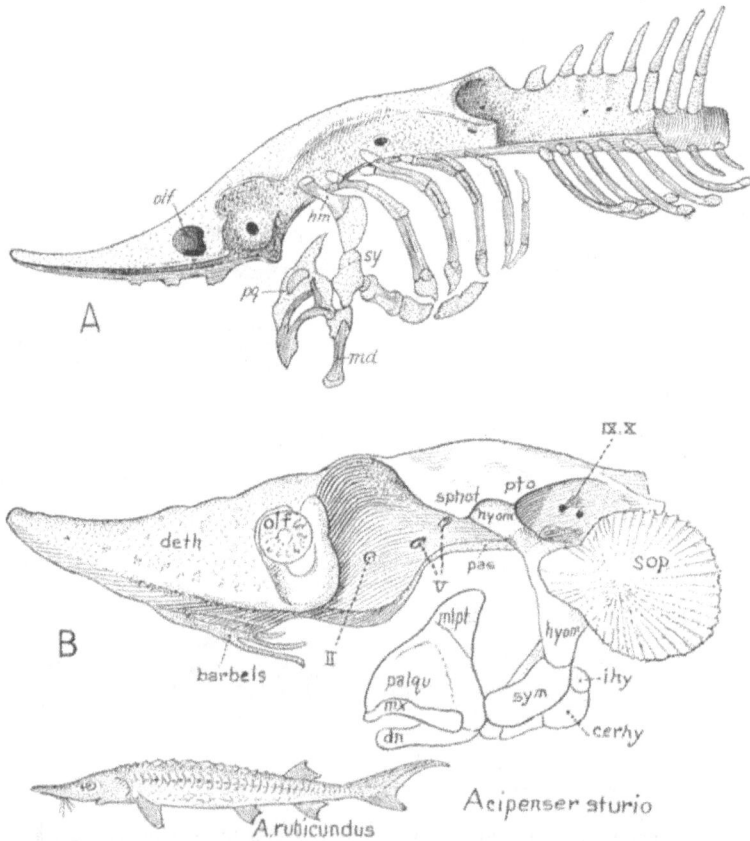

Fig. 20.   A. Skull of larval sturgeon (after Wiedersheim), showing downward bending of anterior end of skull upon the vertebral column.   B. Skull of larval sturgeon (after W. K. Parker).

adult.   Here belong such points as the concrescence of many vertebral segments with the occiput, the curious downward bending of the fore part of the skull upon the occipital portion (Fig. 20), the lack of a sharp cranial flexure in the early stages, the presence of barbels, the foreshadowing of the suctorial characters of the adult mouth, the peculiar features of the immense hyomandibular and large symplectic, the loss of the true opercular and the substitution for it of an enlarged subopercular.

The bony scutes of the sturgeons, including those in the rostrum, seem to me to be obviously secondary characters developed after the loss of the ganoine, analogous to the ossified exoskeleton of certain highly specialized teleosts, or to the fragmented rostrum of certain pycnodonts.

In brief, I fail to see in the embryo strugeon any especially elasmobranch characters not shown in other fish embryos, or any which would imply the derivation of the sturgeons and spoonbills from some post-elasmobranch type that stood widely apart from the lines running to the palæoniscids and their allies.

Moreover, many of the peculiar characters of the sturgeons are foreshadowed by the Jurassic *Chondrosteus* (Fig. 19*B*), which on the other hand retains features that are clearly inherited from a palæoniscoid stock, as well noted by A. S. Woodward (1895, p. viii). Watson (1925, p. 831) has already shown the annectant character of the Chondrosteidæ between the palæoniscids and the sturgeons and concludes as follows:

"Thus such new information as I can add only emphasizes that resemblance between *Chondrosteus* and the Palæoniscids which Traquair long ago pointed out, and shows how untenable is the view of Bridge, adopted by Sewertzoff, that the Acipenseroides are the most primitive of bony fish and owe many of their peculiarities to a persistence of Elasmobranch structures."

Protospóndyli (Semionotiids, Pycnodonts, Garpikes, Macrosemiids)

Long ago (1895) Smith Woodward, developing the results of Agassiz, Traquair and others but adding greatly thereto from his own observations, traced the morphological evolution of the vertebræ, scales, caudal fin, skull, mandible, etc., of the Paleozoic and Mesozoic ganoids. Except in a few instances all subsequent palæontological investigations have tended to confirm his conclusions. He showed also that while the known fossils often afford a good morphological series indicating the evolution of the different organs and parts, they do not give us a direct ancestral series, and that there are great breaks in the record as we pass from a lower major group to the next higher one. Thus, as we have seen above, the chondrostean series, very primitive at first, radiates in many directions, most of the branches becoming extinct during the Mesozoic era but a few of them giving rise to the modern spoonbills and sturgeons. The catopterid branch (Fig. 21*A*), while progressing toward the next higher order (the Protospondyli or Holostei), never attained that goal but became extinct long after the early members of that group had been derived from some still undiscovered stock. Hence we cannot yet affirm positively either that the small-mouthed Semionotidæ, which are the oldest known branch of the Protospondyli, were derived from large-mouthed palæoniscoid ancestors, or that they again branched into large and small-mouthed descendants; although that still seems to be by far the most probable view.

The inference that mouths that were once large became small and that some of the latter again became large, does not necessarily involve a violation of the supposed law of "irreversibility of evolution," for the new large mouths probably became so by a different method from that which first led from small to large; that is, different elements were involved in different ways, as will presently be shown.

We need not be surprised that there are still many gaps in our records if we realize how relatively few fish-bearing horizons are known throughout the five hundred million

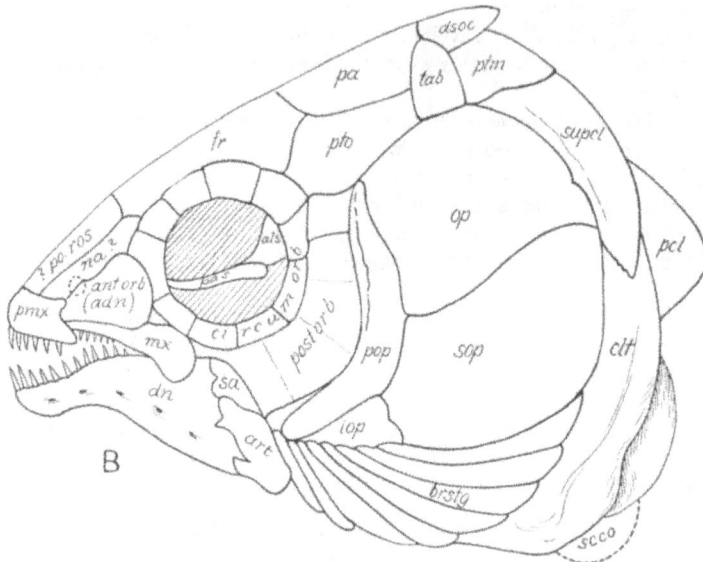

Acentrophorus varians

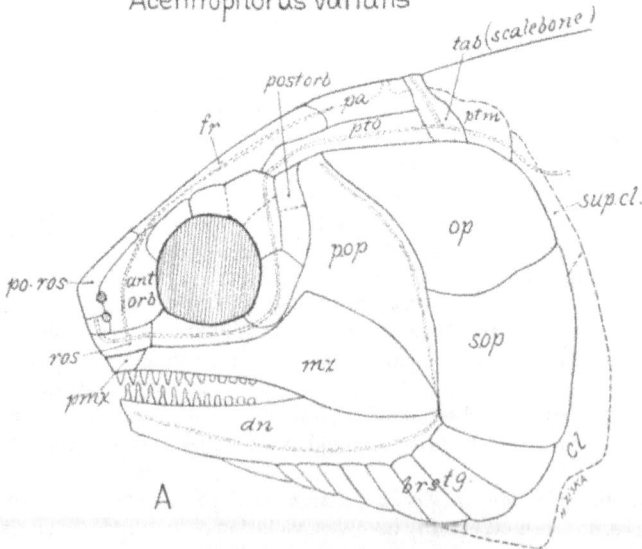

Perleidus woodwardi

FIG. 21. Comparison of the most progressive known palæoniscoid *Perleidus* (A) of the Triassic family Catopteridæ (after Stensiö) with the most primitive semionotid, *Acentrophorus* (B) of the Permian and later periods (after E. L. Gill).

In the advanced palæoniscoid the suspensorium is intermediate in position and the postorbital series is beginning. In the primitive semionotid the suspensorium is inclined forward and the postorbital series is well developed.

years or so of piscine history. The chondrostean series appears in the Middle Devonian and dominates the inland waters through the later Paleozoic, thereafter giving way before the higher ganoids and their descendants and being represented today only by the specialized and in some respects depauperate sturgeons and spoonbills. The Holostei, or Protospondyli, first appear in the records in the closing epoch of the Paleozoic, become dominant in the early Mesozoic and then give place before the increasing hosts of their own descendants. At present the Protospondyli are represented by the garpikes (Lepidosteidæ), a specialized branch of their oldest stage, and by *Amia*, an offshoot from the later Protospondyli that stands near the base of the teleost stem.

Semionotids.—The first of the Holostei, or Protospondyli, was the Permian *Acentrophorus* (Fig. 21*B*), which has the characters typical of the family Semionotidæ. In these fish the body is short and deep and so is the head, which, as described by E. L. Gill, conforms to the body contours. The skull pattern contrasts widely with that of the palæoniscids: the mouth is small and of the nibbling type with delicate styliform teeth; the suspensorium is vertical or inclined forward; the eye is large, located partly behind and above the very small upper jaw. In addition to the sclerotic plates there are two rows of large orbital plates, called here the circumorbitals and the postorbitals, the latter more or less parallel to the curved opercular and branchiostegal series. Above the eyes there are several supraorbital plates. The more or less concentric arrangement of the bones of the face in these fish, while contrasting widely with the oblique arrangement of the cheek-plates of the Chondrostei appears to be one of the "basic patents," so to speak, for the skulls of all higher fishes. The probable origin of this arrangement is discussed below.

The same type of skull pattern (Fig. 22) is preserved in the Triassic and later genera, *Dapedius*, *Semionotus*, *Lepidotus* and their relatives. All are more or less orbicular to fusiform in body, with small nibbling mouths and concentric arrangement of circumorbitals, postorbitals, etc. The circular arrangement thus seems, in fact, to be especially correlated with an orbicular body-form and small mouth. Hence when later in longer-bodied fish we find clear traces of the circular pattern, the inference seems probable that such body-forms have been derived by lengthening of a sub-orbicular type, as has clearly been the case in several other families (Pholidophoridæ, Characidæ, Carangidæ). All bony plates of the typical teleost skull are present and in addition others that become reduced or rare, especially in higher teleosts. In this category belong the several extra elements in the inner or circumorbital series, the entire row of "postorbitals," the antorbitals and the several sheathing bones (so-called splenials and coronoids) on the mesal side of the mandible.

Most of the elements in the "circular" pattern of the skull can be identified readily with those of the oblique or asymmetrical pattern of the primitive palæoniscoid skull (Fig. 12). Nevertheless it is difficult to be sure in certain cases, especially where the number of elements in each of the series is different in the two groups, as in the circumorbital and postorbital series.

The circular pattern may be supposed to have been derived from that of the palæoniscoids, partly as a result of the marked shortening of the jaw and the consequent forward migration of the joint between the mandible and the quadrate. This would drag the attached subopercular downward, perhaps causing the lower end of the subopercular to be fractured and pulled forward out of its place to form the interopercular, which now appears for the first time as distinct from the subopercular (Tate Regan, 1929, pp. 31, 313). The

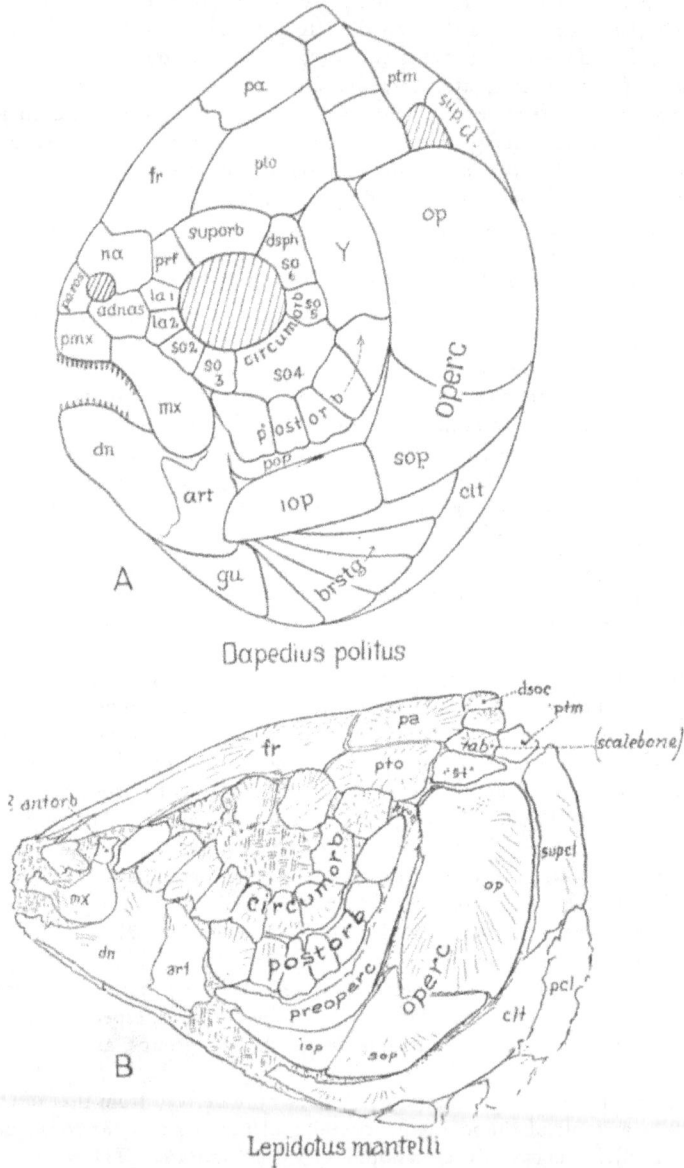

Fig. 22. Comparison of semionotid skulls of semi-orbicular type, (A) *Dapedius*, with more normal type (B), *Lepidotus*. In both the suspensorium is produced well forward. The postorbital row seems to have arisen through the down-growth and budding of plate Y. The interopercular (absent as such in the palæoniscoids) has been derived from the produced lower end of the subopercular (Tate Regan).

forward displacement of the quadrate-articular joint would also give increased importance to the lower part of the preopercular in its function of stiffening the posterior side of the quadrate. The same movement would give space for the small postorbitals of a form like *Oxygnathus* to grow downward behind the first circumorbital row. This postorbital series is of course not to be confused with the true postorbital of the tetrapods, which is equivalent merely with a single one of the inner or circumorbital series of fishes.

Many of the Semionotidæ tended to parallel the Sparidæ, or breams, among teleosts in the development of pebble-like teeth, presumably adapted to crushing mollusc shells. Some of the more normal members of this series gave rise to the very long-lived and successful genus *Lepidotus* (Fig. 22*B*). Others, becoming still more specialized, gave rise to the peculiar Pyncodontidæ, with rows of pebble-like teeth clustered on a median cylinder in the upper jaw and on the inner sides of the lower jaw. The skull pattern of these forms became aberrantly modified, chiefly through consolidation and strengthening of its parts, but as the family has nothing to do with the line of ascent to later families, it may here be passed by.

Lepidosteids.—The garpike (Fig. 23*A*, 24) of the existing ganoids preserves the basic heritage of the semionotid skull, including the large circumorbital plates and the forwardly inclined suspensorium, but it has become specialized in the great elongation of the snout, in the fragmentation of the maxilla into numerous plates, and in many other features. Goodrich (1909, pp. 342–344) has shown that in many features the garpikes agree with *Lepidotus*, and suggests that very possibly *Lepidosteus* is merely a specialized late remnant of the family Semionotidæ.

Recent authorities differ as to the identification of the bone beneath the opercular. Tate Regan (1923*a*) identifies it as an interopercular, partly on account of its relations with both the subopercular and the angular of the mandible. Mayhew (1924), on the other hand, identifies this bone as the preopercular because it carries part of the operculo-mandibular sensory canal. He also applies the term interopercular to the small bone that lies behind and beneath the quadrate. This is also the identification of these elements adopted by Goodrich (1909, p. 342).

A dried skeleton of *Lepidosteus* (Figs. 23, 24) with all these elements in their nearly natural positions shows without doubt that the large curved bone called by Regan the interopercular has all the proper connections for the preopercular. Thus it articulates with the lateral surface of the hyomandibular, running to the top of that element and receiving the preopercular branch of the lateral line system. Then it runs downward and forward, expanding greatly on its posterior and inferior borders so as to afford the main outer brace for the suspensorium. This part of the bone evidently overlapped and then thrust itself between the subopercular and the interopercular. The dorso-medial surface of the ascending branch of the bone evidently affords a secure origin for the lateral parts of the adductor mandibulæ muscle. The inner border of this area has a contact with the head of the interhyal, as has the preopercular of *Salmo*. Continuing forward, the bone in question forms also the lateral brace for the lower end of the symplectic and for the palatal extension of the metapterygoid. At its antero-inferior end it overlaps the true interopercular, which as usual is fastened by ligament to the derm-angular. One reason why the bone in question does not overlap the lateral posterior border of the quadrate in the normal manner of a preopercular is that the latter is too small and has been crowded out by the enlarged metapterygoid. In short the preopercular of *Lepidosteus*, while retaining its normal relations lateral to the hyomandibular and symplectic has expanded on its posterior and inferior borders, and has lost its contact with the forwardly shifted quadrate. Meanwhile the interopercular, retaining its ligamentary connection with the derm-angular, has moved forward, losing its normal connection with the subopercular and usurping the position of the preopercular as the lateral brace of the quadrate.

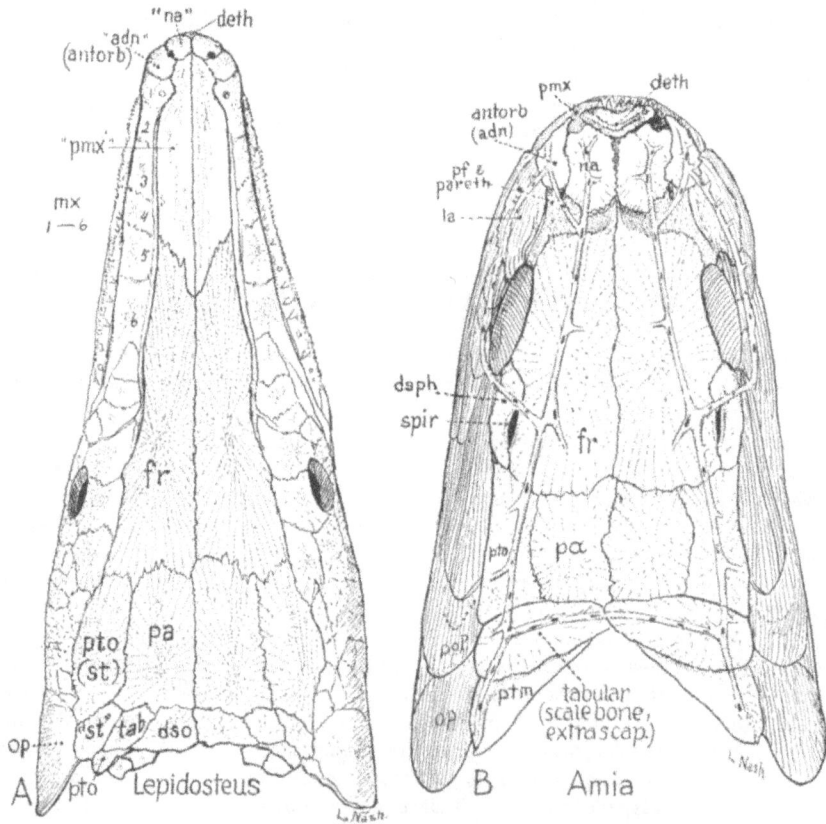

FIG. 23. Comparison of *Lepidosteus* and *Amia* skulls. Top view.
A. *Lepidosteus tristoechus*. B. *Amia calva*.

The long bones on the dorsal surface of the anterior part of the snout are regarded as homologous with the ascending processes of the premaxillæ of *Amia*, as shown by Tate Regan. Mayhew calls them nasopremaxillaries. The small ethmoid appears to represent the rostral or postrostral of palæoniscids, and the mesethmoid (dermethmoid) of teleosts. The nasal and adnasal together retain the relations of the antorbital of palæoniscids except that by the elongation of the snout they are far removed from the eye. The lacrymal has

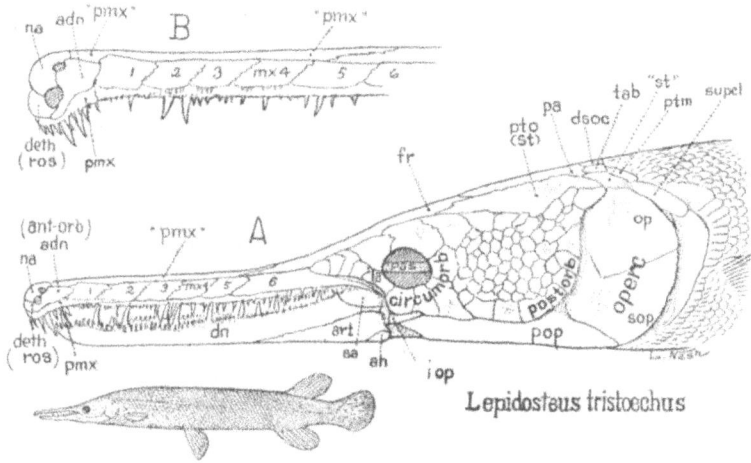

FIG. 24. *Lepidosteus.*
A. Side view.   B. Detail of rostrum.

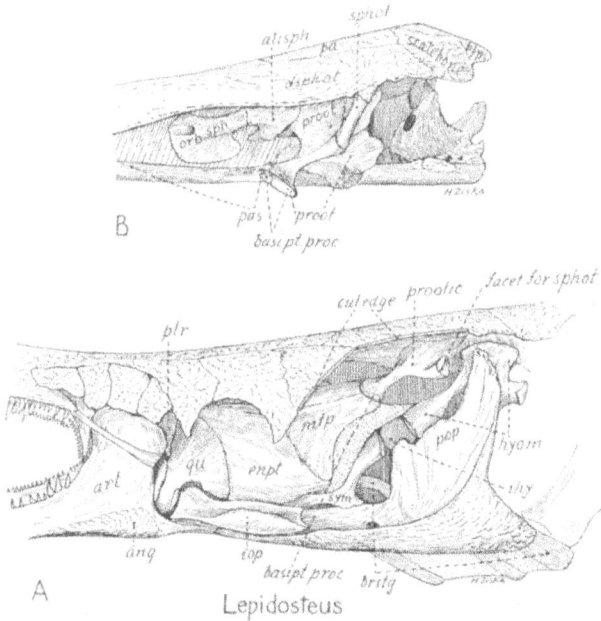

FIG. 25. *Lepidosteus.*
A. Relations of preopercular and adjoining parts.
B. Slightly oblique side view of braincase, showing "basipterygoid" facet on parasphenoid and proötic.

grown forward into the space left between the maxilla and the orbit, and has become subdivided into several bones.

The mosaic of small bones on the cheek surely represents a fragmentation of the post-orbital series of *Lepidotus*. That this fragmentation is secondary is indicated by the presence of a plate corresponding in position to the normal undivided maxilla but here sub-divided into six or more pieces which are fused with lateral line plates representing a forward continuation of the suborbital series (Allis, 1905).

The most notable features of the palate are the paired condition of the vomers (pre-vomers) and the presence of paired, transversely extended articular facets (Fig. 25) on either side of the mid-line, just in front of the cranial vault; the inner parts of these facets are supplied by the parasphenoid, the outer by the proötics. They serve for the attach-ment of the metapterygoids behind the mesopterygoids. These paired facets tie in the long palate posteriorly and permit slight movements without loss of strength.

The entire construction of lepidosteids indicates that these fish are descended from short-bodied, short-mouthed forms in which both body and jaws became elongated antero-posteriorly, thus transforming a peaceful nibbler into a predatory pike. Possibly the labyrinthodont-like teeth of *Lepidosteus* may have been derived from stout, pebble-like teeth which acquired pointed tips.

The principal objection to this derivation is the fact that in the lepidosteids the vertebræ are highly perfected, completely ossified and opisthocœlus, while in the Mesozoic protospondyls the notochord is persistent and the centra are at most ring-like. For this reason several authors have placed *Lepidosteus* in an order by itself, the Lepidosteoidei. But the students of amphibian and reptilian centra have shown, for example, that a procœlous vertebra may become opisthocœlous by annexing the ossified intercentral ball of the vertebra in front of it. At any rate, the perfection and specialization of the *Lepidos-*

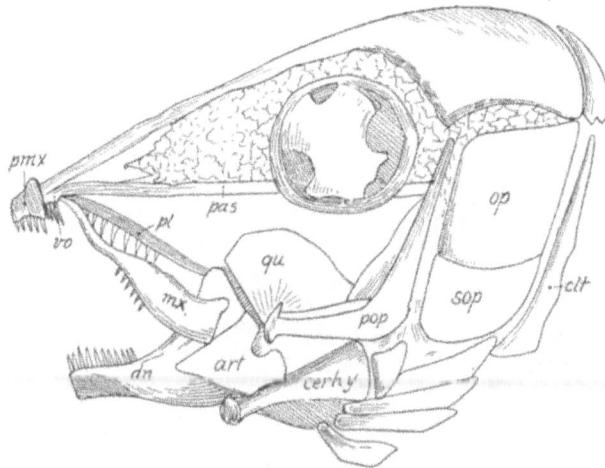

Macrosemius

FIG. 26. *Macrosemius.* After Smith Woodward.

teus vertebræ seem quite easily derivable from the generalized condition of the column in the Semionotidæ, especially in the light of the comparison of skull patterns.

Macrosemiids.—The incompletely known skull pattern (Fig. 26) reveals a few important morphological details. In this Triassic ganoid the eyes had already assumed the dominance characteristic of teleosts in comparison with sharks. The large backwardly-positioned eyes contrast widely with small forwardly-placed eyes of the primitive chondrosteans. The base of the skull was stiffened by a strong parasphenoid. The suspensorium was inclined forward, the preopercular being sharply bent and ridged in contrast with the gently curved cheek-plate preopercular of palæoniscids. The jaws are nearly as short as in the Semionotidæ and there is a sharp ascending ramus of the mandible. Both the small premaxillæ and short maxillæ bear pointed teeth. The opercular is short and deep. The skull gives the impression of having been derived from one that was shorter and deeper. The elongate dorsal fin of the Macrosemiidæ suggests a secondary elongation of the body from a more deep-bodied ancestral stock, perhaps not far from the Permian *Acentrophorus*.

AMIOIDEI

In the Triassic of Europe appears the first known of the great group of amioids, which as a whole stand between the Palæozoic ganoids and the late Mesozoic teleosts. These primitive amioids have inherited from some older ganoid stock the complete ground-plan of their skeletons, the microscopic structure of the scales, the abbreviate heterocercal tail, the general pattern of the skull bones. But they are far more progressive than any known earlier ganoid group in the swift-swimming type of body with its large forked hemiheterocercal tail, as well as in numerous details of skull structure. During the Jurassic and Cretaceous periods they exhibit a wide adaptive radiation into many different body-forms. Most of these end in specialized side lines; one gives rise to the existing *Amia*, while the leptolepids finally attain the teleost grade and appear to be directly ancestral to the order Isospondyli at least.

As to the exact point of origin of this entire series, even the prolonged researches of Smith Woodward, Watson and others have failed to reveal it. On the whole, evidence seems to indicate that in spite of the marked difference in habits and habitus the primitive pike-like amioids may have been derived from a relatively short-bodied, small-mouthed form like the Permian *Acentrophorous* rather than directly from the Devonian pike-like palæoniscids. This view has indeed been more or less independently suggested by Smith Woodward (1895, p. ix), Tate Regan (1923a, p. 456; 1929, p. 312) and the present writer (1923, p. 239).

Eugnathids.—Whatever its precise derivation may have been, the skull (Fig. 27) of *Eugnathus* (A. S. Woodward, 1895, Pls. IV, V) gives the impression that the eye has migrated backward, thus giving rise, through some non-Lamarckian principle, to confusion and irregularity in the circumorbital plates, while the snout has grown forward, the two movements conditioning the marked elongation and downward pitch of the lacrymal and antorbital elements, a feature unknown in earlier ganoids, but highly characteristic in the amioids and teleosts. The long preorbital plate is subdivided longitudinally into two parallel plates, of which the dorsal corresponds to the antorbital, the lower to the lacrymal of *Amia*. A close scrutiny of the original specimen shows that at least parts of the suture

5

separating the antorbital from the lacrymal are precisely like other and certain sutures near by.

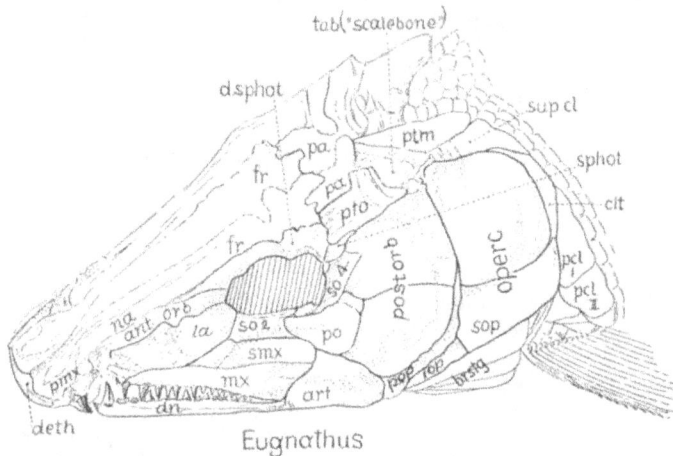

FIG. 27. *Eugnathus.* After Smith Woodward.

The beautiful figure of this skull published by A. S. Woodward gives a very accurate record of the facts, especially as regards sutures. The postorbital plates, the beginnings of which have been seen in the palæoniscids, are very large but now few in number and there is a much higher degree of differentiation between adjacent elements of the surface pattern than was the case in the primitive semionotids. The operculars and branchiostegals all obviously form part of a single opercular fold. They are all wider antero-posteriorly than those of primitive palæoniscids. They agree, however, with the semionotid type in the important fact that the interopercular is in horizontal alignment with the subopercular, thus lending weight to Tate Regan's suggestion that the interopercular represents a separated extension of the antero-inferior corner of the subopercular. Both the sub- and inter-opercular lie immediately dorsal to the long first branchiostegal. Examination of this fine specimen of *Eugnathus* reveals a small transversely-extended bone lying above the premaxillæ and below the broad nasals, which seems to be the median ethmoid, as it corresponds to a similar bone in *Lepidosteus* and *Amia*. The broad plates in front of the frontals have the appearance of being homologous with the nasals of *Amia* and the teleosts and possibly also with the paired postrostrals of palæoniscoids (*cf.* p. 113).

The hyomandibular of *Eugnathus* is directed at first gently backward and then forward, essentially like those of the Semionotidæ on the one hand and of the teleosts on the other, and unlike the more sharply inclined suspensorium of the primitive palæoniscids. This contrast is related with the fact that in *Eugnathus* the snout is relatively elongate and the eye far backward, while in the palæoniscids the opposite conditions are found. Also correlated with this difference is the profound contrast in the preopercular, which in palæoniscids is large, inclined forward and spread over the upper part of the cheek, but in the primitive amioids small, vertical, inclined slightly backward and limited to the lower part of the cheek. The maxillæ are now freed at their posterior end from the primary upper jaw as in teleosts and each bears a single elongate supramaxilla. Tate Regan (1923a, p. 456) suggests that they have abstracted this element from the outer circumorbital plates in both *Caturus* and *Eugnathus*, as well as in *Amia*. Possibly this has happened in response to the pull on the maxilla of the superior branch of the adductor mandibulæ muscle.

On the cranial roof the frontals have assumed the importance that is characteristic of teleosts, while the parietals are shortened anteroposteriorly. The other surface elements also approach the primitive soft-rayed teleost type.

The skull of *Caturus* with its broad smoothly-rounded operculars and clean-lined contours foreshadows the mackerel type, as also in the progressive thinning of the surface plates and scales, but these resemblances are far more probably due to convergence than to direct phylogenetic relationship. The skull-roof of *Caturus* has been figured by A. S. Woodward (1897, Pl. VIII). It is flat and *Amia*-like, with very large frontals and small parietals, which however meet in the midline, the bony supraoccipital not yet having made its appearance on the skull-roof.

In the more advanced amioids such as *Hypsocormus* the skull foreshadows that of *Amia* in the fact that the posterior circumorbitals are already much elongated anteroposteriorly, although very irregular in contour. The postorbital series of cheek plates, however, is greatly enlarged, a point of wide contrast with *Amia*.

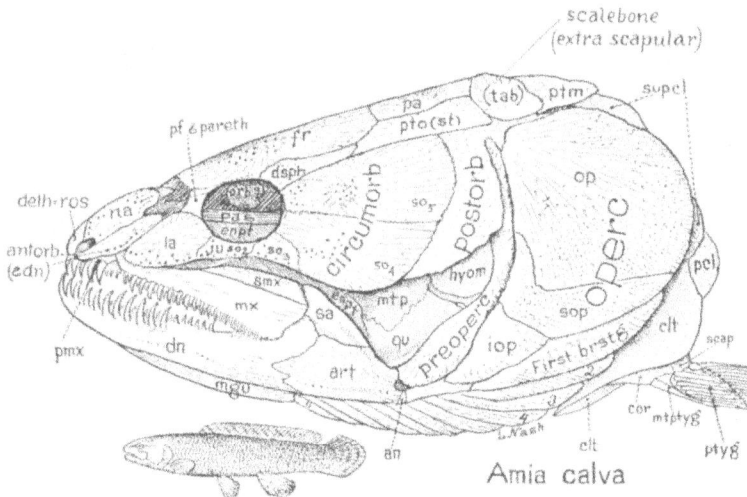

FIG. 28. *Amia calva.* Side view.

Amia.—In *Amia* (Figs. 23B, 28) the first circumorbital series has become greatly enlarged, while the cheek plates (postorbitals) have either entirely disappeared or at most are represented by several irregular ossicles, apparently first noted by Shufeldt and recognized by A. S. Woodward (1895, p. 369) as being the representatives of the cheek plates of the older ganoids.

Taken as a whole, the skull of *Amia* has advanced far toward the primitive teleost type.   All the derm bones have lost completely the ganoine covering that was so characteristic of their Mesozoic ancestors, while, as above noted, the postorbital or second series of plates has practically disappeared.   The ethmoid region is extended anteroposteriorly and includes distinct lateral ethmoid ossifications.   A special point of resemblance to the clupeoid isospondyls is the presence of posttemporal fenestræ in the occipital surface for the insertion of prolongations from the trunk musculature.   The neurocranium, as described by Allis (1895, 1897a), also approaches the primitive teleost type.   There is a distinct though small myodome and the parasphenoid is prolonged backward beneath the occiput.

At the same time *Amia* retains a number of ancient ganoid characters, especially: the small size of the mesethmoid, the large size of the paired bones in front of the frontals (probably rightly called nasals), the paired condition of the "vomers" (= prevomers), and the presence of an antorbital plate and of so-called splenial (= coronoid and prearticular) plates in the mandible.

Very probably the presence of so much unossified cartilage in the chondrocranium is not a primitive character but a retrogressive feature and a retention of larval characters in the adult.

In short, *Amia*, in respect to its skull structure as well as in most other parts of its anatomy, stands much nearer to the base of the teleosts than it does to *Lepidosteus*, with which it was formerly grouped as a "ganoid."

The skull of *Caturus* as figured by Smith Woodward (1897, Pls. VIII, IX) appears to afford an ideal structural ancestor to the *Amia* type, and approaches the latter in many characters of the skull-top face and mandible.

ISOSPONDYLI (PRIMITIVE TELEOSTS)

CONTEMPORARY with the lepidosteoids and amioids of the Triassic and Jurassic were several families that gradually approached and finally attained the basal teleost grade of organization. As noted by Smith Woodward (1895, xix–xxi) and others, their originally rhombic and dense ganoid scales lose the peg-and-socket articulations and become cycloid, overlapping, thin and more or less horny; their tails pass from the hemiheterocercal type, with unexpanded hæmal rods, to the primitive homocercal type, immediately prior to the expansion of the hypural bones; the thin, imperfect bony rings of their vertebral column gradually give rise to well ossified centra; meanwhile, the mandible attains simplification by the progressive reduction of the median gular plate and elimination of the so-called splenial (prearticular); the anterior borders of the fins gradually eliminate the fulcral scales; intermuscular bones appear only in the later members of the series. Thus the old distinctions between ganoids and teleosts are gradually effaced.

Opercular region of primitive teleosts.—In an excellent comparative study of the bones of the opercular series of fishes, Hubbs (1919, p. 63) writes:

"The Isospondyli, comprising the oldest and most primitive of the teleosts, retain certain generalized features of the opercular series. Thus, in *Elops* an intergular plate is developed, and in *Albula*, although the plate itself is lacking, the intergular fold remains. The branchiostegals of the typical Isospondyli (at least the upper ones), persist as thin wide plates. The uppermost and widest ray (which may be termed the branchioperculum, as it seems to be homologous with the plate in *Amia* to which that name is here applied) is attached closely to the inner margin of the sub- and interoperculum; not having become concealed under these bones, it remains visible from the side. The whole series, in fact, remaining scarcely at all folded together after the fashion of a fan, is visible from below, though the branchial membranes are separate (as they usually are). The plates of the opercular series in the isospondylous fishes differ from those of *Amia* in the following respects: the reduction of the suboperculum, so that the interoperculum and operculum are in contact anteriorly; the proximal (or anterior) attachment of branchioperculum and branchiopercular fold to the hyoid arch; the more complete imbrication of all the rays; the attachment of branchiostegals to the epihyal as well as to the ceratohyal; the frequent reduction of the rays below the main hyoid suture to rather slender rods, and the occasional attachment of these reduced rays to the edge of the ceratohyal, rather than to its outer face. These last two features are apparently caused by the strong development of the musculus geniohyoideus of the lower jaw, which is attached to the hyoid arch near the suture separating the ceratohyal from the epihyal. The number of the larger and flatter rays attached to the outer surface of the epihyal (the lowermost sometimes on the suture) varies widely in the Isospondyli and related orders [one to ten]; . . . ."

". . . The total number of branchiostegals is three in the Cyprinidæ and others, twenty-four to thirty-six in the several species of *Elops*. Many other figures might be added, but these are enough to illustrate clearly the inconstancy of the number of branchiostegal rays in the generalized malacopterygian fishes."

Pholidophorids.—Near the base of the teleost series stands the Triassic and Jurassic

135

Pholidophoridæ, at the summit stand all the Jurassic and Cretaceous Leptolepidæ, while the Jurassic and Cretaceous Oligopleuridæ are a side branch.

The skull of *Pholidophorus* (Fig. 29) agrees with that of the holostean *Macrosemius* in the marked forward inclination of the suspensorium of the lower jaw, but the mouth is directed partly upward instead of forward, the jaws are longer, teeth more numerous and

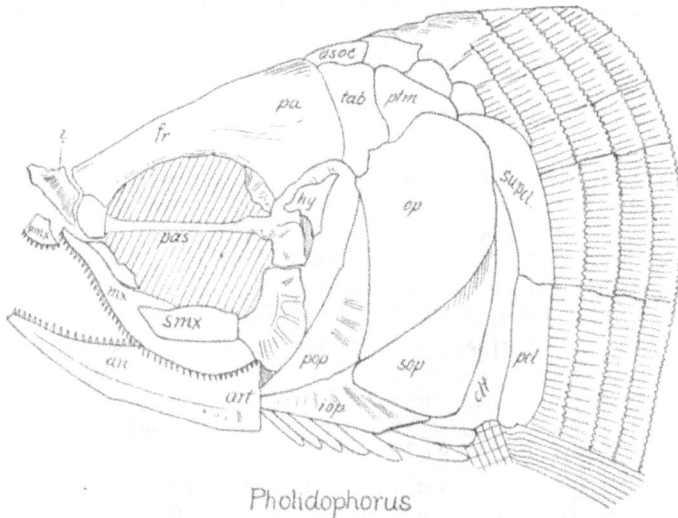

FIG. 29. *Pholidophorus macrocephalus.* After Zittel, from Smith Woodward.

less styloid, the eye much larger; the body as a whole is "elegantly fusiform," while that of *Macrosemius* is more robust and. deep. All these features contribute to the herring-like appearance of *Pholidophorus* and mark its contrast with the far more primitive *Macrosemius*. A further conspicuous agreement with the primitive isospondylous teleosts is seen in the small size of the premaxilla and the convex oral margin of the maxilla, which is loosely attached and bears typically two well developed supramaxillary plates (A. S. Woodward, 1895, p. 446) as in the Elopidæ, Albulidæ, Clupeidæ, etc. The supramaxillæ appear to support Tate Regan's suggestion that these elements in teleosts originally belonged to the suborbital series, from which they were abstracted by the maxillæ. Inspection of A. S. Woodward's figure of *Pholidophorus limbatus* (1895, Pt. III, Pl. XII, Fig. 7) suggests that the maxilla was already free at its posterior end and that the three supramaxillæ were derived from the second or postorbital row of plates rather than from the inner circum-orbital series. The skull top appears to be primitive. According to Watson (1925, p. 866) there were a pair of large tabular bones. The supraoccipital had not yet begun to move forward between the parietals.

Classification of the Isospondyli.—The existing families of isospondylous fishes have branched in many directions and the problem of their phylogenetic relationships with each other and with their Cretaceous forerunners, although essayed by many authors, is still only partly solved.

For the purposes of this study the Isospondyli may be divided into six superfamilies by modifying slightly the arrangement adopted by Tate Regan (1929, p. 313).

Superfamily Clupeoidea: Leptolepidæ, Elopidæ, Albulidæ, Chirocentridæ (Saurodontidæ), Clupeidæ, Ctenothrissidæ, Alepocephalidæ, Chanidæ, Kneriidæ, Phractolæmidæ, Cromeriidæ.

Superfamily Salmonoidea: Salmonidæ, Microstomidæ, Argentinidæ, Opisthoproctidæ, Osmeridæ, Salangidæ, Retropinnatidæ, Haplochitonidæ, Galaxiidæ.

Superfamily Stomiatoidea: Gonostomatidæ (including *Maurolicus, Gonostoma*, etc.), Sternoptychidæ, Astronesthidæ, Chauliodontidæ, Stomiatidæ.

Superfamily Osteoglossoidea: Osteoglossidæ, Pantodontidæ.

Superfamily Mormyroidea: Hyodontidæ, Notopteridæ, Mormyridæ, Gymnarchidæ.

Superfamily Gonorhynchoidea: Gonorhynchidæ.

It is not necessary to define these tentative superfamilies. They are recognized simply for convenience in expressing the apparent interrelationships of this enormously variable series of families.

The researches of G. Allan Frost (1925–1930) on the otoliths of the Neopterygian Fishes add an important and practically new set of criteria for estimating the relationships of the suborders and families of teleosts. Many of Mr. Frost's conclusions are noticed below.

We may now make a brief survey of the skull structure of these groups, supplementing our own very limited material by constant reference to the works chiefly of Smith Woodward and Ridewood for the cranial osteology of the leading fossil and recent types.

CLUPEOIDEA (ALBULIDS, TARPON, HERRINGS, ETC.)

Leptolepids.—In the Jurassic and Cretaceous *Leptolepis* (A. S. Woodward, 1895, p. 501) the isospondyl skull is seen in its most typical and primitive form, without any of the aberrant specializations of later types (Fig. 30). Ganoidean reminiscences, however, are not wanting in the more or less enamelled condition of the delicate membrane bones of the head; the centra, though well ossified, are pierced by the notochord, the scales, though thin, cycloidal and deeply imbricating, usually remain ganoid in structure on their exposed portion. The maxillæ, bearing two prominent supramaxillæ, as in the clupeoid group, are now wholly freed at the posterior end and loosely articulated anteriorly.

The large head and large eye, the somewhat upwardly directed and fairly large mouth bordered with minute teeth, suggest that these fishes were herring-like also in habits, feeding in schools on the plankton of the Jurassic and Cretaceous seas. The broad, rounded preopercular and large circumorbital plates suggest the Elopidæ. According to Frost (1925a, p. 153; 1926, pp. 82, 83), the otoliths of the Elopidæ "are so similar to those of their Jurassic prototypes, the Leptolepidæ, that they confirm in a striking manner Mr. Tate Regan's opinion that these should be placed together as one family." The otolith named by Frost "*Otolithus (Leptolepidarum) rostratus* n. sp." from the Upper Jurassic of England resembles the otoliths of the Elopidæ in general appearance, but differs in certain details.

Elopids.—The Elopidæ, including the "Ten-pounder" (*Elops*) and the tarpon (*Megalops, Tarpon*), are regarded by both Woodward and Ridewood as "the most archaic of existing Teleosteans." *Elops* and *Megalops* both date from the London Clay (Lower

Eocene (Woodward)). The family Elopidæ, as recognized by Woodward (1901, Pt. IV, p. 7; 1907, p. 112), falls into two sections: the first is characterized by the fact that the parietals meet above the supraoccipital, which has, however, already extended forward some distance beneath the parietals. This section includes *Elops*, *Megalops* and several Cretaceous genera. All these Cretaceous genera resemble the existing Elopidæ and each other in their generally primitive isospondyl skeletal characters. They show also the transition from the leptolepid to the isospondyl stage in the evolution of the hypural bones of the tail, three of which in *Spaniodon* are beginning to be expanded. The chief differences

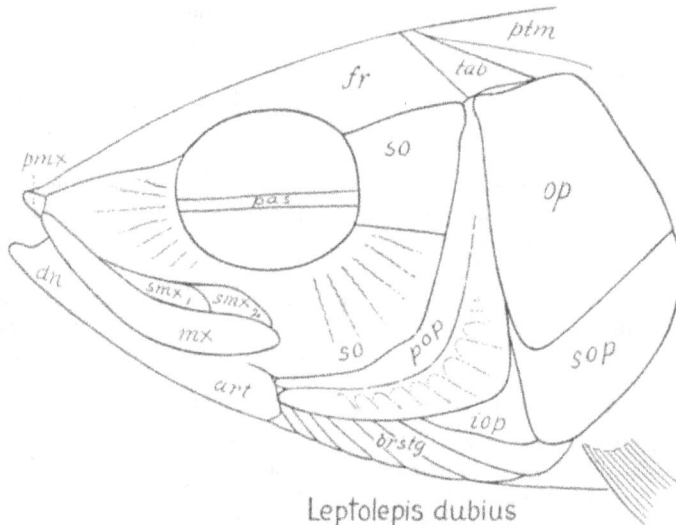

FIG. 30. *Leptolepis dubius*. After Smith Woodward.

among these genera are in the characters of the teeth and jaws: *Elopopsis* was evidently a predaceous form with large mouth and large teeth; *Osmeroides* had minute teeth and smaller mouth. In the existing genera the teeth are very minute and clustered on the margins of the rather large mouth.

In the second section of the family (which does not appear to be either closely related to the first group or very homogeneous in itself) the parietal bones are relatively small and do not meet above in the mid-line but are definitely thrust apart by the supraoccipital. In this division fall the genera *Thrissopater*, *Spaniodon* and several others. In *Thrissopater* the small conical teeth extend right to the distal end of the long and slender maxilla, as they do in the existing anchovy (*Engraulis*) among the clupeoid fishes and in the Gonostomidæ and other deep-sea isospondyls.

The skull of *Elops*, as figured by Ridewood, is more primitive than that of the tarpon in not having a strongly upturned mouth but is otherwise fundamentally similar. The tarpon skull (Fig. 31) is notable for its depth in the opercular region and for the marked uptilting of the nearly edentulous jaws. This is brought about through the relative shortness of the snout and the depth and forward position of the quadrate-articular joint.

The cranium (neurocranium) of a large tarpon (Fig. 32) shows the family characters as well as many primitive isospondyl characters. The cranium as a whole is wedge-like from front to rear, the enlarged dense vomer receiving the thrusts from the upper jaw as well as from the skull-roof. The large orbit is supported anteriorly by the prominent lateral ethmoid (parethmoid), above which is a thin prefrontal. The latter bears a branch of the supraorbital lateral-line canal.

The posterior end of the thin skull-roof is lifted high above the level of the brain tube by the intrusion of the trapezius and the dorsal muscles of the flanks, which extend forward

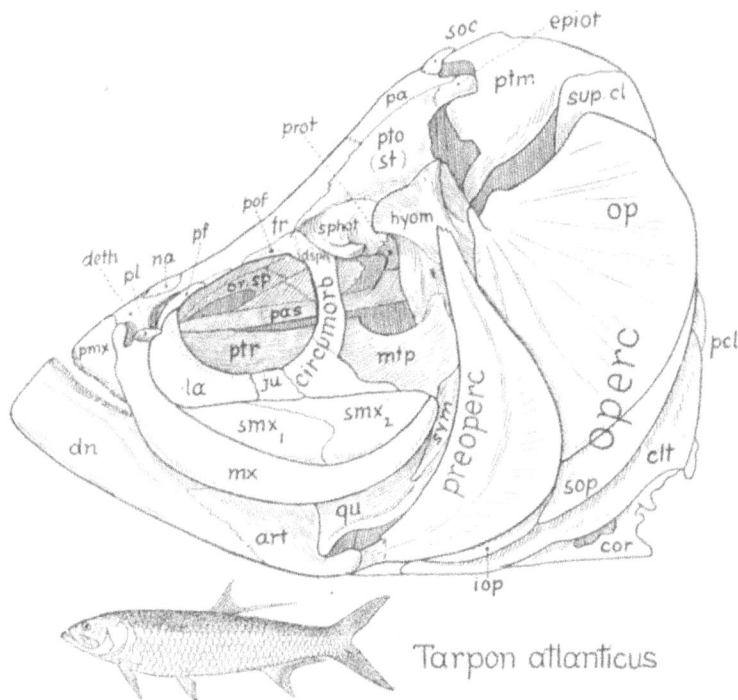

FIG. 31. *Tarpon atlanticus.*

N.B. The "scale bone" (= supratemporal of Owen and Starks and Ridewood = extrascapular in part of Allis) has been removed, exposing the posttemporal and pterotic.

through the large posttemporal fossæ. The skull-roof is supported by vertical plates of the supraoccipital and epiotics, which are prolongations of the otic part of the chondro-cranium. In the tarpon the dorsal plate of the supraoccipital extends forward a short distance beneath the posterior end of the parietals, but is widely separated from the frontals in the top view (Fig. 33). In *Elops*, however, Ridewood (1904a, p. 37) notes that the dorsal supraoccipital extends well forward beneath the posterior parts of the frontals.

The brain trough beneath the skull-roof is stoutly built, the posterior part being flanked by the greatly enlarged otic elements. The ali- and orbito-sphenoids are widened

transversely into broad wings. The basisphenoid is squeezed between the eye muscles
into a narrow but strong median brace, which extends obliquely downward and forward
to support the parasphenoid. Behind this junction the myodome or eye-muscle chamber
is very capacious. It extends back beneath the otic region but does not open posteriorly.

The lateral surfaces of the proötic, pterotic and sphenotic together bear the usual wide
facet for the head of the hyomandibular (Figs. 31, 32). On the inner side the pterotic and

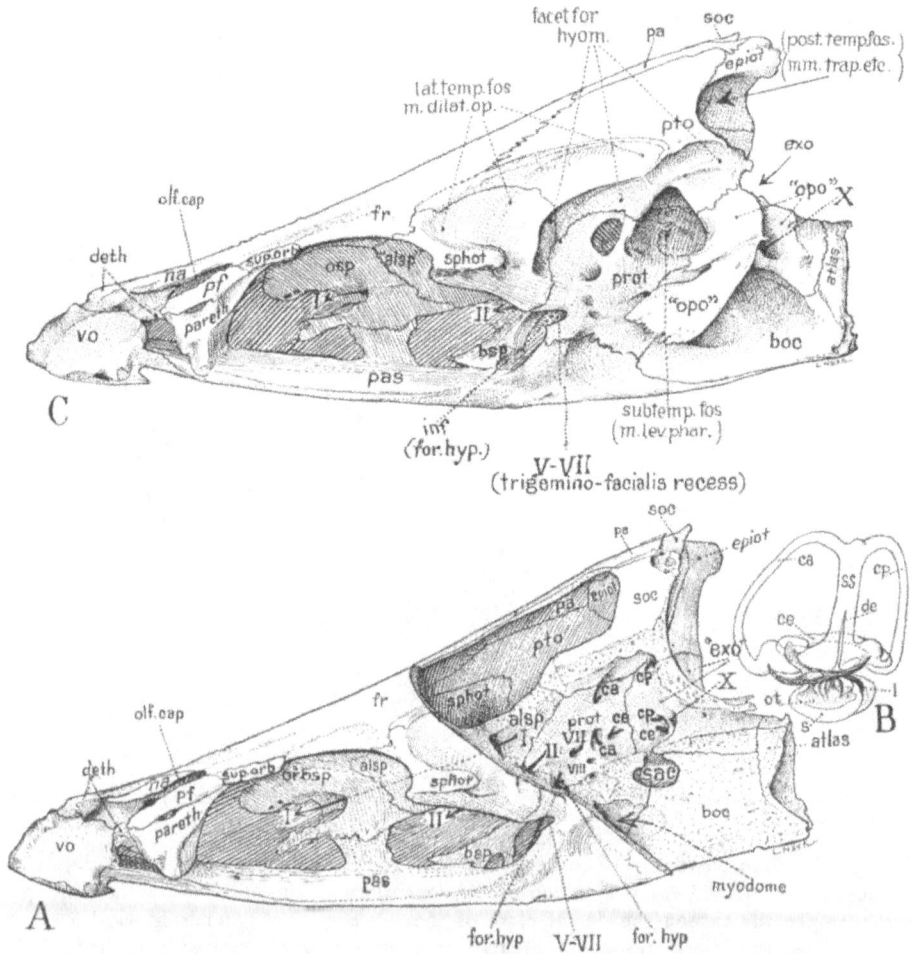

Fig. 32. *Tarpon atlanticus.* Neurocranium.

A. Neurocranium with sagittal section of cranial vault.
B. Diagram of medial aspect of membranous labyrinth, based on data from Retzius.
C. Neurocranium, outer surface.

sphenotic both have the appearance of being endosteal bones, like the adjacent otic and sphenoid elements, but on their dorsal surfaces (Fig. 33) they appear to be of the same texture as the ectosteal parietals and frontals.

On the dorso-lateral surface of the sphenotic is a well marked depression or fossa, called by ·Ridewood (1904a, p. 61) and others the "lateral temporal groove or fossa,"

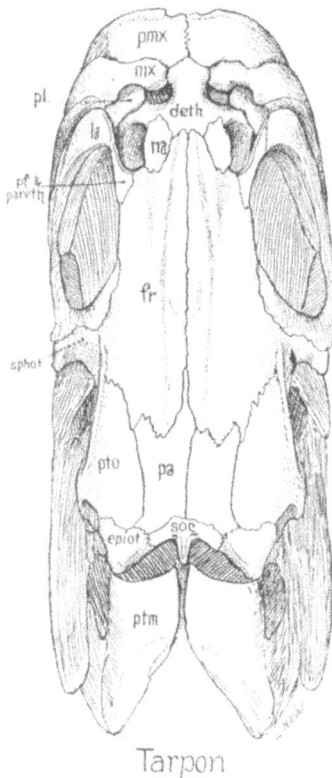

Fig. 33. *Tarpon atlanticus.* Top view.

In this relatively primitive isospondyl the exposed portion of the supraoccipital is small, the parietals are still in contact with the frontals, the mesethmoid (dermethmoid) is small.

The nasals, now well advanced beyond the ganoid stage, have become reduced in size and widely separated across the mid-line.

which gives origin to the dilatator operculi muscle. On the lateral surface of the otic capsule is a deep fossa, bounded by the proötic, the pterotic and the opisthotic, which is called by Ridewood (1904a, p. 62) the "subtemporal fossa." He notes that Sagemehl (1884) found that a similar but larger fossa in the cyprinoid fishes serves for the lodgment of the great muscles which, by pulling up the inferior pharyngeal bones (fifth ceratobranchials), bring the teeth upon those bones forcibly against the callous pad that is carried on the under surface of the basioccipital bone. In the tarpon this fossa, which is likewise prominently developed, may also lodge some of the muscles of the branchial apparatus.

The most massively developed part of the entire cranium is the basioccipital, which serves as a base for the converging lines of stresses from the rest of the skull. The cranial base is further strengthened posteriorly by the incorporation of the first vertebra into the occiput (Fig. 32*A*).

Chirocentrids.—In the modern *Elops* and the tarpons the maxilla is overlapped at its proximal end by a large mallet-like process of the palatine (Figs. 31, 33). This character is emphasized in the existing *Chirocentrus* (Fig. 34) and in the Cretaceous *Ichthyodectes* (Fig. 35), *Portheus* (Fig. 36) and allied genera. These are all referred by Woodward to the family Chirocentridæ (Saurodontidæ), which he regards as being related on the one hand to the older Leptolepidæ and to the Clupeidæ on the other.

The skull of the modern *Chirocentrus dorab* (Fig. 34) has been shown by Ridewood (1904*b*, pp. 448–453, 491–492) to agree closely in many details with the clupeid type. This again tends to strengthen the bonds between the Elopidæ and the Clupeidæ. It is true that the teeth of the Cretaceous chirocentrids (saurodontids) are implanted in distinct sockets in the bone, while those of the modern *Chirocentrus* are merely ankylosed to the

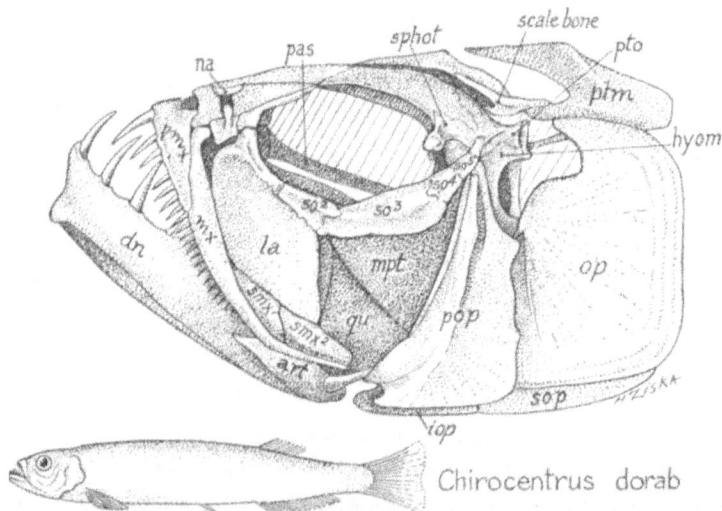

Fig. 34. *Chirocentrus dorab.* After Ridewood.

bone; but a similar difference separates the Cretaceous *Pachyrhizodus* from other genera of the Elopidæ (Woodward, 1901, p. 37). In *Chirocentrus* also this last feature may very well be a specialization and it hardly outweighs the striking resemblances, noted by Boulenger and Woodward, between *Chirocentrus* and the Cretaceous saurodonts. For example, Woodward in his memoir "Fossil Fishes of the English Chalk" (Pt. II, 1903, pp. 93–95) describes and figures the crania of *Chirocentrus dorab* and *Ichthyodectes* sp.,

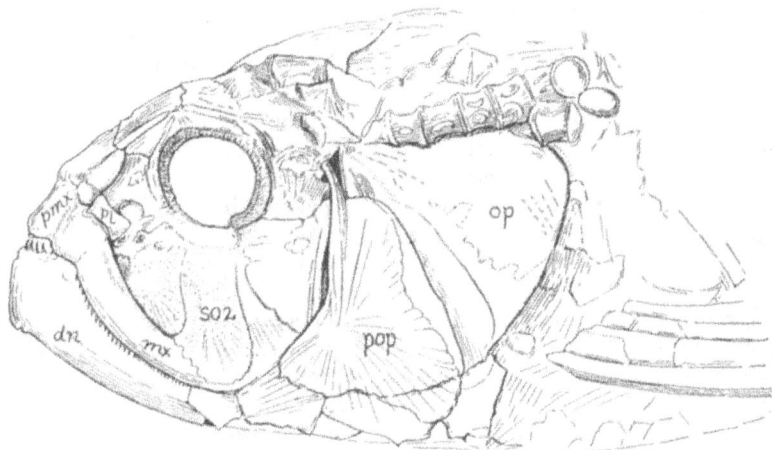

Ichthyodectes serridens

FIG. 35. *Ichthyodectes serridens*, Upper Cretaceous. After Smith Woodward.

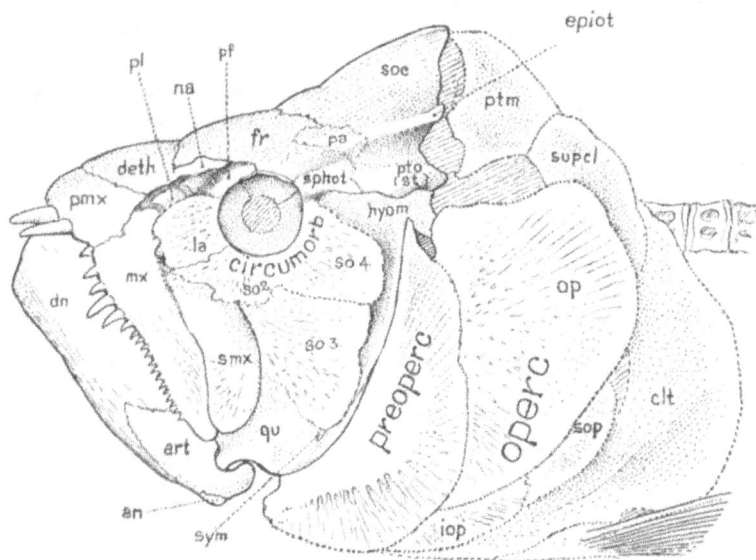

Portheus molossus

FIG. 36. *Portheus molossus*. Composite, based on specimens in the American Museum of Natural History, with certain details from Crook (1892) and from a photograph of a specimen kindly lent by Mr. George F. Sternberg.

showing that apart from minor differences in the proportional development of certain parts the two skulls exhibit the most arresting evidences of close relationship. He notes also (*op. cit.*, Pt. VII, 1911, p. 253), that *Ichthyodectes* belongs among the distinctly synthetic types of Cretaceous genera. "The skull of *Ichthyodectes*," he writes, "is mainly similar to that of the surviving *Chirocentrus*, which belongs in the same or a closely related family; but it differs in exhibiting a pit in the side of the otic region, which is now found, not in the Chirocentridæ, but in the Elopidæ and Clupeidæ."

In *Portheus* the *Ichthyodectes* type becomes of gigantic size. The sharp upturning of the mouth is due to a combination of a short snout with a depressed and anteriorly-placed quadrate-articular joint. The circumorbital bones are large, the supramaxilla exceptionally so. Mr. Sternberg's specimen indicates that the posttemporal was very large. The posterior borders of the bones of the opercular region are not defined and the bones were probably continued into a thin web. The chondrocranium of *Portheus* as described by Hay (1903*a*) was fundamentally similar to that of the tarpon.

Albulids.—The Albulidæ (Fig. 37) are regarded by Woodward as "merely Elopine fishes with a forwardly-inclined suspensorium, a small mouth and reduced branchiostegal apparatus" (1901, p. vi). Ridewood (1904), after extended and intensive comparisons of the skulls of *Albula* and the modern Elopidæ, has shown that the two families agree in possessing many primitive isospondyl characters but have few peculiar specializations in common. *Albula* itself dates back to the Lower Eocene and, according to Woodward (1901, p. 61), is related to the Cretaceous genera *Anogmius* and some others that have small or minute teeth clustered on the margins of the jaws and on the parasphenoid and other bones within the mouth.

The small size of the mouth in *Albula* appears to be a specialization which has involved the reduction of the marginal teeth on both jaws and the enlargement of the premaxillæ, which, as in many more advanced teleosts, have crowded the maxillæ out of the gape. The dwindling of the mouth and the simultaneous elongation of the snout have, as it were, dragged the lower end of the suspensorium forward and with it the attached inter- and pre-operculars. The coronoid process of the dentary rises steeply, as in other short-jawed isospondyls. The lacrymal also has been extended forward in correlation with the marked increase in length of the snout. The surface of the snout is deeply pitted by enlarged organs representing the rostral branch of the lateral-line canal; similar organs have left a raised shelf along the upper suborbital border.

The otolith of *Albula vulpes*, according to Frost (1925*a*, p. 155), is the most aberrant form among the otoliths of Clupeoidea. However, it is approached in a certain peculiar feature by that of *Engraulis mystax*.

The Cretaceous genus *Isteus*, which is placed by Smith Woodward in the Albulidæ, is regarded by him (1901, p. vii) as being "essentially identical with an imperfectly known fish still surviving in the deep sea (*Bathythrissa*)."

"The Cretaceous Clupeoids," writes Smith Woodward (1912, p. 254), "are chiefly of interest on account of their precocious development. They do not differ much from some of the Jurassic Leptolepidæ but it is remarkable that so far back as the Lower Cretaceous, both in Switzerland and in Brazil, some of them had already acquired the row of sharp ventral ridge scales which are so peculiar a feature of the surviving *Clupea* and allied genera."

A related family of Cretaceous clupeoids is the Ctenothrissidæ (see p. 149 below), which had already advanced in the direction of the spiny-finned fishes. "With these specialized Clupeoids there are others," Smith Woodward continues, "of a more generalized grade," including *Crossognathus* and *Syllæmus*.

Clupeids.—The skulls of the modern Clupeidæ (Figs. 38–41), which have been so

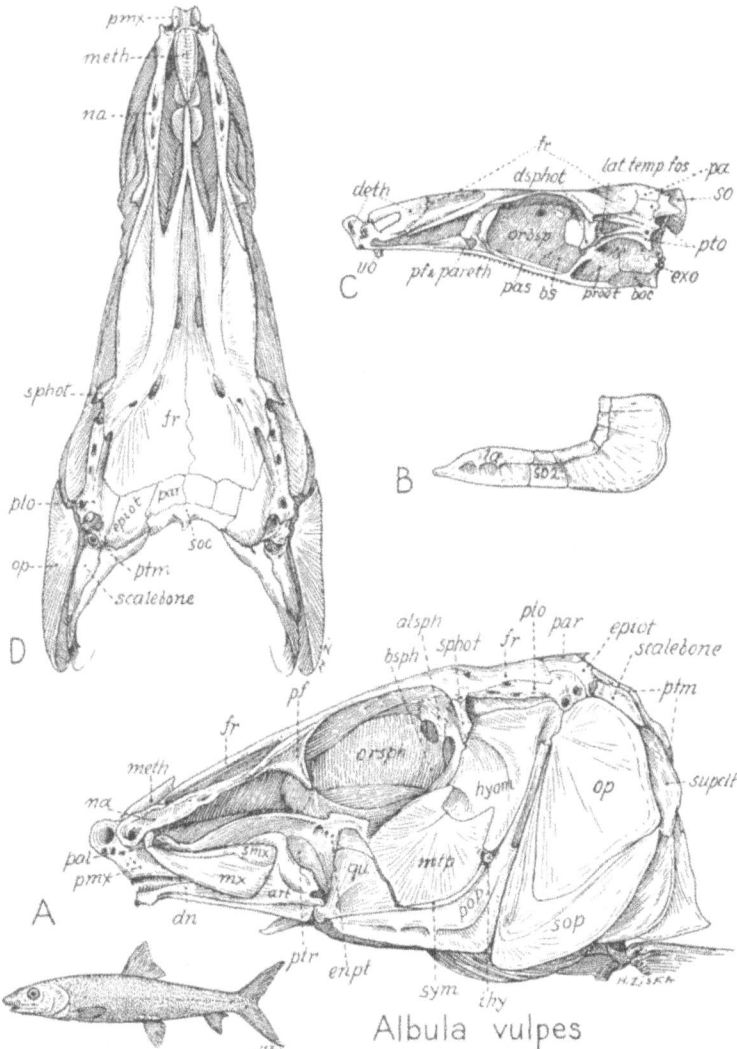

Fig. 37. A, B, D. *Albula vulpes.* C. *Albula conorhynchus.* After Ridewood.

thoroughly investigated by Ridewood (1904a, b), show an extraordinary diversity in general habitus joined with a fundamental identity in family heritage. The typical plankton-feeding genus *Clupea* (Fig. 38) has very delicate teeth with jaws of moderate length; the

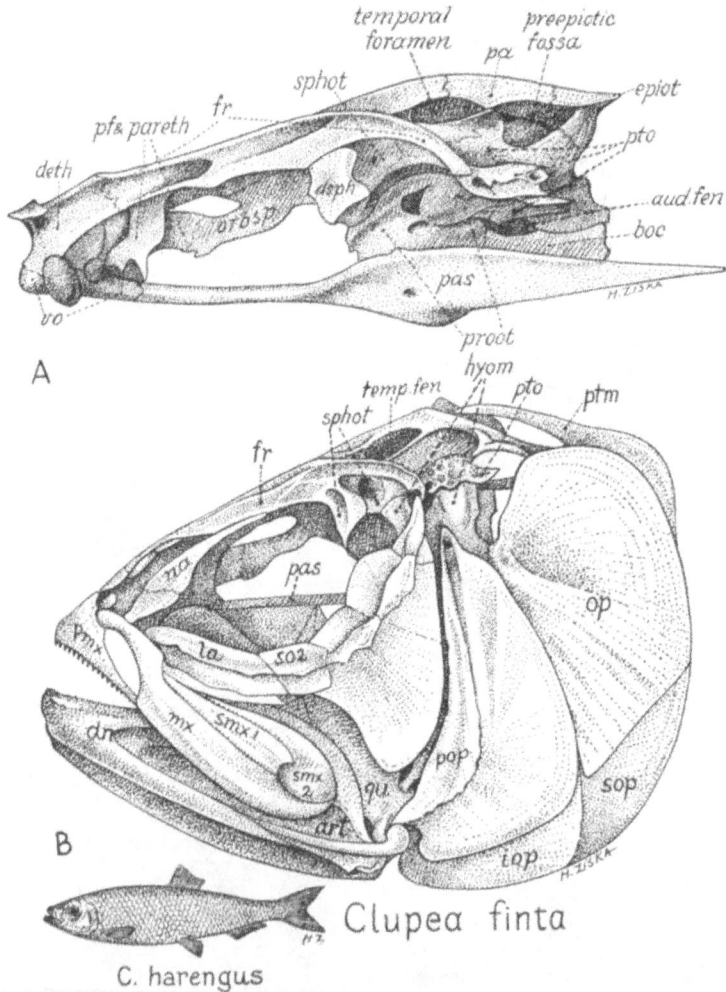

Fig. 38.   *Clupea finta.*   After Ridewood.   A. Neurocranium.   B. Syncranium.

general form of the skull also is normal in appearance. The large maxilla forms the side of the gape. In the predaceous *Chirocentrus* (Fig. 34) (whose skull characters ally it with this family) the strong upturned mandible supports a few very large recurved teeth; in the toothless *Chatoëssus* (Fig. 39) the head is very short and dorso-ventrally deep, the gape very small and bounded above only by the premaxilla, the small maxilla being excluded from the gape. The forwardly-produced suspensorium ends in front of the lower border of the orbit and the mandible is directed sharply upward.

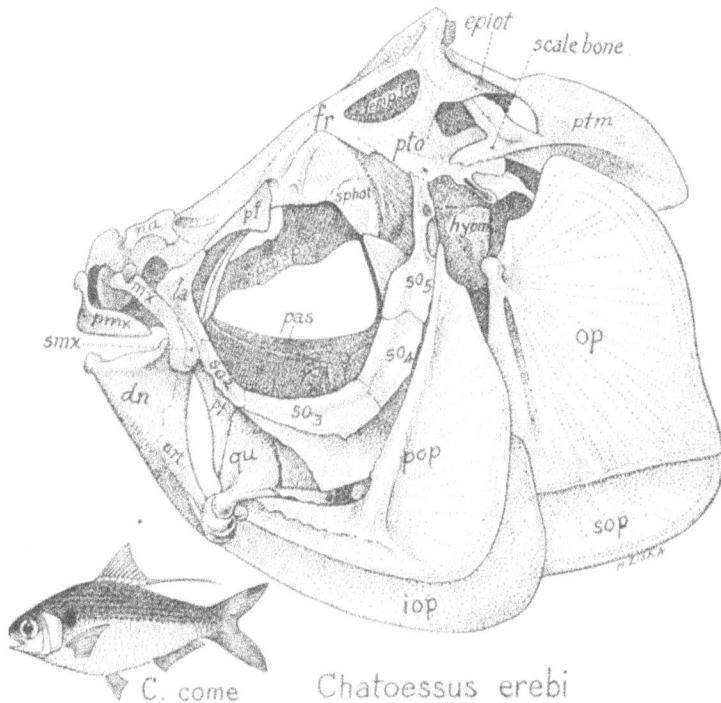

FIG. 39. *Chatoëssus erebi.* After Ridewood.

In *Engraulis* (Fig. 40), on the other hand, the suspensorium is directed obliquely backward, the quadrate-articular joint being well behind the orbit; the jaws are long and slender but wholly below the round protruding snout. In *Dussumieria*, with forwardly-inclined suspensorium and short jaws, the cranium as seen from above is excessively narrow, with very large orbits. In *Coilia* (Fig. 41), with backwardly-inclined suspensorium and long jaws, the maxilla is produced backward far behind the skull into a long, narrow denticulate rod, the orbits are small and the cranium very broad between the orbits. According to Ridewood (1904b, p. 478) the hyomandibular has no articulation with the pterotic but only with the sphenotic,—a most unusual arrangement.

The family heritage beneath this diversity is especially well revealed in the detailed characters of the cranium (Fig. 38A). A few of the outstanding cranial characters recorded by Ridewood (1904a, b) are as follows: (1) in this family the primitive posttemporal fenestræ of the Elopidæ are lacking, or represented by a large groove between the epiotic and the pterotic ("squamosal"); (2) there are cæcal diverticula of the swim-bladder con-

6

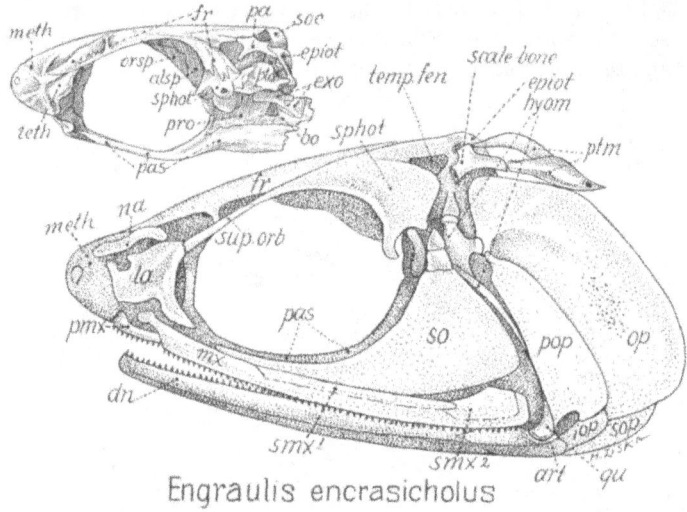

FIG. 40. *Engraulis encrasicholus*. After Ridewood.

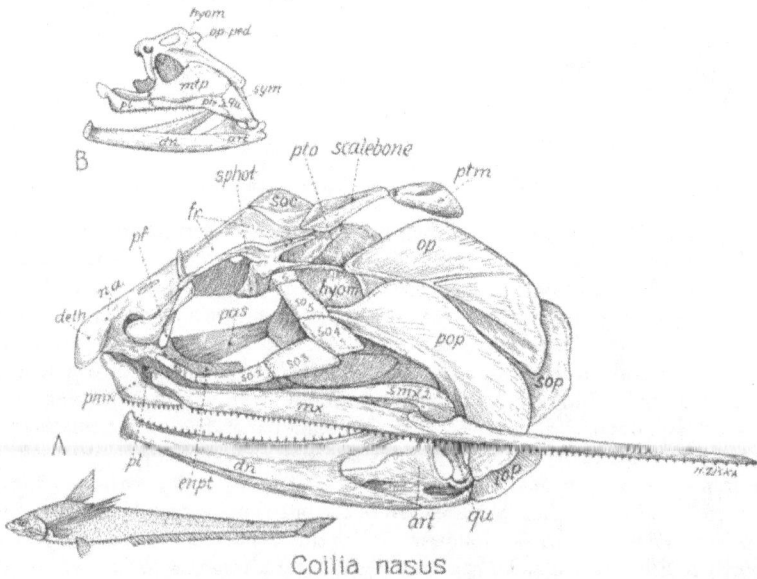

FIG. 41. *Coilia nasus*. A. Side view. B. Inner view of primary jaw. After Ridewood.

tained in the squamosal and proötic bones in many clupeids examined by Ridewood (1904*a*, p. 62); (3) according to the same authority, in many clupeids there is an aperture, the "temporal foramen" in the side of the cranium, bounded by the parietal and frontal bones. "This in life is occupied by a fatty mass and in the dried skull leads directly from the posterior temporal groove to the cavum cranii (1904*a*, p. 61); (4) a short distance behind this is a lateral depression, the "pre-epiotic fossa," situated immediately in front of the epiotic bone and bounded by the parietal, squamosal and epiotic. "The bottom of the depression is composed of cartilage in *Dussumieria* and in *Clupea harengus*. . . ." These details are of importance in the problem of the relationships of the Clupeidæ with other isospondyl families.

Thus the skulls of the Clupeidæ afford numerous examples of what might be called a general principle of the morphology of the vertebrate skeleton, namely, that "the holes are more important than the bones"; that is, the form and position of the bony tracts are largely determined by the form and position of the sensory vesicles, blood-vessels, nerves, muscles, etc.; the strengthening ridges and eminences appear between and around the openings caused by the presence of the various parts mentioned above.

Frost (1925*a*, p. 156) concludes that the otoliths of the Clupeoidea (from which he excludes the Salmonoidea) appear to divide themselves into three groups which he names the "Elopine," the "Clupeid" and the "Engrauline" types.

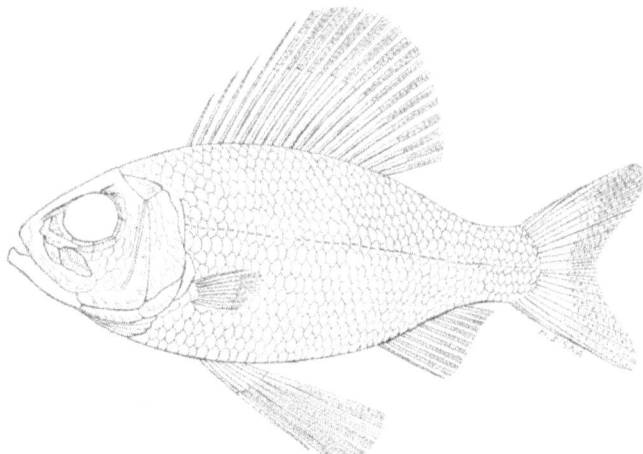

Ctenothrissa radians

FIG. 42. *Ctenothrissa radians.* After Smith Woodward.

Ctenothrissa.—The Upper Cretaceous genus *Ctenothrissa* (Fig. 42), which is the type of a family referred to the clupeoid division of the Isospondyli, is thus referred to by Smith Woodward (1901, pp. vii, viii): ". . . Most of the Cretaceous forms are typical Clupeidæ, and they have scarcely changed during subsequent epochs. A few, however, discovered only in Cretaceous rocks, are of special interest as exhibiting the precocious development of a character which was never permanently acquired by fishes with so primitive a skull, but soon became the common feature of the spiny-finned or acanthopterygian families. These are the Ctenothrissidæ, which have hitherto been mistaken for Berycoids because they display the character in question, namely, the forward displacement of their pelvic fins, which are situated more or less directly beneath the pectoral pair.   The few undivided rays in front of their fins, however, are always articulated distally and never form true spines."   Tate Regan (1907a, p. 642) also concludes that "the Beryciformes [the most primitive of the spiny-finned groups] may have evolved directly from Malacopterygii, such as the Cretaceous *Ctenothrissa* and *Pseudoberyx*, to which they bear considerable resemblance."

Smith Woodward in his paper on "The Antiquity of the Deep Sea Fish Fauna" (1898; also 1912, p. 254) has noted that many of the Cretaceous isospondyls find their nearest relatives today in the deep-sea fauna, and that these modern relicts "probably migrated to the ocean depths during the Tertiary period as the competition from newer types of fishes has increased."   Through the courtesy of Dr. William Beebe I have been enabled to examine the skull structure of the principal types of deep-sea isospondyls.   Our observa-

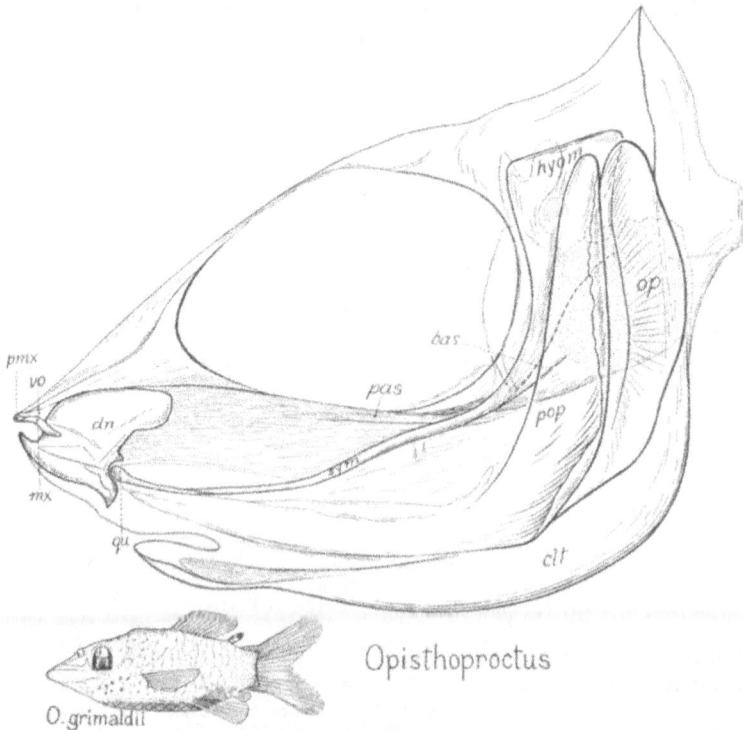

FIG. 43. *Opisthoproctus.*

Diagram of skull of specimen kindly loaned by Dr. William Beebe, who will publish a full description of this rare specimen

tions will be more fully recorded in later papers but for the purposes of the present report the following comments may be made.

Opisthoproctus.—This deep-sea fish (Fig. 43) is strongly reminiscent of the Cretaceous *Ctenothrissa microcephala*, as figured by Smith Woodward (1901, Pl. X). To put it the other way around, the ancient genus approaches the modern in its short deep body, broad caudal peduncle, very large thin cycloid scales, very short deep head with very large orbit, as well as in its forwardly-produced suspensorium, small narrow upturned mouth with very high ascending process of dentary; the preopercular likewise has a boomerang-like elbow and the opercular is deep.

On the other hand, the Cretaceous fish differs from its modern analogue in retaining such primitive characters as two supramaxillæ and in the more normal form of the orbits. A noteworthy difference is the anterior position of the pelvic fins, which lie beneath the cleithra in *Ctenothrissa* but are abdominal in *Opisthoproctus*. *Ctenothrissa* also lacks the flattened abdomen and the pocket lying between the lower border of the mandible and the anteriorly prolonged branches of the cleithra, which is one of the most peculiar features of *Opisthoproctus*. These are obviously specializations, perhaps of relatively late date. An alternate possibility is that *Opisthoproctus*, which retains the adipose dorsal fin that is so characteristic of the Salmonidæ and their allies, may be derived from some small *Argentina*-like form having very large eyes and a small mouth set at the end of a slightly elongate oropharyngeal tunnel.

Chanos.—The peculiar genus *Chanos* (Fig. 44) differs from the Clupeidæ in many

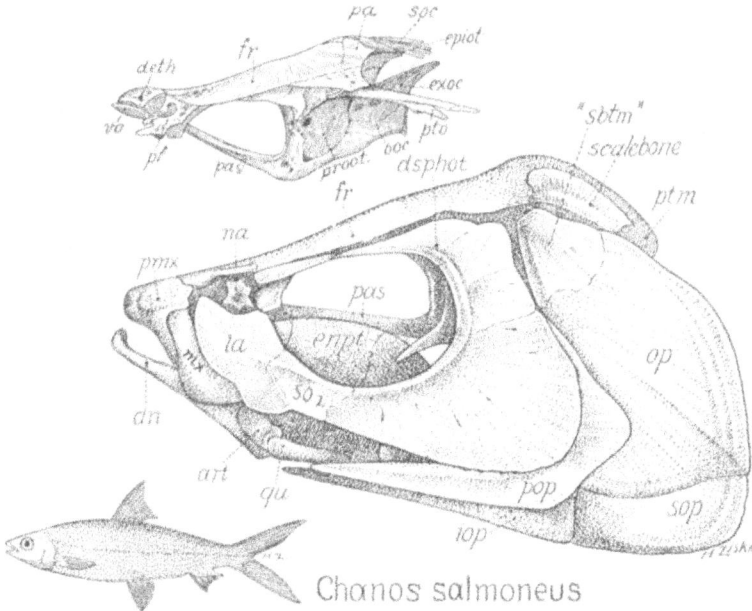

FIG. 44. *Chanos salmoneus.* After Ridewood.

important skull characters noted by Ridewood (1904*b*, pp. 488–493).   It is more primitive than the Clupeidæ in retaining roofed posterior temporal fossæ, a well marked lateral temporal groove and many other features.   Smith Woodward included *Chanos* in the

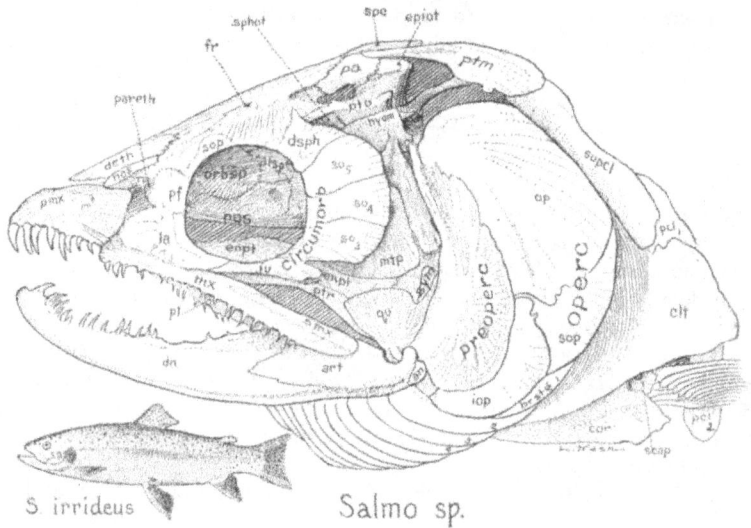

FIG. 45.   *Salmo* sp.   Side view.

family Albulidæ, but Ridewood was not able to confirm this allocation and states that the evidence of the skull favors the view of separating the genus from the Clupeidæ and of according it a family rank.   He notes, however, that it shares a number of points in

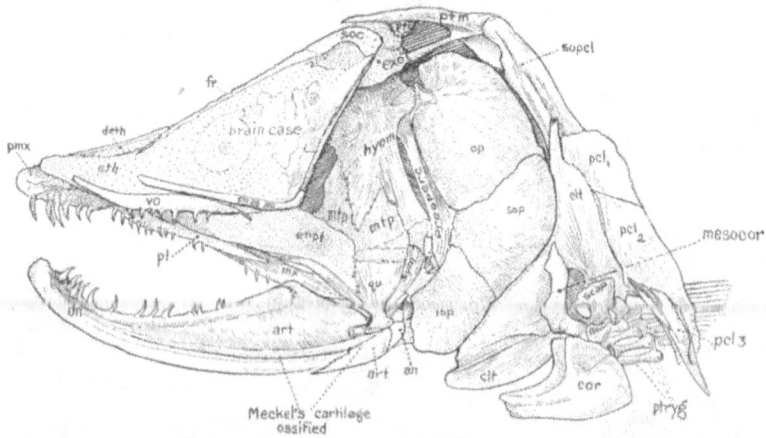

FIG. 46.   *Salmo* sp.   Medial aspect of right half of skull, with sagittal section of front part of braincase.

common with the Albulidæ. In general appearance its skull suggests relationships with the Salmonidæ.

SALMONOIDEA (SALMON, SMELT, ETC.)

Salmonids.—The typical salmon skull (Figs. 45–48) is probably retrogressive or pædogenetic in the high degree to which its endocranium has become cartilaginous; but it is

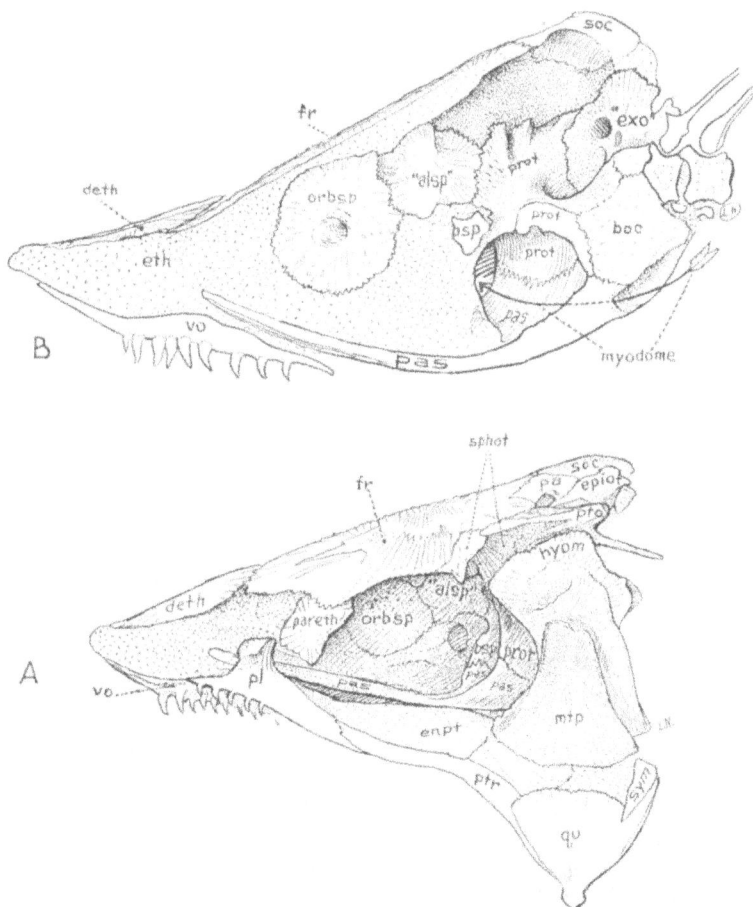

FIG. 47. *Salmo* sp. A. Lateral view of endocranium, with attached suspensorium and primary upper jaw. B. Median sagittal section of endocranium, showing its largely cartilaginous nature.

distinctly progressive in the fact that the supraoccipital is broadly in contact with the frontals and has thrust aside the reduced parietals. Few if any of the numerous highly peculiar features of the clupeid neurocranium are visible in the salmon. The ethmoid cartilage is large, as well as the ectosteal ethmoid ("mesethmoid") above it.

A "subtemporal bone," lying above the opercular, is recorded by Ridewood in the salmon as well as in *Chanos*. Perhaps this bone, which is of unknown origin, is indicated in the Cretaceous elopid genus *Osmeroides* (Woodward, A. S., 1901, Pl. II, Fig. 1). The origin of the "subtemporal bone" is discussed below (p. 166). The maxilla is thin and elongate, recalling that of the Cretaceous *Thrissopater*.

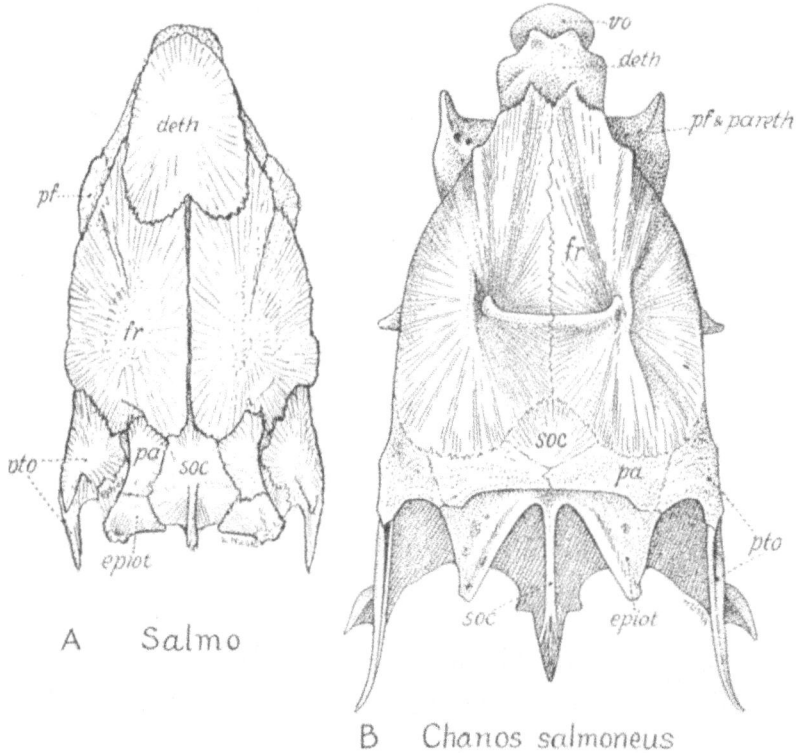

Fig. 48. *Salmo* sp. (A) and *Chanos salmoneus* (B). Comparison of skull tops.

The salmon family (in the broad sense) dates only from the Miocene epoch but Agassiz and subsequent authorities have recognized its relationship with the Clupeidæ.

Starks notes (1926a, p. 150) that in *Argyrosomus hoyi*, representing the Coregonidæ, which are generally admitted to be closely allied with the Salmonidæ, the ethmoid region in general resembles that of the fishes of the family Clupeidæ.

Frost (1925a, p. 157) notes that the Salmonoidea agree with the Clupeoidea in the prominent development of the saccular otolith (sagitta) and show in their otoliths their relationship to the Elopidæ.

In his recent article on the phyletic classification of the teleosts, Professor Garstang (1932, pp. 253, 257, 258) removes the Salmonidæ very widely from the Clupeidæ, assigning them to his first grand division (Haplophysi) of the Teleostei, while the Clupeidæ are referred to the second grand division (Otophysi). This main division rests on the entirely unproved assumption that the cæcal diverticula of the air bladder, which in the Clupeidæ extend forward into the pterotic and proötic bones, are fully homologous with somewhat similar diverticula .found in the Ostariophysi. But this assumption has already been disputed for apparently good reasons by Ridewood (1904c, p. 214). Moreover, numerous figures of the skulls of Clupeidæ, *Chanos*, *Engraulis*, *Salmo*, etc., assembled in the present work, support the more conservative conclusion that in spite of their retention of an adipose dorsal fin the Salmonidæ are a modern offshoot of the old clupeid-elopid stock, which runs back through the leptolepids into early Mesozoic times.

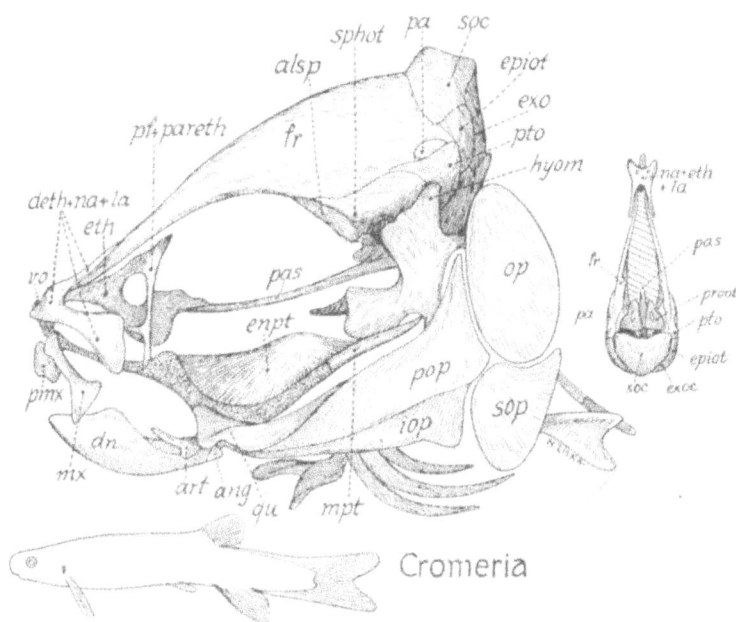

FIG. 49. *Cromeria nilotica.* After Swinnerton.

Cromeria.—This is a scaleless, diminutive fish from the Nile, of uncertain affinities; its osteology has been carefully described by Swinnerton (1903). That it is an isospondyl of some sort is certain, as shown by many features, including the retention of a mesocoracoid arch and of an air bladder and pneumatic duct similar to that of *Salmo;* it also lacks Weberian ossicles (Swinnerton, 1903, p. 59). The general characters of its cranium, jaws and opercular system also conform to the isospondyl type.

Many of the skull characters are highly specialized. The suspensorium extends well forward and the mouth is small and edentulous, with very small premaxillæ and laterally placed maxillæ forming the greater part of the oral border. The hyomandibular has an extraordinarily large anterior process out of all proportion with the reduced metapterygoid. The symplectic is absent. The large lacrymal is fused with the mesethmoid and nasals. Seen from above, the skull is elongate, with greatly swollen, rounded braincase, very suggestive of larval conditions. The elongate frontals are only on the sides of the brain and there is a very large elongate median fontanelle occupied by the swollen brain. The supraoccipital is wide and flat, widely separated from the parietals; parietals much reduced; the alisphenoids are large and the orbitosphenoid absent.

The skull as a whole is extremely unlike that of *Galaxias* and Swinnerton rightly concludes (1903, p. 70) that this fish is not related to the Galaxiidæ (to which it had originally been referred by Boulenger), but is a specialized member of some other offshoot of the malacopterygian stock. Tate Regan (1929, p. 313) puts it near the Chanidæ.

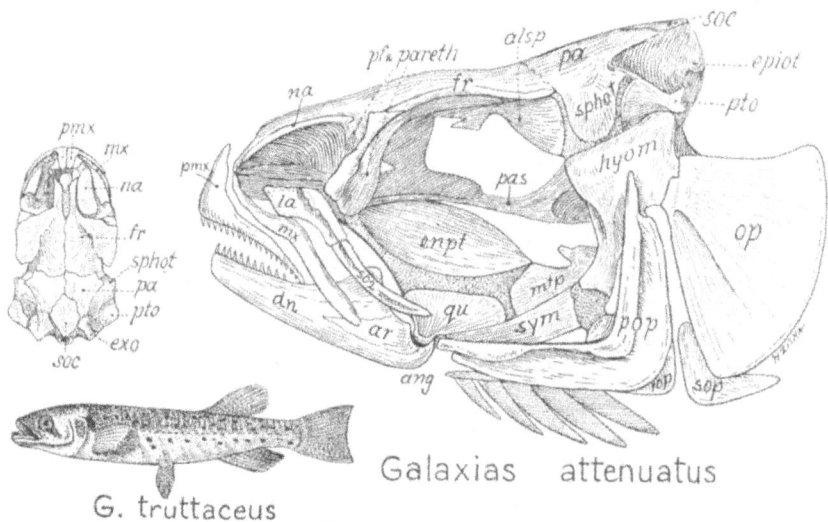

FIG. 50. *Galaxias attenuatus.* After Swinnerton.

Galaxias.—A wide-ranging genus in New Zealand, Australia, South America and South Africa. All but one species (*G. attenuatus*) live in fresh water, but this form is marine and widely distributed in the southern hemisphere. Boulenger (1910, p. 607) interpreted this fact as an indication that the group was originally marine and had invaded the fresh waters from the sea. MacFarlane (1923, p. 360), on the other hand, regards the Galaxiidæ as indigenous to an ancient antarctic hemisphere, which later sank and of which the existing lands of the far south are but fragments. In either case, *Galaxias* (Fig. 50) is on all accounts an important archaic type in the history of fishes.

The osteology of *Galaxias attenuatus* has been described by Swinnerton (1903). The absence of the mesocoracoid arch, the simple unforked shape of the posttemporal, suggest relationship rather with the Iniomi and Haplomi than with the Isospondyli, and Boulenger refers them to his comprehensive order Haplomi.

The cranial table is widened, the flat supraoccipital being well separated from the frontals by the large parietals. The frontals are wide, orbits large, nasals exceptionally large and flat; the prefrontal (parethmoid) with large preorbital buttress and the olfactory fossa unusually large. The jaws are of fair size with conical teeth, the well developed premaxillæ not quite excluding the rod-like maxillæ from the corners of the mouth. The premaxillæ have short ascending processes and the conditions represent the initial phase in the protrusility of the premaxillæ. The suspensorium and opercular series present little that is unusual. The orbitosphenoids are absent (see also Starks, 1908c, p. 414). The dermal mesethmoid as seen from above is circular, much as in *Fundulus*.

Thus the leading characters of *Galaxias* appear to indicate its relationships with the Haplomi and their relatives. Tate Regan, however, in his revised Classification of the Teleostean Fishes (1909a, p. 82) referred the Haplochitonidæ and the Galaxiidæ to the Isospondyli in the following passage (p. 82):

"In some external characters *Retropinna* is intermediate between *Osmerus* and *Prototroctes*. *Retropinna*, *Salanx*, and *Microstoma* are Argentinidæ which have no mesocoracoid. The Argentinidæ, Haplochitonidæ and Galaxiidæ are extremely similar in osteology, dentition, and in the absence of oviducts, and are undoubtedly closely related.

"It is possible to maintain the order Isospondyli, with the addition of the Haplochitonidæ and Galaxiidæ, by taking into consideration the mouth structure, the maxillary entering the gape to a greater or less extent (almost excluded in the Haplochitonidæ) and the unpaired ethmoid. As thus defined, the Haplomi, Iniomi, and Microcyprini are excluded." Here then we have added evidence of the shadowy nature of the boundaries between the Isospondyli, Iniomi, Haplomi and Microcyprini, as long ago noted by Smith Woodward and others.

STOMIATOIDEA

Alepocephalus is placed by Tate Regan next to the Ctenothrissidæ among the clupeoids and it shares with that assemblage the diagnostic character of two supramaxillaries. With the salmons it shares the secondary development of the cartilaginous chondrocranium. It illustrates an early stage of the effect of abyssal life on a branch of the clupeoid stock and might be a structural ancestor of the stomiatoids, at least in many respects (Fig. 51).

In *Maurolicus* and *Ichthyoccus*, which are generally referred to the Gonostomidæ, the photophores are arranged much as in the short-bodied Sternoptychidæ, but the body is of moderate length. *Maurolicus* (Fig. 52C) in fact appears to be the descendant of an ancient common stock which diverged, on the one hand, into such excessively deep-bodied forms as the Sternoptychidæ and, on the other, into the long-bodied Astronesthidæ, Chauliodontidæ and Stomiatidæ. According to Tate Regan (1923b, p. 613) the Gonostomatidæ are near the Elopidæ. "Comparing *Photichthys* with *Elops*," he writes, "I find a striking agreement in the head-skeleton, the general form of the skull and the relations of the bones being almost exactly the same. In *Photichthys* the orbitosphenoid appears to be absent and the posterior temporal fossæ are somewhat smaller than in *Elops*, but there are no other differences of importance."

Smith Woodward notes (1908, p. 138) that in the extinct *Tomognathus mordax* from the English Chalk the skull and dentition are in some respects "suggestive of those of the Stomiatidæ and their allies, which exist in the deep sea." This form, like *Astronesthes*, has a quite short head with the orbit very large and far forward; the jaws are fairly short with very strong pointed teeth in front.

In the narrow and deep-bodied Sternoptychidæ (Fig. 52) the suspensorium is inclined forward progressively as we pass from *Argyropelecus* to *Sternoptyx* to such a degree that the preoperculars are finally lateral to the postero-external part of the huge upturned eyes and the quadrate-articular joint lies below and in front of the orbit.   On the other hand, in *Astronesthes, Gonostoma* (Fig. 53), *Cyclothone* (Fig. 54), and the Stomiatidæ the suspensorium extends backward as in the morays, but as the snout is very short the mouth slopes some- what upward.   This effect is heightened in *Chauliodus* (Fig. 55) by the extreme depression

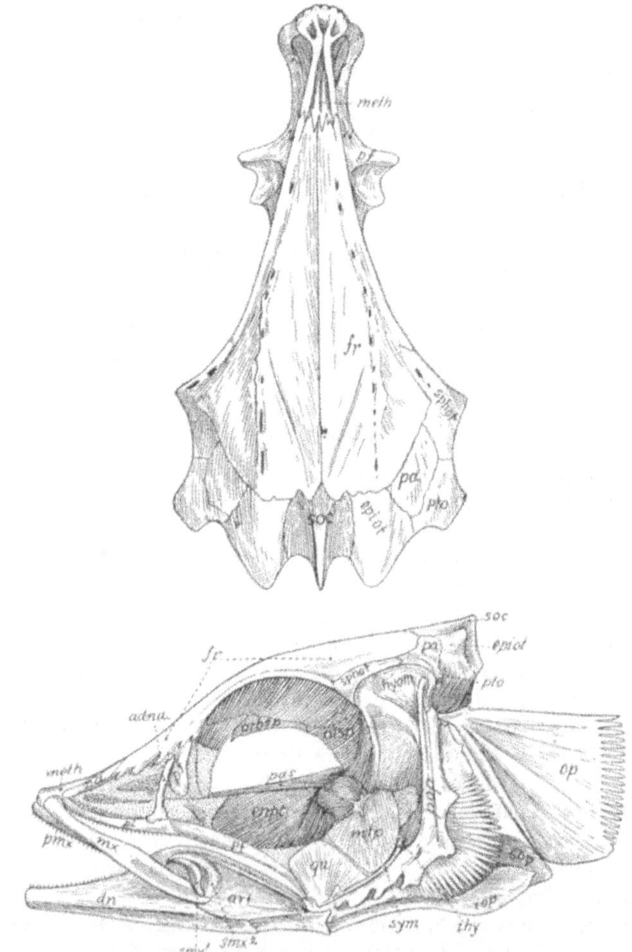

Fig. 51.   *Alepocephalus rostratus*.   After Gegenbaur.

of the quadrate-articular joint almost directly beneath the orbit. In *Cyclothone* the jaws have become a huge expanded trap, with all the long bones reduced to slender bars.

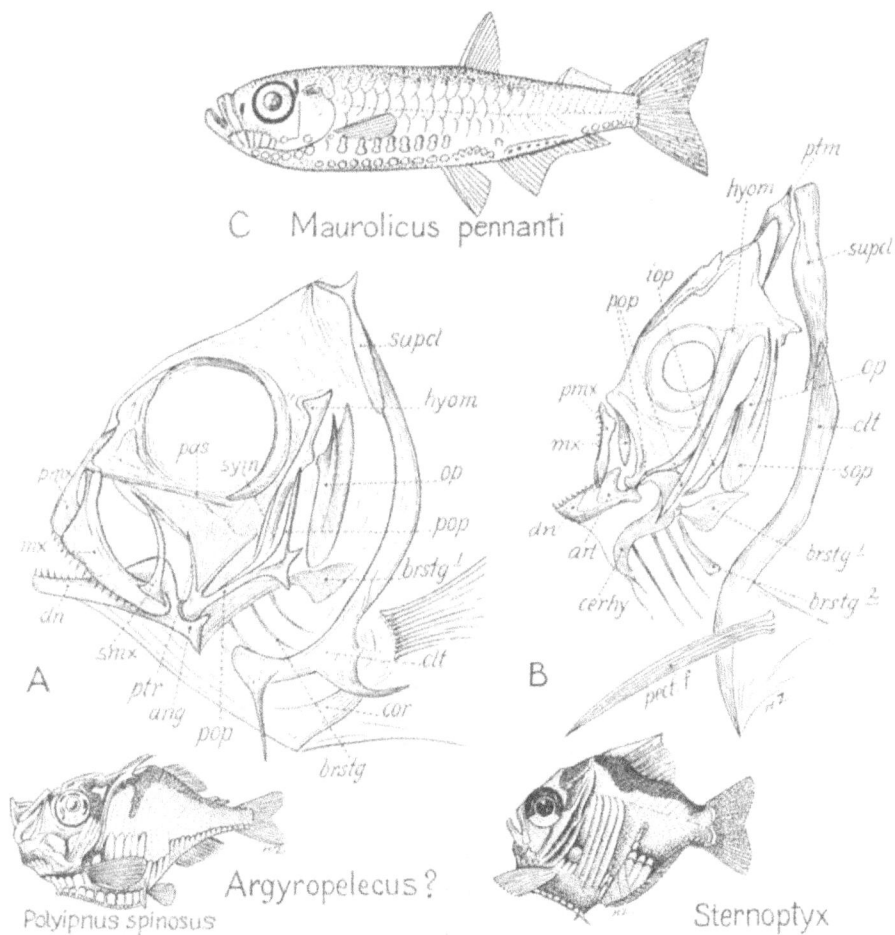

FIG. 52. A. *Argyropelecus* and B. *Sternoptyx*. Enlarged from young specimens cleared and stained by Miss Gloria Hollister for Dr. William Beebe. C. *Maurolicus pennanti*. After Bigelow and Welsh.

Regan and Trewavas (1930, p. 44) show that in certain of the Stomiatidæ, "the loose attachment of the palatine in front and the ectopterygoid behind permit their movement forward and backward, and that of the upper jaw which is rigidly connected with them." In these forms (*Eustomias*) the cervical portion of the column is bent into one or two loops and the vertebræ are more or less replaced by an incompletely ossified region which acst as a shock absorber during the protraction of the jaws in wrestling with large prey (p. 46).

The cranium is strongly built.   In the Chauliodontidæ (Regan and Trewavas, 1929, p. 31) the first vertebra is greatly enlarged and serves as a fulcrum for the head, which in turn carries relatively short jaws with enormously long fangs.

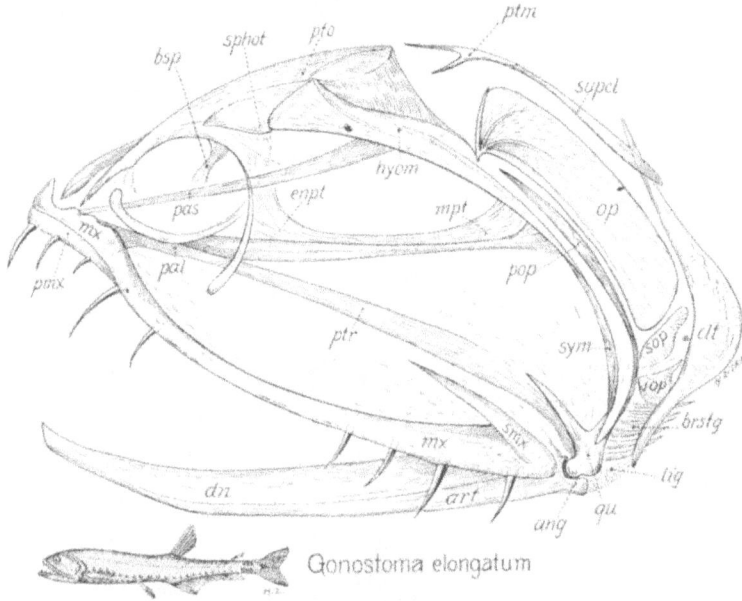

FIG. 53. *Gonostoma elongatum.*   Specimen cleared and stained by Miss Gloria Hollister for Dr. William Beebe.

Parr (1927a, 1930b) has traced the divergent evolution of three suborders as follows: —(1) the Lepidophotodermi, including *Stomias*, which retains scales and has free premaxillæ, and possibly the Chauliodontidæ; (2) the Gymnophotodermi, including the Astronesthidæ and Melanostomiatidæ (these have no scales and the premaxillæ are fastened to the maxillæ, with highly differentiated luminous organs); (3) the Heterophotodermi, including the Gonostomatidæ and the Sternoptychidæ.

In conclusion, while some of the stomiatoid fishes approach certain of the Iniomi in the development of long piercing teeth and of photophores, their retention in the Isospondyli (in the restricted sense) rather than their inclusion in the Iniomi would seem to be justified by the fact that they retain the primitive characters of a dentigerous maxilla, a supramaxilla, a mesocoracoid arch, typically abdominal ventrals and wholly soft-rayed fins. Their connection with the Isospondyli seems also to be supported by the characters of their otoliths, as Frost (1925, pp. 159, 160) notes that the otoliths of *Gonostoma denudatum* of the suborder Stomiatoidea, "resemble in their general lines those of *Argentina*, but are more specialized in some respects. . . . The rostrum of this otolith is more produced than in any other species of the order Isospondyli. . . ."

Garstang (1932, p. 258) unites the stomiatoids with the scopeloids under the name Lampadephori, believing that the common possession of photophores indicates a community of origin of the two groups.   They may indeed have sprung from not distantly related families of Isospondyli, but photophores have also been developed, presumably independently, in *Halosauropsis* of the order Heteromi (see Boulenger, 1904, p. 621). There are indeed, as Garstang observes, some noteworthy resemblances in the upper jaws

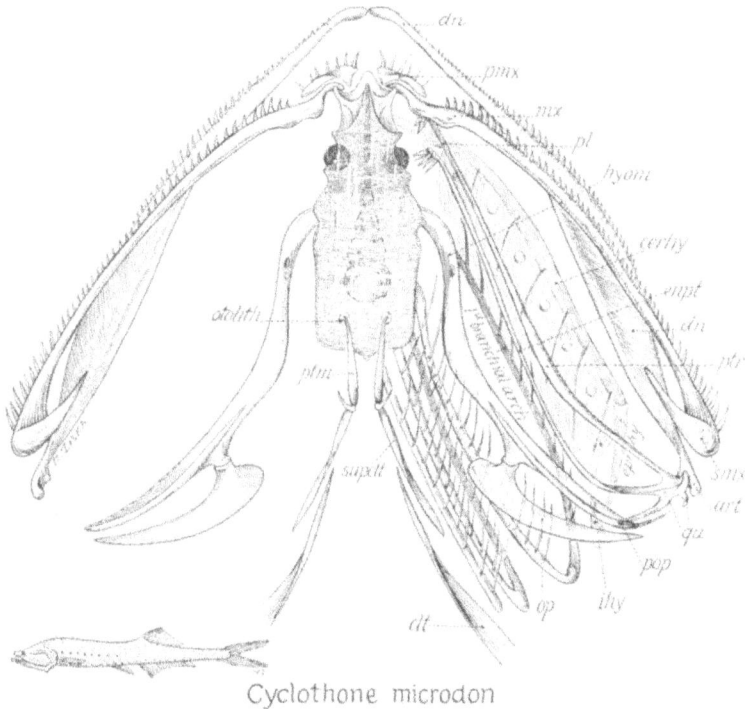

Cyclothone microdon

FIG. 54. *Cyclothone microdon.*   Top view.   Specimen cleared and stained by Miss Gloria Hollister for Dr. William Beebe.

between stomiatoids and Iniomi, but the differences between them, pointed out by Boulenger (1904, pp. 570, 611), still seem to outweigh the resemblances and to indicate that the stomiatoids are an older branch derived perhaps from *Argentina*-like salmonoids, while the Iniomi seem to have sprung from elopine Isospondyli.

Osteoglossoidea

The superfamily consisting of the Osteoglossidæ and the Pantodontidæ at the present time has representatives only in the fresh waters of tropical America, Australia, the East Indies and Africa; but in Lower Eocene times the group apparently had its headquarters in the northern hemisphere, fossil Osteoglossidæ being known from the Eocene of Wyoming and England.   The skulls of the recent representatives combine archaic characters with certain marked specializations, but on the whole the degree of specialization is far less than

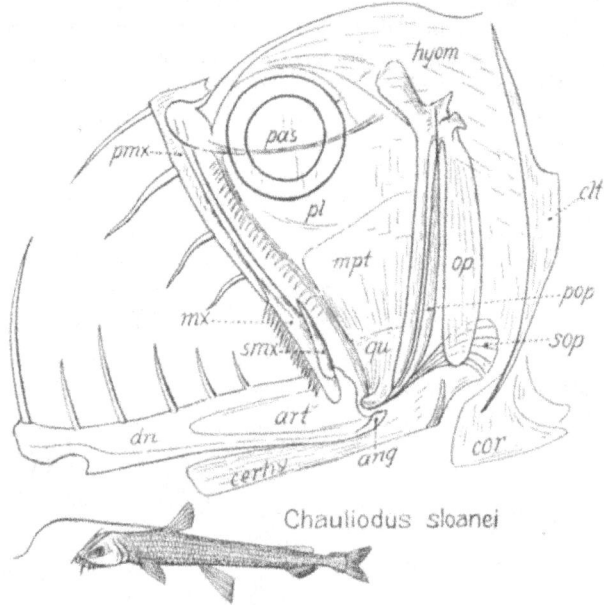

FIG. 55. *Chauliodus sloanei.*
Sketch of young specimen much enlarged.   Cleared and stained by Miss Gloria Hollister for Dr. William Beebe.

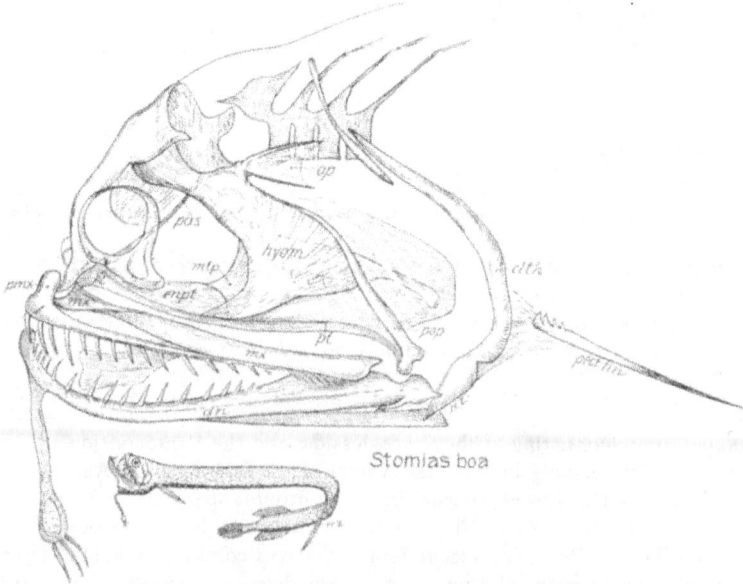

FIG. 56. *Stomias boa.* Sketch of very young specimen much enlarged.   Cleared and stained by Miss Gloria Hollister for Dr. William Beebe.

it is in either the Mormyridæ or the Clupeidæ. The skulls have been very fully described and figured by Ridewood (1905a).

The otolith of *Osteoglossum bicirrhosum*, according to Frost (1925a, p. 159), is pentangular in shape, in this respect resembling the Jurassic fossil *Otolithus* (*Leptolepidarum*) *pentangulatus* and that of *Argentina sphyraena*. The front part of this otolith resembles that of *Salmo trutta*. In another member of the Osteoglossidæ, *Heterotis niloticus*, the otolith is more elopine in character, but with certain differences (p. 159).

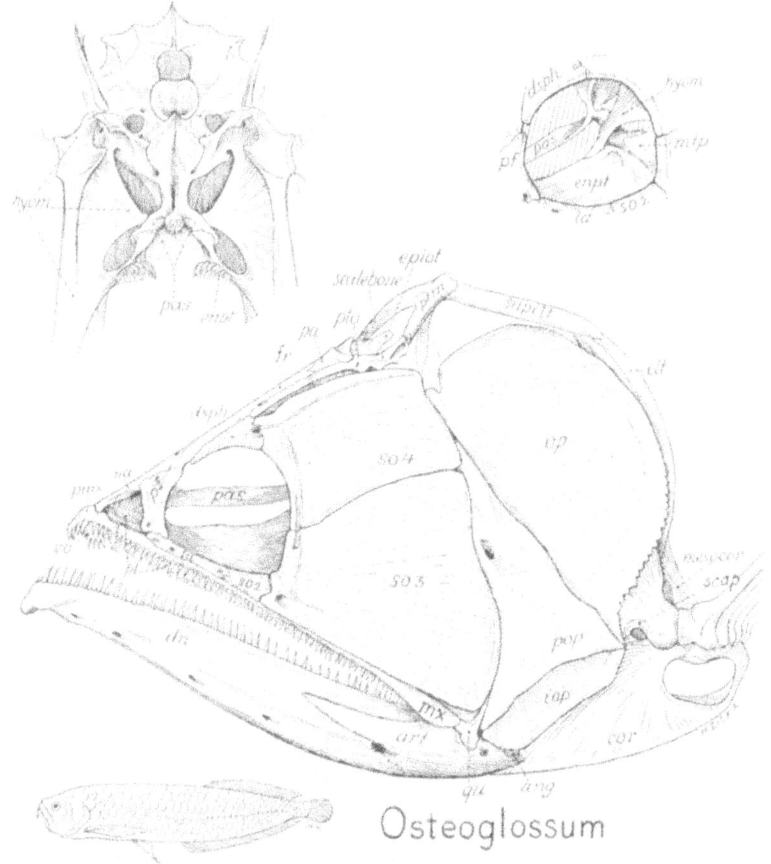

Osteoglossum

FIG. 57. *Osteoglossum.*

In *Osteoglossum*, the three species of which are found respectively in Brazil and Guiana, Borneo and Sumatra, and Queensland, the skull is fairly primitive in external appearance. The surface bones (Fig. 57) are sculptured, the parietals meet in the mid-line, the nasals are well developed, all the opercular elements are present and the occiput retains the

7

paired posttemporal fenestræ. "One of the most striking and characteristic features of the skull of Osteoglossid fishes," writes Ridewood, "is the occurrence of a paired lateral peg of the parasphenoid for articulation with the hyopalatine arch, described by Bridge in *Osteoglossum formosum* (Proc. Zoöl. Soc. London, 1895, pp. 302–310)." Bridge found that "the linear series of obliquely set teeth in the two mesopterygoids of this fish become opposable in the median line of the oral cavity and, in conjunction with the mesial teeth on the parasphenoid, form part of an additional oral masticating mechanism." Bridge compares this arrangement with the somewhat analogous conditions in the palate of *Lepidosteus*, where there is a secondary facet for the metapterygoid, at the junction of processes from the parasphenoid and the proötic. Ridewood found the articulation in question not only in all three species of *Osteoglossum* but also in *Arapaima* and *Heterotis* (Fig. 58). Its presence in *Pantodon* constituted a striking evidence of the relationship of

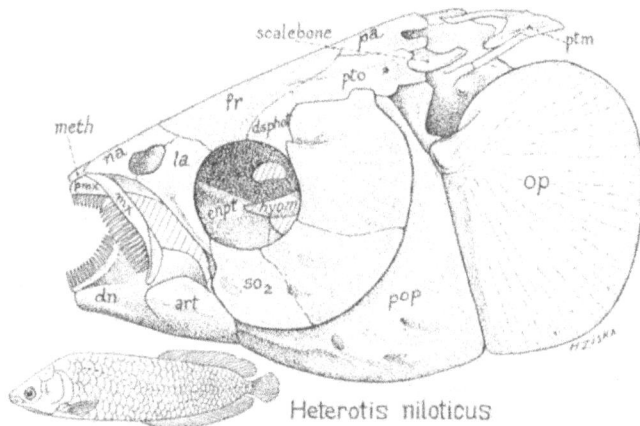

Fig. 58. *Heterotis niloticus.* After Ridewood.

that form to the Osteoglossidæ. The rear view of the skull of *Osteoglossum vandelii* (Fig. 57) shows that these diverging processes from the parasphenoid are also analogous with the basipterygoid processes of primitive amphibians and reptiles. They brace the ascending dentigerous processes of the mesopterygoids and transmit the upward thrusts from the palate to the neurocranium. They are also attached to a pair of long, special and very peculiar processes from the hyomandibulars, the existence of which is difficult to account for. Probably they represent an elongation of that process of the hyomandibular which in other isospondyls is often found above the metapterygoid. We may assume perhaps that as the quadrate-articular joint grew downward, this supra-metapterygoid process of the hyomandibular grew downward and forward, as if to brace the hyomandibular, until it passed over the junction of the ento- (meso-) pterygoid and the ascending processes of the parasphenoid.

Another marked characteristic of all the Osteoglossidæ is the fact, noted by Ridewood, that the nasals are large bones which meet in a long suture in the mid-dorsal line and widely separate the frontals from the mesethmoids, as in *Amia*. Examination of a parasagittal section of an *Arapaima* skull (Fig. 59) shows that even though the frontals extended forward as a thin sheet for some distance beneath the posterior ends of the nasals, they were still widely removed from the small ethmoids, which were lodged in a terminal notch between the nasals and the premaxillæ.   In view of the fact that in all other known primitive

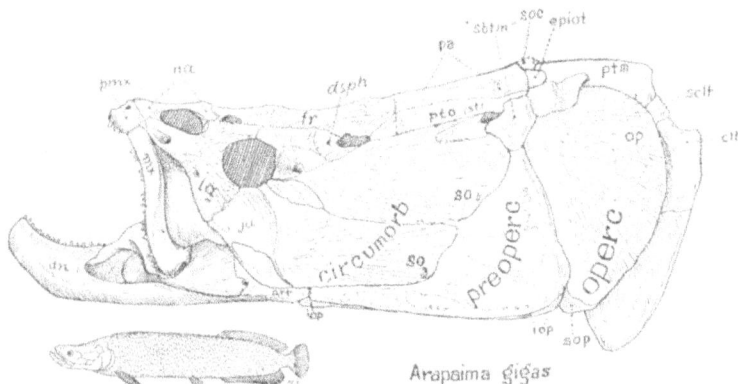

FIG. 59. *Arapaima gigas.*

teleosts the mesethmoids are larger and more important than the nasals, it would seem that the opposite conditions in the osteoglossids, with the consequent resemblances to the early ganoids, is due to convergence, like the sculpturing of the surface bones.   And this inference becomes more probable in view of the many other losses and specializations of the *Arapaima* skull, for instance, the loss of the basisphenoid, of the supramaxillæ, the spreading of the symplectic, the marked reduction of the subopercular and the enlargement of the interopercular, which is seen only in the medial aspect of the preopercular.

The osteoglossids, like other primitive isospondyls, have a well developed "scale bone" (the "supratemporal" of Owen, Ridewood, Starks).   In my specimen of *Arapaima*, however, there are two surface bones (Fig. 59), either one of which might be named the scale bone or lateral extrascapula.   The first lies in the usual position immediately below and in front of the posttemporal; the second lies in front of the first and articulates with the pterotic, with the enlarged fourth suborbital, with the preopercular and the opercular. Comparison with Ridewood's figures of osteoglossids shows that he has applied the name supratemporal, in all three genera, to the second of these two, the first not being shown in the figures.   Comparison with his figures of the supratemporal region of other isospondyls, however, reveals the fact that in *Elops* the large "supratemporal" was nearly divided in two by a deep posterior notch and that in the clupeoid genus *Chanos* there is, in addition to the "supratemporal," a separate bone to which he applies the name "subtemporal," and which has exactly the same position and connections as the above noted "second scale bone" of *Arapaima*.   Ridewood (1904*b*, p. 485) notes that this element in *Chanos* carries a branch of the sensory canal that passes downward to the preopercular; also that this fact "taken in conjunction with the position of the bone below the squamosal [pterotic] and above the preopercular, points to the conclusion that the bone is the homologue of that which, in the Salmon, Parker (Phil. Trans. Roy. Soc., clxiii, 1873, p. 99 and Pl. VI, Fig. 1, *st.*) erroneously called the supratemporal." In the Eocene species of the genera *Megalops* and *Notelops*, Smith Woodward states that the "operculum is subdivided by a transverse suture" (1901, pp. 24, 27) and he notes the presence in a certain specimen of *Notelops brama* of a "separate plate above the operculum" (p. 28). Hence it seems not improbable that in primitive isospondyls a "subtemporal" was present either as a separate element or as the lower and anterior moiety of the "supratemporal" scale bone or tabular.

With regard to the relationships of the family with other Cretaceous isospondyls, Smith Woodward concluded that the "Osteoglossidæ with a curiously thickened skull, also seem to be closely related to the early Albulidæ" (1901, p. vii). Several Cretaceous and Eocene genera having small teeth on the parasphenoid and other bones within the mouth may possibly mark the transition from the Albulidæ to the Osteoglossidæ. The several species of *Plethodus*, for example, bore concave dental plates on the base of the cranium, which were opposed to a similar but convex plate supported probably by the basihyal bone. This at least suggests the prominent patch of teeth on the parasphenoid of *Arapaima*, which were opposed to the lingual teeth. Bridge (1895) inferred that in *Osteoglossum* there was a lateral movement of the teeth on the mesial edges of the two entopterygoids, but Ridewood (1905*a*), referring to the fact that although these upper teeth are obliquely set, their tips point downward, concludes that "There can be no question that these teeth act in a vertical direction and are opposed to the lingual teeth borne upon the bone that covers the glossohyal cartilage and the basibranchials." Thus in certain Cretaceous Albulidæ and Osteoglossidæ, as well as in their modern relatives, there was a dental apparatus within the mouth analogous in some respects with those which have been developed independently and in various ways, in pycnodonts, wrasses, synentognaths, etc.

Returning to the consideration of Ridewood's analysis of the skull characters of the Osteoglossidæ, he did not find any very conspicuous specializations peculiar to osteoglossids and albulids apart from the tendency to develop a masticatory apparatus within the mouth, as above noted. Nevertheless he remarks, in discussing the various arrangements of the families of malacopterygian fishes proposed by Günther, Gill, Smith Woodward, Boulenger, that "I should be disposed to associate the Osteoglossidæ with the Pantodontidæ for the reasons given on page 276, and to regard the next nearest family to be the Albulidæ. The conclusion is arrived at by a consideration of the craniological features mainly, but the characters of the other parts of the skeleton and of the soft parts of the body, so far as they are known to me, do not militate against the suggestion that the Osteoglossidæ and the Albulidæ have descended from a common stock."

Garstang (1932, p. 245) considers that the "feeding mechanism of parasphenoidal and hyoidean teeth" in the osteoglossoids constitutes an important link with the elopines, that their skull and jaw characters entitle them to membership in the "archicraniote" section of the "Otophysi" (pp. 253, 256), and that the presence of air vesicles in the temporal fossæ link them with the Ostariophysi in the grand division Otophysi. I, however, am now rather more impressed by the differences from the albulids and elopines, even in essential details of the median teeth on the floor of the branchial region; while the total absence of Weberian ossicles, together with the lack of essential resemblances to the characins even in the Eocene osteoglossids, indicate that the connection with the Ostariophysi is, at best, very remote.

MORMYROIDEA

In the third superfamily, Mormyroidea, specialization and instability in certain features finally result in bizarre forms of body and skull.

Hyodon.—As described by Ridewood (1904c, pp. 206–210), the least aberrant is the genus *Hyodon* of North America, which retains fairly normal toothed jaws and body form.

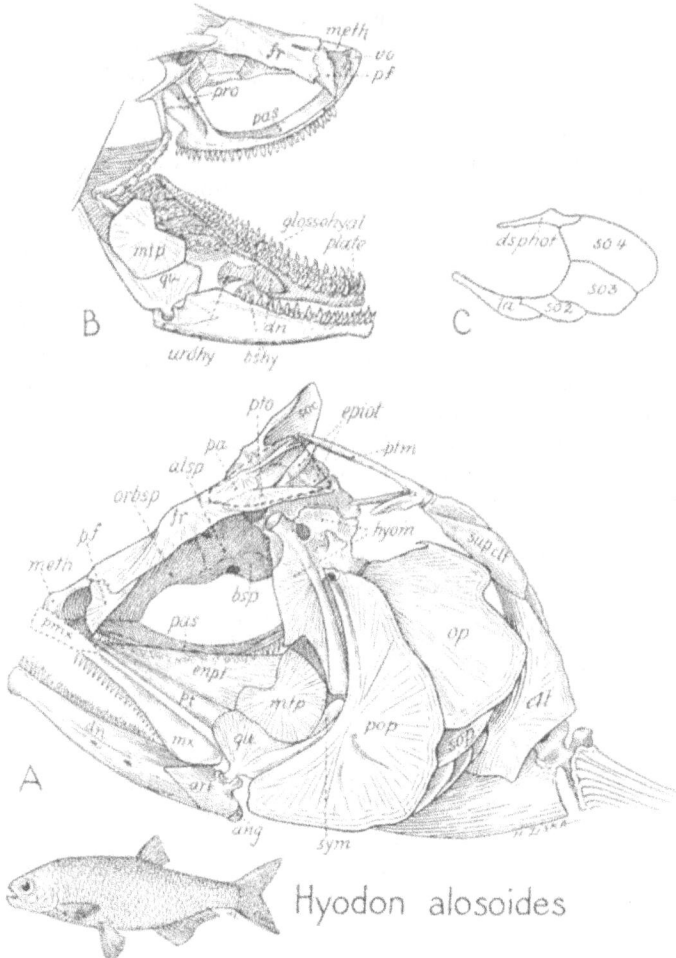

FIG. 60. *Hyodon alosoides.* After Ridewood.

The most conspicuous specialization of the rather short skull (Fig. 60) is the relatively prominent dentition on the narrow parasphenoid, which is bowed downward far below its normal level, perhaps in order to bring its teeth near to those on the tongue and floor of the mouth. The skull top is short and fairly broad, with a high projecting supraoccipital crest, which is flattened and thick on top. The broad supraoccipital base has grown forward, thrusting apart the parietals, or overgrowing them, in the rear; but they still retain a short contact in the mid-line in front of the supraoccipital, separating the latter from the broad frontals. In Ridewood's specimen but not in mine the large "supra-temporal" or scale bone covered over an oval tract of cartilage corresponding perhaps with the pre-epiotic fossa of clupeids, and perhaps related morphologically to the lateral cranial foramen of mormyrids. At the side of the base of the cranium and below the level of the horizontal ridge on the pterotic and opisthotic is a great vesicle of the swim-bladder. "Its outer wall," writes Ridewood, "is composed of fibrous tissue, which is attached to the cranium along the line marked with dots in Fig. 20. Its inner wall is formed by the exoccipital and basioccipital and its anterior wall is formed by a vertical lamina of the pro-otic. Between the exoccipital, basioccipital and pro-otic is a fairly large auditory fenestra, opening into the perilymphatic cavity and traversed vertically by the pro-otic lamina just mentioned." Ridewood points out that the auditory fenestra is a clupeoid feature and that its occurrence in *Hyodon* is of some interest.

The circumorbital plates are fairly normal but the interopercular is concealed in the lateral view by the broad preopercular. In contrast with the Osteoglossidæ, the nasal bones are very slender and laterally placed, the prominent mesethmoid being in contact with the frontals. There are no posttemporal fenestræ, no supramaxilla, no angular bone in the mandible. A peculiar feature is that the ectosteal articular remains suturally distinct from the endosteal or true articular (Ridewood, 1904c, p. 208).

At first sight the skull of *Hyodon* suggests relationships with that of *Osteoglossum*. In both the circumorbital series is reduced in front of the orbit and much enlarged behind it. Both lack the supramaxillæ and have downwardly projecting teeth on the parasphenoid. In both the interopercular is concealed from the outer side by the preopercular. But along with these and other resemblances there are many differences: the marked reduction of the nasals in *Hyodon*, the development of prominent pegs on the parasphenoid for articulation with the entopterygoid in *Osteoglossum*, and so forth.

Notwithstanding these and other differences noted by Ridewood, the construction of the pectoral girdle in *Hyodon* shows the most unmistakable marks of affinity with that of *Osteoglossum*.

Notopterus.—This genus affords a fine example of mutual adjustment of skull form and body form. In *Notopterus chitala* (Fig. 61) the concave dorsal contour of the skull rises steeply toward the high hump on the back, the great size of which must aid in the balance of the body when propelled by the undulations of the much elongated anal fin. The rear of the skull has thus increased greatly in height. As the orbits have moved forward to the front of the head, while the hyomandibular has as usual retained its articulation at the back part of the lateral wall of the braincase, the middle part of the skull behind the orbits has become much elongated, more so in this species than in *Notopterus kapirat*, as figured by Ridewood.

In top view the skull is long and narrow, surmounted in its posterior half by several crests that mark the boundaries between forwardly-extended strips of body muscles. The long supraoccipital is separated from the frontals by the short parietals.

In the side view, primitive isospondyl features, such as the extension of the lateral teeth to near the posterior end of the maxilla, the normal appearance of the circumorbital plates, the presence of an orbitosphenoid and the retention of a symplectic, exist side by side with specializations and losses, such as the close connection of the quadrate with the

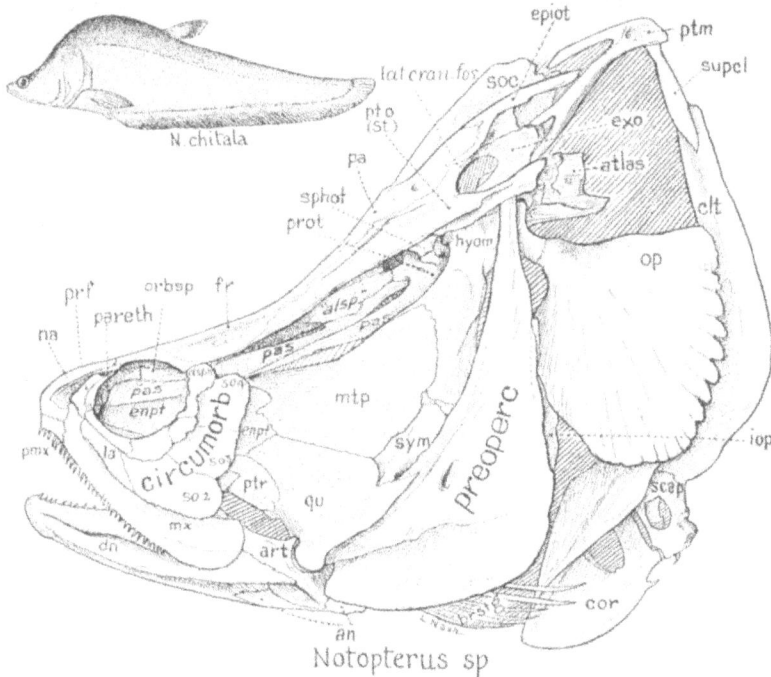

Fig. 61. *Notopterus* sp.

preopercular by means of two serrated ridges, the incipient reduction of the opercular, which is membranous around the margin, and the loss of the subopercular.

Above the pterotic is a large "lateral cranial foramen," leading in the dried skull directly into the cranial cavity. This is a marked point of resemblance to the mormyrids, in which this foramen is occupied by a thick-walled vesicle (Ridewood, 1904c, p. 189). "The base of the skull is inflated," writes Ridewood (*op. cit.*, pp. 202, 203), "the bulla being formed by the pro-otic and basioccipital at the side and by the posterior end of the parasphenoid below. Behind the bony swelling is a ventro-lateral vacuity bounded above by the opisthotic and pro-otic, and internally by the basioccipital. This vacuity lodges the inner and upper portion of a rather large air-vesicle, the outer and lower walls of which are fibrous, and are consequently wanting in a macerated skull. The anatomy of this diverticulum of the swim-bladder has been minutely described by Bridge [1900], who terms it the 'auditory cæcum.'" Thus the notopterid air-bladder, like that of the mormyrids, clupeids, cyprinids and other malacopterygian groups, gives the impression of having sent out various exploratory diverticula which in the different families succeeded in penetrating the auditory chamber by different routes.

Ridewood's thorough studies indicate that the notopterids, while constituting a specialized side branch, have come off from near the stem of the mormyrids.   On the other hand Frost (1925, p. 160) has shown that in *Notopterus kapirat* the very peculiar swollen otolith with a styliform rostrum is very unlike those of the typical clupeoids but might be derived from a type similar to that of *Gonostoma denudatum.*

Mormyrids.—The mormyrids of tropical Africa are well known for their remarkable and peculiar specializations.   The least specialized mormyrid skull is that of *Mormyrops*

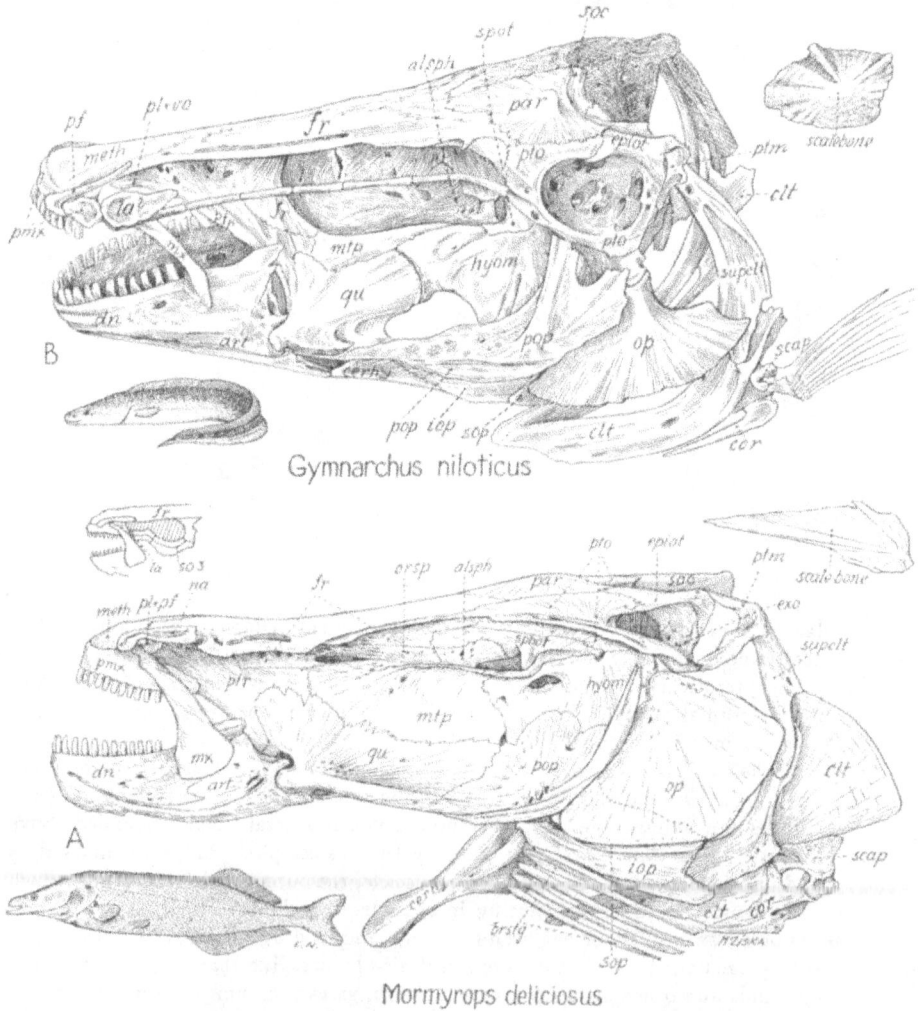

Gymnarchus niloticus

Mormyrops deliciosus

FIG. 62.   A. *Mormyrops deliciosus.*   B. *Gymnarchus niloticus.*   After Hyrtl.

(Fig. 62), in which the middle part of the skull is already well elongated. The stout hyopalatine arches articulate closely with the lower lateral borders of the neurocranium, concealing the parasphenoid in the side view of the skull. The symplectic has been lost and the broad preopercular articulates by a long irregular suture with the wide hyomandibular. The stout opposite premaxillæ are fused at the proximal end. They are placed horizontally and bear small teeth. The stout curved maxilla is toothless and is displaced to the outer angle of the gape. The dentary is rather short, thick, and must be capable of giving a strong bite, since it is moved by powerful adductor mandibulæ muscles, which have an extensive area of origin on the elongated tract behind the mandible. Teeth are wanting from the vomer, palatine and pterygoid bones. The subopercular is hidden beneath the opercular and the interopercular is produced into a long narrow rod. The large lateral cranial foramen (for the reception of a vesicle that penetrates the cranial cavity) is covered laterally by a thin scale bone ("supratemporal," extrascapular).

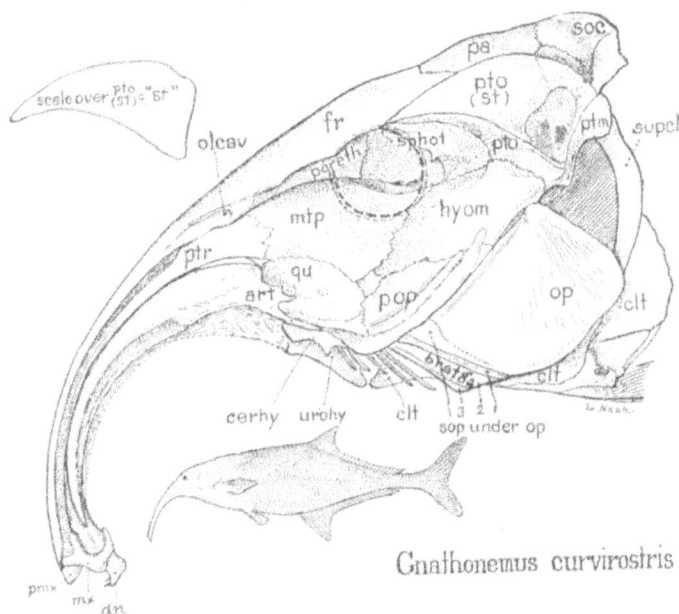

FIG. 63. *Gnathonemus curvirostris.*

In *Gnathonemus curvirostris* (Fig. 63) the typical mormyrid specializations are much further emphasized. The skull could be derived from that of *Mormyrops* by the inflation of the braincase and by the pulling out of the ethmoid, pterygoid and articular bones into a long decurved trunk. The terminal mouth has become minute but is doubtless capable of acting like powerful small pincers.

In *Petrocephalus* (Fig. 64) it would seem that mormyrid specialization had reached its present limit. As figured by Ridewood, the skull appears to me to have been derived from a long-skulled mormyrid type by a relatively rapid shortening of an already down-turned "proboscis." This hypothesis would account for the following otherwise inexplicable facts: (1) the nasal bones and mesethmoid of *Petrocephalus* are turned very sharply downward as they would be if derived from those of more typical mormyrids; (2) the form and detailed relations of the premaxillæ, maxillæ and dentary strongly suggest the conditions in typical mormyrids and the same is true of the form and relations of the opercular, preopercular and hyopalatine series; (3) the hypothesis of a trunk curved convexly in front

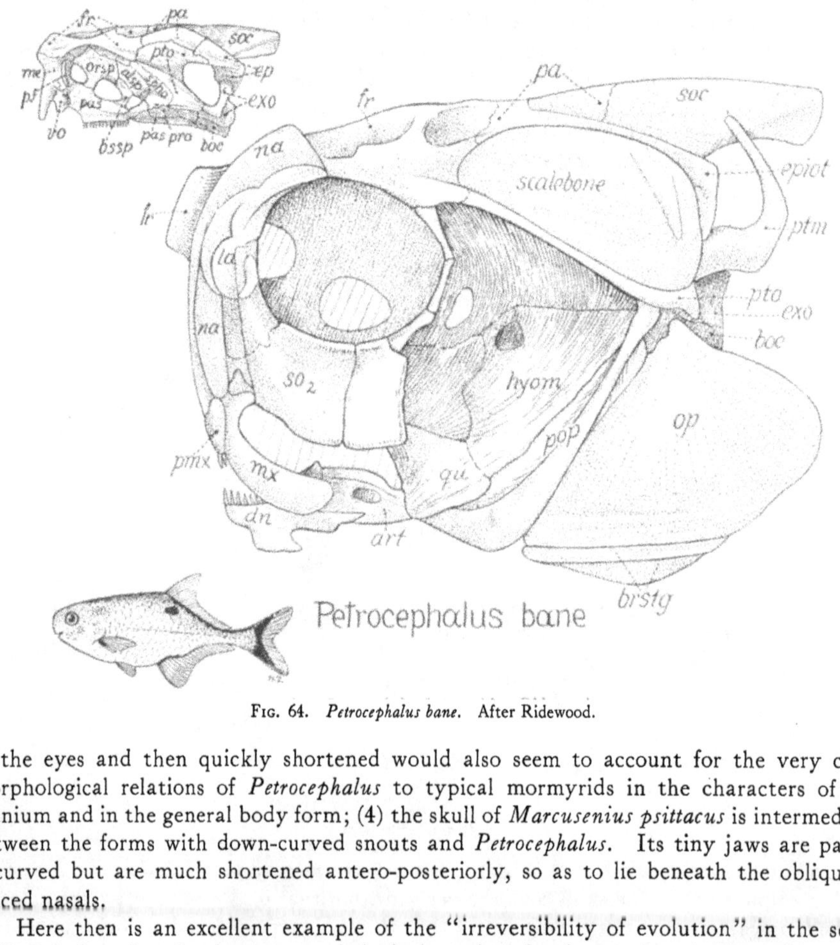

Fig. 64. *Petrocephalus bane.* After Ridewood.

of the eyes and then quickly shortened would also seem to account for the very close morphological relations of *Petrocephalus* to typical mormyrids in the characters of the cranium and in the general body form; (4) the skull of *Marcusenius psittacus* is intermediate between the forms with down-curved snouts and *Petrocephalus*. Its tiny jaws are partly decurved but are much shortened antero-posteriorly, so as to lie beneath the obliquely-placed nasals.

Here then is an excellent example of the "irreversibility of evolution" in the sense of Dollo's law, for the deeper morphological results of a former lengthening of the mid portion of the skull and of the development of a trunk are still evident even after the dwindling away of this trunk. But the same material affords an equally clear example of a change in the trend of evolution; for the former tendency toward the development of a "trunk" has been arrested and a rapid secondary shortening of the bony tract between the dentary and the quadrate has brought the mouth beneath the orbit instead of far in front of it.

Still more specialized is the eel-shaped *Gymnarchus*, which is said to propel itself through the water entirely by the action of its elongate dorsal fin, forward and backward with equal facility (Boulenger, 1904, p. 552). Its strange skull, which has been well figured by Erdl (1847, Pl. V) and especially by Hyrtl (1856, Pls. I, II), might well be derived from that of *Mormyrops* (Fig. 62), which it resembles in the emphasis of the olfactory and otic regions and the retention of strong mandibles, which likewise have an even row of stout teeth; the eye also is small and placed far forward, above and slightly behind the quadrate-articular joint; the lower part of the hyomandibular-quadrate is horizontal. The skull of *Gymnarchus* is more specialized than that of *Mormyrops* in being secondarily shorter, in having a large fenestra below, a stout sutural junction of the quadrate and hyomandibular; also in the almost horizontal position of the preopercular, in the forward displacement of the opercular and other features. In a classification based on skull structure *Gymnarchus* would be placed following *Mormyrops*, its eel-like adaptations being undoubtedly secondary.

The chondrocranium of a forty-third-day larva of *Gymnarchus niloticus*, as figured by Assheton in the Budgett Memorial Volume (Pl. XX), shows the usual foreshadowing of adult conditions, except that the cranium is far less elongate than in the adult stage. The semicircular canals are very large and thick. In the midst of them is a large "temporal vacuity" on the lateral wall of the cranium, which receives the "anterior bulb" of the air-bladder (Pl. XVIII, Fig. 31) much as in *Notopterus* (*cf.* Bridge, 1900, Pl. XXXVII).

As to the phylogenetic relationships of the families just treated, Ridewood (1904c, pp. 212, 213) writes as follows:

"On the whole, the study of the craniological characters impels one to the conclusion that the families Mormyridæ, Notopteridæ and Hyodontidæ, though more closely related *inter se* than is either family with any other family of Malacopterygian fishes, are not more intimately related with one another than was previously assumed to be the case . . . the cranial characters of the families are so conflicting that any phylogenetic arrangement based upon them is out of the question. The three families must remain, as hitherto, the terminals of a radiating system."

As Boulenger points out (1898, p. 778), the Mormyridæ cannot rightly be grouped with the Ostariophysi of Sagemehl (1885, p. 22), as Jordan and Evermann have done (1896, p. 114), since they possess no Weberian ossicles nor other modification of the anterior vertebræ. It is true that Garstang (1932) has recently assumed that the common possession of otic diverticula of the swim-bladder in mormyroids and Ostariophysi is alone sufficient to warrant the bracketing of these two groups along with some others in a superordinal division Otophysi; but the profound differences of mormyroids and characins in skull structure at least lend no support to such a procedure. The absence of a separate symplectic in both mormyroids and siluroids is evidently nothing more than a matter of convergence; the symplectic is absent also in the eels. The study of the skull of mormyroids shows also that they can have no close affinity with the Esocidæ, with which family Johannes Müller associated them. According to Boulenger (1898), the nearest allies of the mormyroids are to be found in the Albulidæ, as suggested by Valenciennes in 1846. Ridewood then states that he regrets that the study of the skull brings forward no evidence in favor of this conclusion and then goes on (1903–1906, pp. 214, 215) to give a careful and judicious evaluation of the craniological resemblances and differences between the mormyroids and the Albulidæ, to which the interested reader is referred.   Consequently it cannot be said that the point of origin of the whole mormyroid series from primitive Cretaceous isospondyls has been determined with any degree of certainty, since they share important characters with such diverse families as the Albulidæ and Clupeidæ.   The otoliths of mormyroids, as noted by Frost (1925, p. 160), are highly peculiar.   The saccular otolith (sagitta) is small and ill-formed, while the asteriscus and lapillus are unusually large. In the sagitta the sulcus widens out behind and is large and rounded posteriorly; in this and certain other features the sagitta resembles that of *Osteoglossum*, but Frost does not place much value on this resemblance; he concludes (p. 162) that a comparison of the otoliths suggests that the Mormyridæ represent a highly specialized group, and that, so far as the otoliths are concerned, they have little in common with the remainder of the Isospondyli.

"Concerning the genera *Notopterus* and *Hyodon*," concludes Ridewood (1904c, p. 214), "there is but little to be said, except that the latter possesses a greater proportion of primitive characters.   Of the forms described in the present paper, there can be little doubt that the Mormyridæ are the most specialized, and *Hyodon* the least specialized; but the close study of the skulls of these fishes does not lend support to a view of relationship recently expressed by Boulenger.   On page 116 of his book ' *Les Poissons du Bassin du Congo*,' 1901[b], he writes:—' *Les Notoptérides me semblent occuper vis-à-vis des Hyodontides une position analogue à celle qu'occupent les Mormyrides vis-à-vis des Albulides, c'est à dire qu'ils peuvent en être considérés comme modification excentrique.*'   In considering the possibility of evolution of the Notopteridæ from the ancestral *Hyodon*, one must not lose sight of the fact that *Notopterus*—in the presence of the large lateral cranial foramen bounded by the squamosal, epiotic and exoccipital, in the attempt (a futile one, it is true) of the thin scale-like supratemporal to cover it, and in the presence of a paired tendon-bone of considerable size projecting down from the side of the second basibranchial—exhibits characters strikingly constant in the Mormyridæ, but not possessed by *Hyodon*.

"Although in both *Notopterus* and *Hyodon* there are vesicles of the swim-bladder on the lateral face of the otic region of the cranium, it does not necessarily follow that these structures have had a common origin.   The connection between the swim-bladder and the ear must not be relied upon too implicitly as indicating close relationship between such fishes as possess it.   That it has arisen independently in different groups is evident from the remarkable difference between the methods by which the result is arrived at.   Compare, for instance, *Clupea* on the one hand and the Ostariophysi on the other.   Stannius (Handb. d. Anat. d. Wirbelth., i, p. 2) mentions that there is a connection between the swim-bladder and the ear in the Macruridæ among the Anacanthini, and in the Berycidæ and Gerridæ among the Acanthopteri; while Sagemehl (Morph. Jahrb., x, 1885, p. 51, footnote) observes that it occurs in the Gadidoids *Physiculus* and *Uraleptus*, and in the Scleroderm *Balistes*" (1904c, p. 215).   This passage, I think, offers a sufficient answer to the recently expressed views of Garstang (1932) concerning the supposed relationship of the mormyroids and other isospondyl "Otophysi" to the Ostariophysi.

Thus Ridewood's detailed analysis of the craniological characters of the families of malacopterygian fishes has revealed the great extent and complexity of the phylogenetic problem and contributed greatly to clarify the issues and distinguish the well established from the merely plausible. The whole work also contrasts favorably with the ordinary taxonomic monograph, the chief object of which is to expose the differences between species.

Gonorhynchoidea

Gonorhynchus.—The single existing species (*Gonorhynchus greyi*) of this isolated group inhabits the seas off Japan, South Africa, Australia and New Zealand. It has an elongate cylindrical body and a sturgeon-like head with a pointed snout, a small inferiorly-placed mouth and a rostral barbel. The body and head are covered with small spiny scales. The dorsal, ventral and anal fins are in the posterior half of the more or less pike-like body.

To this family are referred the genera *Notogoneus* from the freshwater Eocene beds of France and North America, and *Charitosomus* from the Upper Cretaceous of Westphalia and Syria. According to Woodward (1901, p. ix), the "Gonorhynchidæ are only slightly modified Scopeloids, and are now shown to date back to the Cretaceous period, when all the characteristic features of *Gonorhynchus*, except the extension of the scales over the head, seem to have been already acquired."

In the list of family characters of Scopelidæ and Gonorhynchidæ as given by Woodward (1901, pp. 235, 271), the agreements far outnumber the differences, but it must be confessed they are of a rather general character, such as the exclusion of the maxilla from the oral margin by the premaxilla, the forward extension of the supraoccipital between the parietals, the lack of a precoracoid process in the pectoral arch, the loss of the air-bladder, and the like. On the other hand, the abdominal vertebræ of gonorhynchids, while well ossified, are usually pierced by the notochord and have robust parapophyses ankylosed with the centra (Tate Regan, 1929, p. 313) and bearing delicate ribs (Woodward, 1901, p. 271), while in the Scopelidæ the vertebral centra, though also well ossified, lack transverse processes, the ribs being sessile (Boulenger, 1910, p. 611). To Woodward the differences between the Gonorhynchidæ and Scopelidæ are less important from a phylogenetic viewpoint than the resemblances, and he accordingly refers the two families to his very broad order Isospondyli, following the Enchodontidæ, a Cretaceous family that appears to be related to the scopeloids (Iniomi). Boulenger (1910, p. 572) puts the Gonorhynchidæ at the end of the "suborder Malacopterygii" (= Isospondyli in part), while grouping the scopeloids near the Enchodontidæ with the Haplomi (p. 611). Tate Regan (1929, p. 313) treats the Gonorhynchidæ as the closing section of the Isospondyli, referring the Enchodontidæ to another division of the same order and the scopeloids to the order Iniomi.

From this and much similar evidence it appears that the "orders" Isospondyli, Iniomi and Haplomi are more or less artificial groups between the extremes of which lie many combinations of intermediate characters.

A monographic comparative study of the skull of *Gonorhynchus greyi* has been made by Ridewood (1905b) in continuation of his studies on other isospondyls. The skull (Fig. 65) parallels that of the Albulidæ in its small mouth and forwardly produced suspensorium, but after a searching analysis of the cranial characters, Ridewood concludes (p. 370) that although the ancestral elopids and albulids were upon the line of descent of the gonorhynchids, the relationship is not nearer.

From the excellent figures by Ridewood we may now attempt a new functional analysis of the chief characters of this skull. The elongation of the snout has apparently involved the pulling out of the frontals into a long narrow tract (Fig. 65). The length and narrowness of the interorbital bridge and bony rostrum would perhaps be a source of weakness if these very long frontals were not coalesced in the mid-line. The rostrum is also braced on the ventral side by the parasphenoid, which in turn is fastened posteriorly by strong

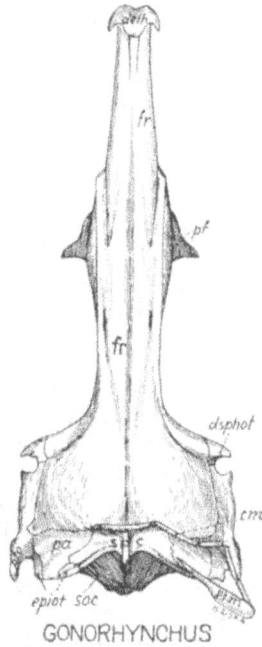

FIG. 65. *Gonorhynchus greyi.* Top view. After Ridewood.

ascending processes to the low cranial vault. The down-bending of the rostrum and the vaulting of the roof of the oral chamber have probably conditioned the marked shortening of the prefrontal (parethmoid) in the vertical plane. The anterior preorbital (lacrymal) is also involved in this prolongation and so are the slender nasals, but to a far less extent, while the mesethmoid is represented on the dorsal surface only by a small terminal bone.

As in the sturgeon, the rostrum is curved downward so as to project in front of the very small mouth, while the palatopterygoid arch is curved upward and then downward and forward to form the roof of a special cavity for the small mouth parts (Fig. 66). The premaxillæ are small rods only about half the length of the rod-like maxillæ. They lie below and in front of the maxillæ. The mandible is relatively stout and much longer than the maxilla, but only the front part of the mandible enters into the gape, owing to the fact that the ascending ramus of the dentary is tilted upward at a sharp angle to the anterior process of the articular. In other words, "The lower margin of the gape is nearly at right angles to the long axis of the mandibular ramus" (Ridewood, 1905*b*, p. 365). From the relative robusticity of the mandible and the presence of strong ridges on the hyomandibular and preopercular it may be inferred that the adductor mandibulæ muscles are fairly strong, so that in spite of its weak premaxillæ the fish ought to be able to pluck up small creatures living on the bottom, such food habits being also indicated by the downwardly-turned mouth, down-bent rostrum and sensory barbel.

It is obvious then that some of the peculiar characters of the skull of this fish are connected with the presence of a blunt projecting rostrum and with the development of a small and peculiarly specialized downwardly-directed mouth. Thus if we were to endeavor to make a pictorial restoration of the skull of the remote ancestor of this fish prior to these

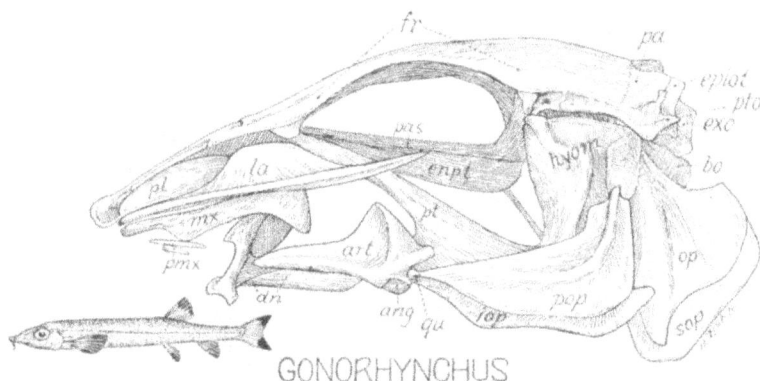

FIG. 66. *Gonorhynchus greyi.* Side view. After Ridewood.

specializations, we should have to shorten greatly the rostrum, free the opposite frontals on the mid-line, greatly enlarge the jaws and bring the mouth forward to its primitive anterior position. This would give us a generalized isospondylous skull, not unlike that of *Leptolepis.*

But there are other specializations in this skull whose functional interpretation is not difficult. On the dorsal surface of the second basibranchial bone Ridewood figures (p. 366) a circular patch of "about twenty strong blunt teeth, which engage with the teeth on the entopterygoids, and with the latter constitute the entire dentition of the animal." Thus *Gonorhynchus* like various other isospondyls chose, as it were, to develop teeth on the floor and roof of the mouth and to sacrifice the teeth on the jaws, and in so doing it evolved rounded peg-like teeth. But in this case these were located exclusively on the second basibranchial and opposite entopterygoids instead of being spread over several bones on the floor and roof of the mouth. The upward thrust of these basibranchial teeth on the roof of the mouth, under the pull of the branchial levator muscles, would tend to disrupt the ordinary edge-to-edge contact between the thin entopterygoid and the thin edge of the hyomandibular. Hence it is not surprising to find that in this fish the metapterygoid is represented by a thin rod of bone, running from the lower end of the hyomandibular upward and forward to the entopterygoid. Thus this rod together with its ligaments would be in an excellent position to check the upward thrust of the basibranchial teeth, while the rod-like symplectic would perform a like office for the quadrate, steadying it against the wrenching movements caused by the pull of the adductor mandibulæ muscles on the relatively strong mandible. The marked forward displacement of the mandible and the vertical depression of the head as a whole have caused the lower part of the pre-opercular to project at right angles to the hyomandibular and to assume more than its ordinary share in the work of bracing the quadrate.   The preopercular is stiffened by a sharp boomerang-like ridge on the outer surface, which doubtless also resists the pull of the superficial sheet of the adductor mandibulæ.   A similar strong ridge on the hyoman-dibular above the preopercular probably has a similar significance.   The hyomandibular has the usual two articular heads, connected, as in many isospondyls, by a web of bone. The dilatator operculi fossa, above the hyomandibular, is a narrow groove on the lateral edge of the sphenotic and pterotic.   The adductor hyomandibularis was doubtless attached on the concave under surface of the projecting pterotic.   The opercular and subopercular are small and leave a wide gap above the opercular, which was doubtless filled by the retractor hyomandibularis muscle.

The cranial vault is remarkably wide in the top view and shallow dorso-ventrally. The body musculature evidently did not extend forward, there being no posttemporal fossæ and no longitudinal crests on top of the stoutly built cranium.   The commissural lateral-line canal passes through a slender bony canal lying above the surface of the trans-versely widened parietals and small supraoccipital; the front end of the latter is in contact with the much broadened, coalesced frontals, which, as above noted, are the dominant elements of the skull, extending almost from the occiput to the front part of the elongate snout.   The posttemporal is reduced to a slender fork, its third or opisthotic fork being apparently represented by a partly ossified ligament or intermuscular bone (Ridewood, 1905b, p. 364).   A true though small opisthotic is present on the back of the lateral extension of the exoccipital.   This element is absent in many fishes but its presence here may not safely be assumed as a primitive character on account of the unusual specializations of the flattened occiput.

A very remarkable feature is the presence of a pseudo-occipital condyle with a convex rather than a concave posterior surface.   This condyle, according to Ridewood (1905b, p. 363), represents "a portion of a vertebral centrum fused with the basioccipital and lower parts of the exoccipitals."   This type of articulation is in wide contrast with the tripartite occipital condyles of typical percomorphs.   The convexity of the occipital articulation, notes Ridewood, is not peculiar to Gonorhynchus but is found also in Fistularia and a few other specialized fishes.   Very possibly the presence of this comparatively small and weak articulation between the skull and the vertebral column may be connected with the loss of the ordinary extension of the dorsal axial muscles above the occiput.   Thus there would be less danger of dorsal displacement of the column and consequent strangulation of the spinal cord.   On the other hand, the wide lateral extension of the occipital surface and the flexibility of the functional occipital condyle seem to imply that the column was flexed rather widely in the horizontal plane from the broad occiput as a base.   Thus we see that when considered as a natural mechanism this rather peculiar isospondylous type of fish skull is of no little interest.

Finally, from the taxonomic and phylogenetic side, Ridewood's penetrating analysis (1905b, pp. 363–371) seems to exclude one after another of the long list of isospondyl families from close relationships and to justify his main conclusions, which are as follows:

"Of the two remaining families which I propose to consider—the Alepocephalidæ and Salmonidæ—the former is to a certain extent specialized in relation with its deep-sea habits, but in some respects remains more primitive than the latter. It has no opisthotic, no teeth on the maxilla, an eye-muscle canal closed behind, and an opercular bone very narrow in front; but, on the other hand, it possesses two surmaxillæ and an ossified first pharyngobranchial in addition to the spicular. *Alepocephalus* resembles *Gonorhynchus* in possessing an epibranchial organ, borne by the fourth and fifth arches, and in possessing a cartilage which may be identified as the fifth epibranchial; but the list of resemblances is soon exhausted.

"On the other hand, the Salmonidæ, though offering no close resemblances to the Gonorhynchidæ, consist of a variety of forms but little specialized and highly plastic. For the purposes of comparison the genus *Salmo* is less suitable than such a form as *Coregonus*, for the Salmons have an excess of cartilage, presumably of secondary origin, in the cranium, and no membranous interorbital septum such as *Coregonus* has. It may be pointed out that within the family Salmonidæ there are forms, such as *Coregonus oxyrhynchus*, with prominent snout and reduced mouth with no teeth.

"Although a study of the cranial osteology of the Gonorhynchidæ and Salmonidæ cannot bring forward direct evidence of affinity between these families, the hypothesis of the descent of the Gonorhynchidæ from the Salmonoid stock is open to little objection of any serious import."

Notogoneus.—A comparison of Ridewood's figures of the skull of the existing *Gonorhynchus greyi* with Smith Woodward's figures (1896) of the extinct *Notogoneus osculus* from the Eocene of Wyoming and of *Notogoneus squamosus* from the Upper Eocene of France shows that the ancient form had already acquired the chief characteristics of the gonorhynchid mouth parts, although slightly less specialized in details. The ascending or coronoid process of the dentary and the coronoid process of the articular-angular are both higher than in the recent type, the snout less elongate, the opercular less reduced, the subopercular large, with four deep clefts on its hinder border. According to Smith Woodward (1901, p. 275) and Ridewood (1905b, p. 363) there are no pterygoid and lingual teeth in *Notogoneus*. The skull as a whole seems relatively less elongate and depressed than in the recent genus. Even as far back as the Upper Cretaceous of Mt. Lebanon and Westphalia, according to Woodward (1901, p. 273), the genus *Charitosomus* much resembled *Gonorhynchus* not only in general characters but even in the form of the maxilla. A patch of bluntly conical teeth in the throat are found "just above the ceratohyal as if they had been fixed upon the hyoid arch." It seems not impossible that these were really located on the basibranchial as in *Gonorhynchus* and that the ceratohyal had merely pressed against them, while the opposing set above them appears to correspond to the entopterygoid teeth of *Gonorhynchus*. Thus the isolation of the specialized family of Gonorhynchidæ among the Isospondyli, even in Upper Cretaceous times suggests that their line of ancestry may run back to the Jurassic Leptolepidæ, which are truly generalized isospondyls.

This inference seems to be supported by the characters of the otolith of *Gonorhynchus*, which as described by Frost (1925, pp. 161, 162) conforms in general to the elopine type, while approaching the mormyroid type in certain features.

In his great work on the "Classification of Fishes" (1923, p. 120), Jordan makes the following remark: "It is doubtful whether the extinct forms of this group (*Notogoneus*, *Charitostomus*) really belong to the same family as the living *Gonorhynchus*." If this means that *Notogoneus* is not at least closely related to the direct ancestry of *Gonorhynchus*, then it seems that Dr. Jordan must have overlooked or underestimated the striking and detailed resemblances in the jaw parts and skull as a whole between the modern and the Eocene genera, as clearly figured by Woodward. In this connection Jordan cites Dr. Cockerell's view (*in lit.* May 13, 1922) as follows: ". . . I suppose the Gonorhynchid type of scales, originating during the Mesozoic, may have persisted in several distinct branches of the original stem, of which our modern Gonorhynchidæ constitute one only. One of the other branches would be represented by *Notogoneus*. I think the resemblance is too close for convergence from entirely different stems."

OSTARIOPHYSI

Characins, Gymnotids, Carps and Catfishes

As is well known, this immense assemblage of some five thousand species, mostly of freshwater fishes, constituting the order Ostariophysi, exhibits the utmost diversity of habitus in body- and skull-form, but is shown to be a natural group by the common possession of the elaborate Weberian apparatus connecting the swim-bladder with the inner ear, which is essentially the same in all the families of the order. These ossicles, which are believed to be derived from the ribs and neural arches of the four anterior vertebræ, transmit vibrations to the membraneous labyrinth of the ear and probably serve to increase the sense of hearing (Tate Regan, 1929, p. 315). It seems curious that the importance and significance of this unique apparatus seem to have been greatly underestimated by Garstang (1932, p. 241), who refers depreciatively to the "one peculiar specialization of a few anterior vertebræ" as being inadequate for excluding the Ostariophysi from a place "on the top shelf" of the Teleostei, along with the Isospondyli. But it is doubtful whether there is any taxonomic and phylogenetic group in the whole series of vertebrates that is more deeply stamped than is the Ostariophysi with the mark of unitary origin through the possession of an elaborate "basic patent" and basic pattern. If this is not a strictly circumscribed natural group, where in all the animal kingdom shall one be found? And how is it possible to speak of "one peculiar specialization of a few anterior vertebræ," when one contemplates the extraordinary complexity of this apparatus and of the relations of its numerous parts to each other, to the swim-bladder and to the otic region of the skull?

However, all this specialization has not prevented Sagemehl, Smith Woodward, Goodrich, Tate Regan or any other author that I know of, from recognizing the probably high antiquity of the Ostariophysi nor the primitive nature of many features of the jaws and skull of the less specialized characins.

Indeed if it were not for the possession of this apparatus the least specialized of the Ostariophysi might well be classed among the Isospondyli. It is not impossible that the family Lycopteridæ from the Jurassic of China and Siberia, which are closely related to the Leptolepidæ, may stand in or near the line of ancestry to the carps, since the microscopic characters of their scales, as studied by Cockerell (1925), approach the carp type. The skeleton of the short-skulled *Lycopterus sinensis*, as restored by A. S. Woodward (1901, Pt. IV, p. 3), would seem indeed to be a favorable starting-point for this order. According to Cockerell, the anterior vertebræ of *Lycoptera middendorffi* are not modified, so that the Weberian apparatus had not yet been acquired in this very generalized forerunner of the Cyprinidæ.

As might be expected, the otoliths of this order are widely different from those of normal teleosts. Frost (1925, p. 553) states that in the suborder Cyprinoidea (of Regan, including the characins, gymnotids and carps) the lagenar otolith, the asteriscus, is generally the most developed, often (except in the Cyprinidæ) occupying lateral capsules or open recesses in the cranial cavity above the floor supporting the brain. The utricular otolith, the lapillus, is small in the Characiformes but well developed in the Gymnotiformes and Cypriniformes. In the Cyprinidæ the lapillus may be remarkably different in form in closely related species in which the lagenar otoliths are closely similar or almost indistinguishable. The sagitta, which in most orders affords diagnostic characters, in this suborder is small and attenuated.

The number of branchiostegal rays, according to Hubbs (1919, p. 65), varies widely in the Ostariophysi. In cyprinids it is never more than three, in characins from three to five, in the Nematognathi, six, seven, nine, eleven. The low number of branchiostegals in certain malacopterygian fishes is usually correlated with the broad union of the branchial membranes and with a freshwater habitat (Hubbs).

HETEROGNATHI (CHARACINS)

One of the least specialized skulls in the entire order is possessed by the characin *Erythrinus* (Fig. 67) of the primitive family Erythrininæ. Indeed in this connection Sagemehl (1885, p. 117) came to the following important conclusion:

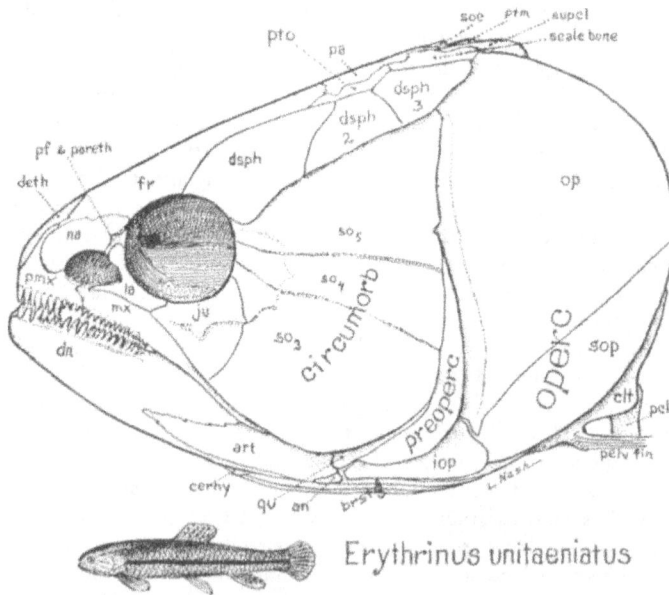

FIG. 67. *Erythrinus unitæniatus.* Side view.

"Wie es sich klar ergiebt, lassen sich die Characiniden in den meisten Organisationsverhältnissen direkt an die bei Amia bestehenden anschliessen, und zwar ist es die Gruppe der Erythrininen, welche die grösste Uebereinstimmung erkennen lässt. Nur in wenigen Punkten ist ein Anschluss nicht möglich und müssen wir in diesen Fällen auf tiefer stehende Formen als Amia zurückgehen. Jedenfalls stand die Stammform der Characiniden nicht fern von Amia."

It seems more probable, however, that at least part of this resemblance is convergent, since the two genera plainly belong to different orders, which must have begun to diverge perhaps even before the Lower Jurassic. The skull of *Erythrinus* in the side and top views resembles that of *Amia* in being covered with broad flat bony plates of somewhat similar sculpturing, but in *Erythrinus* the epiotic seems to be secondarily reduced in size and the pterotic is much crowded by a row of three bones which include the dermosphenotic and what appear to be posterior extensions of the dermosphenotic, or perhaps anterior extensions of the scale bone or so-called supratemporal.   The nasals (Fig. 68), as in most teleosts, are widely separated by the large and progressive mesethmoid (dermethmoid) but they might be conceived to be in a transitional stage in which they were withdrawing from the mid-line toward the lateral borders of the blunt snout.

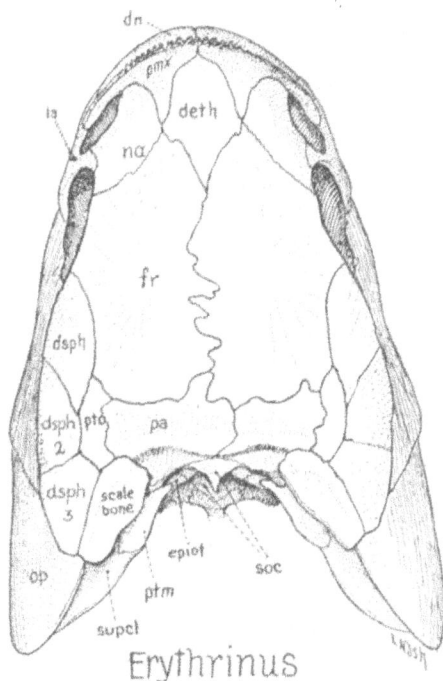

Fig. 68. *Erythrinus unitæniatus.*   Top view.

As the eye is relatively far forward and of only moderate size, the cheek plates are very large and simulate those of *Amia*. The lacrymal, on the other hand, is small and I find no trace of an adnasal in this genus although there seems to be one in *Distichodus* (Fig. 65).   The nasal has much the position and appearance of the nasal of the Semionotidæ save that it is now overspread dorsally by the premaxilla and the dermethmoid.   Perhaps the greatest objection to Sagemehl's idea of a close relationship between the primitive characins and *Amia* lies in the fact that the former possess a fully developed Weberian apparatus and that no known member of the amioid group shows the slightest tendency toward the development of these highly complex ossicles.

Neither in *Erythrinus* nor in the carps do the cranium and surface bones give any reliable suggestions of close relationships with the osteoglossids and mormyrids, which are isospondyls that merely parallel the Ostariophysi in the extension of diverticula of the

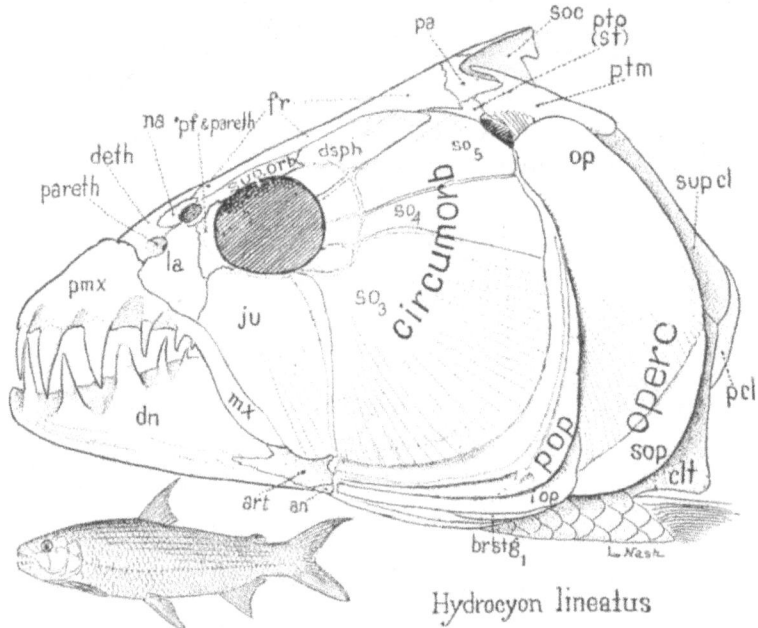

FIG. 69. *Hydrocyon lineatus.* Side view.

swim-bladder into the bony chamber of the inner ear.   But in none of these isospondyl families is there any suggestion either of the beginning of a Weberian apparatus or of the development of diverticula which would have the relations with the surrounding parts that obtain in the Ostariophysi.

The same general type of skull seen in *Erythrinus* is also shown with minor modifications in the other three characins here figured.   The typical characins (*Serrasalmo, Hydrocyon*) are famous for their ferocity and for the severity of their bites.   But although many are predaceous, with sharp teeth, none are of the gulping open-mouthed type; all have relatively short, powerful jaws, typically with close-set, even, sharply cuspidate teeth (Fig. 70), capable of tearing off pieces of the food.   The premaxillæ are usually strong and well set, the maxillary typically bears teeth, often to the posterior end, but it may become toothless and even be excluded from the mouth (Boulenger, 1910, p. 576).   It would seem that the ancestral stock of the characins must have been a form much like *Erythrinus*, with moderately short but strongly attached jaws.

In *Distichodus langi* (Fig. 71) the highly specialized mouth is very short anteroposteriorly but wide transversely, bordered by two concentric rows of fine styloid, pointed teeth.   The mouth is directed partly downward and there is a peculiar accessory joint

between the articular and the dentary, due to the buckling of the lower border of the
mandible at the junction of these two bones. The effect of the contraction of the large
adductor mandibulæ must be to convert a long, nearly horizontal, into a strong, short
vertical, movement of the tooth row, while posterior dislocation of the dentary is prevented
by the high ascending bar of the articular. In this arrangement there is a certain resemblance to the double-jointed mandible of the parrot fishes (Fig. 134).

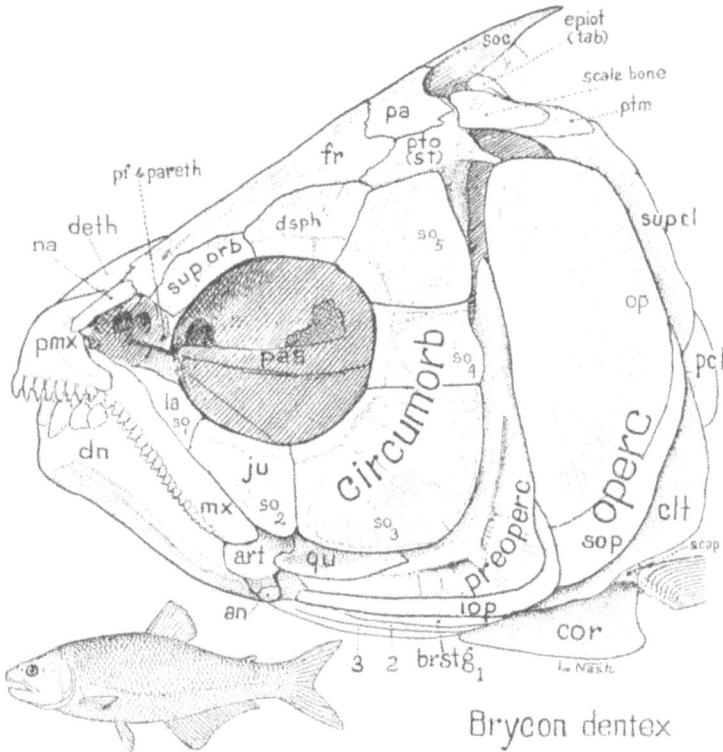

FIG. 70. *Brycon dentex.* Side view.

Gasteropelecus.—In the "fresh-water flying fish" of South America the skeleton as a
whole conforms to the characin type, but the opposite coracoids are enormously enlarged
into a fan-like keel for the support of the muscles of flight. In the excellent figure of the
skeleton given by Ridewood (1913, Pl. XVI) the skull is seen to be compactly built and to
be crowded forward, so to speak, by the enormous coracoids. The mouth is small and
sharply upturned, but with stoutly built jaws and strong short teeth. The broad suspensorium curves forward beneath the large orbit. The opercular is shortened vertically.
The skull-roof is low but strongly built.

GLANENCHELI (GYMNOTIDS)

The gymnotids or "electric eels" represent a group of four closely related families of Ostariophysi from South and Central America, which collectively seem to deserve the rank of a separate suborder (Glanencheli). They are all more or less eel-like in external appearance but are peculiar in having no dorsal fin, a greatly elongate anal fin and a shortened body cavity with the vent on the throat. Of the nine known genera only *Electrophorus* is provided with electric organs, which are extended along the sides of the body.

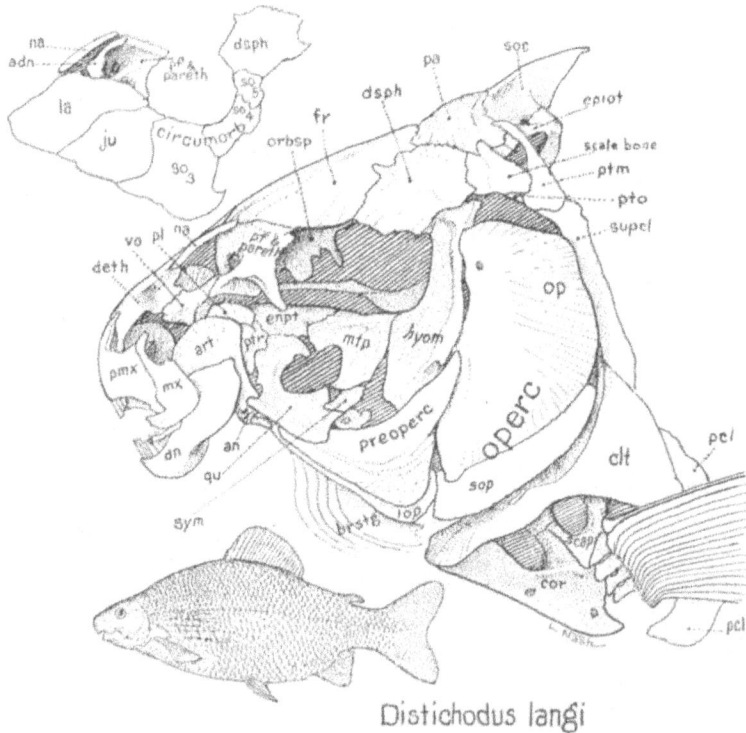

FIG. 71. *Distichodus langi.* Side view.

It was long since shown that this fish is not a true eel but a strongly modified, degraded characin (Boulenger, 1904, p. 579). The form of the skull differs very widely within the group, as shown in the monograph by Ellis (1913). In an American Museum specimen of *Eigenmannia macrops* the head (Fig. 72*A*) is fairly deep and the skull top narrow, with a long narrow median fontanelle extending forward to the very narrow, downwardly sloping mesethmoid. The latter ends below in a transversely widened facet for the small transversely placed premaxillæ. The mouth (Fig. 73*A*) is small, not protrusile, and bordered laterally by the toothless maxillæ. The premaxillæ and dentaries bear minute teeth. A cylindrical bone that seems to be the prefrontal forms a prominent, nearly vertical pillar, which spreads out below on the palatopterygoid arch. The orbit is of fair size and the same is true of the preopercular and opercular. Apparently this type of skull could be derived from that of such a characin as *Distichodus.*

Very different indeed in appearance is the skull of *Electrophorus (Gymnotus) electricus* (Figs. 72B, 73B). This is wide and depressed, with flattened cranium, no inter-frontal

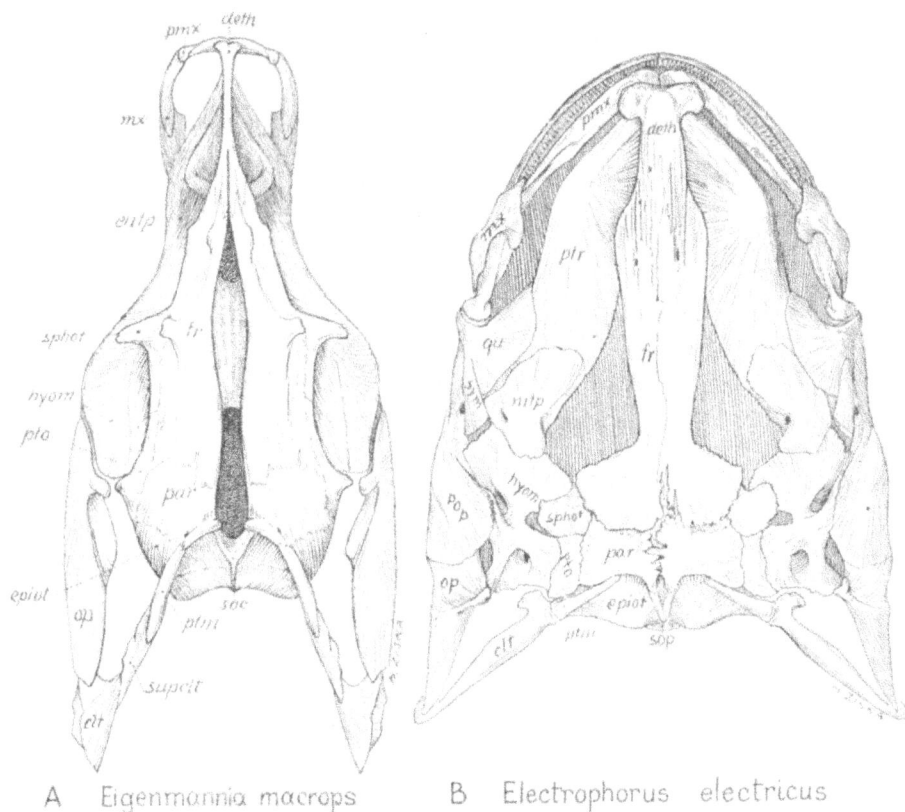

FIG. 72. A. *Eigenmannia macrops.* B. *Electrophorus electricus.* Top views.

fontanelle, a long narrow interorbital bridge and flattened mesethmoid, the latter expanding distally into two rounded plates above the premaxillæ. The mouth is relatively large, bordered chiefly by the premaxillæ and provided with thick-set small teeth with recurved tips. The maxillæ are reduced to small triangular bony flaps at the corners of the mouth. The large dentary has a strong ascending process. The prefrontals (parethmoids) appear to be absent and the roof of the oral cavity is formed chiefly by the greatly expanded entopterygoids, the palatines and ectopterygoids being absent. The symplectic is unusually large and prominent, the hyomandibular much widened anteroposteriorly, especially at the top. One large and two smaller fenestræ perforate its upper part, above the preoper-

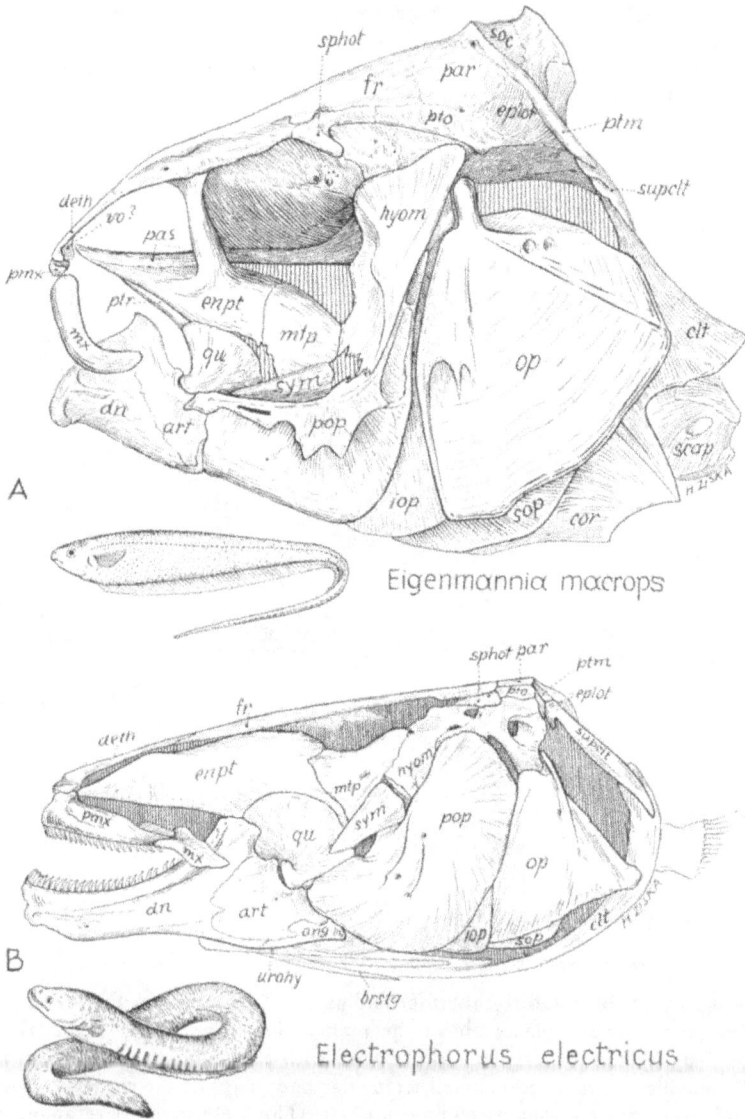

FIG. 73.   A. *Eigenmannia macrops.*   B. *Electrophorus electricus.*   Side views.

cular. The latter is somewhat inclined forward. It is much larger than the opercular which, however, has a large glenoid process for the pedicle of the hyomandibular.

This very peculiar skull exhibits no obvious marks of near relationship with that of *Eigenmannia*, but somewhat intermediate conditions are found in *Gymnotus carapo* (Ellis, Pl. XVI) and *Eigenmannia virescens* (Pl. XVIII). In *Rhamphichthys rostratus* (Pl. XVII) the snout is very long, the suspensorium much produced and the mouth very small, the whole recalling somewhat the form of the head and snout in certain mormyrids. The elongate preopercular is almost horizontal.

The subhorizontal position of the preopercular in such short-headed forms as *Eigenmannia*, together with the fact that the hyomandibular process of the opercular points upward rather than forward, suggest that the primitive gymnotid had the middle part of the skull at least fairly elongated, and that the shortening of the skull in *Eigenmannia* and many other genera is quite secondary. This view is also consistent with the graphic diagram of generic relationships given by Ellis (Pl. XV).

The amazingly wide range in skull structure in this group and the indubitable evidence of relatively close relationship of all its genera and species indicate that it is now undergoing a rapid evolutionary expansion, along with its parent group the characins.

With regard to the otoliths, Frost (1925*b*, p. 556) notes that "The otoliths of the Gymnotiformes are not far removed from those of the typical Characiformes, from which type they have apparently been derived."

Eventognathi (Carps, Suckers, etc.)

The typical carp skull (Fig. 74) is more specialized than that of the primitive characin in the edentulous and highly protrusile character of the upper jaw. The mechanism of the carp's mouth was carefully described and figured in 1886 by Vitus Graber, whose work is cited by Thilo (1920). The latter gives a very clear brief account of the origin and "degeneration" of this mechanism. Sagemehl (1891, p. 583) described the circumoral structures of the carps, suckers and loaches chiefly from the morphological viewpoint, while Starks (1926*a*) gives many important descriptive details of the ethmoid region, including the attachments of the "submaxillaries" in these families. L. F. Edwards (1926) has recently published an excellent description of the protractile apparatus of the mouth of the catostomids. The following brief description is based primarily on the above mentioned papers but with constant reference to preserved specimens and skulls of various cyprinids and catostomids.

As noted by Sagemehl (1891, p. 583), the protrusile condition of the jaws has doutbless been acquired independently of that of the percoid fishes, from which it differs in important details.

In the typical "suckers" (*Catostomus*) the toothless premaxillæ and maxillæ are hidden in thick circular lips. In the carp (Fig. 74) the lips are not so thick but the edentulous jaws are equally protrusile. The premaxillæ have ascending processes to which are attached by ligament a slender median tracker, the "rostral bone" of Starks (1926*a*), or "preëthmoid bone" of Edwards (1926). This bony tracker in turn is attached posteriorly by a long flexible ligament to the anterior end of the mesethmoid just above the notch (Fig. 75*A*) between the anterior forks of the vomers (Starks, 1926*a*, p. 173). When the jaws are retracted and closed, the bony tracker and its posterior ligament are folded up into a sharp U and pressed vertically against the fork of the vomer and the median cavity of the mesethmoid. In the oblique front view (Fig. 76) the opposite maxillæ are seen to end dorsally in two sharply diverging, rounded processes. The opposite pairs of maxillary processes surround the vertically placed premaxillary tracker or rostral bone, meeting both above and below it. The rounded dorsal processes fit dorso-posteriorly into thick cartila-

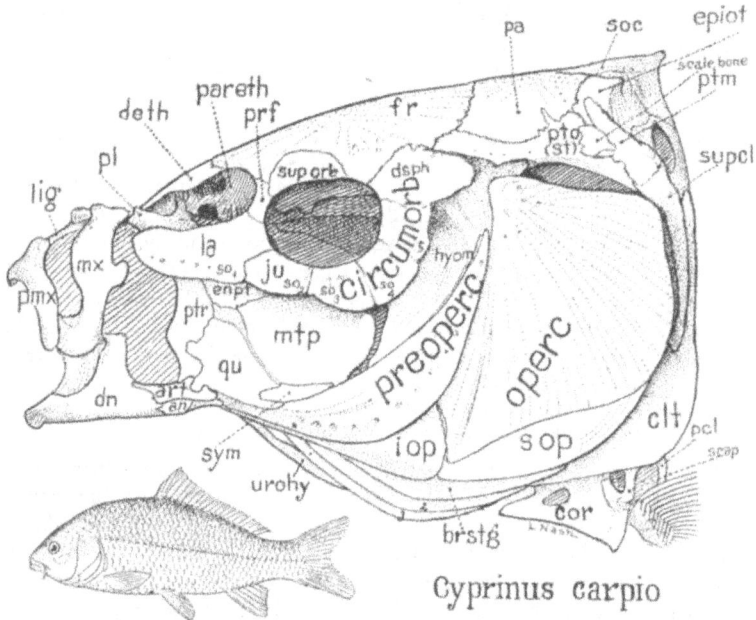

Fig. 74. *Cyprinus carpio.* Side view.

ginous or bony caps called the "submaxillary cartilages" by Starks and "cartilaginous rods" by Edwards, which are more or less fully ossified. These fit against smoothly rounded, condylar-like "preëthmoid" projections from the anterior forks of the vomer. By means of these rolling condylar surfaces of the vomer and preëthmoids the maxillaries can rock back and forth as the mouth is protruded and withdrawn.

The parts described above are protected from lateral dislocation by an anterior or maxillary process of the short palatine, which is lateral to the proximal end of the maxilla; from its front end a ligament passes forward to the outer side of the dorsal process of the maxilla. In the closed position of the mouth the lower border of the mandible is directed partly upward. Depression of the mandible by the longitudinal throat muscles (genio-hyoideus and sternohyoideus) causes the ascending process of the dentary, which is very far forward, to move forward and downward together with the fold of skin that passes to the lower border of the maxilla. Since the ends of the premaxillæ are tied by ligaments to the lower end of the maxillæ, and since the latter are fastened to the ascending process of the dentary, depression of the mandible also tends to pull the premaxillæ downward. The forward movement is checked at the end by the median tracker and paired ligaments.

The thrusting forward of the upper lip is, according to Edwards, effected partly by the outward and forward swing of the entire upper jaw and suspensorium, which also carries with it the mandible. This movement is caused by the contraction of the strong protractor hyomandibularis muscle. Thus the ascending rami of the opposite dentaries push the lower ends of the maxillæ before them, rocking the upper ends of the maxillæ backward and apparently releasing the premaxillæ.

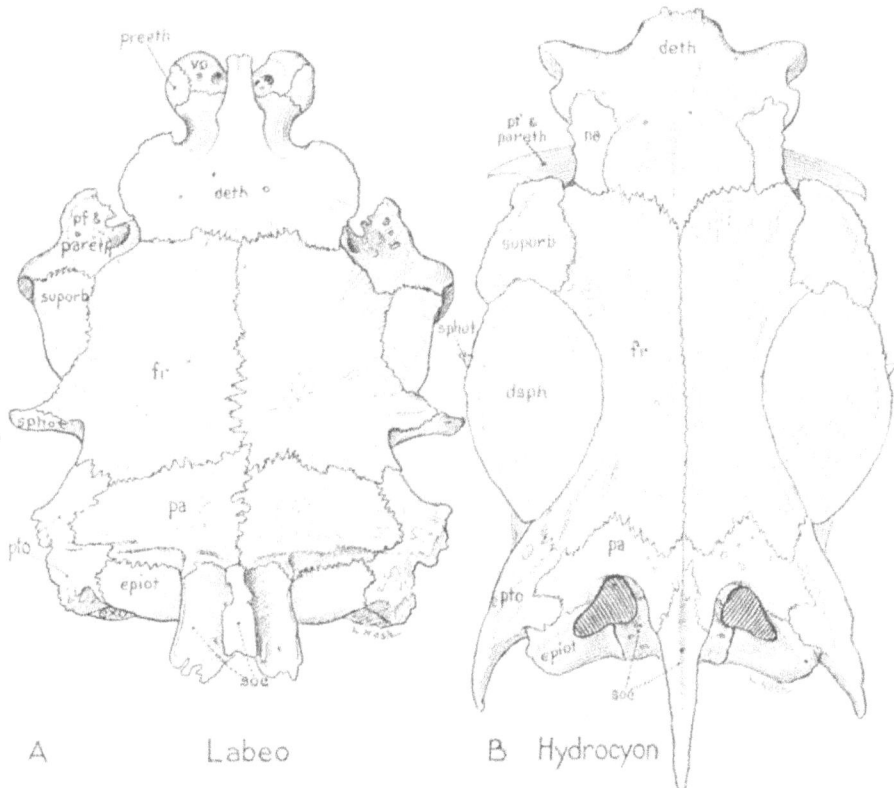

FIG. 75. Comparison of (A) *Labeo* sp. and (B) *Hydrocyon*. Top views.

Food is ingested by suction through the alternate enlargement and constriction of the oral cavity. Expansion is accomplished largely by the coöperation of the protractor hyomandibularis and of the levator arcus palatini muscles with the ventral muscles of the mandible, contraction, by the adductor hyomandibularis, intermandibularis and adductor mandibulæ (Edwards, 1926). By stimulating the geniohyoideus and sternohyoideus of the live fish, Edwards (p. 268) obtained a forward and downward thrusting of the pre- maxillæ. On the other hand, when the adductor mandibulæ was stimulated the jaws were retracted.

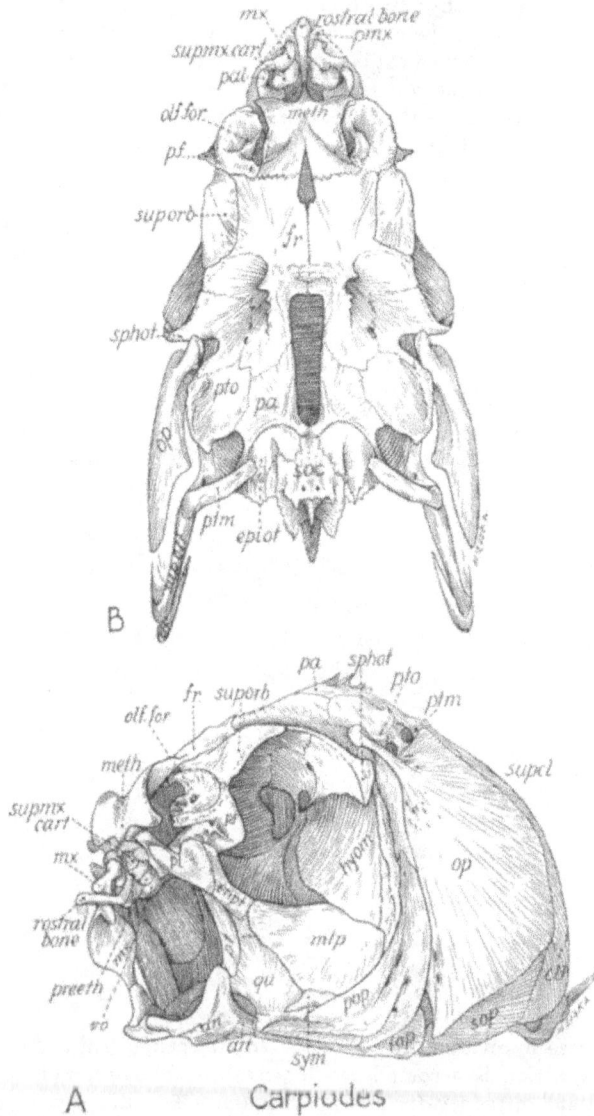

FIG. 76. *Carpiodes* sp.   A. Oblique front view of mouth, after removal of both premaxillæ and of the left maxilla, showing proximal fork of right maxilla, rod-like tracker bone, ball-like preëthmoid and dried cartilaginous rods or "submaxillary cartilages" of Starks.   B. Top view.

In the cyprinid *Aspius rapax*, which lives on small fishes, the typical carp mouth has been modified toward predatory habits. According to Thilo (1920, p. 221), the jaws are now much larger, with sharp cutting edges; on the anterior end of the mandible is a sharp tip, which fits into a corresponding excision of the upper jaw. The ascending process of the dentary is further back toward the fulcrum and the long premaxillæ exclude the maxillæ from the border of the mouth. Nevertheless, Thilo's description shows that the mechanism as a whole is the same as that in the typical carp and the conditions in *Aspius* plainly represent only a convergence toward the moderately protrusile jaws of the percoid fishes.

From the evidence reviewed by Sagemehl (1891) it seems probable that the most primitive members of the cyprinoid group are not the free-swimming types, with large mouths, but the more sedentary or slow-moving, bottom-living forms, derived eventually perhaps from some small-mouthed characin like *Distichodus*, which developed thick lips and became edentulous.

The absence of teeth on the jaws of cyprinoids is more than compensated by the presence of tooth-like processes on the fifth ceratobranchials or lower pharyngeals, arranged in one to three rows in different subfamilies. These teeth, being drawn upward by the powerful action of muscles that are attached to the lateral fossa of the cranium, oppose a prominent horny pad which rests on a raised bony projection from the basioccipital. Of these pharyngeal teeth, Boulenger writes that "adapted to various requirements, [they] may be conical, hooked, spoon-shaped, molariform, etc."

In the suckers (Catostomidæ), writes Tate Regan (1929, p. 315), "the premaxillaries are small, the lips fleshy, the pharyngeal teeth in a single series often numerous, and the pharyngeal processes of the basioccipital united to form an expanded perforated lamella, rolled up at.the edges and not covered by a horny sheath." These fishes feed on small aquatic animals, weeds and mud.

A median interparietal fontanelle of varying size and anteroposterior extent is found in various genera of the different subfamilies of the cyprinoid group. Sagemehl (1891, p. 506) was unable to find any physiologic significance in this variable feature. He could not accept E. H. Weber's suggestion that this fontanelle might afford access of sound waves to the labyrinth, for the reason that it is always closed by a thick, fatty membrane, and that between it and the nearest parts of the labyrinth lies a considerable mass of fatty interdural tissue. On account of the presence of a similar interparietal fontanelle in certain characins, and of its sporadic occurrence in the cyprinids, Sagemehl (1891, p. 495) was inclined to regard it as an ancient hereditary or reversional feature of the whole cyprinoid stem. In conclusion, Sagemehl (1891, pp. 580–594) has abundantly shown that the cyprinoid skull as a whole gives evidence of derivation from characinid ancestors; but I cannot accept his conclusion that such resemblances as there may be between the skulls of cyprinoids and characins and that of *Amia* are indicative of the derivation of the former two from the latter.

The extraordinary idea expressed by Garstang (1932) that the presence of an adipose fin in various Ostariophysi, Isospondyli and Iniomi compels us to carry the Teleostei back to osteolepid-like ancestors, finds no support, I think, in the study of their skull structure.

The otoliths of the cyprinoid families afford some interesting evidence as to relationship. In the Cyprinidæ, according to Frost (1925b, p. 561), the otoliths of *Barilius* "present certain Characid features. They are very flat compared with those of the remainder of

the family, and the lower line of the sulcus is domed as in the Characiformes. Another
Characid feature is the position of the asteriscus in the cranial cavity. These points confirm Mr. Regan's contention that *Barilius* is the most primitive of living Cyprinoids.

"The otoliths of other species of the Cyprinidæ, while resembling the Characid type
in certain features, such as the radiating furrows of the outer side and the serrated edges,

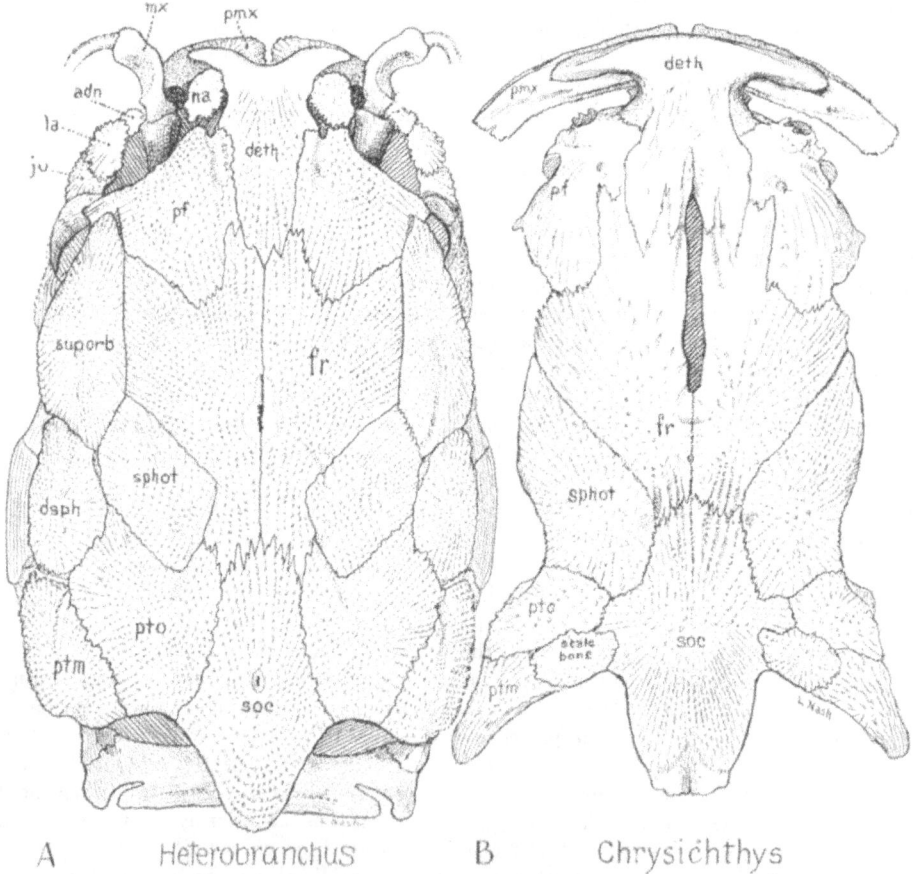

Fig. 77.   Skull tops of two catfishes.
A. *Heterobranchus longifilis.*   Family Clariidæ.
B. *Chrysichthys* sp.   Family Bagridæ.

are generally elongated and also differ in the shape of the sulcus, owing to the lower line
being either straight or distended ventrally instead of being domed.

"In the Cobitidæ and Homalopteridæ, the asteriscus, which throughout the remainder
of the suborder appears to be invariably the largest of the three otoliths, gives place in
importance to the lapillus, this feature occurring elsewhere only in the succeeding suborder,
the Siluroidea."

NEMATOGNATHI (CATFISHES)

In the catfishes (Nematognathi) of the various families, specialization in skull structure goes to far greater lengths than in the characins and carps. In the typical forms the skull-roof (Fig. 77) is widened out into a sort of cephalic shield to which the broadened supra-orbital, dermosphenotic, sphenotic, pterotic and posttemporal plates contribute. In addition to this, a nuchal shield is often formed by the expansion and coalescence of the bony

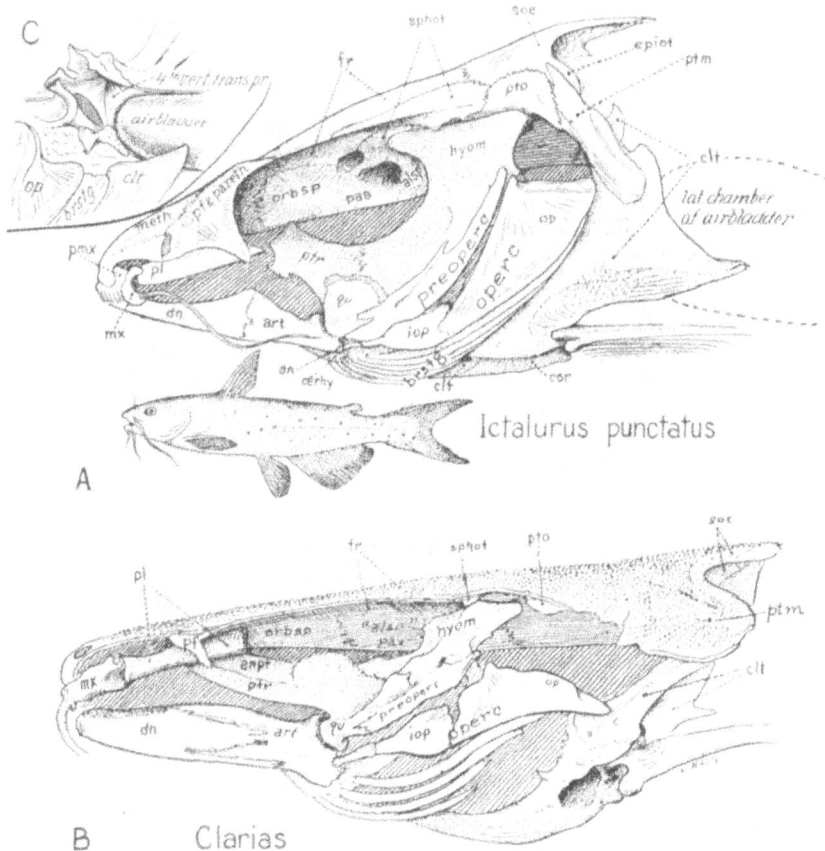

FIG. 78. Skulls of catfishes. A. *Ictalurus punctatus*. B. *Clarias* sp. C. Relations of the air-bladder to the fourth vertebra in a typical nematognath, *Ictalurus* sp.

supports of the first three rays of the dorsal fin, together with the posterior process of the supraoccipital. Another sign of high specialization is the apparent loss of the parietals, opisthotic, symplectic and subopercular. The jaws are peculiarly specialized in most siluroids in the reduction of the maxillæ to small bones that support the barbels. *Diplomystes*, however, the most primitive catfish, has a well developed maxillary, expanded

9

distally and toothed (Tate Regan, 1929, p. 316).   In the typical forms (Fig. 78) the palatine has become rod-like and freed from the pterygoid but retains its attachment to the maxilla (cf. Starks, 1926, p. 185).   The metapterygoid, according to Regan (1911ƒ (Ostariophysi), p. 563) has moved forward over the top of the quadrate and usurped the place of the pterygoid, while the pterygoid itself has disappeared, except in the Bagridæ, where it forms a small plate behind the palatine.   No trace of these primitive conditions is evident in the chondrocranium of a 10 mm. *Amiurus* figured by Kindred (1919, Pl. I).   Here the hyomandibular bears a deep sinus on its front border in the place where a metapterygoid would be looked for, while the functional pterygoid bar is in its usual position connected with the anterior inferior border of the quadrate.   The palatine has already become long and rod-like, the orbits are very small and far forward; in general this foetal chondrocranium rather fully foreshadows the adult skull.   The circumorbital series is often reduced to a thread or absent, the preopercular and opercular apparatus variously reduced.   The post-temporal has been practically annexed to the skull and apparently the supracleithrum has disappeared.   In the midst of these and other specializations the orbitosphenoid element is strongly developed, perhaps because it is needed to stiffen the immense skull and nuchal spine.

Still greater specializations appear in connection with the peculiar modifications of the air-bladder and of the Weberian apparatus.   Perhaps in order to support the massive skull the first vertebra forms a disc "rigidly united to the basioccipital and to the second, third and fourth vertebræ, which are ankylosed to form a complex to which the fifth is rigidly attached and with the parapophyses ankylosed to the centra" (Tate Regan, 1929, p. 316).

The branches of the transverse processes of the complex vertebræ have different relations with the parts of the air-bladder and of the "supracleithrum" (= supratemporal) in the different families (Tate Regan, 1929).

In passing we may allude to the well known fact that the under side of the skull (Fig. 79) of many catfishes bears a certain resemblance to a crucifix (Gudger, E. W., 1925). The crown is formed by the opposite tripus of the Weberian apparatus.   The arms of the cross are formed by the transverse processes of the complex vertebræ.   The arms of the figure on the cross are the ascending processes of the parasphenoid.   The "inscription" across the base of the cross is furnished by the vomerine tooth-patch.   No better example perhaps could be found of a class of fortuitous resemblances between wholly unrelated objects, which the late Professor Bashford Dean called "Unnatural History Resemblances" (1908).   Is it any wonder then that still closer resemblances are often produced between form patterns that *are* related?

The resemblances of the cephalic shield of the loricariid catfishes (Fig. 80) to those of the cephalaspid ostracoderms are evidently to be referred to "convergence."   In these heavily armored forms the siluroid skull attains its highest specialization.   In *Plecostomus* (Fig. 80) the small, almost tadpole-like mouth bears several rows of minute, curved, rod-like teeth in the premaxillæ and dentaries.   The width across the quadrate-articular joints is about twice the length of the jaw itself.   This is braced posteriorly by the transversely widened pterygo-entopterygoid, which articulates firmly at two contacts with the side wall of the ethmoid, below the large olfactory fossa.   The palatine extends forward as a long process overlapping the reduced maxilla.   The very large plate-like supratemporal overrides the reduced ascending limb of the cleithrum and acts as a cover for a bony tunnel (that probably housed a lateral branch of the air-bladder) formed by the transverse process of the so-called fourth vertebra. The very small opercular still overlaps the anterior border of the reduced ascending limb of the cleithrum. The preopercular is reduced laterally to an oblique splint but it extends inward under the adductor muscles and con-

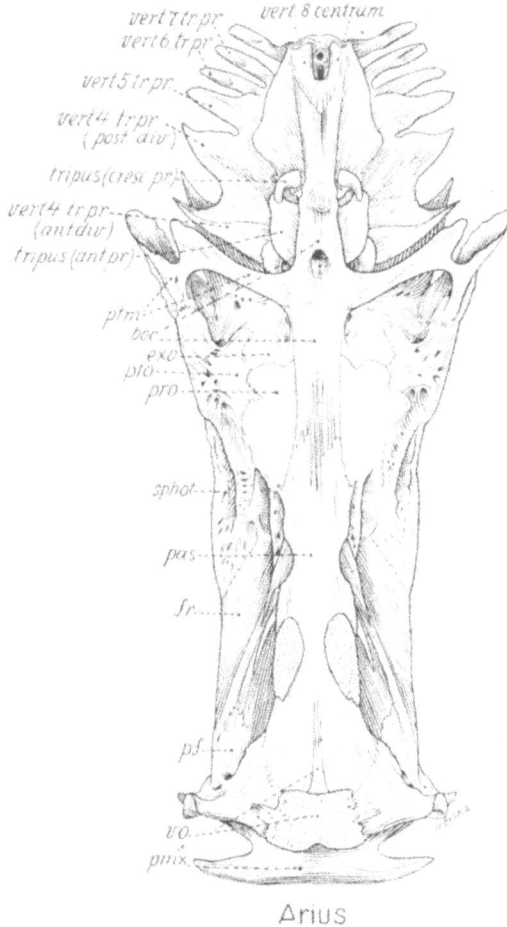

FIG. 79. *Arius* sp. "Crucifix" fish.

tinues to form a significant lateral brace of the suspensorium. The very broad hyomandibular is covered postero-laterally by the expanded supratemporal, which has largely usurped the place and appearance of an opercular. The row of fused anterior vertebræ is fully incorporated with the occiput, so that the tunnel-like transverse processes of the fourth vertebra already mentioned appear to spring from the sides of the occiput, like the paroc-

cipital processes in dinosaur skulls.   Moreover the enlarged supraoccipital plate lies immediately above this newly annexed region.   The under side of the skull also exhibits a high
degree of specialization, as shown in Fig. 80C.

The otoliths of the siluroid fishes have yielded to Mr. G. Allan Frost (1925c, p. 445)
the following significant conclusion:

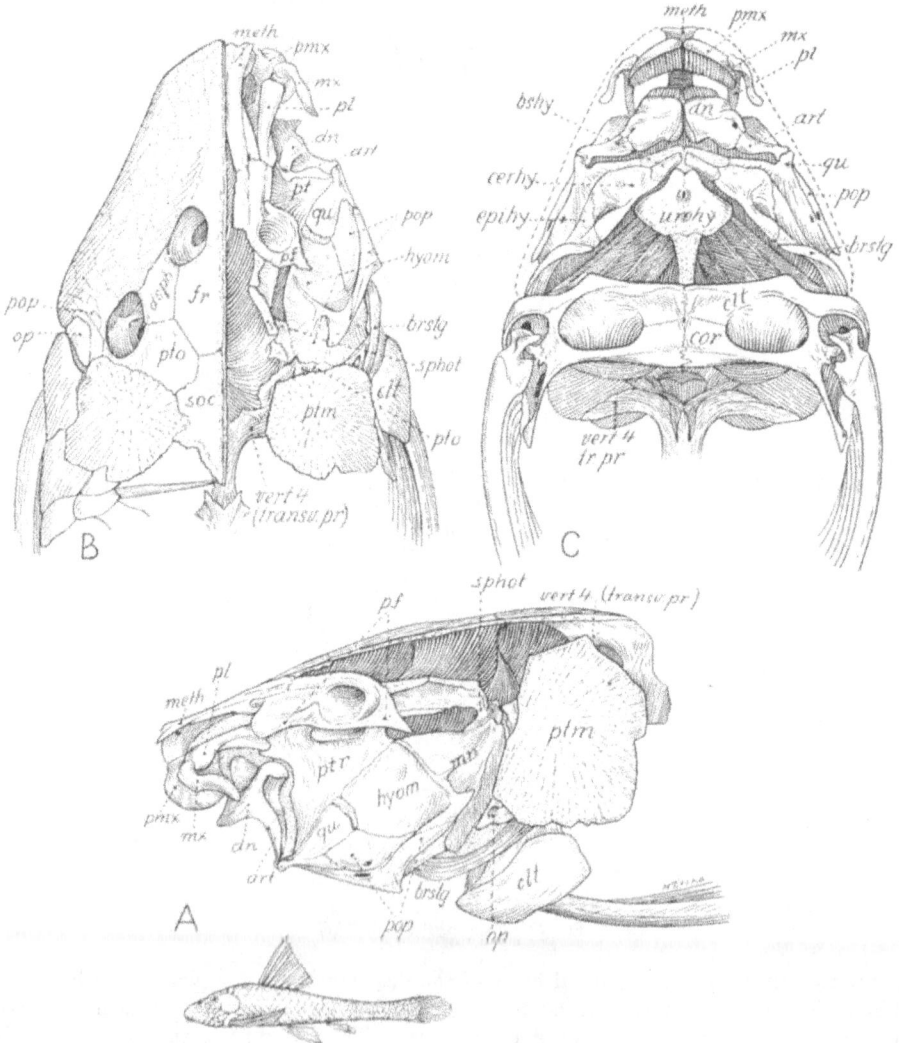

Plecostomus commersonii

FIG. 80.   *Plecostomus.*   A. Side view.   B. Top view.   C. Underside.

"In the Ostariophysi the saccular otolith, the sagitta, which is generally the principal otolith in other fishes, is attenuated and diminutive. In the Cyprinoids and *Diplomystes* the asteriscus is the largest otolith, in the Siluroids (except *Diplomystes* and a few South American species) the lapillus. It seems possible that the reduction of the sagitta may be related to the development of the Weberian mechanism, and that the great development of the lapillus or utricular otolith, in the Siluroids, may in muddy waters compensate for the decreased use of the eyes for maintaining equilibrium."

The otoliths also throw light on the interrelationships of the siluroid families with each other and with the primitive characins (1925c, pp. 445–446).

A utricular otolith resembling those of the siluroids has been recorded by Frost (1926a, p. 83) from the Upper Jurassic of England. Here is another suggestion of the relative antiquity of the ostariophysial fishes.

HETEROMI

HALOSAURS, NOTOCANTHS

Halosaurs.—These are a family of deep-sea fishes with elongate bodies and long tails tapering to a point, without caudal fin and with the anal much elongated. The dorsal fin is short, located on the middle of the back above the short abdominal fins. The short rostrum is pointed, projecting beyond the mouth, which is of fair size. According to Boulenger (1910, p. 621) the air-bladder has no trace of an open duct (at least in *Halo-sauropsis*, so that the fish is literally a physoclist, but primitive isospondyl characters are the absence of spines in the fins, the cycloid scales, the abdominal ventrals, the parietal bones separating the frontals. On the other hand, the mesocoracoid arch is absent (as in Iniomi, Haplomi, Apodes, etc.). As to the otoliths, the sagitta of *Halosaurus macrochir*, according to Frost (1926, p. 466), is of the elopine type, resembling that of *Elops* in certain features and that of *Megalops* in others.

According to Smith Woodward (1901, Pt. IV, p. 162; 1903, Pt. II, pp. 74–76), the family is represented in the Upper Cretaceous of England, Westphalia and Mt. Lebanon by the genus *Enchelurus*. In this form as described by Woodward (1903, p. 76) the cranial

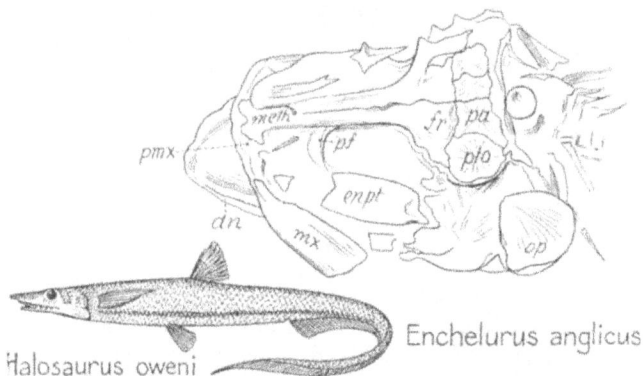

FIG. 81. *Enchelurus.* After Smith Woodward. *Halosaurus oweni.*

vault (Fig. 81) is wide and flat-roofed, the interorbital bridge elongate and narrow, with distally forked ethmoid. The paired parietals and pterotics ("squamosals") form a trans-verse row of four rather small flat bones, which widely separate the supraoccipital from the frontals. The premaxillæ are relatively small, with a delicate rod-like extension behind, bearing minute teeth. The maxilla is very large and closely similar to that of *Halosaurus* in general shape. The gape of the mouth must have been small, and the relatively large inferior limb of the preopercular extends forward to the articulation of the mandible. The opercular is as deep as broad, rounded behind and quite smooth on its outer surface. In the allied Upper Cretaceous *Echidnacephalus* the enlarged suborbital plates bear a well developed slime canal (Woodward, 1901, p. 162). In the modern *Halosaurus*, according

200

to Boulenger (1910, p. 623), the premaxillæ and maxillæ both enter the border of the moderate-sized mouth; the preopercular is rudimentary.

Notacanths.—Another deep-sea fish, *Notacanthus*, differs from the halosaurs in having a very blunt snout, a small mouth and the dorsal fin represented by six separated spines; but Boulenger (1910, pp. 622, 624) showed that a third form, *Lipogenys*, is intermediate in structure between the Halosauridæ and the Notacanthidæ. According to Smith Woodward (1901, p. 168), the notacanths were represented in the Upper Cretaceous by *Pronotacanthus*.

Tate Regan (1909a, p. 82) states that "In skeletal characters *Halosaurus* and *Notacanthus* agree in that the orbito-rostral part of the cranium is elongate, the parietals meet, opisthotics, basisphenoid, alisphenoids and orbitosphenoid are absent, the parasphenoid unites with the sphenotic (postfrontal) in front of the proötic, the posttemporal is simple or ligamentous, ..." The otolith (sagitta) of *Notacanthus* is described by Frost (1926, p. 466) as being quite different from that of *Halosaurus* in its elliptical contour and unusual thickness; its sulcus resembles that of *Halosaurus* in length and position but differs in other details. Hence the evidence from the otoliths indicates merely that *Notacanthus* is more highly specialized than *Halosaurus* but possibly related to it.

Dercetids.—This Upper Cretaceous family is also regarded by Smith Woodward (1901, Pt. VIII, pp. 162–189; 1903, pp. 64–74) as apparently related to the Halosauridæ and Notacanthidæ. *Leptotrachelus* as restored by Smith Woodward (1903, p. 69) is an elongate sagittiform fish with a large homocercal tail, an elongate dorsal fin, small abdominal ventrals and a long pointed head. A horizontal row of small scutes extends along the flanks from the head to the tail and there were other rod-like scutes on the flanks. In general it is rather suggestive of the sand-lance (*Ammodytes*). The latter, however, differs chiefly in having more prolonged dorsal and anal fins, no ventral fins, highly protrusile premaxillæ, smaller tail. Thus the resemblances between these two types are very probably convergent, especially as Tate Regan has cited good evidence for a quite different allocation of *Ammodytes* (see below, p. 354).

As to the interrelationships of the Halosauridæ, Notacanthidæ and Dercetidæ, Smith Woodward states (1901, p. viii) that "Of all the Cretaceous Isospondyli, three families of eel-shaped fishes are the most difficult to understand. They are all characterized by a primitive cranium of the Jurassic type; but they exhibit the new specialization by which the extending premaxilla gradually excludes the maxilla from the upper border of the mouth. Their elongated shape is alone indicative of high specialization but no intermediate forms are yet known to afford a clue to their more normally-shaped ancestors. The Dercetidæ are interesting as being the earliest type of fish in which evidence of a distensible stomach has been observed (Woodward, 1901, p. 177). Their fins are less specialized than those of the two families just mentioned; and their trunk is provided with paired longitudinal series of enlarged scutes."

Tate Regan (1911a, p. 120) holds that: "The Dercetidæ are of uncertain relationships, but the orbital and postorbital parts of the skull and the posttemporals show considerable resemblance to *Evermanella* (*Odontostomus*) whilst the ethmoid region and jaws are more like those of *Alepidosaurus*." To this may be added that the general arrangement and form of all the fins in *Leptotrachelus* is assuredly more favorable to relationship with *Alepidosaurus* than with any of the halosaurs, notacanths or allied forms. Hence it seems that on such evidence as we have the Dercetidæ had better be transferred to the Iniomi.

APODES

Eels, Morays

It is well known that the eel-like body-form has been acquired independently in many groups of fishes and other vertebrates. The outstanding morphological feature of the eel skull (Fig. 82A) is the reduction of the dorsal part of the opercular and the freeing

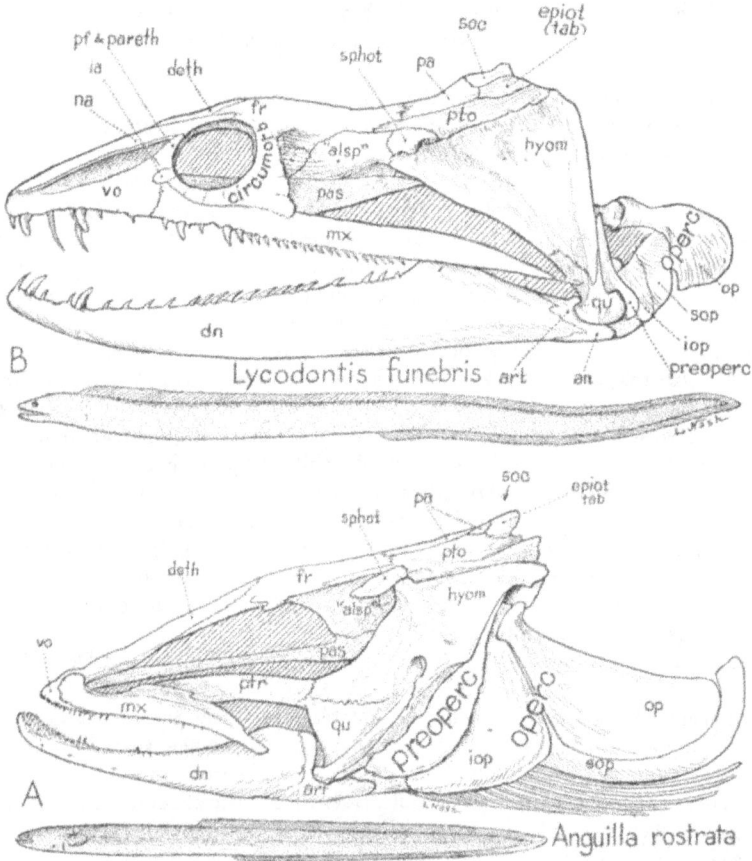

Fig. 82. Skulls of Apodes. A. *Anguilla rostrata*. B. *Lycodontis funebris*.

of the pectoral girdle from the skull by the loss of the posttemporal. This result has been conditioned by the great development of the muscles that dilate the branchial chamber. This tendency is carried to an extreme in the moray (Fig. 82B), in which the opercular apparatus has become almost vestigial. The gills too have become much reduced in size, perhaps because a smaller aërating surface is necessary on account of the "forced draught" of oxygen-bearing water. Or it may be that the "forced draught" apparatus is a response to a diminishing gill area.

As to the branchiostegal rays, Hubbs (1919, p. 65) notes that in the Apodes (eels), Heteromi and Lyopomi, the rays are all slender, usually numerous and long and frequently curved upward posteriorly about the free margin of the opercular bones.

The increased size of the respiratory muscles have apparently conditioned the powerful development of the hyomandibular and of the whole roof of the cranium; the latter has to resist the wrenching action of these muscles and afford a firm lodging for the brain, as well as a strong base for the heavily muscled jaws.

The eel represents a predaceous adaptation in an early stage, in which the suspensorium is directed forward as in primitive teleosts. The moderate length of the jaw is obtained by a moderate lengthening of the postorbital region. The moray represents an advanced stage in which the very strong hyomandibular has become directed backward, thus lengthening the jaws posteriorly.

The main upper jaw-bone of the morays appears to be the maxillary and not the pterygoid, as Boulenger (1904, p. 599) held, for the reasons that, as seen in a fresh moray head, this bone lies definitely outside of the powerful adductor mandibulæ muscles and that its posterior tip passes to the outer side of the quadrate. Obviously the pterygoid bone could not pass through the jaw muscles, which always lie lateral to it. We may also suppose that the great increase in cross-section of the adductor mandibulæ muscle could not have taken place if the pterygo-quadrate bar had remained in the position which it has in the eel, and that it was no longer needed for the bracing of the suspensorium on account of the great increase in size of the hyomandibular. The backward growth of the latter would also tend to pull the posterior end of the palatopterygoid out into the thin thread of bone which Tate Regan (1912b, p. 378) records as present in the Murænidæ.

Leighton Kesteven (1926a, p. 133) adopts the identification by Owen and Richardson of the main dentigerous upper jaw bones of the murænids with the palatines, chiefly on the ground that these bones in the morays are not "lip-bones," like those of ordinary fishes, but belong to the inner row. But if so, how is it that the posterior ends of these bones in the morays pass outside of the jaw muscles?

As to the derivation of the eels, Smith Woodward (1901, p. x) noted that relatively primitive representatives of the order were already in existence in the Cretaceous period, and that they cannot be regarded as degenerate members of any group of Cretaceous "Teleostei" hitherto discovered but have been derived directly from some of the Mesozoic fishes which would be termed "Ganoidei" by some authors. On the other hand, the "leptocephalus" larvæ of the eels is very similar to that of the isospondyl *Albula*, while the skull of eels seems to be merely a highly specialized derivative of some large-mouthed Cretaceous isospondyl type. *Thrissopater* (A. S. Woodward, 1909) might be such a form, except that the supraoccipital is in contact with the frontals, while in the eels it is separated from them by the parietals. The recent *Engraulis* among the clupeoids shows that the hyomandibular may easily become secondarily directed backward.

The otoliths (sagittæ) of Apodes, as studied by Frost (1926b, p. 99), fall into three types: the "Anguillid," the "Congrid" and the "Heterenchelid," of which the first resembles those of the Clupeoidea with certain added features. The lapillus of *Anguilla* resembles the conchoidal lapilli of the Ariidæ; the asteriscus is slight and upright as in the Elopidæ but is slightly different in form (1926, p. 100). On the other hand, it seems not impossible that the Apodes may also stand as a specialized offshoot from near the base of the Iniomi.

MESICHTHYES [1] (INTERMEDIATE TELEOSTS)

INIOMI (SCOPELOIDS)

THE scopeloids (Iniomi) are much further advanced than the typical isospondyls in the predominance of the premaxilla, in the frequent loss of the supramaxilla, and in the generally more pike-like skull characters, including the frequent predominance of the inner over the outer upper tooth rows. The mesocoracoid arch is lost.

The Cretaceous genus *Enchodus* (*Eurypholis*) of the family Enchodontidæ, as described and figured by Smith Woodward (1901, pp. 6, 189–234), has a pike-like skull (Fig. 83) with large mouth and long, sharp, piercing teeth on the palatine and ectopterygoid, with

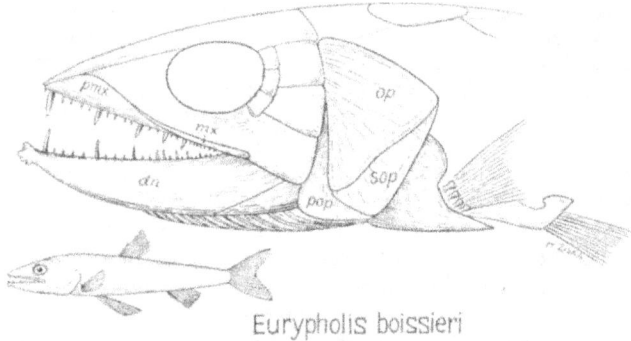

FIG. 83. *Eurypholis boissieri.* After Smith Woodward.

small teeth on the premaxilla; the maxilla is either finely toothed or toothless at the oral border. The teeth are firmly fused with the supporting bone, not implanted in sockets (Woodward). The enlargement of the premaxilla, which in many forms finally crowded the maxilla completely away from the oral border, is a process which took place independently in this family as in others among the Cretaceous fishes. The genus *Halec* of the same family was less advanced in this respect than *Enchodus*, since its maxilla still entered the gape behind, where it bore a spaced series of relatively large conical teeth pointing slightly forward (Woodward, 1901, p. 212; 1902, p. 51), recalling the conditions in the stomiatoid fishes. The presence of longitudinal rows of long sharp teeth on the pterygo-palatine arch and mandible is a point of resemblance to such scopeloids as *Alepisaurus* and *Omosudis* (Fig. 89), and a comparison of skulls of the latter two with that of *Enchodus* reveals a striking resemblance in many parts, as implied by Woodward (1901, p. 189).

A conspicuous, posteriorly-directed spine at the lower end of the very narrow and deep preopercular (Woodward, 1902, p. 42) recalls the conditions in the Astronesthidæ and the Sternoptychidæ.

According to Woodward's restoration of *Enchodus boissieri* (1901, p. 206), there were two rows of plates behind the eye, an inner smaller set marked *co* and a much broader

[1] O. P. Hay, 1902, p. 254.

posterior set marked *so*. It seems more probable that the part marked *co* was merely the raised and reflected orbital part of a single plate, as in *Albula*, *Notopterus* and others.

According to Tate Regan (1909*a*, p. 82; 1911*a*, p. 120), the Enchodontidæ "fall into the division Stomiatoidei [of the Isospondyli]; they agree with the Stomiatidæ in the structure of the skull and of the mouth." Smith Woodward (1902, p. 37), however, regards the enchodonts as "closely related to the existing Scopelidæ, Odontostomidæ and Alepisauridæ," but distinguished from all of these by having the margin of the jaw formed partly by the maxilla.

Thus the Cretaceous Enchodontidæ have been referred by one leading authority to the stomiatoid division of the Isospondyli and by another to the scopeloid division of the Iniomi—a fact that is in line with other evidence of the close interrelationships of these orders.

Sardinoides.—This Cretaceous genus (Fig. 84) is referred by Smith Woodward (1902, p. 33) to the Scopelidæ; it is in a general way intermediate in appearance between the

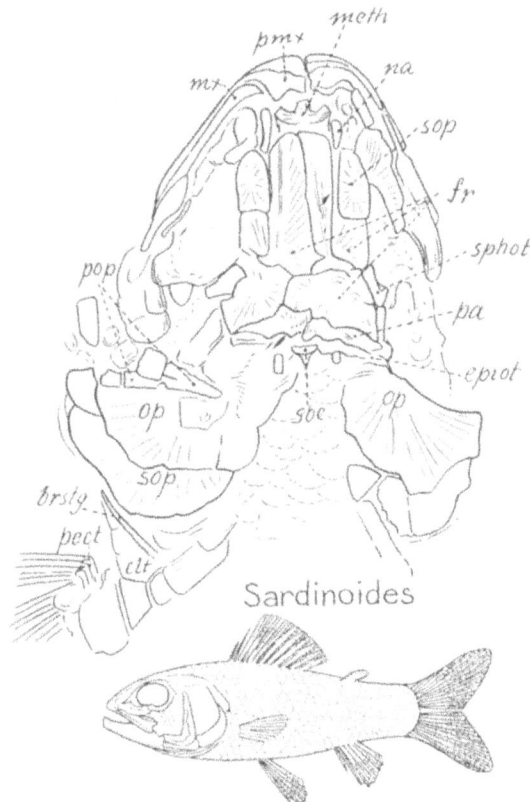

Fig. 84. *Sardinoides crassicauda.*

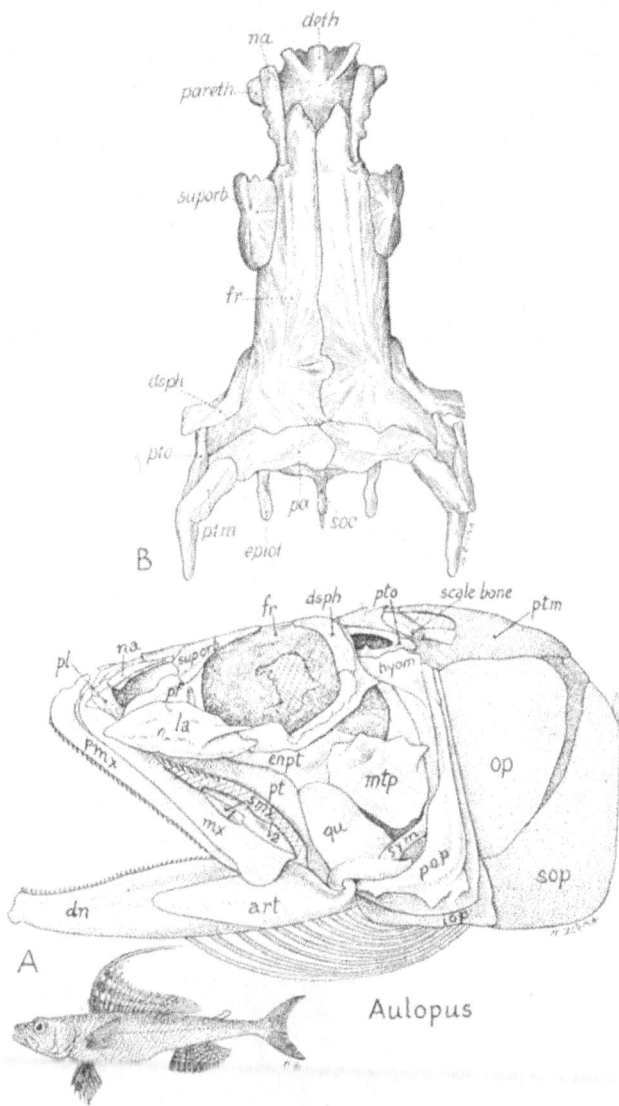

FIG. 85. *Aulopus.*

salmonoid isospondyls that retain an adipose fin and such generalized Cretaceous acanthopterygians as *Aipichthys* (Smith Woodward, 1912, p. 254). It distinctly approaches the existing scopelid *Chlorophthalmus*, but is more primitive in several respects (Smith Woodward, 1902, p. 33). Tate Regan also (1911a, p. 120) agrees that *Sardinoides crassicauda* and *illustrans* certainly belong to the Aulopidæ (*cf.* Fig. 85), the most primitive of the existing scopeloids, with which they agree closely in skull structure. On the whole, the skull characters of *Sardinoides* and *Aulopus* are very primitive. The mouth is fairly large, bordered with fine teeth, which however do not extend to the maxillary. The latter bears two supramaxillæ, a very primitive character. The parietals, although short anteroposteriorly, meet in the mid-line and exclude the small supraoccipital from contact with the frontals. The orbitosphenoid is well developed, all the opercular bones are present and it is only in minor details that the skulls differ from those of such primitive isospondyls as *Elops* (Regan). The subopercular forms the border of the gill cover.

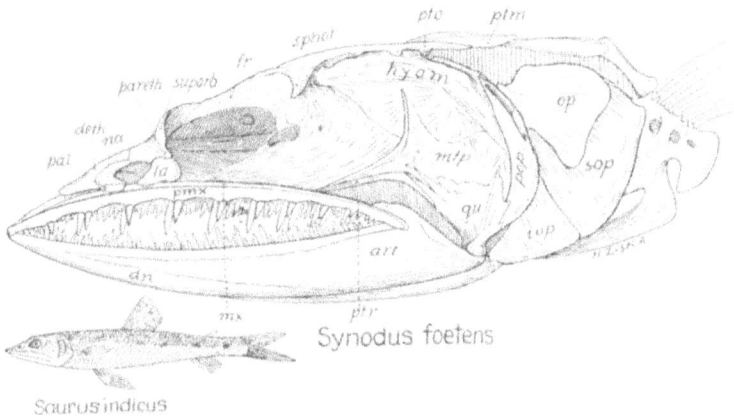

FIG. 86. *Synodus foetens.*

Synodus.—*Synodus* (Fig. 86) and its allies are more specialized than *Aulopus* in the backward inclination of the suspensorium, with the presence of more or less laniary clustered teeth. In *Synodus foetens*, according to Starks (1926a, p. 155), the vomer is entirely absent, as it is also in the allied *Trachinocephalus myops* (Fig. 87), whereas in *Saurida argyrophanes*, which is very much like *Synodus* in other respects, the vomer is well developed.

A very comprehensive and thorough analysis of the osteology of the Iniomi with special reference to the phylogenetic relationships and divergent trends of the families was published by Parr in 1929, from which the following passages may be quoted:

"The line of differentiations leading from *Chlorophthalmus* through *Bathysudis* to the Omosudidæ is characterized by a strong reduction of the lateral ethmoids, with complete obliteration of their transverse process and a corresponding shifting forward of the main attachment of the palatines to the naso-ethmoidal region. Further, by the reduction of the suborbital bones and by the anterior fusion of the nasals with the mesethmoid. The parietals remain separate. The posterior temporal fossæ are entirely unroofed. There is

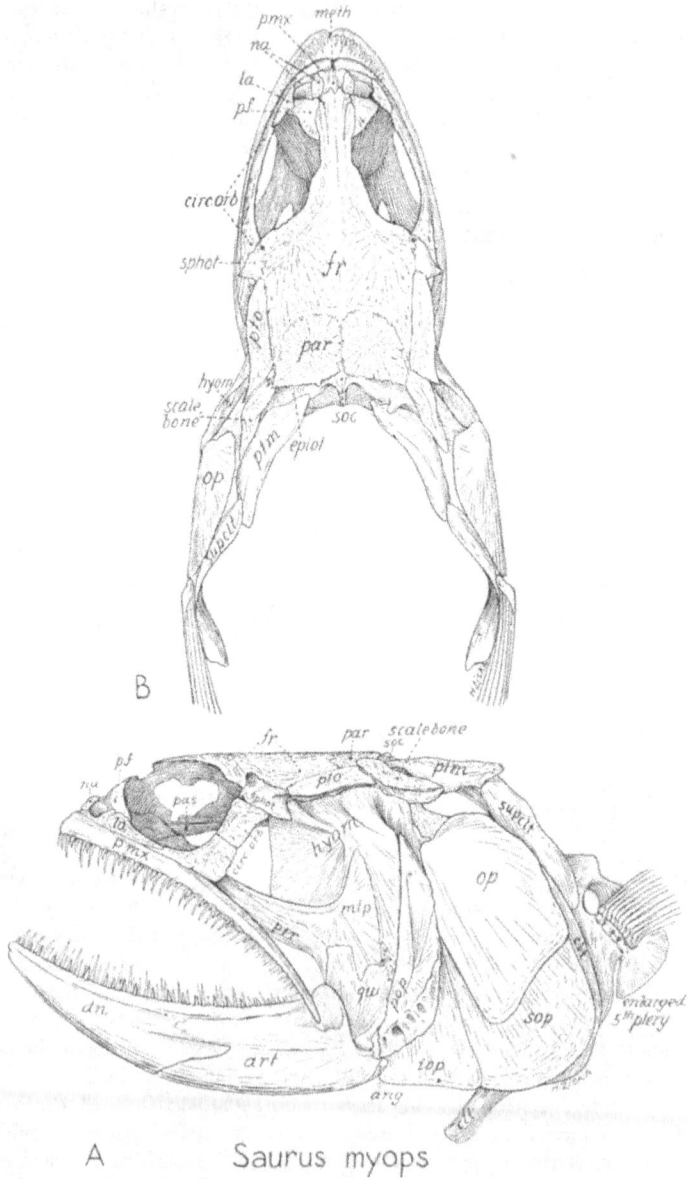

FIG. 87. *Trachinocephalus* (" *Saurus*") *myops.*

no orbitosphenoid and the basisphenoid is absent or reduced to a mere vestigeal ossification. In all these respects the Osmosudidæ [Figs. 88, 89] show a perfect continuation of the series already started by the subgenus *Bathysudis* (gen. *Lestidium*, see Parr, 1928, d. 42). . . .

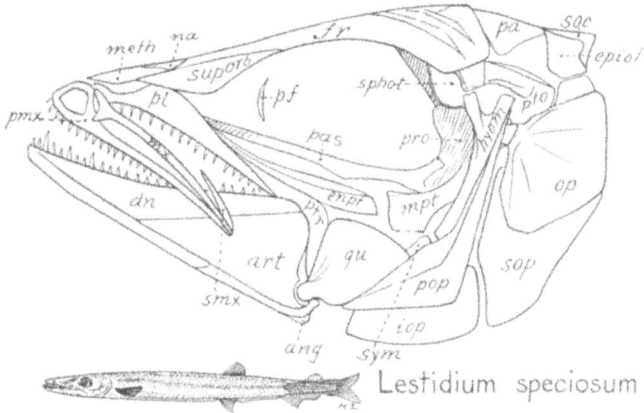

FIG. 88. *Lestidium speciosum.* After Parr.

"The evolution of the phylogenetic branch represented by the Scopelarchidæ, Evermannellidæ, and probably also by the Cetomimidæ has followed along lines entirely opposite to those of the just considered forms derived from a *Chlorophthalmus*-like ancestral type.

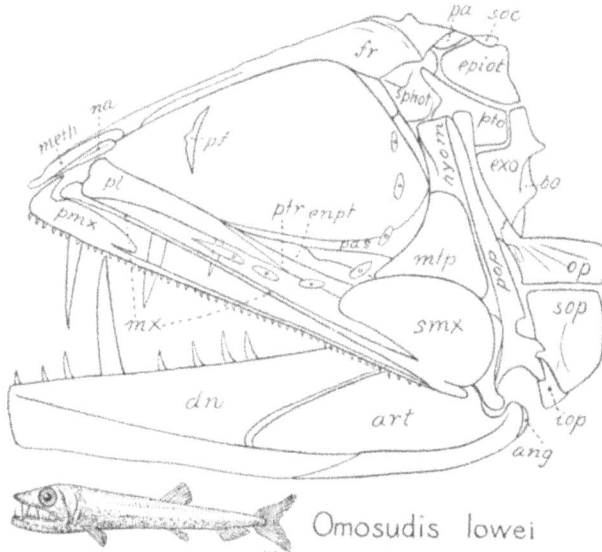

FIG. 89. *Omosudis lowei.* After Parr.

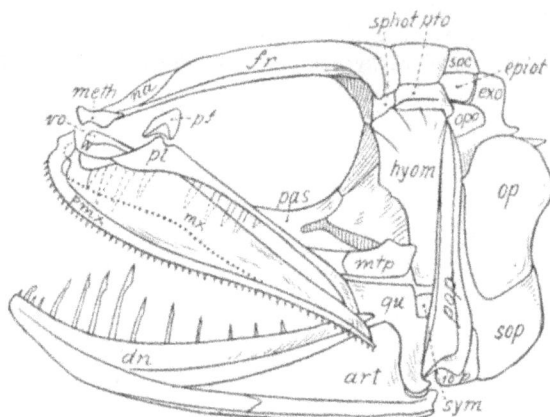

FIG. 90.   *Scopelarchus anale.*   After Parr.

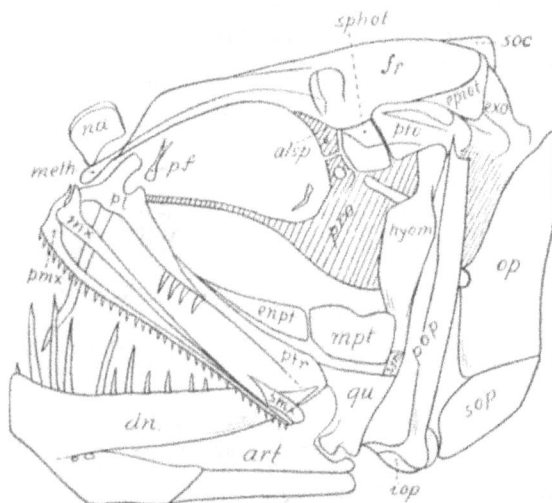

FIG. 91.   *Evermanella atrata.*   After Parr.   *Odontostomus hyalinus.*

In the Scopelarchidæ [Fig. 90] and Evermannellidæ [Fig. 91] the suborbital bones are strongly developed, instead of reduced, and a posteroventral production of the infraorbital along the upper margin of the maxillary has appeared. The lateral ethmoids and their transverse processes for attachment of the palatines are well developed throughout the series. The nasals remain entirely separate from the mesethmoid, being only attached by ligament to the anterior margins of the frontals. The parietals have become completely fused to the frontals. This is, as far as our knowledge goes, an entirely unique feature among the Iniomi. A fundamental difference is finally contributed by the fact that the Scopelarchidæ and, to some extent at least, also the Evermannellidæ, show a distinct and well developed roof over the anterior parts of their posterior temporal fossæ. According to Regan's description this roof has already disappeared in *Chlorophthalmus* and no traces are found in any of the Sudidæ and Omosudidæ examined by the author. As the feature

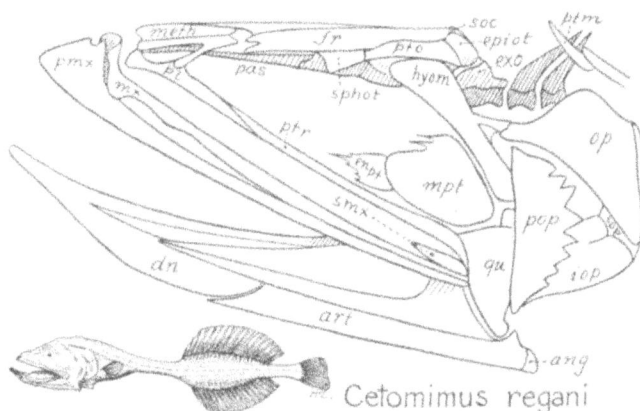

FIG. 92. *Cetomimus regani.* After Parr.

is presumably to be regarded as a primitive character, the Scopelarchidæ can therefore not be derived from a *Chlorophthalmus*-like ancestral type, and the various features of specialization exhibited already by the most primitive forms of the *Scopelarchus*-branch (parietals fused with frontals, etc.) make it equally impossible to derive *Chlorophthalmus* from a *Scopelarchus*-like ancestral type. The two phylogenetic sub-branches of the Iniomi now under discussion must therefore have separated before the stage of either of these two primitive types had been reached, as shown in the diagram on the opposite page. . . .

"The family Cetomimidæ [Fig. 92] comprises a group of highly differentiated or degenerate deep-sea forms, which, in spite of their degeneration, carry evidence of similar fundamental phylogenetic tendencies as those expressed in the skeletal structures of the Scopelarchidæ. They thus agree with the latter family in having the parietals fused with the frontals and in having the lateral ethmoids strongly developed, with prominent transverse processes serving as main support and attachment for the palatines. The author is therefore, particularly on account of the first mentioned of these features, inclined to regard the Cetomimidæ as derived from the same root as the Scopelarchidæ, and it may be added that the general aspect of the skull and visceral skeleton seems to offer no obstacles to this

10

view, when the secondary effects (such as loss of myodome and posterior temporal fossæ) of the great depression of the skull in *Cetomimus* are taken into due consideration."

A few of Doctor Parr's numerous figures of the skulls of the above mentioned families are here reproduced, with his kind permission. It will be seen that in this order, as in the Isospondyli, there is a very wide range in the inclination of the suspensorium, which is inclined far backward in *Cetomimus* but well forward in *Lestidium*.

The otoliths of the Iniomi, as described and classified by Frost (1926c, p. 466), fall into two types, called respectively the "Elopine" and the "Scopelid." In *Aulopus* the sagitta

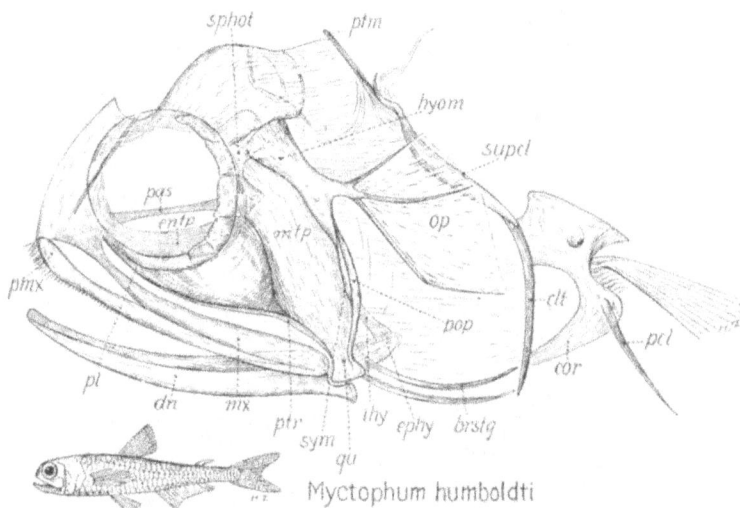

Fig. 93. *Myctophum humboldti.* Larval skull, stained by Miss Gloria Hollister for Dr. William Beebe.

is of the "Elopine" type and also resembles that of *Osmerus*, but differs in certain details. In several species of *Synodus* the otolith is also described as being markedly "Elopine," the shape being elongate and cuneiform. In *Ceratoscopelus maderensis* we have a passage from the "Elopine" to the more specialized "Scopelid" type, seen in the family Myctophidæ, which is distinguished by the heightening of the otolith, the reduction of the rostrum and other characters.

Thus the evidence of the otoliths is in harmony with much other evidence that the Iniomi are derivatives of primitive Isospondyli.

LYOMERI (GULPERS)

THE gulpers (Saccopharyngidæ) are generally regarded as "degraded eels," perhaps because they pass through a 'leptocephalus' larval stage. They have reached the logical extreme of moray-like specializations, since they have enormous mouths and a highly distensible pharynx and stomach. The rod-like maxillæ (Fig. 94) are slung from the very short solid skull by an enormously long backwardly-directed rod composed of a hyomandibular and a quadrate connected by a movable joint (Zugmayer, E., 1911). The mandibular rami are slender, loosely united at the symphysis and connected by a broad distensible membrane (Tate Regan, 1912c, p. 348). Nearly all the usually associated bones have disappeared; there are no palatopterygoids, operculars, branchiostegals, while the gills are much reduced and placed far behind the jaws.

If it were not for the last character and for the fact that the gulpers pass through a 'leptocephalus' larval stage, one might regard them as degraded derivatives of some long-jawed gonostomid like *Cyclothone*. But detailed comparison of the skulls does not support this hypothesis. Also the 'leptocephalus' of the gulpers is very unlike those of the isospondyls and more like those of certain eels. Thus it would seem at first sight that the gulpers must have been derived from some big-mouthed eel type, as Boulenger (1910, p. 603) suggests. In a phylogenetic system of classification the gulpers would in that case be placed in the order Apodes, instead of standing by themselves in the order Lyomeri.

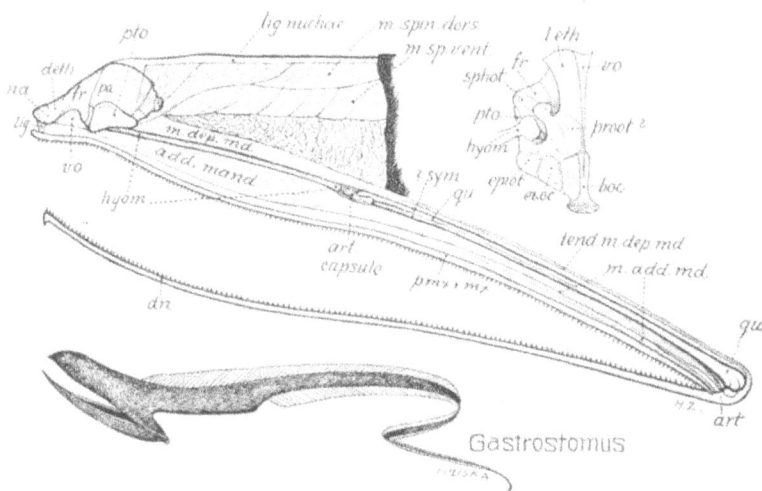

FIG. 94. *Gastrostomus*. Data from Nusbaum-Hilarowicz.

On the other hand, Tate Regan (1912b) has made out a strong case for the derivation of the gulpers from some of the predaceous Iniomi, such as *Synodus*, the skull of which seems to afford an ideal starting-point for the peculiar specialization of the skulls of *Saccopharynx* and *Gastrostomus*. The crania of these forms are progressively widened and made more solid for the attachment of the enormously long jaws and pharynx (Fig. 94). They are also much shortened anteroposteriorly as a result of the forward displacement of the eyes, shortening of the snout and forward and upward thrust of the hyomandibular. The skull of *Cetomimus regani* as described by Parr (1929, pp. 24–27) might indeed make a still better starting-point for that of *Gastrostomus* if we assumed only that the final shortening and widening of the neurocranium took place *pari passu* with the excessive elongation of the jaws. According to Regan (1912c, p. 349) *Saccopharynx* is piscivorous but *Gastrostomus* probably feeds mainly on small invertebrates.

HAPLOMI (PIKES, ETC.)

The pikes (Esocidæ) and their immediate allies the Umbridæ are regarded by Woodward as "essentially freshwater Scopeloids." Starks (1904a, 1926a) holds that the three main families of the group, the Esocidæ, the Umbridæ and the Poeciliidæ "have either widely diverged from each other or are not of the same line of descent"; also (1926a, p. 203, footnote) that "If the order *Cyprinodontes* [Microcyprini] is recognized as distinct from the order *Haplomi*, the family Esocidæ should be raised to co-ordinate rank. It has little in common with the Umbridæ." Starks (1904a) accordingly divided the order Haplomi into three superfamilies: (1) Esocoidea, containing the Esocidæ and the Umbridæ; (2) Poecilioidea, containing the Poeciliidæ; (3) Amblyopsoidea, containing the Amblyopsidæ. That the Haplomi in the restricted sense are well separated from the Microcyprini and closely related to the Isospondyli is indicated by the comparative study of the otoliths by Frost (1926, p. 465). In *Esox* the sagitta resembles that of the isospondyls *Megalops cyprinoides* of the Clupeoidea and *Gonostoma* of the Stomiatoidea but it is more highly specialized, while in the typical Microcyprini this otolith, although widely varied, seems to start from a peculiar "Microcyprinid" type.

Boulenger (1910, p. 606) referred to the order Haplomi a large series of families, many of them having no apparently strong claims to close relationship with the Esocidæ.

Tate Regan (1909a, p. 83) restricted the Haplomi to the Umbridæ, Esocidæ and Dalliidæ, raising the Poeciliidæ and its allies to ordinal rank as Microcyprini.

Evidence of the relatively close relationships of both the Haplomi and the Iniomi to the typical Isospondyli (to which group they were referred by Smith Woodward) is afforded by the form and arrangement of the branchiostegal rays. "In the Haplomi (Esox, Umbra and Dallia)," writes Hubbs (1919, p. 66), "but not in the poecilioid fishes which have been confused with them, the branchiostegals are like those of the Isospondyli. In the Iniomi (the Synodont fishes and their allies) the branchiostegals vary greatly in number (from six to twenty, four to eight attached to the suture between ceratohyal and epihyal, two to twelve below the suture); in Plagyodus the uppermost ray, as in the Isospondyli, is not wholly concealed, but in most of the genera several of the upper rays are covered by the opercula; when the rays are numerous several of the upper ones are closely approximated basally."

Esox.—If we consider the skull of the muskellonge, *Esox musquinoni* (Fig. 95), with special reference to its "adaptive" features, we shall note that the dominant physiological feature is the very large lower jaw with its row of high, well-spaced, great laniary teeth pointing upward and inward, in front of which is a series of small sharp teeth arranged along the upward curve of the undershot jaw. Obviously the big teeth pierce the prey, which is held fast by the *chevaux-de-frise* of smaller inwardly-directed sharp teeth arranged in long rows on the dermopalatines and vomer. The premaxillary teeth are practically vestigial. The maxillary, although edentulous, serves to press the prey against the lower dagger-like teeth. The supramaxillary is a souvenir of much earlier times. As the dominant elements of the lower jaw are the great laniary teeth on the sides, so the most important elements of the upper jaw are the large dentigerous areas on the dermopalatines and entopterygoids, which run in nearly parallel antero-posterior tracts. These are supported and braced dorsally by the large true palatines and posteriorly by the ectopterygoid, quadrate and metapterygoid, which also receive the heavy thrusts from the lower jaw and transmit the whole load to the strongly-braced hyomandibular.

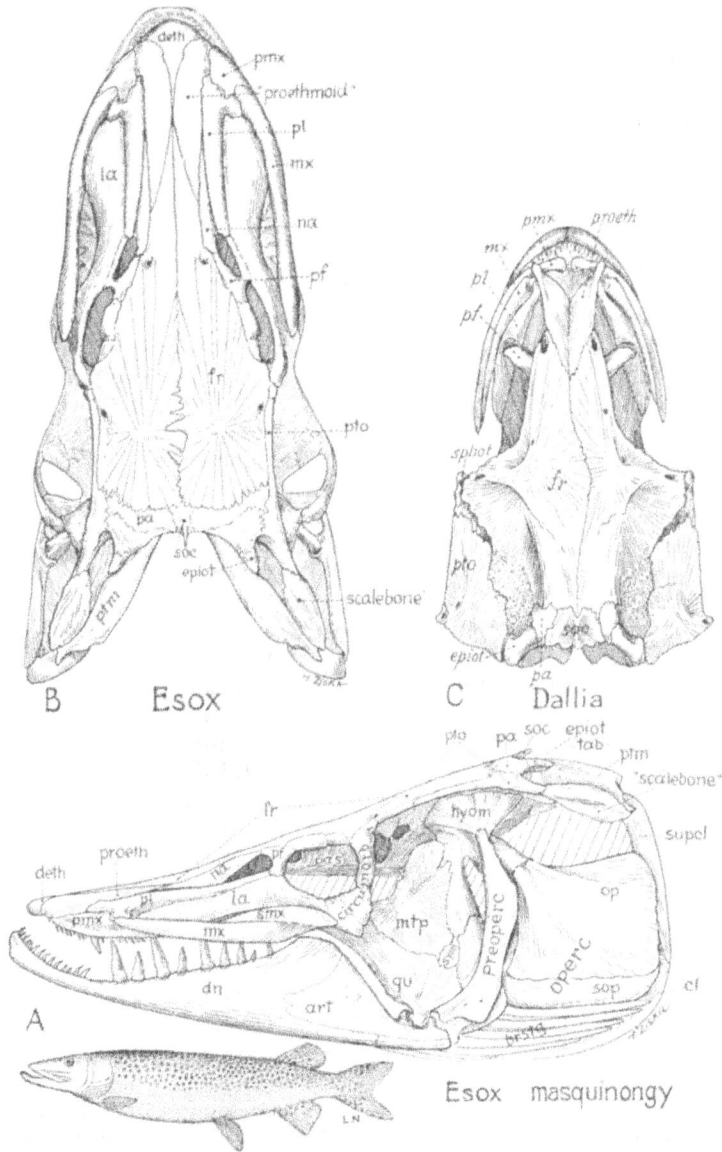

FIG. 95. *Esox masquinongy.* A. Side view. B. Top view. C. *Dallia pectoralis.* Top view.

The opposite palatines, which are exposed in the top view, do not converge toward the front but stretch, as it were, the rostrum into a thin flattened surface. This lateral stretching of the rostrum may somehow be connected with the loss of the mesethmoid and the presence of the two well developed proethmoids, which are of unknown origin.[1]

The long wedge-like cranium is strongly braced on its dorsal side by the immense flat frontals, which extend from near the tip of the snout almost to the occiput. As the body is long and low the supraoccipital crest is small, although the supraoccipital itself forms the well-braced keystone of the occipital arch. The short wide parietals, separated slightly by the supraoccipital above the occiput, form the roof of a pair of deep caverns extending inward between the pterotics and the epiotics and in front of the posttemporal fossæ of isospondyls.

The orbits are small and the myodome not large. The lacrymals are much produced in front to cover the sides of the snout and there are five rather small suborbitals, a preorbital and a dermal prefrontal. The lateral ethmoids (parethmoids) form diverging braces for the palatopterygoid arches.

Other osteological details of the cranium, as recorded by Starks (1904*a*, p. 256), are chiefly of interest in contrasting the Esocidæ with the Umbridæ and Poeciliidæ.

Umbra.—The skull of *Umbra krameri* from Hungary (No. 1013, Brit. Mus. Nat. Hist.) is singularly devoid of marked specializations beyond those noted by Starks (1904*a*). The mouth is fairly small; the maxilla enters the posterior part of the gape. The quadrate is produced forward beyond the middle of the fairly large orbit. As a whole this skull would seem to make a good structural ancestor for the specialized skulls of the cyprinodonts.

The otolith (sagitta) of *Umbra crameri* is described by Frost (1926*c*, p. 465) as being of the salmonid type, but differing in certain details. On the other hand, it resembles that of *Esox* in a number of important features that appear to indicate close relationship.

Dallia pectoralis.—This is a very peculiar Alaskan fish for which Gill erected the order Xenomi, based chiefly on certain peculiar characters of the shoulder-girdle with which we are not especially concerned. Starks (1904*c*) subjected the osteology of this fish to a thorough analysis and showed that it is related to the order Haplomi in the following characters: (1) in having paired dermal ethmoids; (2) four separate superior pharyngeals on each side, those of the anterior two arches toothless, the others with teeth; (3) upper limb of posttemporal attached to the epiotic by a ligament. Characters 1–3 he found elsewhere only in *Umbra* and *Lucius*. In addition to these he found that (4) the palatopterygoid arch was reduced to a single element, as in the Poeciliidæ; (5) that the splanchnic anatomy is very similar to that of *Umbra*. The cranium was largely cartilaginous, many of the usual bones being absent. He redefined the order Xenomi on the characters of its coracoids and actinosts, but pointed out its close relationship to the Haplomi.

Microcyprini (Top-minnows, etc.)

The ordinary *Fundulus* is a stout-bodied, vigorous little fish with broad, quick-darting tail, large rounded head and small upturned bulldog-like mouth adapted for seizing food

[1] Allis' suggestion that they may be the homologue of the mesethmoid is not accepted by Starks (1926*a*, p. 335) for the reason that he found in *Osmerus* and *Thaleichthys* a median mesethmoid in addition to paired proethmoids. On the other hand, Allis' suggestion (1909, p. 24) that these paired ethmoidal elements gave rise to the ascending processes of the premaxillæ of acanthopts has in its favor only an imperfect correspondence in the topographic relations of the respective elements (see p. 96 below).

from below (Figs. 96, 97). The suspensorium, including the symplectic, is produced far forward. The pre- and inter-opercula are also involved in the forward growth of the suspensorium.

More in detail, the mouth is more or less widened transversely, is bordered by small teeth usually simple but sometimes with bifid crowns; teeth borne by the premaxillæ and dentaries, maxillæ excluded from gape in the manner characteristic of spiny-rayed fishes. The premaxillæ lie in an anterior lip-fold which also includes the dentigerous border of the

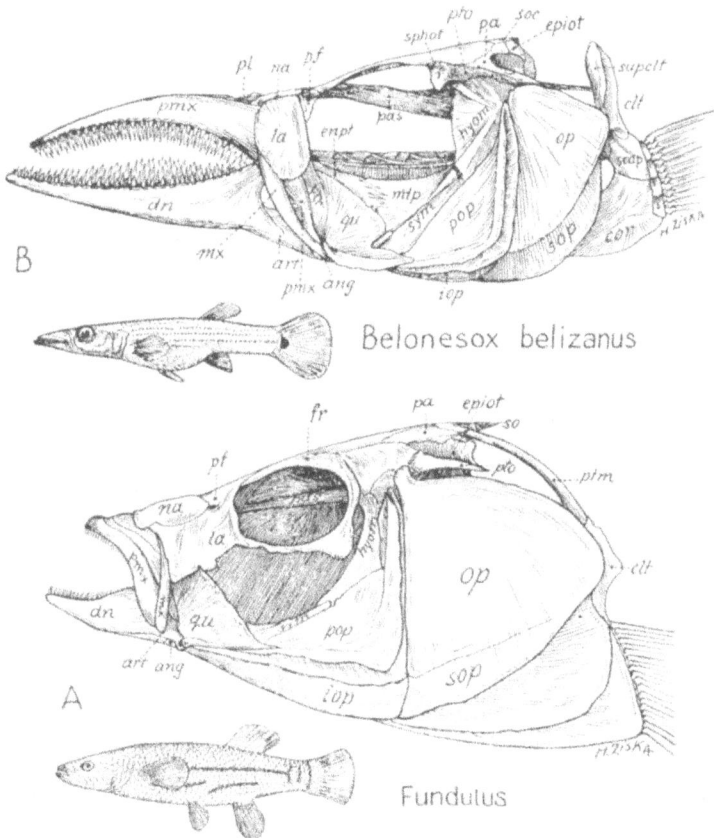

FIG. 96. A. *Fundulus.* Side view. B. *Belonesox belizanus.* Side view.

dentaries, while the maxillæ lie in a postero-lateral lip-fold which overlaps the dentary and is associated with the lacrymal and cheek-folds. A ligament passes from the lower outer part of the maxilla obliquely upward and forward to the lower outer end of the premaxilla. The tendon of the superficial layer of the adductor mandibulæ muscle is inserted on the back of the lower outer part of the maxilla, nearly opposite the tendon that runs to the premaxilla. When the mandible is depressed by the throat muscles the premaxilla is drawn downward and the maxillo-premaxillary ligament also pulls the maxilla downward and forward. When the adductor mandibulæ muscles begin to contract the mandible is drawn upward, the maxillæ backward, the premaxillæ backward and upward. The palatine also is movable, permitting the movement of the maxilla (Starks, 1926a, p. 205).

Belonesox (Figs. 96B, 97C) is essentially a small predatory poeciliid with much enlarged

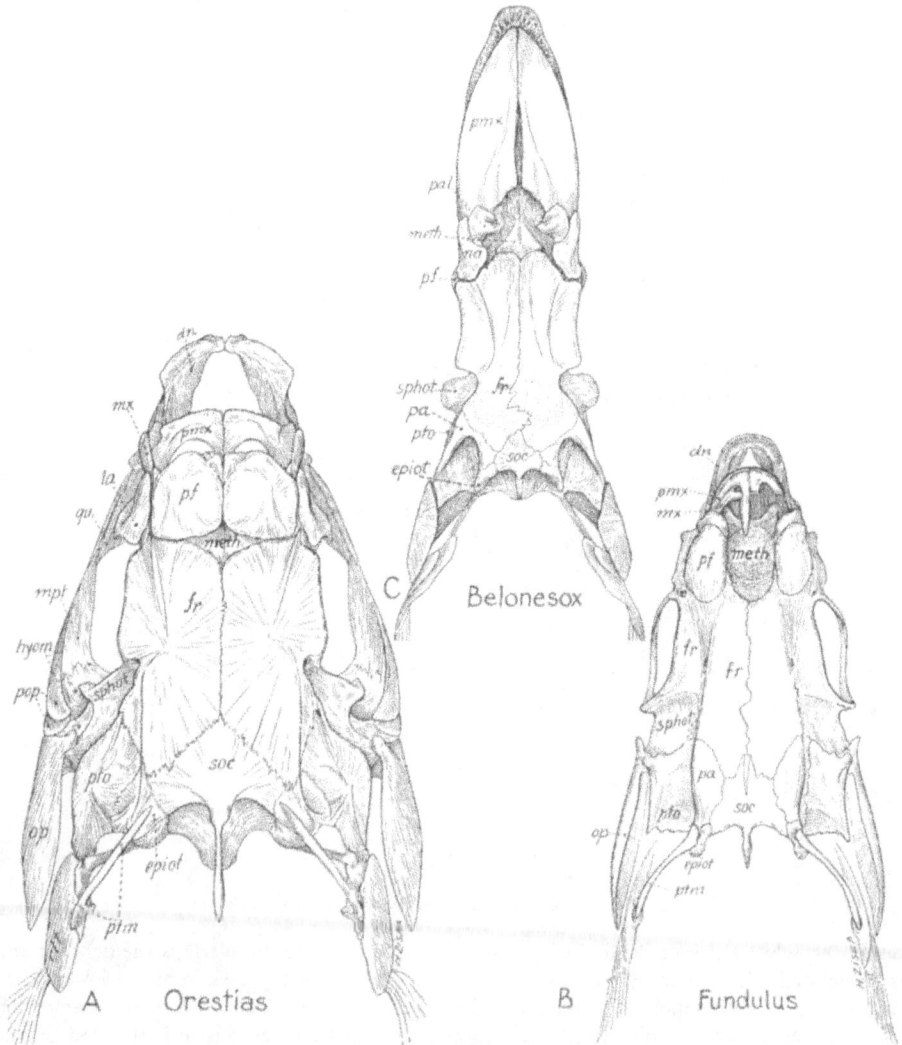

Fig. 97. A. *Orestias* sp. Top view. B. *Fundulus* sp. Top view. C. *Belonesox*. Top view.

curved jaws and fine cardiform teeth arranged in many rows on the premaxillæ and dentaries. These premaxillæ are movably hinged on the mesethmoids and vomer. The maxillæ are produced downward below the level of the dentaries and serve as levers for depressing the premaxillæ. The skull agrees in most essentials with that of *Fundulus*.

Another acanthopterygian resemblance, noted by Hubbs, is to be seen in the form and arrangement of the branchiostegal rays: "The Microcyprini (Poeciliidæ and Amblyopsidæ) were long confused with the Haplomi, but have recently been shown to have a more advanced organization. The structure of the branchiostegal rays in the two groups confirms this view; those of the Haplomi are quite like those of the Isospondyli, whereas those of the Microcyprini are similar to those of the Acanthopteri. In the Poeciliidæ there are six, or fewer, branchiostegals, which are folded up behind the opercular and above its lower margin. The upper four saber-shaped rays are attached to the outer surface of both the ceratohyal and epihyal, postero-superior to the prominent angle of the hyoid arch; the lower rays arise from the inner face of the ceratohyal." (Hubbs, 1919, p. 67.)

In his very thorough investigation of the osteology of the order Haplomi, Starks (1904a, pp. 258–261) records the many technical characters which distinguish the "Superfamily Poeciloidea" from the Esocoidea, or true Haplomi, as well as from the Amblyopsoidea. The ethmoid in the Poeciloidea is represented by a nearly circular scale of bone not unlike that seen in the Synentognathi (see also Starks, 1926a, pp. 206–211). There are no "proethmoids."

The palatopterygoid arch is toothless; "posttemporal is attached to the epiotic without the intervention of a ligament; supraclavicle is a very small scale of bone, scarcely sufficient to separate the posttemporal from the clavicle . . ." (Starks). Starks also shows that the three subfamilies of the Poeciliidæ differ widely from each other in skull characters (1904a, pp. 259–261).

The lower pharyngeal bones in *Fundulus similis* are "joined to each other by a deeply dentate and interlocked suture, together forming a triangular bone with concave sides, covered with teeth similar to those above." In *Fundulus heteroclitus* the opposite pharyngeals were elongate, attached only at their anterior ends and diverging posteriorly. In other species intermediate conditions were observed (Starks, 1904a, p. 259). Similarly in *Fundulus similis* the three superior pharyngeals on each side are "joined (not ankylosed) to form an ovate plate," covered with molar-like to conical teeth; in *Poecilia* the elongate pharyngeal teeth, both upper and lower, are bristle-like. In *Amblyopsis* the upper pharyngeals are separate and bear normal teeth and the lower pharyngeals are but loosely in contact in the mid-line. Hence in these various Microcyprini we may observe the steps by which tooth-bearing pharyngeals approaching the synentognath type are being evolved.

In conclusion, the Microcyprini seem to have paralleled the true Acanthopterygii in a number of characters, including the predominance of the premaxillæ, the exclusion of the maxillæ from the gape, the loss of the orbitosphenoid (Regan, 1909a, p. 78; 1911d, p. 321), the loss of the mesocoracoid arch, the form and arrangement of the branchiostegals (Hubbs). On the other hand, they have lagged behind the Acanthopterygii in retaining soft-rayed fins, abdominal ventrals, unarmed opercles, etc.

According to Starks (1904a, p. 254), *Umbra* (of the true Haplomi) is certainly nearer to the family Poeciliidæ than it is to any family in the order Iniomi, while Tate Regan says (1910a, p. 7) that "the relationship of the Cyprinodontidæ to the Esocidæ is generally recognized." Hubbs, on the other hand (1919, p. 67), says that "The Microcyprini (Poeciliidæ and Amblyopsidæ) were long confused with the Haplomi, but have recently been shown to have a more advanced organization." He then cited the characters of the

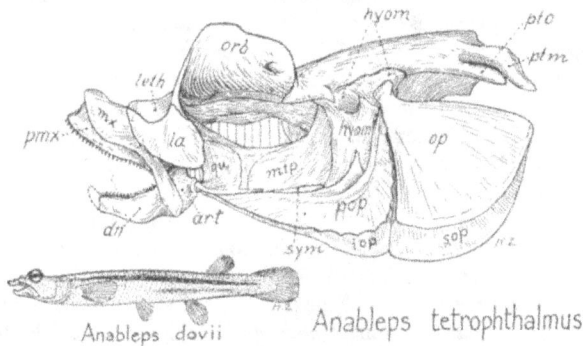

FIG. 98.   *Anableps tetrophthalmus.*

branchiostegals, as noted above.   Thus it appears probable that the Microcyprini represent a specialized offshoot of the Iniomi-Haplomi group, which have evolved in the direction of their cousins the true acanthopts.

This conclusion appears to be in harmony with the evidence from a comparative study of the otoliths by Frost (1926c, p. 474), whose figures show that the sagittæ of the Microcyprini in general vary around a "Microcyprinid" type which is notably different from the central perciform type (Frost, 1926c, Pl. XXI, Figs. 6–31; 1927b, Pl. V).   Moreover, in the Microcyprini the asteriscus, the second ossicle, is well developed and about half the height of the sagitta, its position is upright and its form usually of the characinid type.   "The lapillus is [normally] the smallest otolith, and is often microscopic; it has the form of a bean in many species, and resembles those of the Characinidæ and the primitive cyprinid *Baralius* . . ." (Frost, 1926c, p. 474) while in the Perciformes the asteriscus and lapillus remain diminutive and without special features (1927b, p. 298).   Thus the Microcyprini appear to be entitled to the ordinal rank conferred upon them by Tate Regan.

The skull of *Anableps tetrophthalmus* (Fig. 98) of this order has enormous orbits, the upper borders of which protrude high above the level of the cranial roof.   The very small mouth lies wholly in front of the orbits.   The maxillæ form large, relatively flat flanges which are excluded from the gape by the delicate premaxillæ.   The mandible is minute. In accordance with the forward position of the mouth the quadrate, preopercular and interopercular are likewise produced far forward, and form a sharp angle with the anteroposteriorly extended hyomandibular.   The opercular lacks a spine.   The cranial roof is broad and flat, without crests.   To judge from the characters of the orbits and mouth parts *Anableps tetrophthalmus* ought to be able to swim at the surface and pick up small prey, such as water insects.

The skull top of "*Goodea atripinnis*" (= *Characodon luitpoldii*) has been figured by Tate Regan (1911d, Pl. VIII).   As a whole this skull (Brit. Mus. Nat. Hist., No. 1026B) strongly suggests the *Fundulus* type.   The mandible is extremely small and transversely flattened.

SYNENTOGNATHI (NEEDLE-FISH, FLYING-FISH, ETC.)

The skull of the needle-fish (*Tylosurus*) (Figs. 99, 100) is specialized in many features relating to the excessive elongation of the pointed jaws and of the body, the latter being adapted for leaping from the water. The bill, or upper jaw, is composed largely of the prolonged premaxillæ. This is braced posteriorly by the narrow maxillæ (which barely, if at all, reach the angle of the gape) by the forwardly-produced palatines, the thin vomer, the long parasphenoid, the forwardly-produced naso-proethmoids (Starks, 1926a, p. 208), and by the delicate, minute, disc-like mesethmoid. The cranial table is flat and elongated,

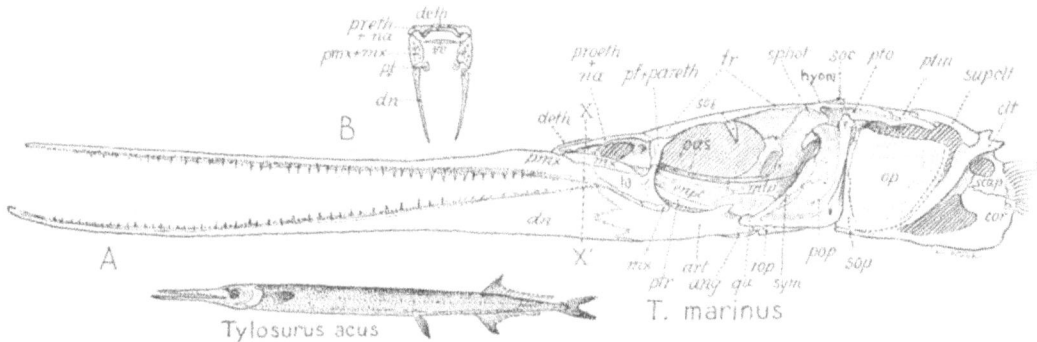

FIG. 99. A. *Tylosurus marinus*. B. Cross-section along a line XX¹.

the frontals being the dominant elements. The orbits are fairly large. The minute supra-occipital meets the frontals and the epiotics are crowded in toward the mid-line. As seen from above the skull ends posteriorly in a pair of divergent horns, formed mostly from the posttemporals and pterotics. The braincase is a rather narrow stiff trough.

The elongate mandible is supported posteriorly by a forwardly-produced suspensorium, which bears stiffening ridges on the quadrate and metapterygoid. The upper limb of the preopercular is inclined slightly forward, the lower limb is horizontal. The roof of the mouth is stiffened by the parasphenoid. As is well known, the lower pharyngeals are coalesced and support a triangular dentigerous surface, the teeth of which can be opposed to those on the upper pharyngeals. Thus stiffness and lightness seem to be the main adaptional features in this long-beaked skull.

The skull of the flying-fish *Halocypselus evolans* (Fig. 101) somewhat recalls that of *Fundulus* in the relatively small, upturned mouth, triangular pharyngeal plate, smooth, rounded operculars, disc-like mesethmoid. But it is obviously nearer phylogenetically to the skull of *Tylosurus*, although contrasting with the latter in its small jaws without a beak. From the excellent studies of Nichols and Breder (1928, p. 439) it even appears probable that the short-jawed *Halocypselus* skull has been derived from one more like that of *Tylosurus* by the rapid shortening and finally by the complete loss of the beak. Possibly reminiscent of a long-jawed stage are the bracing of the mandible by lateral ridges on the quadrate and metapterygoid, the prolongation of the postero-superior border of the mandible, the buttressing of the small upper jaw by the enlarged lateral ethmoids.

The characters of the back of the skull in both *Tylosurus* and *Halocypselus* have been

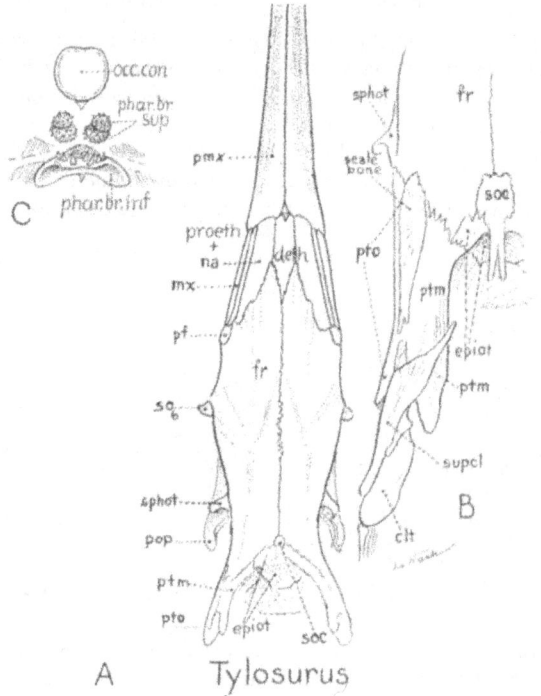

FIG. 100. *Tylosurus acus.* Details of skull. A. Top view. B. Top view of occiput much enlarged. C. Posterior view of pharyngo-branchials.

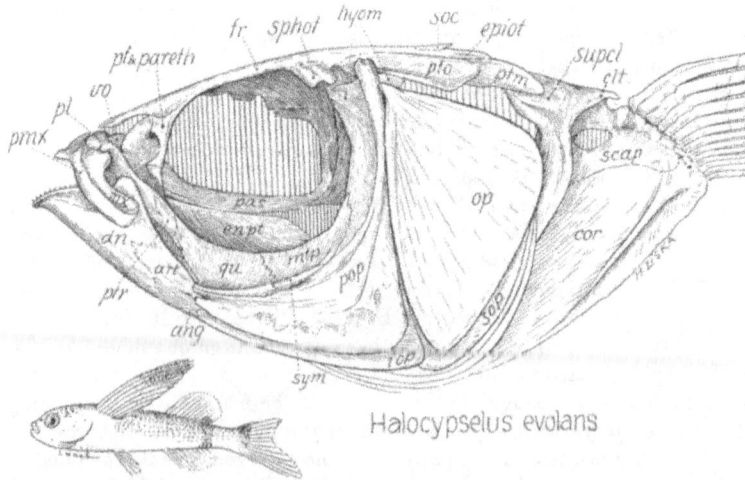

FIG. 101. *Halocypselus evolans.*

influenced by the very high position and large size of the pectoral fins. As a result the supracleithrum has become reduced, while the enlarged posttemporal collaborates with the backwardly-produced pterotic in forming a firm strut for the pectoral girdle. Meanwhile the epiotic has been reduced, especially in *Tylosurus*, and crowded in against the small but well braced supraoccipital. The latter is firmly embraced by the ends of the broad frontals, which are the dominant bones of the skull-roof; the parietals, at least in this form, having been eliminated entirely.

The Synentognathi are unquestionably a natural group. Their characteristic fused lower pharyngeals, which form a toothed plate opposing the more or less enlarged upper pharyngeals, are merely analogous with those of the wrasses and many other groups of the spiny-finned series, but may possibly be related by community of origin with the suturally-united dentigerous lower pharyngeals of some of the cyprinodonts. Boulenger (1910, p. 636) indeed regards the Scombresocidæ (gars, etc.) as being "somewhat related to the Cyprinodonts." A review of the family characters of the two groups as listed by several authors leads at first to the impression that the synentognaths and cyprinodonts are the divergent offshoots of a broad common stock, the cyprinodonts becoming specialized in many features. The practice of internal fertilization, so highly developed in the cyprinodonts, is also inherited by *Hemirhamphus* of the synentognaths (Boulenger, *op. cit.*, p. 638).

A comparison of the surface characters of the skulls of *Fundulus*, representing the cyprinodonts, and of *Tylosurus*, *Hemirhamphus* and *Halocypselus*, representing the synentognaths, reveals a sufficient degree of similarity to suggest at least superordinal relationships. The operculars are large, convex and smooth, without spikes. The not large supraoccipital is in contact with the frontals; the skull-top more or less flat without sharp crests; opposite "prefrontals" nearly in contact medially; mesethmoid more or less disc-like in top view (Starks, 1926a, pp. 205–211); orbits fairly large; lacrymal prominent, overlapping maxilla, other suborbitals absent or represented by narrow rim; suspensorium produced forward, mouth pointing more or less upward, rather broad transversely; premaxillæ more or less beak-like, excluding maxillæ from gape; presence of proethmoids.

The gars, needle-fishes and flying fishes were referred by Boulenger to the "suborder Percesoces," which comprised some twelve families of very diverse-looking fishes. These were grouped together because collectively they showed transitional characters between the soft-rayed fishes with open duct of the swim-bladder, and the spiny-finned groups with closed swim-bladder. But the five characters cited by Boulenger in the definition of the order are each in itself subject to wide variation among teleosts as a whole. Realizing this, Boulenger (1910, p. 636) remarks: "Although this suborder is perhaps only an artificial association, it must be borne in mind that notwithstanding the very wide divergence which exists between the first and last families, and however dissimilar their members may appear to be at first sight, a gradual passage may be traced connecting the most aberrant types."

Tate Regan (1929), however, does not accept this view of the case and distributes the contents of Boulenger's "Percesoces" under a number of different orders. He follows earlier authors in treating the skippers and flying-fishes as a distinct order, Synentognathi, while he refers the typical Percesoces (Sphyrænidæ, Atherinidæ, Mugilidæ, etc.) to the Percomorphi.

In this connection the surface views of the skull (Figs. 100, 101) reveal a close resemblance in family heritage between the needle-fish *Tylosurus* and the flying-fish *Halocypselus*, but only a general and ambiguous resemblance between the latter and *Mugil* (Fig. 138) and *Sphyræna* (Fig. 141) of the true Percesoces. Tate Regan (1929, p. 317) also notes that "the large number of branchiostegals (9–15), the structure of the mouth, the absence of spinous rays, the truly abdominal pelvic fins, etc., indicate their derivation from the Isospondyli." Evidence from the branchiostegal rays is also cited by Hubbs (1919, p. 66) as follows: "The branchiostegals of the Synentognathi (Belonidæ, Scombresocidæ, Hemirhamphidæ, Exocoetidæ) are wholly similar to those of the typical Isospondyli; they are rather numerous (ten in *Euleptorhamphus*), but not constant in number, flat, imbricate plates; the uppermost skirting the lower margins of the opercula, and all with their lower edges exposed. The characters of the branchiostegal rays of the Synentognathi strongly confirm Regan's view that the resemblance between these fishes and the Percesoces is purely fictitious; the group should be placed among the typical soft-rayed fishes."

The otoliths of the Synentognathi, according to G. Allan Frost (1926c, pp. 471–473), retain elopine and clupeoid features in the details of the sagitta, while some species show resemblances to the forms of the orders Salmopercæ and Apodes. On the other hand, these otoliths differ considerably from those of the Mycrocyprini. These facts, added to other evidence, seem to justify Regan in placing the Synentognathi ahead of the Microcyprini and Percomorphi in his ordinal classification.

THORACOSTEI (STICKLEBACKS, TUBE-MOUTHS, SEA-HORSES, ETC.)

The skull (Fig. 102) of the rough-tailed stickleback (*Gasterosteus trachurus*, No. 86, Brit. Mus. Nat. Hist.) represents a primitive stage in this group before the elongation of the preorbital part of the face. The mouth is fairly small, the quadrate-articular joint

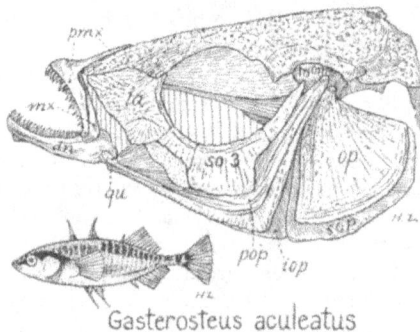

FIG. 102. *Gasterosteus aculeatus.*

lying well in front of the preorbital border. The ascending processes of the premaxillæ are of moderate length, the small maxillæ are closely fastened at the distal end to the premaxillæ and by ligament to the side of the small mandible, so that the mouth must be protrusile, at least to some extent. The lacrymal is fairly large and the third suborbital over-arches the cheek and extends downward and backward to make contact with the preopercular. This analogy with the Scorpænoids is hardly borne out by other skull characters. The opercular lacks a spine and the skull is devoid of crests and is not percoid in general appearance.   On the whole, the skull type suggests rather those of the Umbridæ and Poeciliidæ.

In *Gasterosteus spinachia* (No. 87, Brit. Mus. Nat. Hist.) the preorbital face is decidedly longer.   The premaxillæ have long ascending processes, indicating marked protrusility. The opercular region and the skull top suggest the poeciliid type.

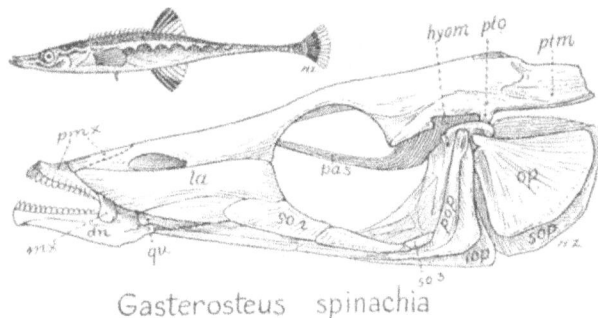

Fig. 103.   *Gasterosteus spinachia.*

The elongation of the preorbital face is carried to an extreme in *Aulostomus* (Fig. 104) and still more in *Fistularia* (Fig. 105).   We have already seen several types of fishes in different orders in which the suspensorium was inordinately produced forward, but nothing approaching the condition in *Fistularia* is known outside its own group.   Much the greater part of the long tube that lies between the mandible and the eye is roofed by the mesethmoid, the remaining part being contributed by the vomer.   Beneath this lies the elongated

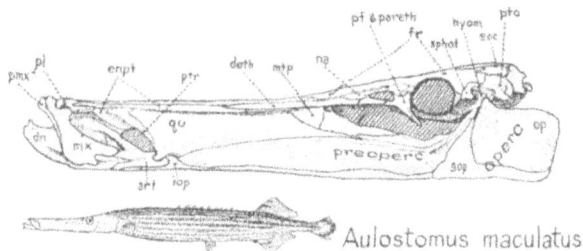

Fig. 104.   *Aulostomus maculatus.*   After Jungersen.

entopterygoid, quadrate, interopercular and preopercular.   In the top view the small supraoccipital is seen to be wedged in between the appressed epiotics.   No parietals are present.   Doubtless in order to strengthen the vertebral column against the adverse leverage of the excessively long skull, the four anterior vertebræ are greatly elongated, with wide parapophyses, the whole complex being fused into a rigid tube.   In *Aulostomus* the specialization of the skull is less advanced than that of *Fistularia*.   In the Syngnathidæ (pipe-fishes and sea-horses) the skull (Fig. 106) is fundamentally the same as in *Fistularia* but even more specialized.

The skull of *Phyllopteryx* (Fig. 106), as figured by Jungersen (1910, Pl. V, Fig. 8) has in fact the leading specializations of the *Fistularia* skull together with others of its own, notably the presence of a row of "antorbital plates" on the sides of the oral tube and the downward bending of the basioccipital, in correlation with the upright posture of the body. The skulls of *Nerophis* (Pl. V, Figs. 9, 10) and of *Siphonostoma* (Pl. V, Figs. 1–5) appear to be less specialized and tend further to connect the syngnathid with the aulostomid stem.

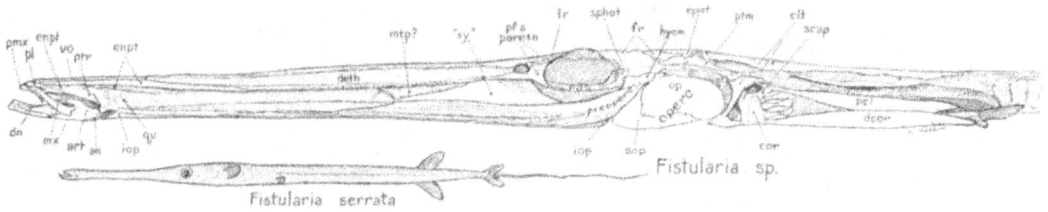

FIG. 105. *Fistularia* sp. After Jungersen.

Nevertheless it seems not improbable that part of the resemblances between the syngnathid and the aulostomid skulls may be due to parallelism, in view of the fact that *Solenostomus* (Fig. 107) tends strongly to connect the syngnathids with the centriscoids.

The centriscoids, in spite of their peculiar specializations, probably indeed stand nearer to the gasterosteoid stem than do the solenostomids. *Centriscus* (Fig. 108) could be de-

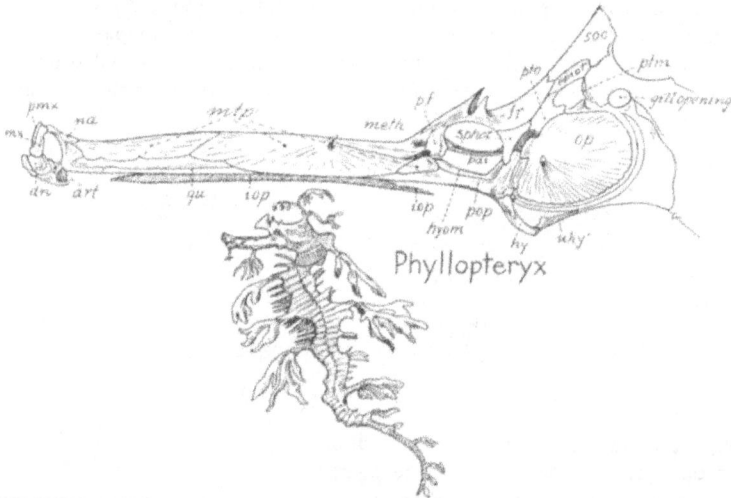

FIG. 106. *Phyllopteryx foliatus.* After Jungersen.

rived from a form like *Spinachia* by the marked elongation of the snout and by the great increase in size of one of the dorsal spikes. The latter is so enormous as to require the magnification and close appression of the first four vertebral spines, epineurals and centra. *Amphisile* (Fig. 109) is obviously only a peculiarly specialized centriscid.

Conceivably the fistulariids might well be derived from a short-bodied form like *Centriscus* by the loss of the dorsal spine, the sudden elongation of the body and the fusion of the already elongated first four vertebræ. Thus the general lines of ascent may be

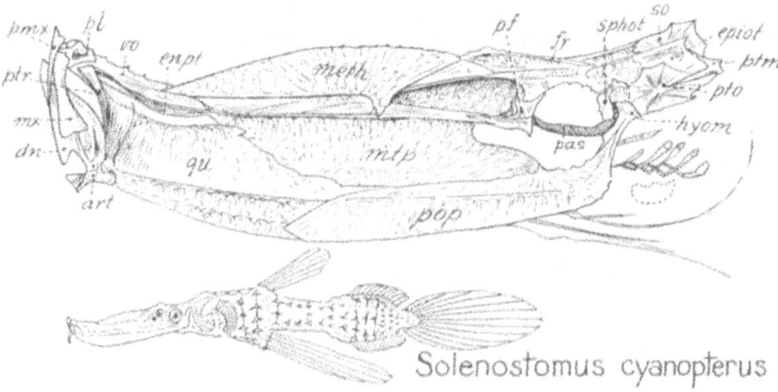

FIG. 107. *Solenostomus.* After Jungersen.

visualized as first from *Gasterosteus* to *Centriscus*, thence on the one hand to the fistulariids, and on the other, through *Solenostomus*, to the syngnathids.

The osteology of the skull and skeleton of the hemibranchiate fishes has been fully described by Starks (1902*a*) and by Jungersen (1908). Up to the time of the publication

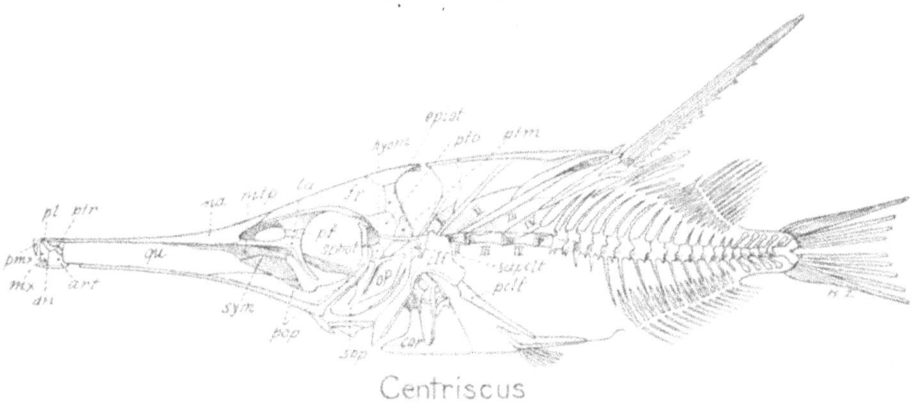

FIG. 108. *Centriscus.* After Jungersen.

of Jungersen's monograph modern ichthyologists had generally accepted the view of Cope, Gill, and Smith Woodward that the sticklebacks, aulostomids, tube-mouths, centriscids, sea-horses and their collateral branches together constituted either a single order or at most two orders, formerly called Hemibranchii and Lophobranchii. Jungersen, however, came

11

to the conclusion that the gasterosteids and aulorhynchids were offshoots of the scorpænoid group and that the aulostomoids, centriscoids, and lophobranchs form a natural group, for which Tate Regan (1906–1908, p. xi) had invented the name *Solenichthyes.* Tate Regan in 1909 (1909*a*, p. 84) could not, however, accept Jungersen's view that the gasterosteids and aulorhynchids belong to the Scorpænoidei, but in 1929 (p. 323) he tentatively accepted this view, stating that in the stickleback the third suborbital bone is extended backward

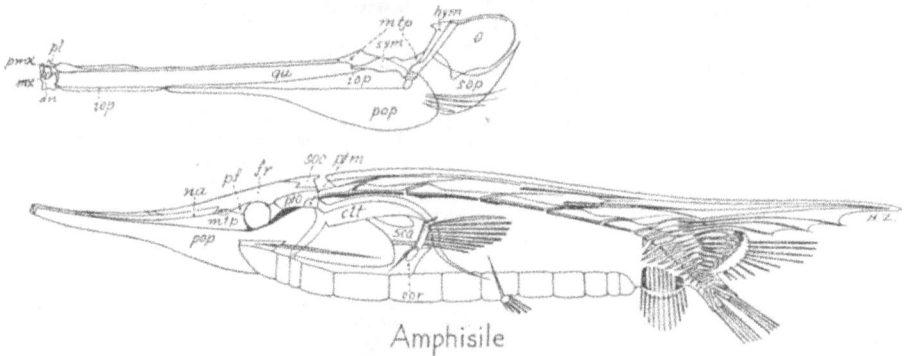

Fig. 109. *Amphisile.* After Jungersen.

over the cheeks as in the scorpænoids and that none of the other characters seem to be inconsistent with this allocation of the group. Starks (1926*a*, pp. 212, 213, footnote; p. 213), on the other hand, dissented from Jungersen's view and felt that the group might better be left in its customary position not far from the synentognaths and Percesoces. In 1902 (1902*a*) Starks wrote as follows (p. 621): "The Hemibranchs certainly do not deserve coördinate rank with the Acanthopteri but should be included as a suborder under them, coördinate with the Percesoces. Probably the Synentognath fishes should also be so included." He then cites (p. 621) a number of osteological characters which they share with the Synentognathi and another set of characters (p. 622) which they share with the Percesoces. He gives extended diagnoses of the Hemibranchii and of its main subdivisions, recognizing the following superfamilies: Gasterosteoidea, Aulostomatoidea, Macrorhamphosoidea and Centriscoidea.

Comparison of the figures, descriptions of the pectoral girdles and family definitions given by Starks, Jungersen, Tate Regan and the other authors cited above, indicate that in a general way the gasterosteoids, the hemibranchs and the lophobranchs form successive grades of organization, and that the relationships of the families within these groups are as follows (*cf.* Goodrich, 1909, pp. 410–413):

I

Gasterosteoidea
Gasterosteidæ
Aulorhynchidæ

II
Hemibranchii

A

Macrorhamphosoidea
Macrorhamphosidæ
Centriscidæ
Amphisilidæ

B

Aulostomatoidea
Aulostomatidæ
Fistulariidæ

III
Lophobranchii
Solenostomidæ
Syngnathidæ

The contrasts in the shoulder-girdle between groups I and II are rather radical and tend to support the views of Jungersen and Starks that we have to do with two distinct orders of tube-mouthed fishes of possibly different origins. Nevertheless both groups have a dermal plate (wrongly called "interclavicle" by early authors but shown by Swinnerton (1905) to be a neomorph in this group), which fuses with the lower border of the coracoid. *Macrorhamphosus* apparently has the least specialized pectoral pterygials, from which those of *Fistularia* and of *Gasterosteus* may have specialized in opposite directions. The lateral line plates over the shoulder-girdle in *Aulorhynchus* in Division I, according to Starks, are homologous with those in *Macrorhamphosus* of Division II.

In short, while there is doubtless a deep gap between Divisions I and II, there seem to be too many indications of ultimate community of origin to justify referring one branch to the scorpænoids and another to a distinct order of quite unknown relationships. After considering the matter with some care, I incline to the opinion that the characters cited by Starks in favor of relationship of the entire series to the synentognaths and Percesoces outweigh the points of resemblance to the scorpænoids cited by Jungersen. In this connection we may cite the following passage from Swinnerton (1902, pp. 580–581):

"To the best of my knowledge, the nearest approach to this order [Thoracostei] among living fishes is made by the Scomberesocidæ. Indeed, so close is this approach that on a consideration of the head skeleton alone one would be almost obliged to place *Belone* in the same sub-order with *Gasterosteus*. Give its cranium an arched instead of a flattened roof, replace its alisphenoid by overlapping frontal and parasphenoid processes, shorten the premaxillæ and mandible to a normal length, elongate the symplectic still further, and it would be extremely difficult to find any feature of importance in which the two crania differed, for in the *Belone* all the roofing bones are sculptured; in spite of its lowly affinities, its opisthotic is absent; the ethmoid, though more cartilaginous, is of the same type; the branchial apparatus is an exact replica of that in *Gasterosteus* in the number and nature of the basibranchials, in the number, shape, and proportional size of its pharyngobranchials, and in all other features except the fusion of the vestigial elements of the fifth arch. Again, the hyomandibular is of the same shape, though its articulations are more generalized; the metapterygoid is equally reduced; one pterygoid line alone is present; the palatine is small, edentulous, and lacks a maxillary process; finally, it presents the acrartete condition.

"The similarity is so great that one may say with considerable truth that the little stickleback is but a slightly specialized *Belone*.

"In the trunk region, however, though the pectorals are raised, the pelvics, abdominal, and the arrangement of the other fins is the same as in Thoracostei; yet the complete absence of bony plates and infra-clavicles gives some excuse for not including the Scomberesoces in the new sub-order."

The peculiar connections of the nasal bones of *Gasterosteus*, as described by Starks (1926a, pp. 212–213), further separate this family from the mail-cheeked fishes. For in *Gasterosteus* the nasals are "closely attached to the frontals by dentate sutures and continue the rugose surface of the frontals forward. They are widely separated, and the area between them drops abruptly to the level of the upper surface of the flat vomer, and forms a deep fossa for the reception of the ascending premaxillary processes. From the lower surface of each nasal a large process turns backward and is firmly attached against the side of the parasphenoid, and against the lower, forward-projecting, plate of the prefrontal. The nasals thus both roof, and (with the help of the prefrontal) floor the deep nasal fossa. This is, as far as known, a unique condition."

The backward extension of the third suborbital bone over the cheek might well be a sign not of relationship with the scorpænoids but of descent from primitive clupeoid isospondyls.

The evidence of the otoliths, as presented by Frost (1929b, p. 263), does not favor the allocation of *Gasterosteus* to the scorpænoids, since its sagitta "shows little resemblance to those of the remainder of the order Scleroparei, but it resembles the Scopelid type." On the other hand, Swinnerton's contention that *Gasterosteus* is related to *Belone* is not supported by a comparison of their otoliths, those of *Belone* and other synentognaths (figured by Frost in 1926c, p. 471) conforming in general to the "Elopine," "Clupeid" and "Biovate" types, while that of *Gasterosteus* is rhomboidal, biconvex and of scopelid type. As to the possible relationships of the sticklebacks to the pipe-fishes, Frost notes (1929b, p. 263) that: "In the example examined of *Spinachia spinachia*, the otoliths were absent or microscopic, as frequently occurs in the Pipe-fishes, which in some features it resembles."

ACANTHOPTERYGII (SPINY-FINNED TELEOSTS)

A PECULIAR arrangement of the branchiostegal rays, characteristic of the spiny-rayed fishes, is thus described by Hubbs (1919, pp. 68–70): "A definite fixed type of branchiostegal structure has been retained, almost without deviation, throughout the great groups of spiny-rayed fishes which flourish so abundantly in the modern seas, and with peculiar constancy in the numerous highly specialized offshoots of the typical Acanthopteri. In fact, it seems safe to assert that none other of the known characters which separate this series from the lower teleosts has been more conservatively maintained throughout the entire group. . . .

"The characteristically stout hyoid arch is strongly angulated at some distance below and before the (typically) dentate suture between the ceratohyal and the epihyal, the angle forming the hinder border of a concavity in which the musculus geniohyoideus is attached. The strong development of this muscle not only modifies the form of the hyoid arch, but also modifies the structure and attachment of the branchiostegal rays, as it also does, usually to a lesser degree and without constancy, in the lower teleosts. The upper four saber-shaped branchiostegals are always attached to the outer surface of both epihyal and ceratohyal, at and above the angle of the arch, and are folded together like a fan above and behind the opercular margins (except in those cases in which the branchiostegal membranes are drawn taut by their broad union ventrally). Below (and before) the angle of the arch, to its edge or inner surface, usually two or three shorter and slenderer rays are attached; these may be reduced to one, or, very rarely, to none, and are increased, in certain berycoids and blennioids to four, but never to a higher number. Thus, the branchiostegals of the Acanthopteri and related groups are usually four plus two or four plus three in number, rarely four plus one or four plus four, and very rarely four plus nought or even three plus nought."

Since this general arrangement is characteristic of the most diversified Acanthopterygii and since clear traces of it persist, even in very highly specialized groups, it may be taken in conjunction with the general perch-like construction of the skull and with other well known characters of the fins as one of the chief characters of the Order Acanthopterygii, even though it does not seem necessary that every fish of this group should still possess all the "ordinal" characters.

As to the division of the spiny-finned fishes into suborders and superfamilies, Tate Regan (1913a, p. 111) admits that "it is largely a matter of opinion whether some of these [his "suborders" of the Percomorphi] may not be regarded as ordinally distinct, or whether others should not rank merely as divisions of the Percoidea." During the course of my studies on the skull of many representative acanthopts I have constantly kept before me his definitions of the families, divisions and suborders of the Percomorphi and related "orders." In the following pages I am not proposing a formally defined classification of the divisions and subdivisions of the spiny-finned series, I am merely using such words as Percoidei, Mugiloidei, etc., as convenient names for more or less well known groups of families, each of which clusters around some typical form. In fact I incline to agree with the late Dr. O. P. Hay that in order to "define" a varying group one need only state what forms are referred to it.

SALMOPERCÆ

In the rivers and streams of Canada and the northeastern United States lives a certain species of small fish, and in the sandy or weedy lagoons along the Columbia lives another and related species, which are of the greatest interest to the student of "missing links," because in their anatomical characters they almost exactly divide the differences between the lower, or soft-rayed, and the higher or spiny-finned teleosts. They have one to four spines on the prominent dorsal and anal fins, the opercular and preopercular both bear a posterior spine, the scales are ctenoid, the ventrals are beneath the pectoral fins, as in acanthopts. On the other hand, the air-bladder retains an open duct and they have an adipose dorsal fin like the salmonids and other isospondyls.

Authorities differ as to the proper place of the percopsids in the system of teleosts. Jordan and Evermann (quoted by Boulenger, 1910, p. 620) concluded that they were "archaic fishes, relics of some earlier fauna, and apparently derived directly from the extinct transitional forms through which the Haplomi and Acanthopteri have descended from allies of the Isospondyli." To Boulenger, however, "an analysis of their characters shows them to belong to the Haplomi, of which they may be regarded as highly specialized members, having evolved in the direction of the Acanthopterygii." To Tate Regan (1929, p. 318) they are "an isolated order, without evident relationships except to the Isospondyli or primitive Iniomi." The branchiostegal rays of these peculiar fishes have afforded important evidence on this question, which has been thus stated by Hubbs (1919, p. 68): "The Salmopercæ, long considered as intermediate between the soft-rayed and spiny-rayed fishes, have six branchiostegals, arranged exactly as in the Acanthopteri. Both of the species usually referred to this group, Percopsis omisco-maycus and Columbia transmontana, have been examined. Aphredoderus sayanus, referred by Regan to the same group, has branchiostegals in all essential respects similar to those of Percopsis and the following groups. . . ."

According to Tate Regan (1929, p. 318) the head of the percopsids, like that of certain berycoids, bears large muciferous channels; those of the frontals are continued forward on the large thin concave nasal bones, which nearly or quite meet in the mid-line. In a skull (No. 981) of *Percopsis guttatus* in the British Museum (Natural History) the orbits are large with raised rings, the suborbitals small with reflected outer border. The olfactory fossa is also relatively large. The mouth is very small, bordered by minute teeth; the rod-like premaxillæ have relatively long ascending processes; the maxillæ are smaller than the premaxillæ and act as levers as do those of acanthopts; no supramaxillæ are visible; the mandible is small, with minute dentary and relatively large articular. Many of these characters indicate that the mouth is protrusile. The opercular region suggests the generalized percoid type, although the point on the opercular is feeble. The large preopercular has a reflected border which bears a moderate spine. The occipital condyles are triple as in the berycoid and percoid groups. The general appearance of the skull suggests an extremely primitive percoid type, but with a minimum of crests and spikes, in accordance with the small size, especially of the mouth.

The otoliths of the Salmopercæ, according to Frost (1926c, p. 470), "closely resemble those of *Ophichthys gomesii* (Order Apodes); in his classification of this order Mr. Tate

Regan notes that it has been suggested by Jordan and Evermann that these fishes are related to the percoid families Percidæ and Centrarchidæ; Mr. Regan adds that this is not confirmed by a study of the anatomy, and the evidence of the otoliths supports his opinion. On the other hand, there is a strong resemblance of the otoliths to those of the percoid genus *Apogon*, which differ from the remainder of the Percoids in the sulcus. . . ." Thus it would seem that the otoliths, like the branchiostegal rays and other characters, lend support to the conclusion that the Salmopercæ are a distinct group intermediate between some ancient soft-rayed forms and the primitive percoids.

BERYCOIDEI

Many of the lower teleosts already noted have foreshadowed the perch type, the "ideally perfect fish," in one or more characters: several of the anterior rays of the dorsal and anal fins may have become more or less spiny, the air-bladder may have lost its duct to the œsophagus, the pelvic fins have perhaps lost the mesocoracoid arch and have moved forward beneath the pectoral fins, the body may have become more or less short and deep (*Cteno-thrissa*), the scales ctenoid, the maxillæ often have withdrawn from the gape before the advancing premaxillæ, the orbitosphenoid may have disappeared, the supraoccipital has often gained contact with the frontals and pushed aside the parietals, and so forth. But except perhaps in the case of *Ctenothrissa*, the few spiny-finned characters thus acquired have been associated with predominant tendencies that led toward other goals and it has gradually become evident that we were not dealing with true ancestors of the spiny-finned fishes but only with originally soft-rayed stocks which had progressed in certain features toward the spiny-finned stage of evolution, the parallelism being on a grand scale and affecting many different lines; but until the true percomorphs themselves appeared in the Upper Cretaceous it never resulted in the complete combination of spiny-finned characters.

In the Cretaceous and recent berycoids, however, we find a group which has almost attained the spiny-finned status and to which Tate Regan (1929, p. 319) has awarded the palm of being an order Berycomorphi, "directly intermediate between the clupeoid Isospondyli and the Percomorphi."

The restoration of the Cretaceous *Hoplopteryx lewesiensis* by Smith Woodward (1901, p. 398) shows a fish (Fig. 110) that in many features recalls the Cretaceous deep-bodied clupeoid *Ctenothrissa*, but which in skull structure is almost a percomorph. The head is short and deep in correlation with the deep compressed form of the body. The premaxillæ have prominent ascending processes and are evidently more or less protrusile; the long alveolar process (which is provided with minute teeth) excludes the stout maxilla from the upper border of the mouth. The maxilla is deepened posteriorly and bears a broad supramaxilla, in front of which is a small triangular anterior supramaxilla (A. S. Woodward, 1902, p. 18). As the suspensorium is nearly vertical and quite long, the articulation of the mandible is brought almost directly beneath the posterior border of the orbit and at a low level. As the snout is also short, the result is that the mouth opening slopes gently upward.

The circumorbital bones have their orbital margins everted, their peripheral areas depressed for the reception of large slime cavities connected with the lateral line system. Similar raised ridges and depressed areas are present on the preopercular and on the lateral surface of the mandible, as in many acanthopts. The relatively large size of the orbits shows that even in Cretaceous times the eyes had already assumed the dominant role in

shaping the brain and its responses that they play in existing teleosts.   In contrast with
the typical isospondyl type, the posterior border of the orbit is so far back that it nearly
touches the hyomandibular.   "The deep and narrow operculum," writes Woodward (1901,
p. 401), "is produced into two short and broad spines at its hinder margin, the upper being

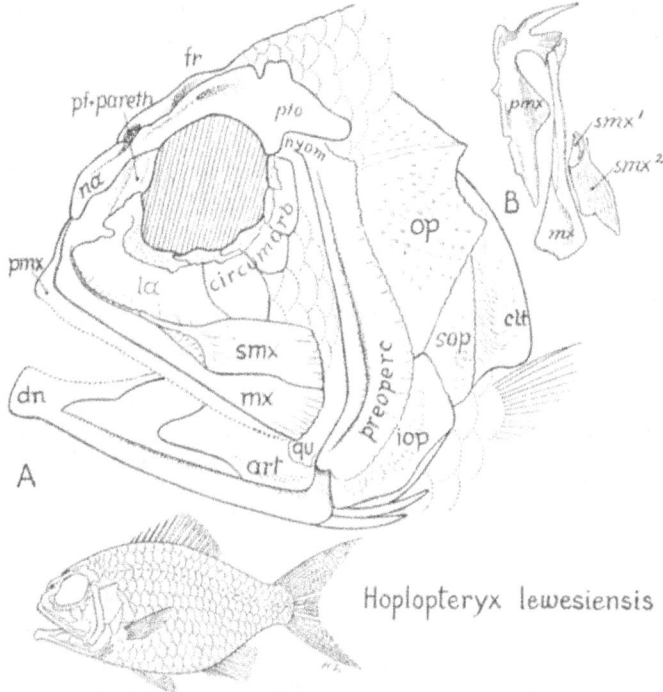

FIG. 110.   *Hoplopteryx lewesiensis.*   After Smith Woodward.
A. Side view of skull.   B. Left premaxilla, maxilla and two supramaxillæ.

connected by a strong ridge with the point of suspension."   Such then, was the beginning
of the "opercular spines" of acanthopts.   The interopercular is well developed and has its
normal connection below with the posterior projection of the mandible.   As to the cranium
proper, Tate Regan (1929, p. 319) states that the orbitosphenoid and Y-shaped "basisphenoid" are retained, two points of inheritance from remote isospondyl ancestors.   In the
recent Berycidæ one or several of the suborbitals emit an internal lamina supporting the
eye (Boulenger, 1910, p. 655).   One of these becomes the "suborbital shelf" of the typical
acanthopts.

A careful investigation of the osteology of six genera of recent berycoid fishes was made
by Starks (1904b).   Among the more constant skull characters of the Berycoidea were the
following (p. 602): orbitosphenoids present, head usually with conspicuous mucous cavities,
a suborbital shelf present on the suborbital ring; maxillary with a large supplemental bone;[1]

---

[1] Tate Regan (1907a, p. 642) records the presence in *Myripristis murdjan* of an anterior but vestigial supramaxillary,
homologous with that of the Cretaceous clupeoid *Ctenothrissa radians.*   Starks (1904b, p. 612) also notes the presence of two
"supplemental bones" on the maxillary of *Holocentrus ascensionis.*

nasals usually very large; occipital articulation tripartite, large exoccipitals meeting broadly beneath the foramen magnum. The osteology and classification of the recent and fossil berycoids were also carefully revised by Tate Regan in 1911. He removed the families Stephanoberycidæ and Melamphaidæ to a new order Xenoberyces, which "differ from typical Berycomorphi in the toothless palate, the absence of a subocular shelf, and the triangular shape of the single supramaxillary but especially in the absence of an orbitosphenoid." Thus the "order" Xenoberyces differs by definition from the Berycomorphi through the loss of a few typical characters.

Myripristis of the family Holocentridæ, the existing species of which live in shallow water in tropical seas, was represented in the Upper Eocene of Monte Bolca, Italy. Its skull (Fig. 111) shows several specializations beyond the primitive Cretaceous berycoid

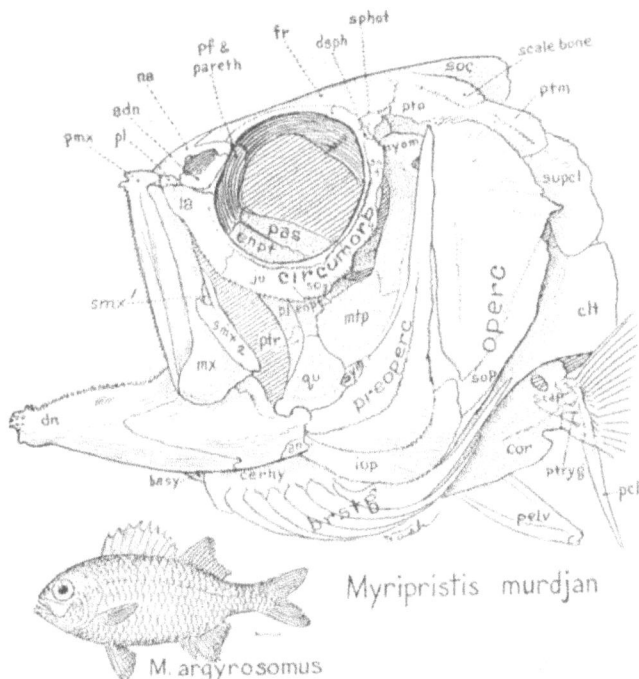

FIG. 111. *Myripristis murdjan.*

type. The mandible is rather massive and its distal end is surmounted by a spiny protuberance which fits into a depression between the opposite premaxillæ. The posterior end of the maxilla is deep and enters into the gape of the mouth, being bordered by serrations. The opercular bears a small spine, the border of the preopercular is finely serrated. The short skull roof is rounded transversely above and behind the orbits.

In the squirrel-fish *Holocentrus ascensionis* the premaxillæ (Fig. 112) are protrusile, with long ascending processes, while the maxillæ have become levers for swinging them.

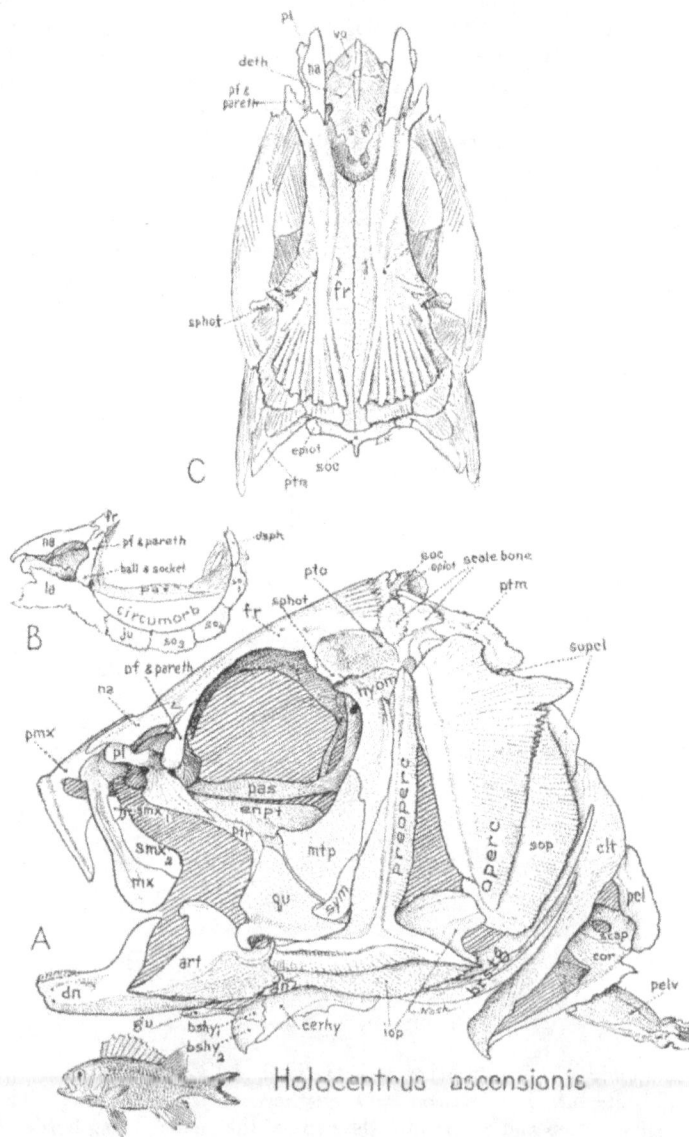

FIG. 112. *Holocentrus ascensionis.*
A. Side view. B. Circumorbitals. C. Top view.

The maxilla bears the primitive two supramaxillæ, while its proximal process, articulating with the vomer, is overlapped dorsally as usual by a strong process of the palatine. The most conspicuous feature of the side view is the very prominent spine at the posteroinferior angle of the strongly ridged preopercular, a feature not seen in fishes below the acanthopt level. The vertically extended opercular also bears a prominent strong spine, below which are sharp serrations. There is a "ball and socket" joint between the dorsal rim of the lacrymal and the lower border of the dermal prefrontal (Fig. 112B).

The cranial roof (Fig. 112) is very dense and solid. Just above the hyomandibular facet it bears a prominent fossa for the dilatator operculi muscles. In the top view (Fig. 112C) the small epiotic and supraoccipitals are depressed beneath the prominent plaited surface of the broad frontals, which are now the dominant elements of the cranial roof. Anteriorly there is a large deep median groove above the mesethmoid, flanked by conspicuous processes formed by the nasals and frontals and serving for the reception of the long ascending processes of the maxillæ. The circumorbital bones number six, as is usually the case, especially in acanthopts, counting the so-called dermosphenotic as the sixth.

Starks has shown (1908b, p. 613) that in *Holocentrus ascensionis* the air-bladder sends forward on each side a diverticulum, the inner membrane of which forms a loose tympanum covering the posterior opening of a tube-like prominence of the otolith chamber. The side of this chamber is formed chiefly by the backwardly prolonged proötic, assisted by the exoccipital and basioccipital. In the species *Holocentrus (Adioryx) suborbitalis* this connection between the air-bladder and the otolith chamber is absent (p. 614). The otoliths of the Berycoidei, as described by Frost (1927a, pp. 440, 444), show a wide range in form from the primitive "Berycoid" type of *Polymixia* to the aberrant form in *Myripristis*. In the Holocentridæ the general form of the sagitta is similar to that of *Elops*, but the sulcus is distinctly percoid in character and the posterior rim is modified. Mr. Frost concludes (p. 440) that these facts tend to confirm Mr. Tate Regan's position that while the berycoids approach the percoids in general structure, they also retain many features which indicate their relationship to primitive clupeoids.

In conclusion, the ancestry of the berycoids and consequently of the entire percomorph series is apparently to be looked for in the neighborhood of the Cretaceous isospondyl *Ctenothrissa*, as held by Regan (1929). Assuredly this form approaches the short-bodied acanthopt type more closely than do any known scopeloids, which are typically long-bodied and more or less advanced on the roads leading to Haplomi, Microcyprini, etc., which are at most pseudo-acanthopts rather than true pre-percomorphs.

PERCOIDEI (BASS, PERCH, SNAPPERS, SPARIDS, CICHLIDS, WRASSES, ETC.)

"The large and varied order Percomorphi," writes Tate Regan (1913a, p. 111) "occupies a central position among the Teleostean fishes. On the one hand it appears to be derived from the Berycomorphi, and on the other it seems to have given rise to a number of specialized offshoots, which may be regarded as ordinally distinct: Scleroparei, Heterosomata, Plectognathi, Discocephali, Xenopterygii, Pediculati, Symbranchii, and Opisthomi." He then proceeds to define the order Percomorphi as follows:

"Symmetrical acanthopterous physoclists with normal dorsal fin, pelvic fins never more than 6-rayed, subabdominal, thoracic, jugular, or mental in position, the pelvic bones typically attached to the cleithra; principal caudal rays not more than 17. No orbitosphenoid. Second suborbital not forming a stay for the præoperculum. Posttemporal more or less distinctly forked."

These of course are merely the characters selected by systematists for the construction of convenient keys for "running down" the taxonomic position of a given specimen. But they are mere isolated fragments of the typical percomorph skeletal pattern, which happen to be easily recognizable in at least most of the members of this large and highly diversified assemblage and to be useful in excluding from the central group many of its own derivatives.

In this chapter we show a few skulls of representatives of this many-branched stock. In the most primitive members of the series the body is short and comparatively deep with only twenty-four vertebræ, recalling the conditions in the ancestral berycoids, but in many more-advanced forms the body becomes more or less elongated while the vertebræ increase in number. The shape of the head as a whole is closely correlated with the general body-form, or to put it more precisely, each one is correlated with the other, so that the skulls pass from relatively short and deep to long and shallow contours. The contours are so adjusted to each other that the "angle of entrance" is greater than that of the "run" or tapering part of the body.

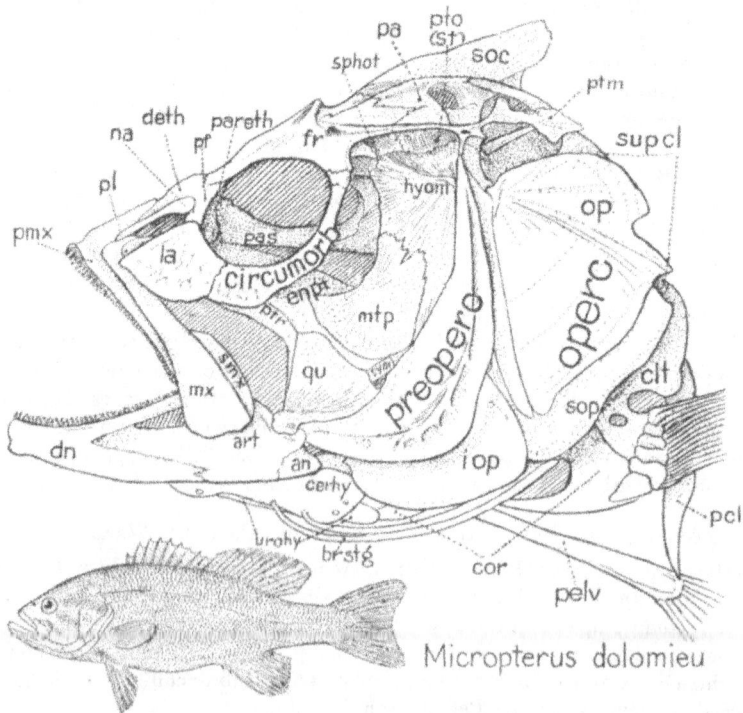

Fig. 113. *Micropterus dolomieu.*

In *Micropterus dolomieu* (Fig. 113) a progressive member of the family Centrachidæ, the skull may be regarded as near the central percoid type. The distance between the posterior border of the orbit and the preopercular is relatively much greater than it is in the primitive berycid (Fig. 110), while in *Luciolates* (Fig. 117) this lengthening is further emphasized.

The otolith (sagitta) of *Micropterus salmonoides*, according to Frost (1927*b*, p. 301), while frail and delicate, in general resembles that of *Perca*.

Mechanism and Evolution of the Protrusile Upper Jaw.—In the percoids the protrusile upper jaw varies from incipient to extreme stages. The phenomenon of protrusility of the mouth has been partly described by Thilo (1920) and by Delsman (1925) from the mechanistic and the morphological viewpoints respectively. There is need, however, of a reconsideration of the subject from the phylogenetic viewpoint, as protrusility has arisen independently in different groups of fishes. In the incipient stage (e.g., in *Lates*, Fig. 114) the

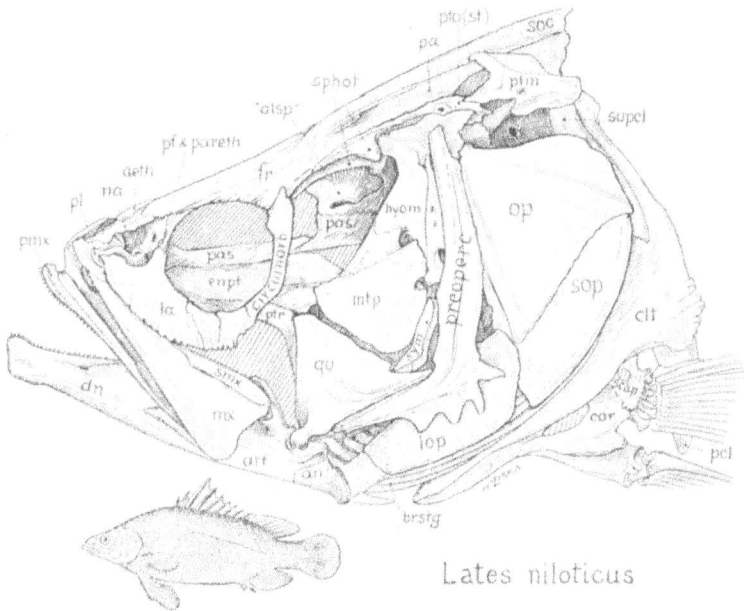

FIG. 114. *Lates niloticus.*

premaxillæ have relatively short ascending processes, which are more or less closely conjoined in the mid-line. Beneath them (Fig. 115) is a median piece of cartilage, the rostral,[1] which slides on the upper surface of the vomer; the latter bears a low keel for the groove on the under side of the rostral cartilage. The conjoined ascending processes and their supporting cartilage form an inverted V, the top of which is flanked by a notch formed by the nasals and upper part of the mesethmoid. Immediately behind and lateral to the ascending process of each premaxilla is a shorter broader process called by Allis (1909, p. 24) the articular process of the premaxillary. This fits posteriorly into the anterior fork of the

---

[1] So named by Allis (1909, p. 28) in the mail-cheeked fishes (Fig. 116). Not to be confused with the surface plate of the same name in Palæozoic ganoids.

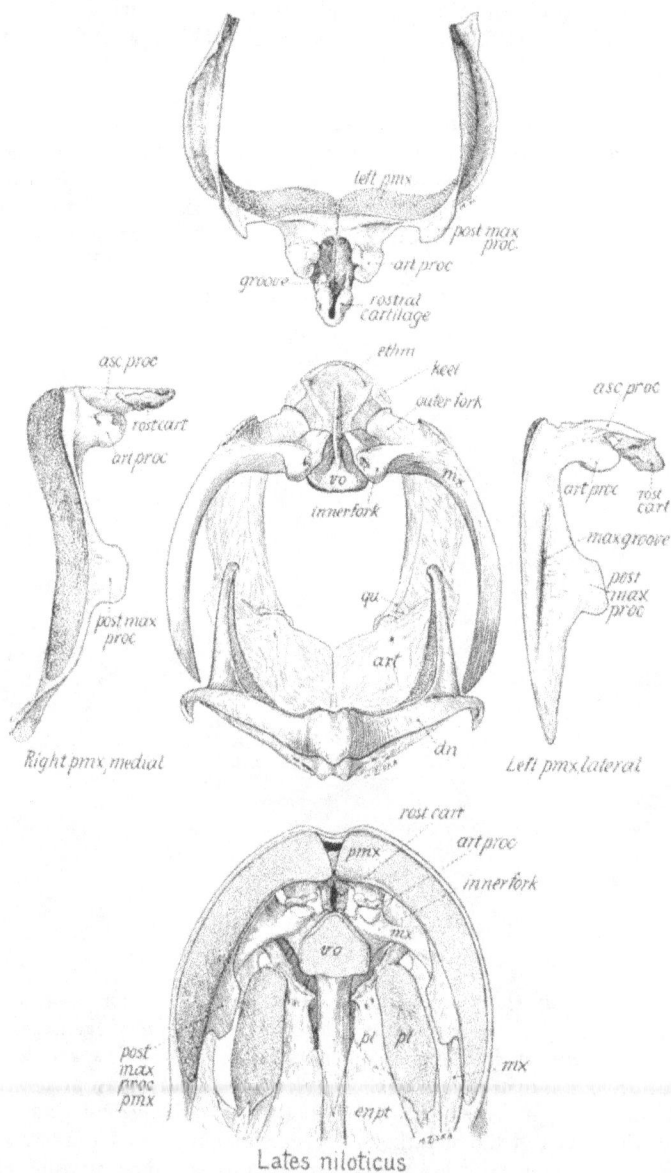

FIG. 115. *Lates niloticus*. Relations of the premaxillæ and maxillæ to the ethmo-vomer block.

maxilla. A third process of the premaxilla is a thin broad low crest on the medial dorsal border of the bone; this crest is called by Allis the postmaxillary process of the premaxilla. It prevents lateral dislocation of the premaxilla with reference to the maxilla.

The maxilla rests in a groove on the postero-lateral face of the premaxilla and is excluded from the gape by the latter. Its narrow proximal end bears a prominent notch. The rounded inner fork slides on the vomer; it carries the ascending process of the premaxilla on its dorso-anterior surface. The outer fork overlaps the articular process of the premaxilla below and joins the inner fork dorsally where both together end in a smooth convex

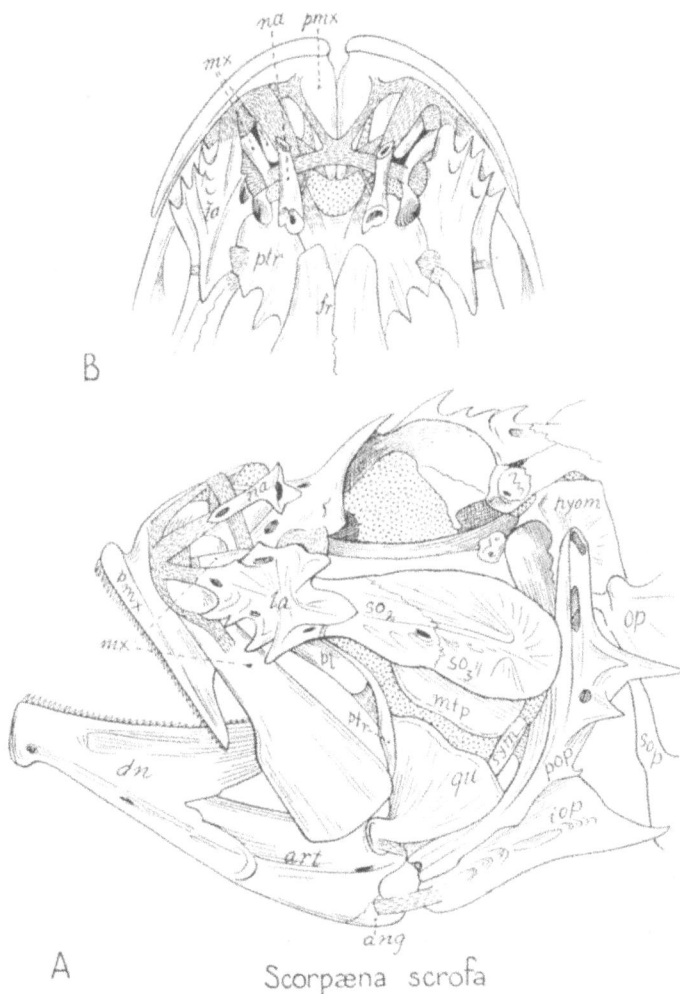

Fig. 116. Connections of the premaxillæ and maxillæ with the rostral cartilage and the skull in a percomorph (*Scorpæna scrofa*). (After Allis.) A. Side view. B. Top view.

surface, which must fit loosely into a socket in the cartilaginous portion of the olfactory capsule.

The relations of the premaxillæ and maxillæ to the ligaments connecting them with the skull are shown in Figure 116, after Allis.

The lower end of the maxilla is tied to the lateral surface of the mandible by a loose fold of skin and connective tissue. The tendon of the superficial portion of the adductor mandibulæ (ad. 1) is inserted on the medial surface of the shaft of the maxilla, behind its articular process and near the fork of the palatine. Contraction of this muscle would tend to pull the maxilla and premaxilla backward and slightly downward and thus help the other adductors to close the mouth (Vetter, 1878, pp. 495, 497). In *Roccus lineatus*, a typical percoid, depression of the mandible by the action of the paired geniohyoideus muscles merely tends to rotate the premaxillæ slightly outward and to protrude them slightly.

Marked protrusion of the premaxillæ can only take place under the following conditions: (1) the gape must be small; (2) the fold of skin attaching the maxilla to the mandible must be moved far in front of the fulcrum in order to exert a downward pull on the premaxillæ; (3) the quadrate-articular fulcrum must be brought much below the level of the closed mouth; (4) at the same time the ascending process of the premaxillæ must be greatly prolonged. Under these circumstances a long downward movement of the mandible under the pull of the geniohyoideus muscles will draw the lower ends of the maxillæ sharply downward, while the premaxillæ are pulled downward and forward, the upper end of the ascending processes remaining on top of the vomer.

Stages preceding the protrusile condition of the jaw may be summarized as follows: (1) the premaxillæ and maxillæ together form the dentigerous outer border of the upper jaw, both being fixed to the primary upper jaw. This stage is well illustrated in the Palæozoic ganoids and in the living *Polypterus*. (2) The posterior end of the maxilla protrudes laterally, but little or no movement is possible; e.g., *Amia*. (3) The proximal end of the maxilla acquires a movable articulation with the vomer, the small premaxilla is closely fastened to the maxilla and moves with it, as in the tarpon. (4) The premaxilla lengthens its alveolar process and finally excludes the maxilla from the gape, and it also begins to send out another ascending process growing upward above the vomer. An early stage in this development may be seen in *Mugil*. (5) The proximal part of the premaxilla becomes differentiated into ascending and articular processes. By this time the maxilla has completely lost its function as a dentigerous element and has become merely a lever for the partial eversion and closure of the mouth. Doctor Allis' view that the ascending process of the teleost premaxilla represents a formerly independent element which has become fused with the main part of that bone is considered at some length on page 96 above.

Mechanism of the Neurocranium.—In *Micropterus* (Fig. 113) the skull is of very primitive percoid type. In *Lates niloticus* (Fig. 114) of the family Centropomidæ specialized features are the forward position of the eye, the sharp snout, the gently concave forehead leading to the high back, and especially the prominent spikes on the angle of the preopercular.

The neurocranium of *Luciolates* (Fig. 117) is perhaps even better adapted than that of the isospondyl tarpon (Fig. 32) to resist the stresses generated within it by the thrusts coming from its own locomotor, masticatory and respiratory apparatus. It is more compact than that of the tarpon in that the anterior prolongations of the spinal muscles in the occipital region pass *above* the dermal skull roof, between the long and well-braced ridges shown in Figure 117, while in the tarpon (Fig. 32) these muscles pass *beneath* the dermal skull roof, entering through the large posttemporal fenestræ and running immediately above the endocranium itself. In both cases the otic region has to receive the downward and backward

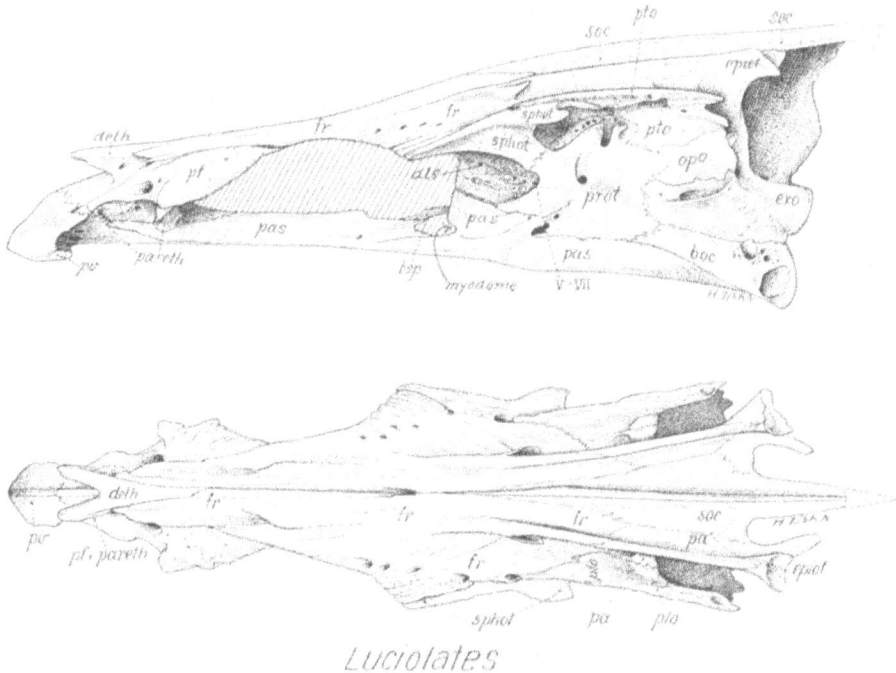

FIG. 117. *Luciolates* sp. Neurocranium, showing strengthening crests and ridges.

thrusts resulting from forward locomotion, etc., but in the tarpon the thin supraoccipital is widely separated from the frontals, while in *Luciolates* the supraoccipital is suturally wedged in between the dominant frontals and braced laterally by stiff ridges on the frontals and pterotics. In fact, the supraoccipital is the keystone of the occipital arch even to a greater extent than in the tarpon. Hardly less important are the backwardly-projecting epiotics, to which are attached one of the horns of the supratemporal, conveying wrenching and pulling forces from the shoulder-girdle. The epiotic is well braced by the parietals, exoccipitals, etc., to meet these demands.

The otic region itself is highly adapted to receive thrusts coming from various directions. The supraorbital frontal roof is beautifully braced by an obliquely-placed buttress on the sphenotic, which when held up to a strong light, in the side view may be seen, as it were, to gather together its trabeculæ and direct them downward and backward across the proötic to the stout basioccipital. Lesser bundles and strands of trabeculæ may be traced in various directions through the occipital region. For instance, in the top view, especially

12

strong bracing is concentrated into a short rounded transverse column at the end of the junction of the parietal and pterotic.   This must resist lateral compression occasioned by the thrusts of the suspensorium of the upper jaw, and it must strengthen the attachment of the high crest on the parietal that divides the trunk muscles into one lower and two upper strands.

The greatest massing of trabeculæ and dense bone occurs, however, near the junction of the occiput with the vertebral column.   Whereas strengthening in this region was effected by the incorporation of the centrum of the first vertebra into the occiput, in *Luciolates* perhaps a still better arrangement is hit upon by having the occipital condyle tripartite, with three inclined condylar articular surfaces, furnished respectively by the basi- and ex-occipitals.   Thus dorso-ventral dislocation, with resulting strangulation of the spinal cord, is

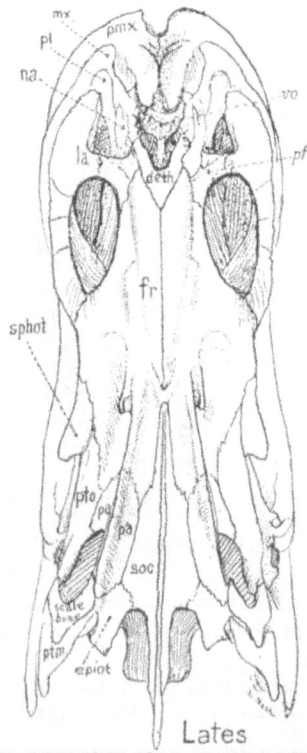

Fig. 118.   *Lates niloticus*.   Top view.

prevented by the overhang of the exoccipital upon the basioccipital segments, while lateral dislocation is prevented by the lateral obliquity of the exoccipital facets.   Here then is an interesting parallel to the tripartite occipito-atlanteal facets of certain extinct reptiles and stegocephalians.

On the lateral surface of the otic region there is a notable absence in *Luciolates* of the deep fossa which in the tarpon and other isospondyls lodged a diverticulum of the swim-bladder.   The fossa for the dilatator operculi muscle, which in the tarpon is conspicuous above the postorbital process of the sphenotic, must be represented in *Luciolates* by the very narrow, nearly horizontal fossa chiefly on the lateral border of the pterotic immediately above the elongate glenoid cavity of the hyomandibular.   The bony brain-trough in front

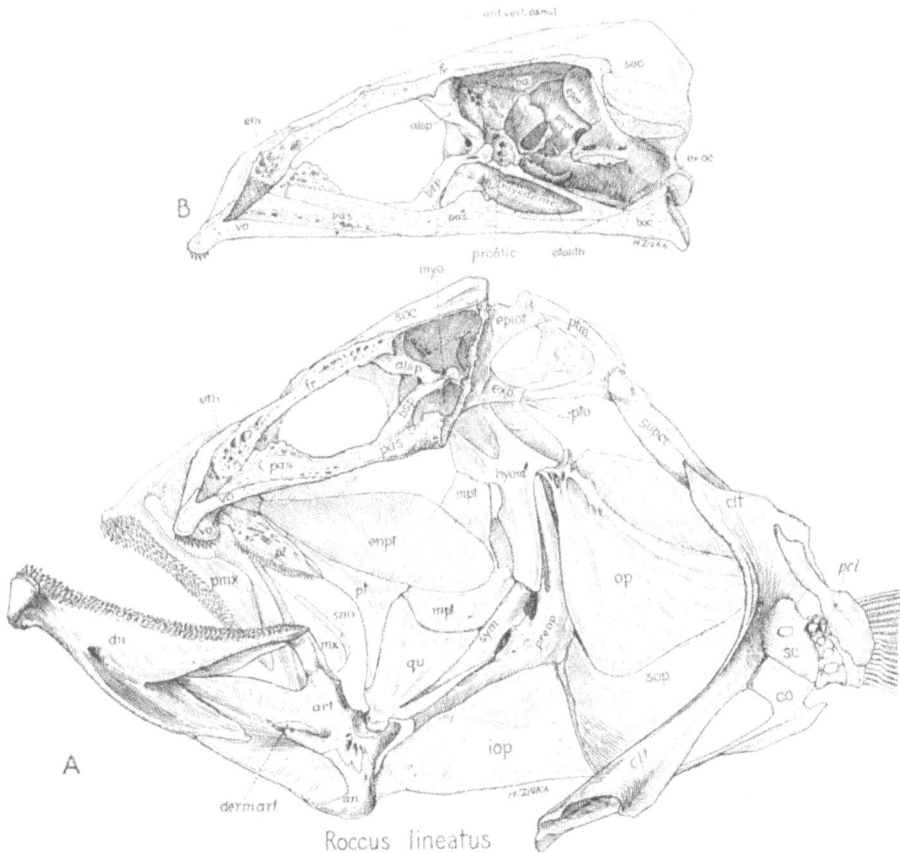

FIG. 119. *Roccus lineatus.*

A. Right half of skull and pectoral girdle, showing the suspension of the jaws and pectoral girdle from the neurocranium.
B. Right half of neurocranium, medial view, showing ethmo-vomer block, keel bone (*pas*), interorbital bridge and cranial vault.

of the sphenotic in *Luciolates* is notably reduced and lightened by the complete elimination of the orbitosphenoid, while the strut from the Y-shaped basisphenoid that comes down to rest on the parasphenoid is here notably weak, so that the parasphenoid, the frontals and the otic bones carry all the load.   However in *Roccus* (Fig. 119*A*) the basisphenoid is still a valuable brace, while the parasphenoid has a strong anterior ascending fork.

In the anterior part of the skull in *Lates, Luciolates* and other percomorphs the top view (Figs. 117, 118) reveals a system of interlocking wedges and abutments.   Thus the vomer abuts against the mesethmoid, the latter is wedged in between the anterior limbs of the frontal, they in turn are wedged in between the prominent lateral ethmoids.   This system of interlocking wedges, which extends also to the occipital region, is well seen also in the top view (Fig. 294*A*) of *Pomatomus.*

Thus in both isospondyl (Fig. 32) and percomorph (Fig. 117) the upper part of the otic region has to bear the load of thrusts from the general forward movements of the body,

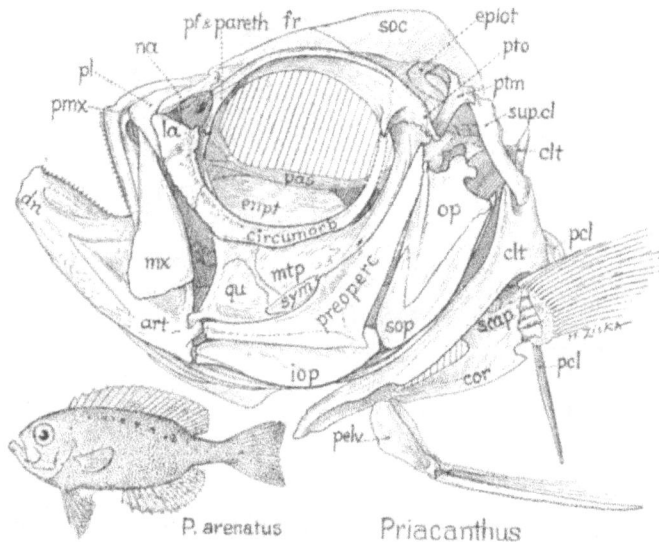

FIG. 120.   *Priacanthus.*

from the powerful jaw muscles (which rest directly and indirectly on the hyomandibular), from the branchial arches, from the opercular apparatus and from the pectoral girdle.   The suspension of the latter by the forked posttemporal is well shown in Figure 119*A*.   As Ridewood (1904*a*, p. 65) has pointed out, the inner fork of the posttemporal represents an intermuscular tendon bone of the obliquely-placed neck muscles.

With regard to the subdivision of the order Percomorphi into suborders and divisions, Tate Regan (1913*a*, p. 111) writes as follows:

"At present I am inclined to recognize thirteen suborders, viz., Percoidea, Trichiuroidea, Scombroidea, Siganoidea, Teuthidoidea, Kurtoidea, Gobioidea, Blennioidea, Stromateoidea, Anabantoidea, Mugiloidea, Polynemoidea.   But it is largely a matter of opinion whether some of these may not be regarded as ordinally distinct, or whether others should not rank merely as divisions of the Percoidea."

He then subdivides the suborder Percoidea into twelve divisions each of which is carefully defined by more or less conspicuous characters.   While for the purposes of the present paper it has not been deemed necessary to follow this arrangement very closely, I have derived much help from it in considering the relationships of the skull-forms described below.

The otoliths of the Perciformes, as described by Frost (1927b) afford valuable hints as to the inter-relationships of some of the families of this enormous assemblage. The sagitta of *Perca fluviatilis*, which is regarded by Frost as the type, is ovate with concave outer and concave inner side. A related type is found in the otoliths of *Serranus* from the Upper Eocene of the Isle of Wight (Frost, 1925c, p. 160).

Basses, Perches, Groupers. The family Serranidæ, the central group of the Percoidea, is the subject of a beautifully illustrated work by Boulenger (1895). Morpho-

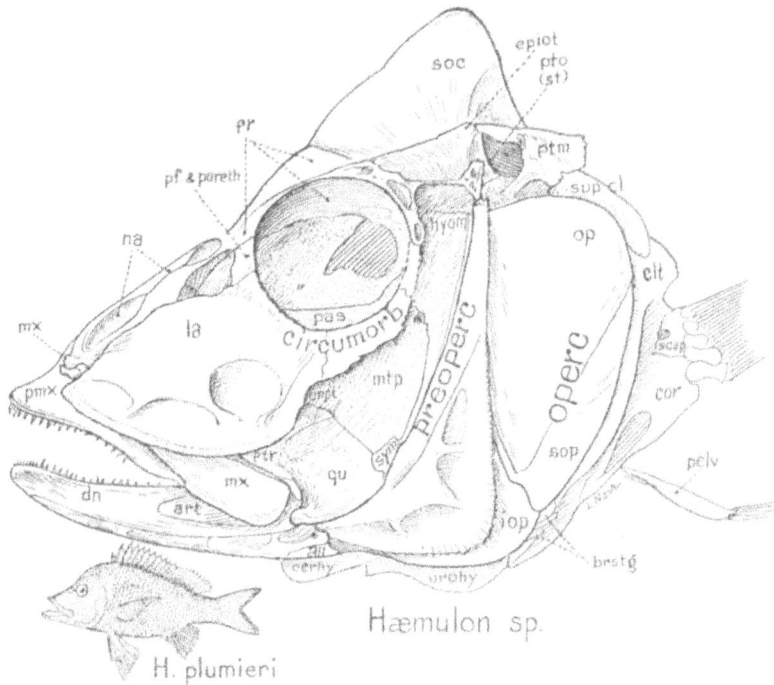

FIG. 121. *Hæmulon* sp.

logical variability of the skull characters within the family seems slight but there are marked differences in the proportional measurements, the skull being markedly elongate and low in *Luciolates*, short and deep in *Cæsioperca lepidoptera* (p. 311).

The cranium of the marine Australian apogonid fish *Dinolestes* (Fig. 140C) as described and figured by Starks (1899a) presents considerable resemblance to that of *Luciolates*. Starks was able to confirm Gill's conclusion that this fish has nothing to do either with *Esox* or with *Sphyræna*, but that it is a true percoid of the family Apogonidæ (Cheilodipteridæ).

Priacanthus.—The most striking feature of this skull (Fig. 120) is the huge size of the orbit and the anteroposterior crowding of the opercular. The mouth, although fairly short, can open upward because the quadrate-articular joint is brought forward to near the anterior border of the eye and because the quadrate is lowered.

Lutianids, Hæmulids, Sparids.—These three families form a progressive series leading
from a predaceous type with sharp teeth on the margins of the jaws to a durophagus type
with blunt molariform teeth on the sides and incisor-like teeth in front.   The pharyngeal
teeth are conical, not very large.   The Lutianidæ are distinguished from the typical Serranidæ by the loss of the supramaxillary and by the broad overlap of the maxillary by the
enlarged preorbital (lacrymal).   The mouth retains the normal percoid protractility and

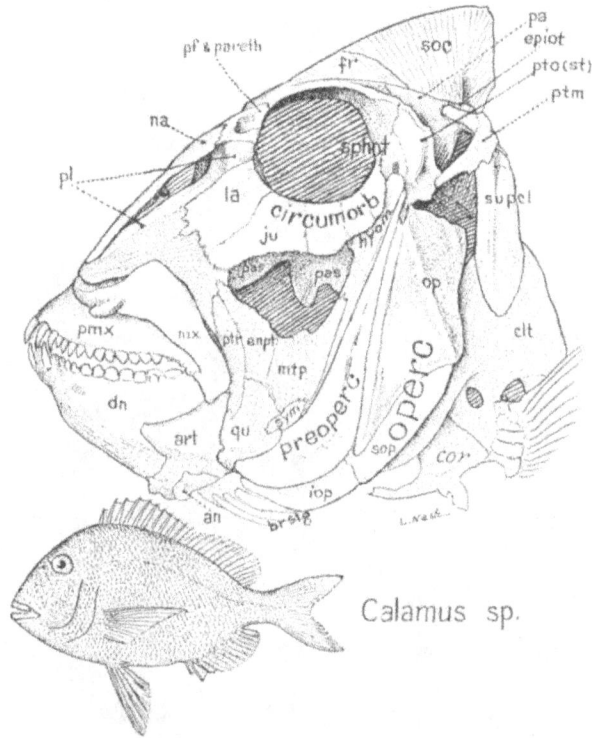

Fig. 122.   *Calamus.*

normal processes of the premaxillæ.   The maxillary broadens distally but is not overlapped
externally by the extremity of the premaxillary (Tate Regan, 1913*a*, p. 120).   The teeth
are villiform or obtusely conical in the jaws and often on the vomer and palatines.
    In *Hæmulon* (Fig. 121) of the family Pomadisidæ, relationship with the Sciænidæ is
suggested by the muciferous cavities on the preopercular and on the lower border of the
mandible; this is probably mere parallelism.   The snout in front of the lacrymal is long;
this brings the tip of the mouth forward and thus makes a more horizontal gape.   The
lacrymal has been enlarged with the ethmoid region; it broadly overlaps the maxilla.   The
premaxillary teeth are delicate.
    In *Calamus* (Fig. 122) of the family Sparidæ, the premaxillary teeth have become

robust, those in the front being more or less conical and recurved, those in the rear, broad and molariform. The large maxilla and the immense overlapping process of the palatine afford a firm base for the massive premaxilla. The palatine in turn is braced by the stout parethmoid and mesethmoid. In *Archosargus* (Fig. 123) the tendencies already noted

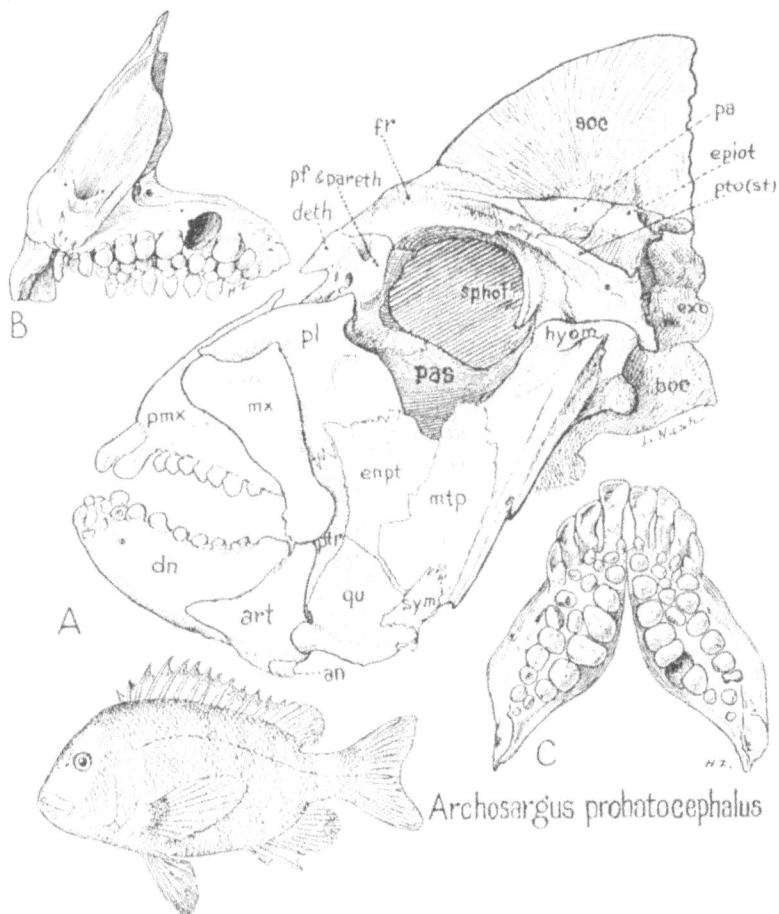

FIG. 123. *Archosargus probatocephalus.* A. Side view. B. Right premaxilla. Inner view. C. Right and left dentaries. Top view.

attain an extreme expression, so that it would be difficult to find a more strongly built skull. Although any form of Lamarckism as an explanation of such facts is of course unthinkable, it is worthy of note that the massiveness of these jaws and their supports is apparently in response to the increasing strength of the jaw muscles as well as to the increasing toughness of the food, and it is a fact that with this apparatus a sheepshead can and does break into live bivalve shells and secure the contents.   Here then the high value of the mechanism in terms of natural selection must be admitted.

Hoplegnathus.—This is the type and sole genus of a peculiar family of percoids that is probably allied with the Lutianidæ (Tate Regan, 1913*a*, p. 130; 1929, p. 321).   The outstanding feature of this skull (Fig. 124) is the coalescence of the teeth on the premaxillæ and dentaries into a powerful beak.   The maxillæ, unlike those of other beaked teleosts, are much reduced but evidently still serve for the insertion of the tendon of the superficial

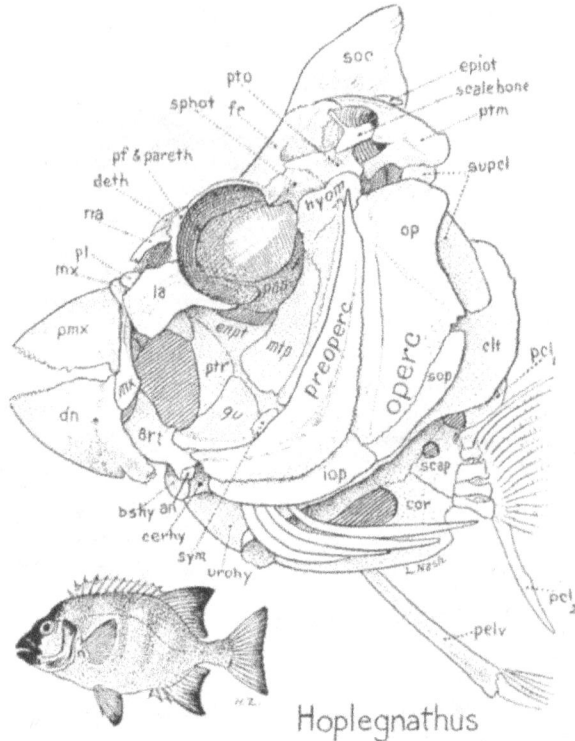

Fig. 124. *Hoplegnathus.*

branch (*A* 1) of the adductor mandibulæ muscle, which would run obliquely downward and backward.   Thus its course would cross that of the deep adductors inserted on the articular. Contraction of the entire muscle mass would thus pull powerfully upward on the mandible and downward on the premaxillæ.

The upper beak rests on the enlarged ethmo-vomer, which is braced principally by the parethmoids, parasphenoid, palatines and frontals.   The dentary, as in other beaked types, tends to form a secondary joint with the articular.   The skull is deep, the supraoccipital towering on top and the opercular region extending downward.   Although the articular-quadrate facet is located far forward, beneath the anterior border of the orbit, the mouth does not open upward but forward. This is brought about by the forward prolongation of the tips of the upper and lower beaks.

Mullidæ.—The "Red Mullets," according to Boulenger (1904, p. 665), "are very nearly related to the Sparidæ, with which they agree in the structure of the vertebral column and the presence of a subocular shelf. They differ in the very weak dentition, the presence of a pair of hyoid barbels, the reduced number (4) of branchiostegal rays, and the double perforation of the scapula. Two short dorsal fins, remote from each other, the anterior with weak spines." Possibly these characters may be related to the bottom-living habits of these fishes. The Mullidæ, according to Tate Regan (1913a, p. 123), though related to the

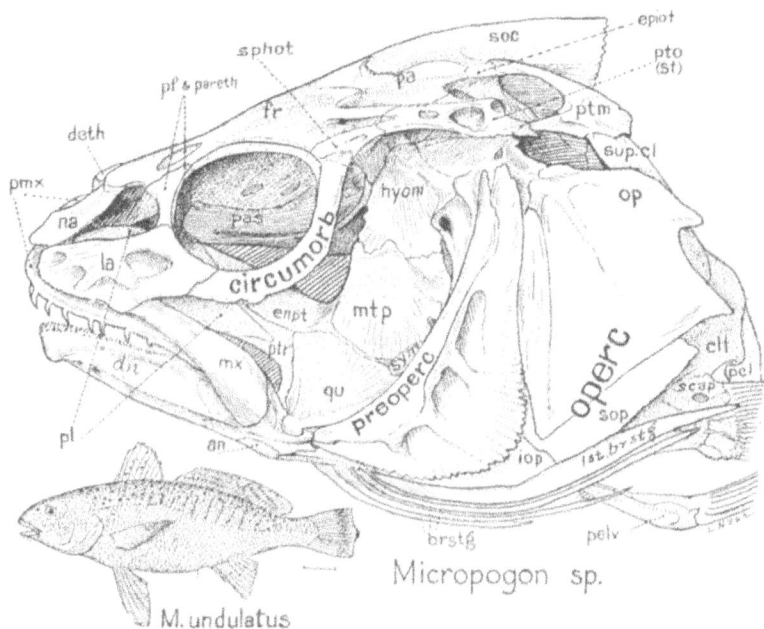

FIG. 125. *Micropogon* sp.

Lutianidæ are more specialized in several characters. As may be inferred from the presence of long barbels attached to the hyoid behind the symphysis of the lower jaw, the mullets rest and feed on the bottom. The mouth is protractile, the alveolar processes of the premaxillary are lacking, the maxillaries not being overlapped externally by the premaxillaries. They are sheathed by the preorbitals (lacrymals). The teeth are villiform in the jaws and often on the vomer and palatines (Regan, 1913a). Tate Regan (1929, p. 320) now classes the Mullidæ among "several families of marine perches," which have "a scaly axillary pelvic process, and the maxillary sheathed by the preorbital," namely, Lutianidæ, Pomadisidæ, Liognathidæ, Sciænidæ, Mullidæ, Sparidæ, etc. He says that the Mullidæ "have a pair of barbels at the chin, used to probe for the small shell-fish, worms, etc., on which they feed." The otolith (sagitta) of a young example of *Mullus barbatus*, according to Frost (1927b, p. 303), was of peculiar type, high and loaf-shaped, with a very short cauda. In aged examples, however, the otolith becomes elongated and the cauda lengthened; in other words, it approaches more normal percoid types.

Sciænids.—In *Micropogon* (Fig. 125), typical of the Sciænidæ, the temporal region, frontals, preopercular, etc., are deeply pitted for muciferous glands of the lateral line system. The opercular has two spikes and the preopercular has a serrate angle as in many other

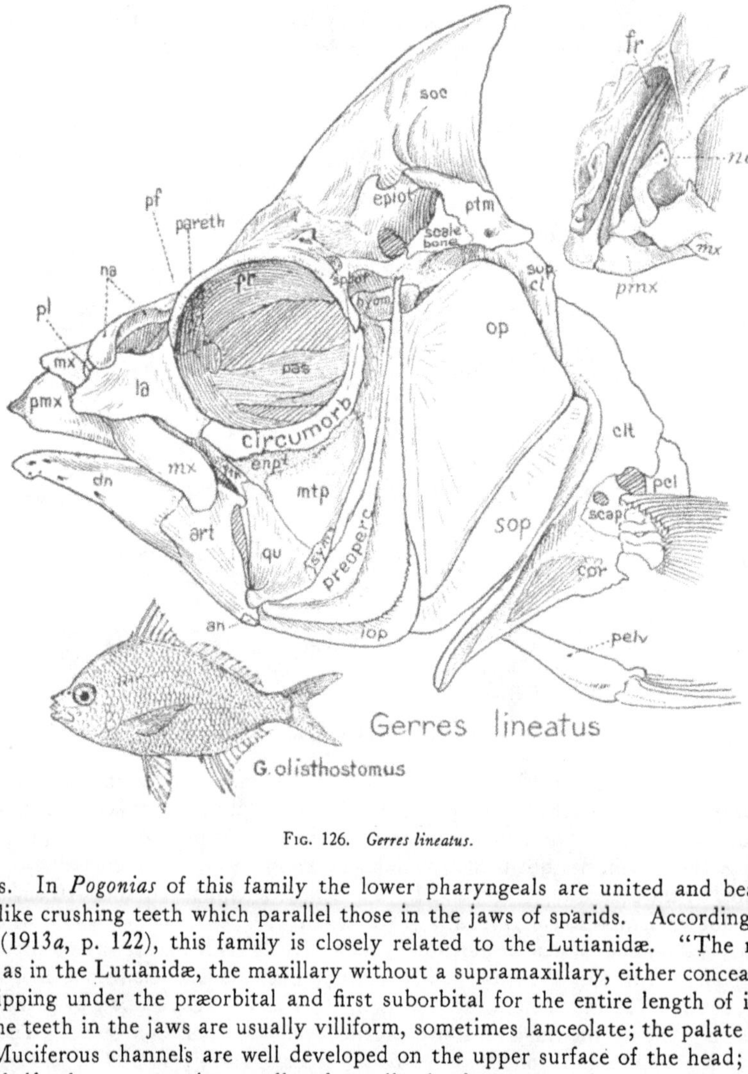

Fig. 126. *Gerres lineatus.*

percoids. In *Pogonias* of this family the lower pharyngeals are united and bear stout, pebble-like crushing teeth which parallel those in the jaws of sparids. According to Tate Regan (1913a, p. 122), this family is closely related to the Lutianidæ. "The mouth is formed as in the Lutianidæ, the maxillary without a supramaxillary, either concealed or at least slipping under the præorbital and first suborbital for the entire length of its upper edge; the teeth in the jaws are usually villiform, sometimes lanceolate; the palate is toothless. Muciferous channels are well developed on the upper surface of the head; the subocular shelf, when present, is a small and usually slender process of the second suborbital."

The otoliths of the family Sciænidæ, according to Frost (1927b, p. 303), are highly specialized, and the sagitta is often of great size and unusually ponderous. Possibly this may be connected with the well-known habit of these fishes of making sounds while in the water.

The main otolith (sagitta) of *Cynoscion nebulosus*, as figured by Frost (1927b, p. 303), is closely related in form to that of *Aplodinotus grunniens* of the same family. This in turn is a very much isolated type which according to Frost has certain features resembling the otoliths of the Berycidæ.

Gerrids.—In *Gerres*, type of the Gerridæ, the mouth (Fig. 126) is extremely protrusile and is withdrawn beneath an overhanging ridge formed by the maxilla and the lacrymal.

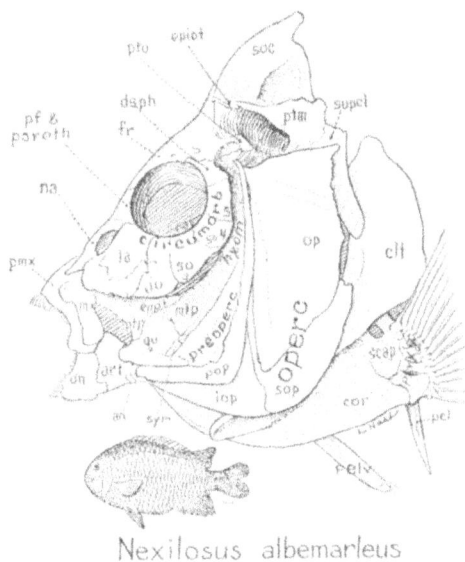

FIG. 127. *Nexilosus albemarleus.*

Boulenger (1910, p. 663) notes that the "premaxillary emits an upward lateral process," and that the lower pharyngeal bones are usually large and more or less completely coalesced as in a number of other families of percomorphs. Tate Regan (1913a, pp. 121–122) notes that in this family the mouth is "very protractile, the long præmaxillary pedicels lying in a groove or chamber formed by the bifurcation of the occipital crest; maxillary variable in form, but always with the anterior edge curved more or less in the shape of an S; distal extremity exposed; no supramaxillary; palate toothless." He regards them as being "closely related to the Lutianidæ." As the body in *Gerres* is deep, the back of the skull rises steeply into the very high supraoccipital.

The main otolith (sagitta) of *Gerres rhombeus*, as figured by Frost (1927b, p. 300), rather closely resembles that of *Lutianus chirtah*. Frost states that it resembles that of *Perca*, although differing in certain features.

Leiognathus.—Boulenger has referred this genus to the Gerridæ, but Starks (1911*a*) after a thorough analysis of its osteology shows that it differs from the Gerridæ in many important skeletal characters.   Nor is it closely allied with the Carangidæ, which it resembles externally.   This fish has a deep compressed body covered with small silvery scales.

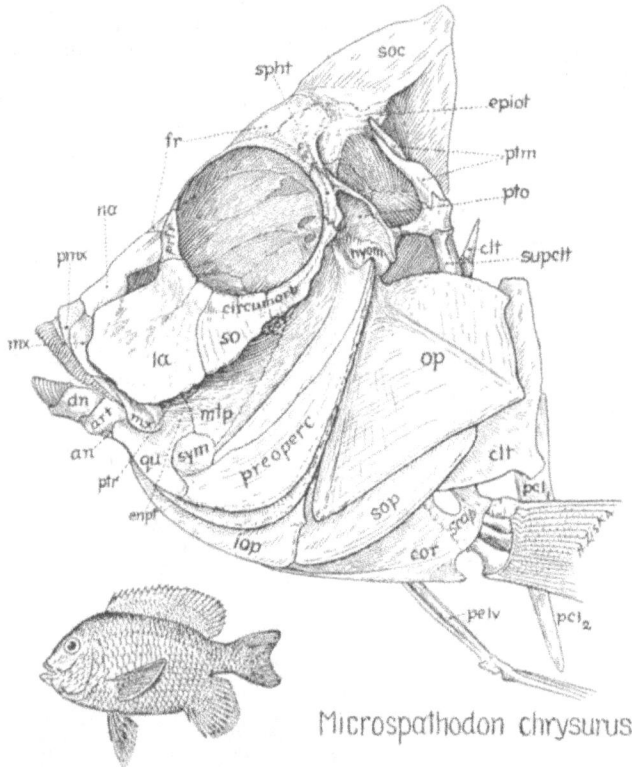

FIG. 128. *Microspathodon chrysurus.*

Its supraoccipital is produced into a very high crest which, however, lies wholly behind the frontals, unlike that of the Carangidæ.   The mouth is small and protrusile with extremely long ascending processes of the premaxillæ, which are received into a triangular area on the skull roof.   Starks considers (1911*a*, p. 8) that "the character of its supraoccipital crest and its deep pelvic girdle may indicate a connection with the scombroid stem near the place where the chætodont fishes branched off."   Nevertheless Tate Regan in 1913 (*a*) united the Gerridæ with *Leiognathus* and its allies in the single family Liognathidæ, while in 1929 (p. 320) he again classes it among those families of marine perches which have "a scaly axillary pelvic process and the maxillary sheathed by the præorbital," namely the Lutianidæ, Pomadisidæ, Sciænidæ, Mullidæ, Sparidæ, etc.

Cichlids.—These numerous freshwater fishes of Africa and South America parallel the wrasses in having strong pharyngeal teeth; but the opposite lower pharyngeals are not coalesced, merely attached by their inner edges or united by suture (Tate Regan, 1913a, p. 130); the ascending processes of the premaxillæ are often, but not always, very long, extending over the roof of the skull to the occiput, this arrangement implying that the mouth is strongly protractile. This is well shown in the figure of the skull of *Petenia splendida* given by Regan (1924, p. 206, Fig. 12A). Regan notes (*op. cit.*) that the fins are usually as in the Serranidæ and other generalized Perciformes.

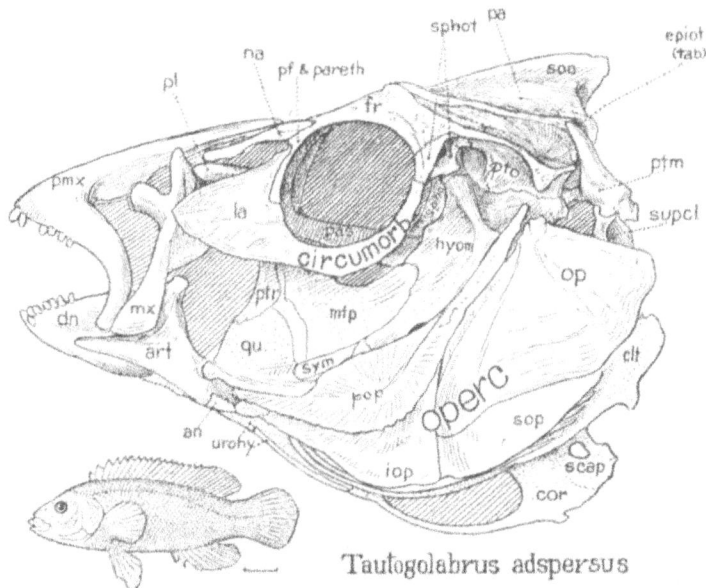

FIG. 129. *Tautogolabrus adspersus.*

According to Frost (1927b, p. 302), the otolith (sagitta) of *Talapia zillii* of the family Cichlidæ resembles that of *Perca* in basic features. This family is classed by Tate Regan in the Perciformes division of the Percomorphi.

Pomacentrids.—The pomacentrids are mostly small-mouthed, deep-bodied percoid fishes that have the lower pharyngeal bones united and tooth-bearing and are in certain characters like the wrasses but differ widely from the latter in the vertebral column and rib attachments. Tate Regan (1913a, p. 131) therefore assigns them to a separate division of the Percoidea, named Pomacentriformes, next to the Labriformes. The two pomacentrid skulls here figured (Figs. 127, 128) show very small mouths of the nibbling type, the solidly-built jaws packed with close-set minute teeth. The position of the quadrate-articular joint in *Microspathodon* is remarkably far forward in front of the anterior border of the orbit; but on account of the downward prolongation of the snout and the mouth, and of the raising of the quadrate joint, the mouth points forward instead of upward. The rest of the skull conforms to the short steep type already seen in other families. Frost (1928a, p. 451) notes that the otolith of *Chromis chromis* is of the percid type, but differs in certain details. It shows no special approach to the labrid type.

Labrids.—The wrasses (Labridæ) are grouped by Tage Regan (1913a, pp. 132–135) with the Odacidæ and the Scaridæ in the division Labriformes.   The jaws are of the protrusile but strong-biting type, with terminal horizontal mouth, again due to the forward and downward elongation of the snout.   The lower pharyngeals are fused into a triangular

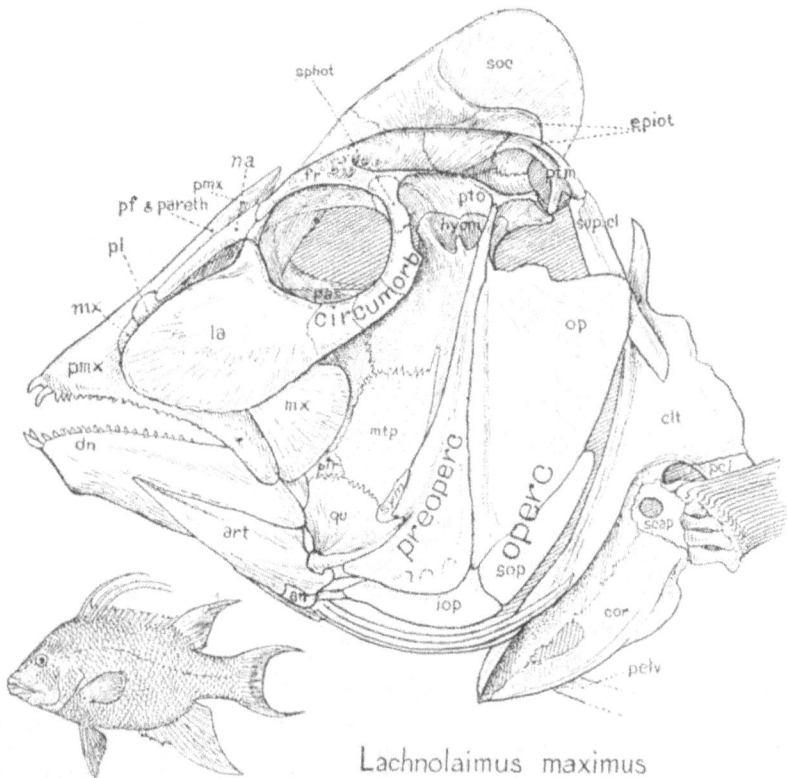

Fɪɢ. 130.   *Lachnolaimus maximus.*

plate and generally bear rounded blunt teeth.   The possession of this pharyngeal dental mill formerly led to the allocation of the wrasses and scarids to a separate order, Pharyngognathi, but they are now generally recognized as a division of the Percomorphi.   Possibly their nearest relatives are not the cichlids and pomacentrids but the hæmulids and sparids. In *Tautogolabrus* (Fig. 129), as in other labrids, the ascending processes of the premaxillæ are longer than the alveolar branch.   The latter is attached at its distal end to the lower end of the maxilla, which in turn is fastened to the side of the mandible.   Hence a lowering in the mandible draws the premaxillæ downward and forward.

In *Lachnolaimus* (Fig. 130) the deepening of the head brings the quadrate-articular joint down to a low level, thus increasing the area of the adductor mandibulæ muscles and lengthening the ascending processes of the premaxillæ. In the highly specialized labrid *Epibulus* as described by Delsman (Fig. 131) the mouth is excessively protrusile. All the conditions mentioned above on page 256 here reach their maximum. Besides this, even the quadrate has acquired a loose joint with the hyomandibular and can now be swung forward to aid still further in the extreme protrusility of the mouth. According to Tate

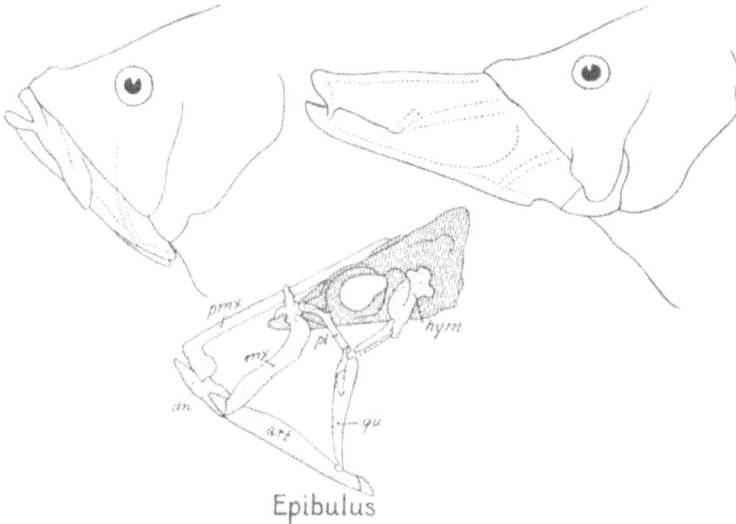

Fig. 131. *Epibulus.* After Delsman.

Regan (1913*a*, p. 133; 1924, p. 205) this form is closely related to the labrid *Cheilinus*, in which the mouth is only moderately protractile.

A peculiar specialization, perhaps from a *Lachnolaimus*-like ancestor, is seen in *Iniistius* (Fig. 132) of the family Coridæ. Here the face has been greatly deepened beneath the orbit. The jaws are short antero-posteriorly and deep vertically, with prominent curved hook-like incisors and small pointed-conical cheek teeth. The opercular, sub- and inter-operculars are membranous. The opposite specialization is seen in the labrid genus *Gomphosus*, which has the face elongated antero-posteriorly into a tubiform snout (*cf.* Jordan and Evermann, 1905, p. 289).

The otoliths of the labriform division, according to Frost (1928*a*, p. 452), are specialized and differ from the percid type. Those of *Iniistius* are very aberrant.

Scarids.—In the parrot-fishes (Scaridæ) the premaxillary and dentary teeth are fused respectively into an upper and a lower beak (Fig. 133). In the more primitive genera the relations of the premaxilla, maxilla and dentary differ from those in the wrasses, chiefly in the fact that the beak is shortened and brought back near the quadrate-articular pivot. In the mandible of certain genera a secondary joint is formed between the dentary and the articular so that a compound lever results, with one pivot at the quadrate and the other at the end of the articular; in the upper jaw the premaxilla is firmly attached to the maxilla by fibrous tissue.   The maxilla is attached below to the ascending process of the dentary by an interarticular disc (Lubosch, 1923, p. 17).   This interarticular disc is a neomorph, or new modification of the tissues normally connecting the lower end of the maxilla and the outer side of the mandible.   Lubosch considers (*op. cit.*, p. 16) that the active factor in this strange arrangement is the strong development of the dentary.   This, how-

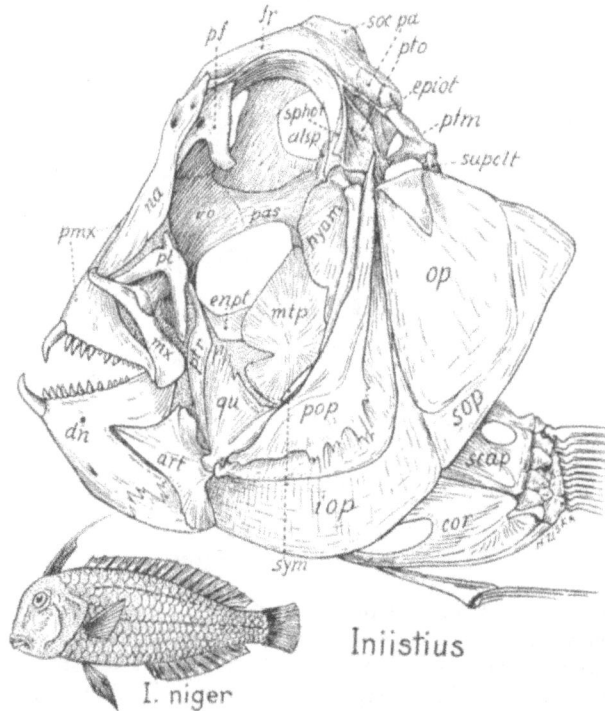

Fig. 132. *Iniistius* sp.

ever, would account only for the initial phase.   The close proximity of the ascending process of the dentary to the maxilla is evidently due to the bending upward of the originally lower border of the articular in order to lessen the adverse leverage against the powerful adductors 2 and 3. Thus the gape of the mouth was reduced but a compound lever of great strength was the result.   It is by means of these strong pincers that the parrot-fishes can tear off the seaweed and other marine growths upon which they feed.

However, the most elaborate part of the dental apparatus of the parrot wrasses lies in the throat, where the fused lower pharyngeals bear a many-toothed plate (Fig. 134) against which works a corresponding but sliding plate borne by the upper pharyngeals. The prolonged base of this upper plate slides back and forth on a pedestal furnished by the parasphenoid, while the lateral processes of the fused lower pharyngeals fit into grooves on the inner side of the cleithra. In each plate the unused teeth lie deep in the alveolar region, ready to come up into use when needed.

The otoliths of the Scaridæ conform to the labrid type, but are more specialized (Frost 1928a, p. 453).

Cirrhitids.—The family Cirrhitidæ (Haplodactylidæ) is regarded by Boulenger (1910, p. 664) as a derivative of the Serranidæ. Tate Regan (1911g, p. 261) however refers this family to the division Cirrhitiformes of the order Percomorphi. It may well be near to the base of the trachinoid series, as suggested by Jordan (1923, p. 203). Possibly it may even

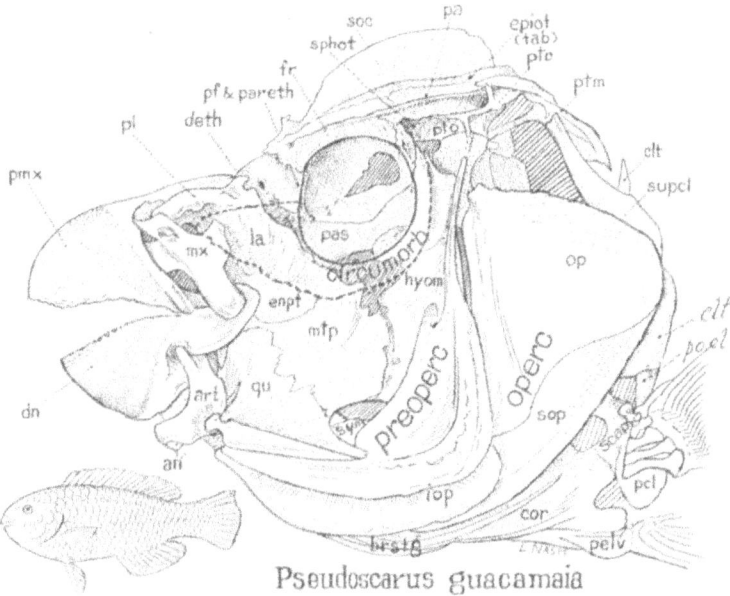

Pseudoscarus guacamaia

Fig. 133. *Pseudoscarus.*

be related on the one hand to the stem of the wrasses and on the other hand to such primitive trachinoids as *Pinguipes* (p. 356). The skull characters noted by Regan as characteristic of the group are as a whole close to the normal percoid type. The skull of *Cirrhitus* (Fig. 135) is a stoutly built derivative of the lutianid type. The third postorbital is tied securely by ligament to the upper end of the preopercular and of the hyomandibular but it does not suggest the stout suborbital stay of the scorpænoids, which is attached further down on the preopercular. The dilatator fossa above this postorbital bar is large and sharply defined. The orbits look partly upward, the opposite prefrontals are stoutly braced against each other in the mid-line and are supported below by the very strong pillars of the palatines. There is an extensive bony floor of the orbit formed by plates from the lacrymal and third suborbital. The opercular spine is membranous and the preopercular is also unarmed. The jaws recall *Pinguipes* and *Parapercis* of the trachinoid series. The upper pharyngeal plates are large but bear fine teeth. The lower pharyngeals are small.

13

Good figures of the skull of *Cheilodactylus spectabilis* are given by Leighton Kesteven (1928, pp. 334–343). Its special interest is that the short deep skull and very short, strong jaws carry the serranid type a long way toward the nibbling specialization seen in the chætodonts, acanthurids and many still more specialized derivatives of the predaceous percoids. Its extremely short dentary forks over a short V from the articular in the manner seen in labrids, scarids and other strong biters. The teeth are small and numerous and could easily give rise either to the chætodont or to the beaked condition. In view of this, it is not surprising to read in Boulenger (1910, p. 664) that these fish "feed chiefly on crustaceans, molluscs, and other invertebrates living among sea-weeds."

In *Chironemus marmoratus*, according to Frost (1928a, p. 451), the sagitta resembles that of *Perca* generally but differs in certain details.

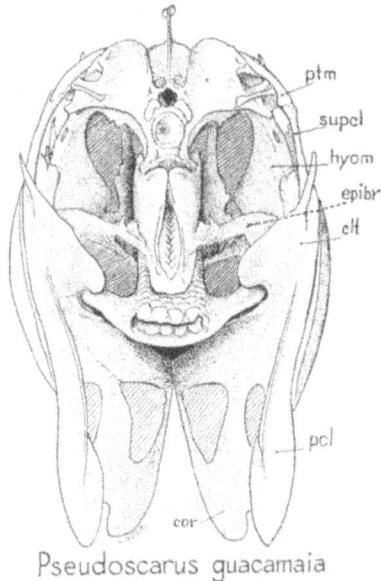

Pseudoscarus guacamaia

Fig. 134. *Pseudoscarus.* Rear view of skull, showing pharyngeal mill.

Chiasmodonts.—These highly voracious deep-sea swallowers have had a checkered taxonomic history, having been pushed about from one order to another; however, the recent thorough study of the osteology by J. R. Norman (1929) seems to have put them definitely in the suborder Percoidea as defined by Tate Regan, but in a new division, Chiasmodontiformes Norman, coordinate with the Perciformes, etc., of Regan. The chief osteological peculiarities are found in the mouth parts, there being an emargination in the upper jaw between the expanded anterior portions of the premaxillæ. Our Figure 137*A* is taken from a young skull of *Chiasmodon* kindly placed at my disposal by Dr. William Beebe. Figures *B, C, D,* are from the paper by Mr. Norman. Analogues with some of the large predatory scopeloids are evident in the slenderness of the premaxillæ, etc. The opercular elements are largely membranous.

PERCESOCES (BARRACUDAS, SILVERSIDES, GRAY MULLETS)

This group, the Percesoces of older authors and the Mugiloidea of Regan, even when restricted to the following three families, includes such diverse-looking fishes as the minute, small-mouthed silversides (Atherinidæ), the round-headed grey mullets (Mugilidæ) and

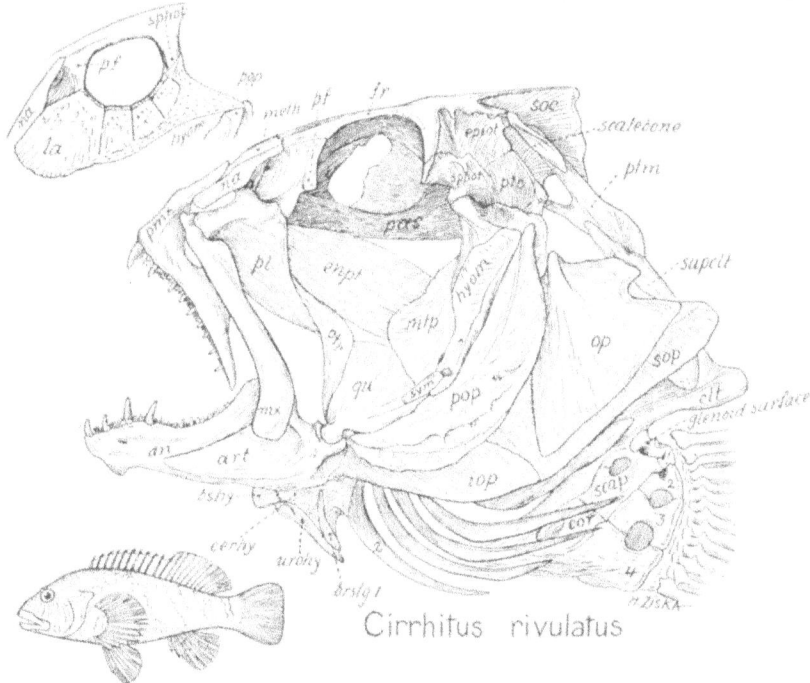

FIG. 135. *Cirrhitus rivulatus.*

the fierce pike-like barracudas (Sphyrænidæ). The researches of Starks (1899b), Jordan and Hubbs (1919), besides those of earlier authors, leave no doubt that the group is a natural one within the percomorph series, but the related problems as to what were the characters of the ancestral group from which the present families have become diversely specialized, and what was the derivation and exact relationship of that group, are still debatable. The group is in some way connected with the Percomorphi, with which it agrees in fundamental skull structure, in the presence of a spinous dorsal fin and in the possession of one spine and five soft rays in each pelvic fin. Even the otoliths of the Mugilidæ and Sphyraenidæ, as described by Frost (1929a, pp. 120–129), resemble generally the serranid and percid types.

The presumption is then that, as Dollo (1909) has suggested, the subabdominal position of the ventral fins in *Sphyræna* is a secondary convergence toward the isospondyl type rather than a direct inheritance. This inference is supported by the following facts:

(*a*) that, at least in *Sphyræna ideastes*, a long ligament runs from the pelvis to the cleithral symphysis (as I noted in dissecting a fresh specimen); (*b*) that in the relatively primitive family of Atherinidæ the pelvic bones are connected with the clavicular symphysis by a ligament (Boulenger, 1910, p. 639), as in all the acanthopterygian fishes.

Each of the existing families is plainly specialized in certain ways. The mullets (Mugilidæ) have a complex arrangement of the pharyngeal bones, a gizzard-like stomach

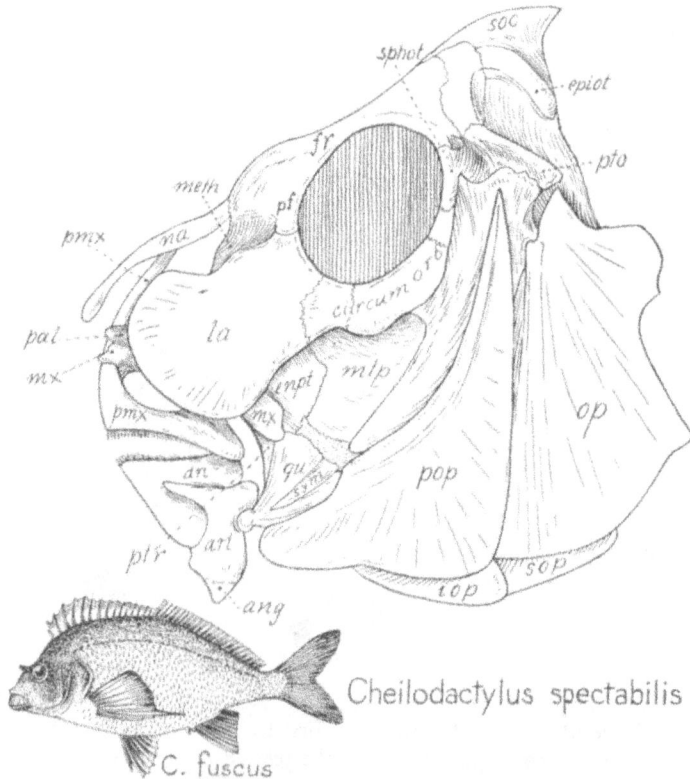

FIG. 136.   *Cheilodactylus.*   After Leighton Kesteven.

and an almost pig-like rim on the protrusile muzzle. The broad rounded skull (Fig. 138) has a vertical concavity in the front wall of the mesethmoid which serves for the reception of the expanded proximal end of the protrusile premaxillæ and limits their posterior movement. The stout parethmoids and paired anterior horns of the vomer support the massive snout and protect the olfactory capsules. The lower edge of the lacrymal is denticulate and overhangs the slender maxillæ. The latter have long medial processes which meet in the mid-line beneath the broad ascending processes of the premaxillæ. The teeth are minute or villiform on the premaxillæ and dentary. The lower or outer end of the rim-like premaxilla is expanded, while that of the maxilla is narrow and rod-like,—the reverse of the usual relations of these elements.

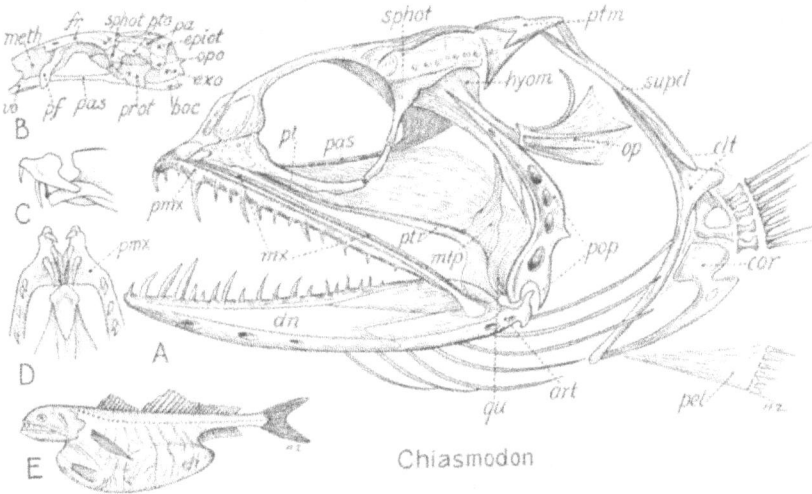

FIG. 137. *Chiasmodon.*

In the skull of *Mugil* (Fig. 138) the marked antero-posterior expansion of the opercular and the elongation of the pterotic and posttemporal may be conditioned by the enlargement of the pharyngeal apparatus, but I do not understand the reason for the backward pro-

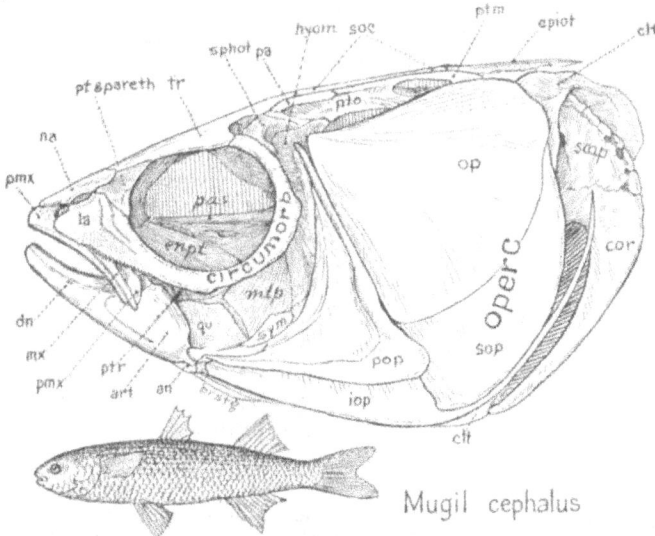

FIG. 138. *Mugil cephalus.*

longation of the preopercular and interopercular.   Probably it is conditioned in some way
by changes in the cerato- and epi-hyal.

In the Atherinidæ the skull and jaws are less specialized, the mouth varies from small
to moderate size, the teeth are usually small.   In the long-bodied *Chirostoma diazi* (Jordan

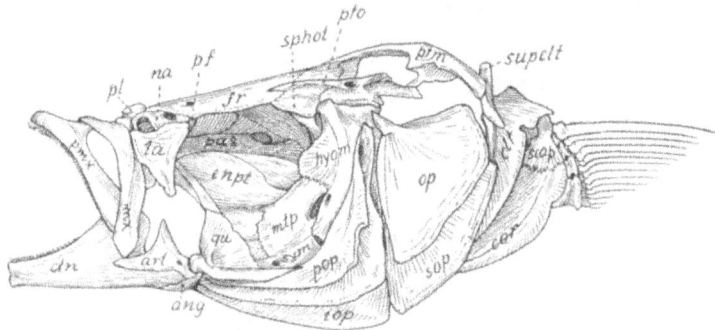

Basilichthys bonariensis

FIG. 139.   *Odontesthes "Basilichthys" bonariensis* Cuv.   Kindness of Dr. Tomas L. Marini and the Museo Nacional de Historia
Natural of Buenos Aires.

and Hubbs, 1919, Pl. VIII, Fig. 32) the whole body, head (Fig. 140*A*) and jaws suggest
that the Sphyrænidæ are merely giant atherinids.   Other atherinids, e.g., *Odontesthes
perugle* (*cf.* Jordan and Hubbs, 1919, Pl. III, Fig. 11) and *Chirostoma chapale* (*op. cit.*, Pl.
VII, Fig. 28) suggest the half-beaks of the order Synentognathi; but Tate Regan (1910*a*,
p. 8) remarks: "The supposed relationship of the Scombresocidæ to the Atherinidæ is
based on a number of resemblances which do not, in my opinion, indicate affinity.   The
Percesoces are more specialized than the Scombresocidæ in that spinous fin-rays are developed, and the features in which the Atherinidæ approximate to the Scombresocidæ
appear to have been evolved within the order Percesoces rather than to be those of the
prototype of the group."

In another direction the *Sphyræna* type is rather closely suggested by the Australian
fish *Dinolestes lewini* (Fig. 140*C*), but the very thorough analysis by Starks (1899*a*) has
shown that the latter is a true percoid of the family Cheilodipteridæ (Apogonidæ) and that
it differs from *Sphyræna* in many important characters, including the structure of the teeth,
which in *Sphyræna* are set in sockets, in *Dinolestes* ankylosed to the bone as in the percoids.

It therefore seems highly probable that the predaceous skull of *Sphyræna* (Fig. 141)
is not the most primitive one of the suborder, that its general resemblances to the *Dinolestes*
percoid type is *partly* an expression of parallelism, that it has been derived from a small,
short-jawed type more like that of the atherinids.   Nevertheless I am loath to give up the
idea suggested by Jordan and Hubbs (1919, pp. 6–8) that the Atherinidæ (and with them
the Percesoces as a whole) have been derived from true percoids, the ancestors of the
Apogonidæ and Ambassidæ, which appear to afford a suitable source for the cycloid scales
of atherinids, for the curiously persistent spinous dorsal of the whole series and for the
percomorph features of the skull.   It is perhaps for such reasons that Tate Regan (1929,

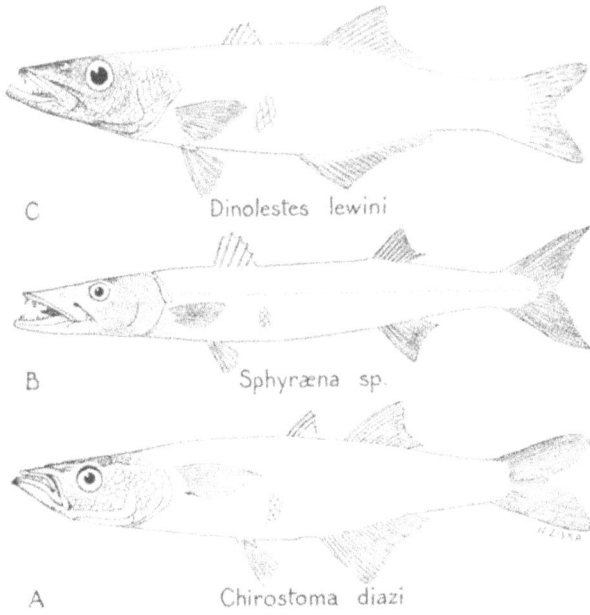

Fig. 140. Comparison of the percoid *Dinolestes* (C) with the Percesoces, *Sphyræna* (B) and *Chirostoma* (A). Scales different.

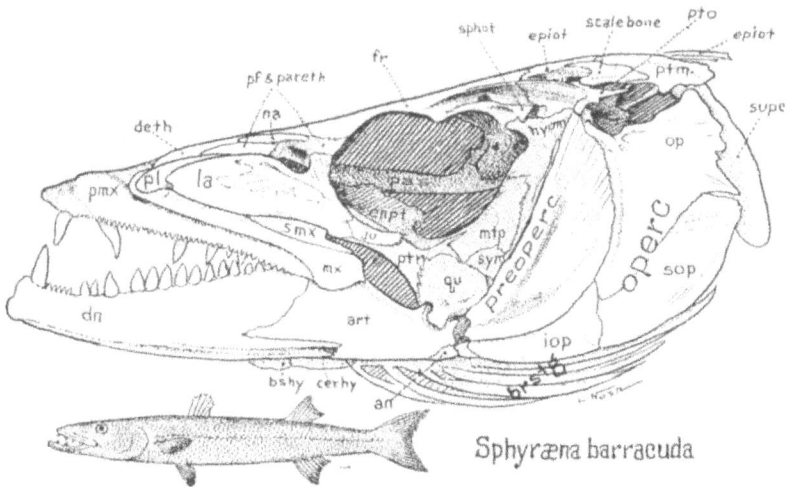

Fig. 141. *Sphyræna barracuda*.

pp. 320, 323) gives up the Percesoces as a separate order and allocates them under the name 'Mugiloidea' to the Percomorphi.

Accordingly it seems probable that *Melanotænia* of Australia (as figured by Weber and de Beaufort, 1922) and its allies are far from being really primitive Percesoces but that they are invaders of the rivers there, derived from marine atherinids.

With this general view of the possible origin of the Percesoces in mind we are perhaps in a better position to evaluate some of the characters of the skull of *Sphyræna* (Fig. 141).

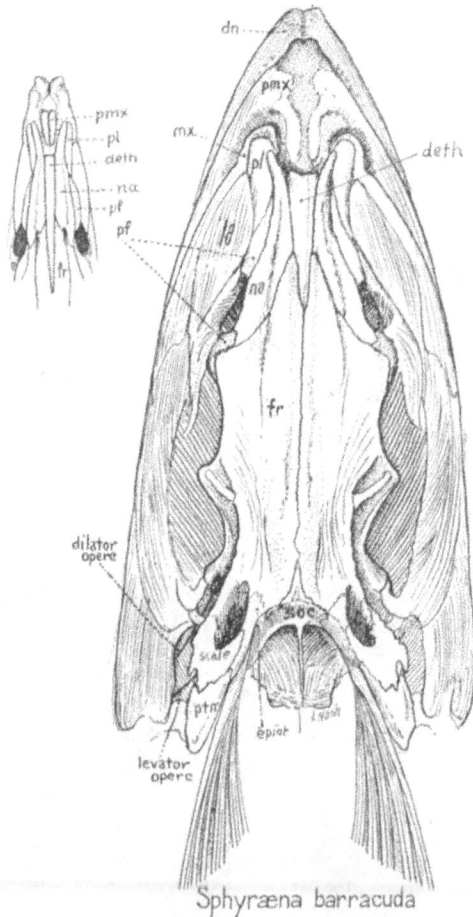

FIG. 142. *Sphyræna barracuda;* top view.

The forward inclination of the suspensorium may be partly due to inheritance from a short-jawed atherinid-like fish, but it may very probably be due in part to the lengthening of the body as a-whole, the rapid forward growth of the underside of the jaws and throat and the flattening down of the top of the head. Again, as in the pikes and eels, with a forwardly inclined suspensorium, the jaws have been lengthened by growing forward, especially at the front end. This in turn has involved a lengthening of the lacrymal and ethmoid region. In the top view (Fig. 142) the ascending rami of the premaxillæ, which are conjoined by a growing cartilaginous-dermal symphysis are blocked posteriorly and braced by the meseth-moid, much as in the Mugilidæ. Detailed comparisons of the premaxillæ and maxillæ

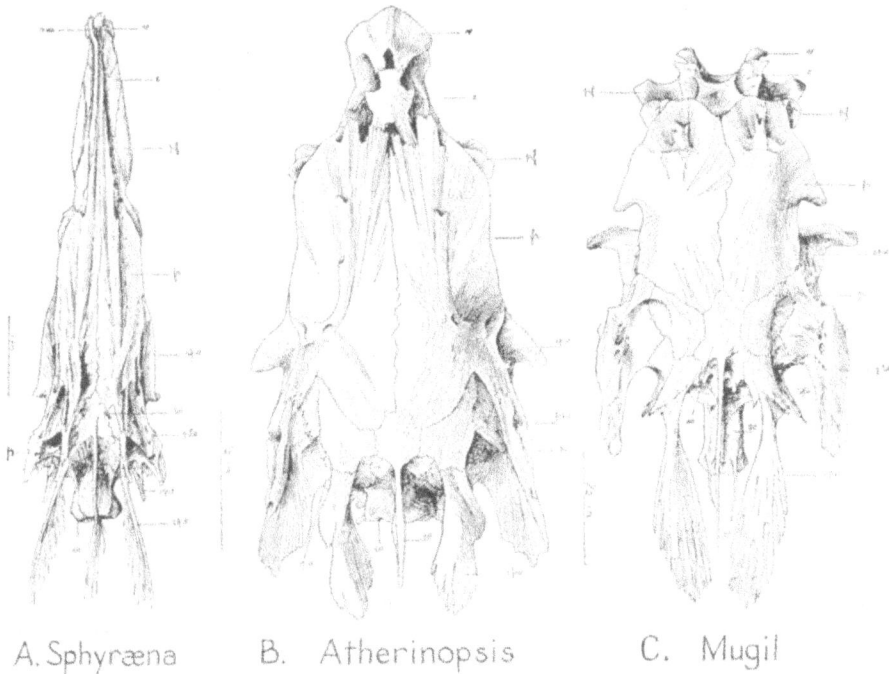

A. Sphyræna   B. Atherinopsis   C. Mugil

Fig. 143. Skulls of *Sphyræna argentea* (A); *Atherinopsis californiensis* (B); *Mugil cephalus* (C). After Starks. Top view.

show that they are plainly of true percomorph type, modified by the presence of the enlarged laniary teeth. The enlargement of the jaws has reduced the protrusility of the premaxillæ almost to zero since the extreme posterior position of the maxillo-mandibular ligament makes it impossible to pull the premaxilla forward. Hence the ascending processes of the premaxillæ are much reduced and the proximal forks of the maxilla firmly clamp the pre-maxillæ. The dentigerous layer is growing forward at the tip, like a predentary or prenasal bone. I cannot agree with Allis that it is this intermaxillary tissue which serves to connect the ascending process of the premaxillæ with the shafts of these bones. In three small specimens of *Sphyræna* there is not a sign of separation between the ascending process and the body of the bone.

The series of excellent comparative views (Fig. 143) of the skulls of *Sphyræna*, *Atherin-opsis* and *Mugil*, which are to be found in the paper by Starks (1899b) on the osteology of the Percesoces, affords a graphic demonstration of unity of subordinal plan, with divergences in proportionate development of parts, in the several families.   While *Sphyræna* is extremely long-headed, in *Atherinopsis* the skull is of normal percoid proportions, and in *Mugil* the skull is short and relatively very wide in front, the last characteristic being plainly correlated with the thickening of the snout and with the peculiar characters of the premaxillæ

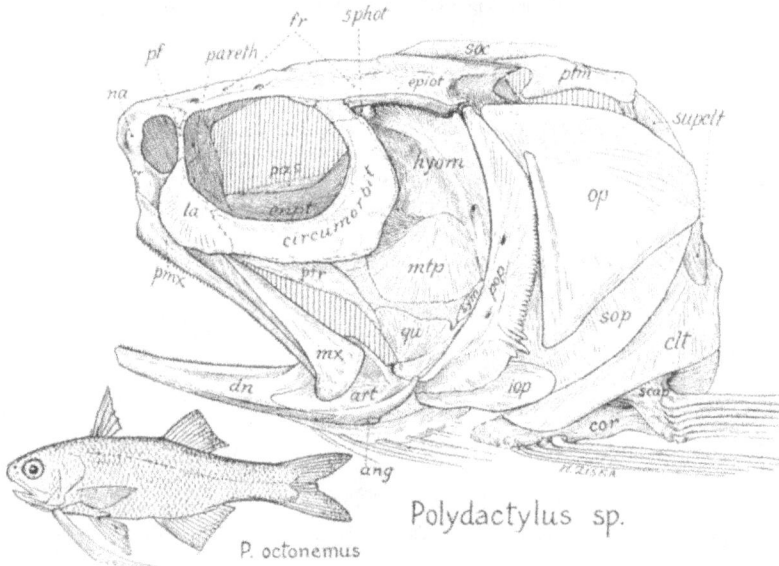

Fig. 144. *Polydactylus* sp.

already noted.   The peculiar bristle-like or branching prolongations of the epiotics and the failure of the supraoccipital crests to project above the level of the skull roof are conspicuous evidences of unity of origin of all three families, when taken in connection with the pervasive resemblances in basic features.

Polynemids.—Most authorities either refer the Polynemidæ to the Percesoces or place them next to that group.   Tate Regan, for instance (1929) gives them a rank coördinate with the "Mugiloidea" as a division of the percoid suborder of the Percomorphi.

The skull of *Polynemus* (Fig. 144) does not closely resemble those of either of the three main types of Percesoces.   The side view gives the impression of a forward and upward displacement of the large eyes, involving the forward shifting of the sphenotic and of the attached anterior head of the hyomandibular, combined with a marked lengthening of the pterygoid and a downward and backward displacement of the quadrate-articular joint to a point almost directly beneath the mid-point of the hyomandibular.   The posterior end of the upper jaw followed the quadrate backward and dragged, as it were, the front end of the premaxilla backward beneath the growing snout, which was covered by the enlarged down-turned nasals.   Thus was produced a convergent resemblance to the head and jaws of primitive palæoniscoids.   Meanwhile the opercular region remained fairly normal, the serrate preopercular being suggestive of percoid relationship.

The otolith (sagitta) of *Polynemus lineatus*, according to Frost (1928b, p. 331), resembles that of *Ophiocephalus* of the labyrinthine series, except in certain details.

LABYRINTHICI (SNAKEHEADS, CLIMBING PERCHES, ETC.)

The ophiocephalids and anabantids were referred to the Percesoces by Boulenger but have been classed as suborders (Ophiocephaloidea and Anabantoidea) of the Percomorphi by Tate Regan. In *Ophiocephalus* the head of the living fish has much more than a vague suggestion of a snake. The presence of an accessory respiratory organ may perhaps be responsible for the great increase in longitudinal diameter from the back of the orbit to the posterior rim of the enlarged subopercular (Fig. 145A). The quadrate-articular joint

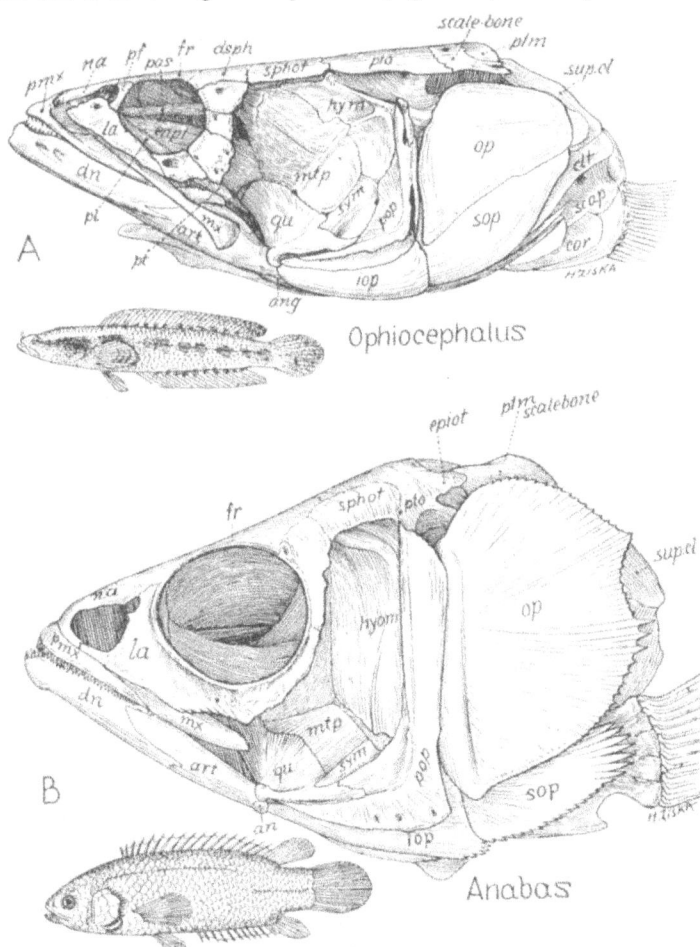

FIG. 145. A. *Ophiocephalus* sp. Side view. B. *Anabas scandens*.

is far behind the eye. No very definite indications of relationship either to the Percesoces or to the typical percoids were noted. The articular processes of the premaxillæ form broad ovals, which resemble those seen in some of the Jugulares. A very peculiar character is the presence of an antero-superior process on the metapterygoid, which process gains contact with the skull roof in front of the very broad hyomandibular. Thus the palato-quadrate arch is technically amphistylic, being supported both by its own process and by the hyomandibular. This is a rare occurrence in teleosts and is doubtless merely a convergent resemblance to the Palæozoic cœlacanths described by Watson and by Stensiö.

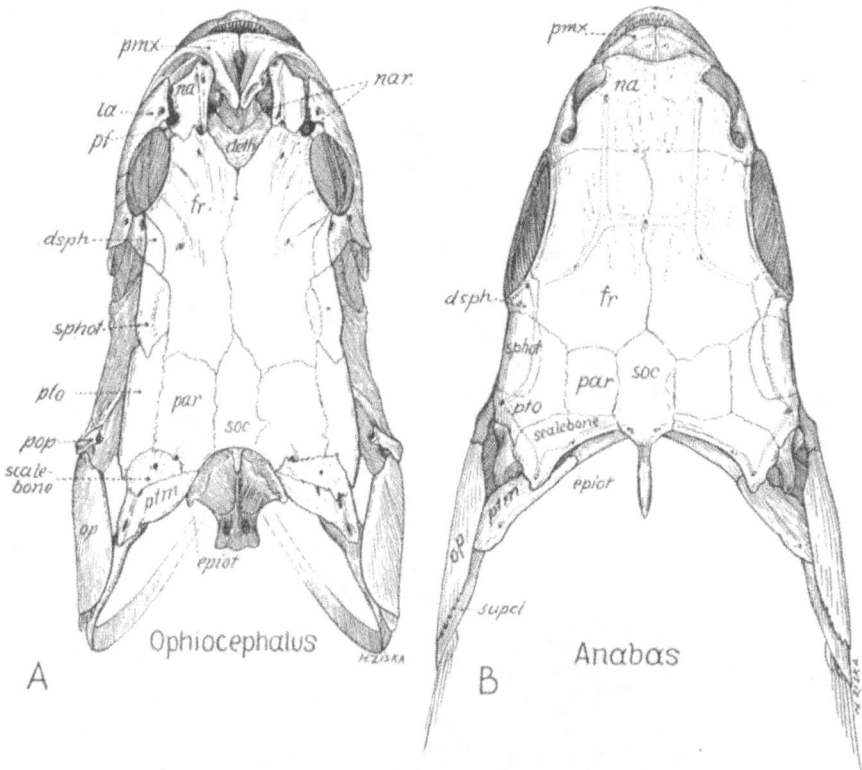

Fig. 146. A. *Ophiocephalus*. Top view. B. *Anabas*.

The deep-bodied anabantoids are probably more primitive than the elongate ophiocephalids and are certainly more percoid in appearance. The genus *Luciocephalus* (Weber and de Beaufort, 1922, p. 369) is long-bodied and pike-like, except for the small mouth. This would seem to indicate that the anabantoids were originally short compressed forms which had given off a long-bodied and pike-like branch. The skull form in general also appears to favor Jordan's view (cited by Tate Regan) that the Ophiocephalidæ are "degraded Anabantidæ" in opposition to Tate Regan's view (1910a, p. 10) that the reverse is the case. Figures 145, 146 show that in many details *Anabas* is more like the cyprinodonts than is *Ophiocephalus*.

The dried skull (Fig. 147) of the gourami (*Osphronemus*, Brit. Mus. Nat. Hist., No. 546) is chiefly notable for the parchment-like skeleton of the labyrinthine respiratory organ,

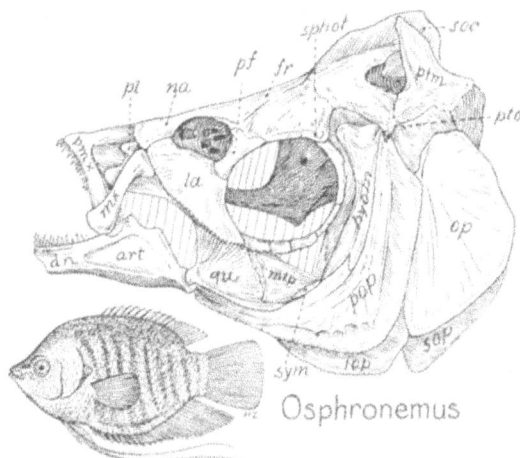

FIG. 147. *Osphronemus*. Side view.

which lies in the upper part of the branchial chamber. The opercular region recalls that of *Anabas*, except that the posterior border of the preopercular is barely, if at all, serrated. The mouth is small and evidently protrusile. The median occipital crest is low.

According to Frost (1928*b*, p. 330), the otolith of *Ophiocephalus lucius* resembles that of the percoid *Centropomus* except in certain details, while that of *Anabas scandens* resembles the serranid type, except in details.

ZEOIDEI (JOHN DORY, BOAR-FISH, ETC.)

The modern *Zeus*, according to the osteological investigations of Starks (1898*b*), is related to the Chætodontidæ, although forming a very distinct family. Tate Regan groups the Zeidæ with the Caproidæ or boar-fishes in an order Zeomorphi, coördinate in rank with, and placed between, the Berycomorphi and the Percomorphi. The Zeidæ have 31 to 46 vertebræ, the Caproidæ only 22. Nevertheless the ordinal characters of the Zeomorphi as defined by Tate Regan seem sufficient to connect the two families at the base of the Berycoidei. In the other direction they appear to be more or less related to the Ephippidæ, which are classed by Tate Regan among the Percomorphi.

The skull of *Zeus* (Fig. 148) is much compressed, with a small braincase, relatively large protrusile mouth, elongate prefrontal-vomerine region, which serves for the support of the long ascending processes of the premaxillæ. The latter also possess the articular and postmaxillary processes which are characteristic of percoid fishes.

The protrusility of the mouth is probably responsible also for the marked upturning of the mandible, the crowding back of the mesopterygoid (which is thrust in between the quadrate and the hyomandibular), the verticality of the pterygo-quadrate border. The roof of the mouth is raised; hence the hyomandibular is much reduced vertically and the opercular and subopercular are remarkably small, while the symplectic and preopercular are prolonged with the downward and forward growth of the quadrate-articular pivot. The cerato-, basi-, and uro-hyals are enlarged vertically, while the glossohyal is very small;

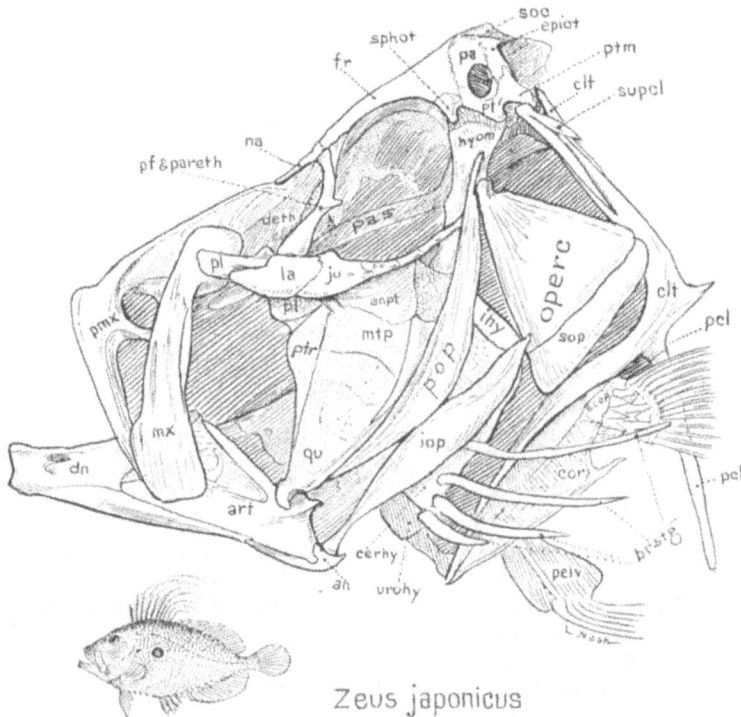

Zeus japonicus

Fig. 148. *Zeus japonicus.*

the lower pharyngeals are rather small and separate; they bear oval vestigial patches (Starks, 1898b, Pl. XXXIV).

Some of the features that are associated with protrusility of the premaxillæ are shared by *Zeus* with the chætodontoids but the relationships are evidently not very close.

Capros. The boar-fish *Capros aper* (Fig. 149A) has a strongly compressed body with a small, extremely protrusile mouth, the ascending processes of the premaxillæ being several times longer than the small dentigerous portion. The maxillary is a bone of unusual complexity, with a broad twisted blade and a great forwardly directed dorsal hook which embraces the ascending process of the premaxilla. The dentary is very short and has a secondary joint with the elongate angular-articular, after the fashion of fishes with small nibbling mouths. The downward movement of the articular, by the action of the longitudinal throat muscles, would pull the mouth downward and forward, and open it. Since the quadrate-articular joint is depressed far below the usual level, the mouth would be directed upward were it not for the great forward and downward prolongation of the premaxillæ and for the sharp downward turn of the dentary on the articular.

In conformity with the rhomboid shape of the body, the cranial vault is very short antero-posteriorly and rises steeply in front into a very high sagittal crest. The nuchal fin muscles are limited anteriorly by a sharp ridge on the supraoccipital crest, which curves downward and is continuous with a curved crest on the frontal and epiotic.

The opercular region is much deepened vertically and shortened antero-posteriorly.

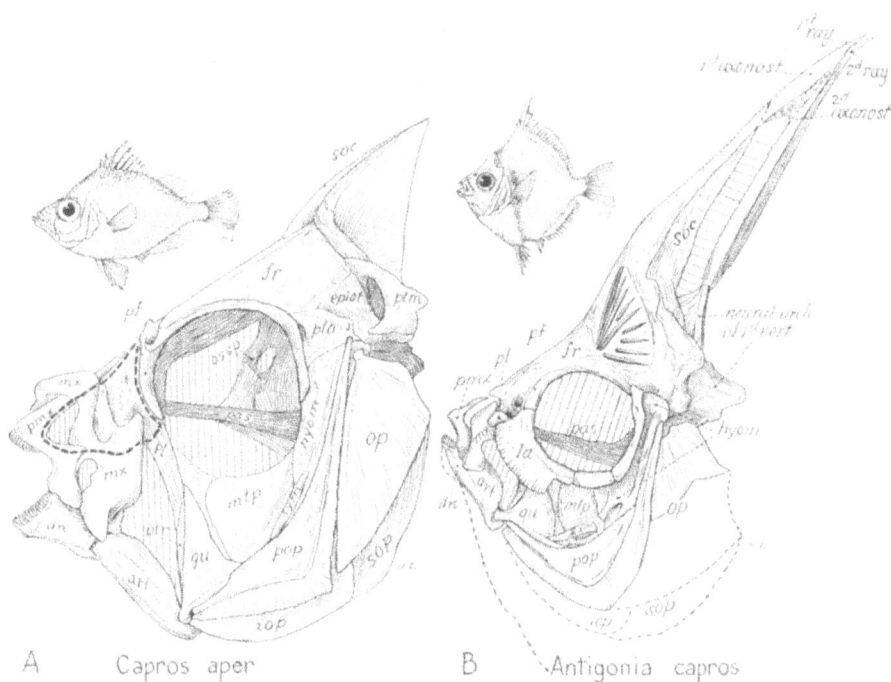

Fig. 149. A. *Capros aper*. B. *Antigonia*.

Antigonia capros.—This form (Fig. 149*B*) is more specialized than *Capros* in the excessive height of the occipital crest, which is conjoined with a still higher crest formed by the greatly enlarged basal rod of the anterior end of the dorsal fin. The neural arch of the first vertebra, which supports this rod, is enlarged and leans against the occiput. The mouth is less protrusile than in *Capros*. The ridge on the preopercular forms a rounded elbow. Starks (1902*b*) cites several characters in which *Antigonia* resembles the chætodonts; they appear also to be related to the Ephippidæ, including *Platax*, which in turn seem to be derived from such deep-bodied percoids as the Scorpididæ. The otoliths of the Zeomorphi, according to Frost (1927*a*, p. 443), while highly aberrant resemble those of the Berycomorphi in certain features.

Chætodontoidei (Butterfly-fishes, Angel-fishes, etc.)

The butterfly-fishes (chætodonts and their allies) show some general resemblances to the pomacentrids but they were regarded by Boulenger (1910, p. 667) as "closely allied to and evidently derived from the more generalized types of the Scorpididæ." Tate Regan (1913a, p. 128) refers them to his rather comprehensive group Perciformes, following the Monodactylidæ, Scorpididæ, Ephippidæ, Drepanidæ and related side-shoots. Starks in his valuable work on "Bones of the Ethmoid Region of the Fish Skull" (1926a, pp. 272–275) emphasizes the close agreement of the chætodonts with the Drepanidæ as showing the very exceptional junction of the opposite prefrontals (lateral ethmoids) in front of the main plate of the mesethmoid.

An examination of skulls of Scorpis, Psettus, Platax, in comparison with those of the ephippids and chætodonts, indicates that Boulenger's conclusion as quoted above is correct,

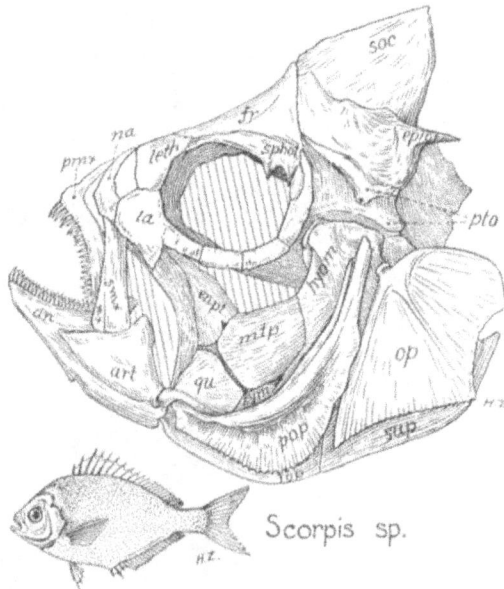

Fig. 150. Scorpis sp.

at least in that the skull of Scorpis (Fig. 150) forms a structural base for the chætodont series and connects it with some deep-bodied primitive percoid with twenty-four vertebræ and pelvic fins with one spine and five soft rays. As shown in Figure 150 the skull of Scorpis georgianus (No. 202, Brit. Mus. Nat. Hist.) is of primitive percoid type, hardly any of the elements showing conspicuous peculiarities. Perhaps in response to the deepening and compression of the body, however, the opposite lateral ethmoids are extended transversely and meet in a broad suture above the parasphenoid, forming a strong anterior pillar of the skull; they are pierced by the olfactory foramina in the normal way. The dorsal crest suggests that of the chætodontoid fishes since its muscle area does not extend forward on the forehead. It also recalls the conditions in the boar-fishes (Caproidæ). The lateral occipital condyles meet below the foramen magnum and above the basioccipital condyle, as in berycoids. The premaxillæ and maxillæ show none of the complexities of those of the boar-fishes.

Psettus.—This excessively deep-bodied fish (Fig. 297*B*) of the family Monodactylidæ may be considered as an aberrant offshoot from near the base of the Scorpididæ. Its skull (No. 460, Brit. Mus. Nat. Hist.) is definitely more specialized than that of *Scorpis* in many features, including the following:

(1) The supraoccipital crest is produced forward, its anterior rim being inclined above the frontals, which also send up paired median crests to meet it. The cranial roof thus forms a platform for the nuchal fin musclés as in *Brama*, *Coryphæna* and *Velifer*.

(2) The mouth is small, the teeth numerous and closely packed.

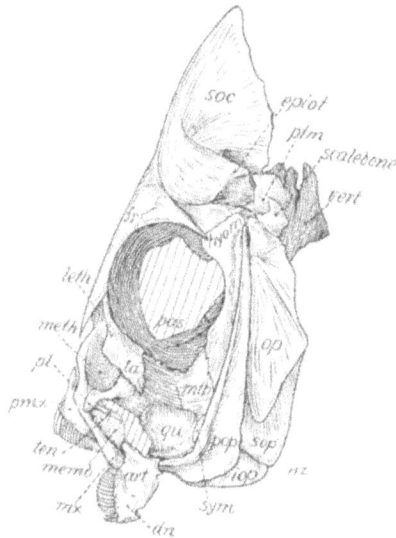

Platax teira

FIG. 151. *Platax teira.*

14

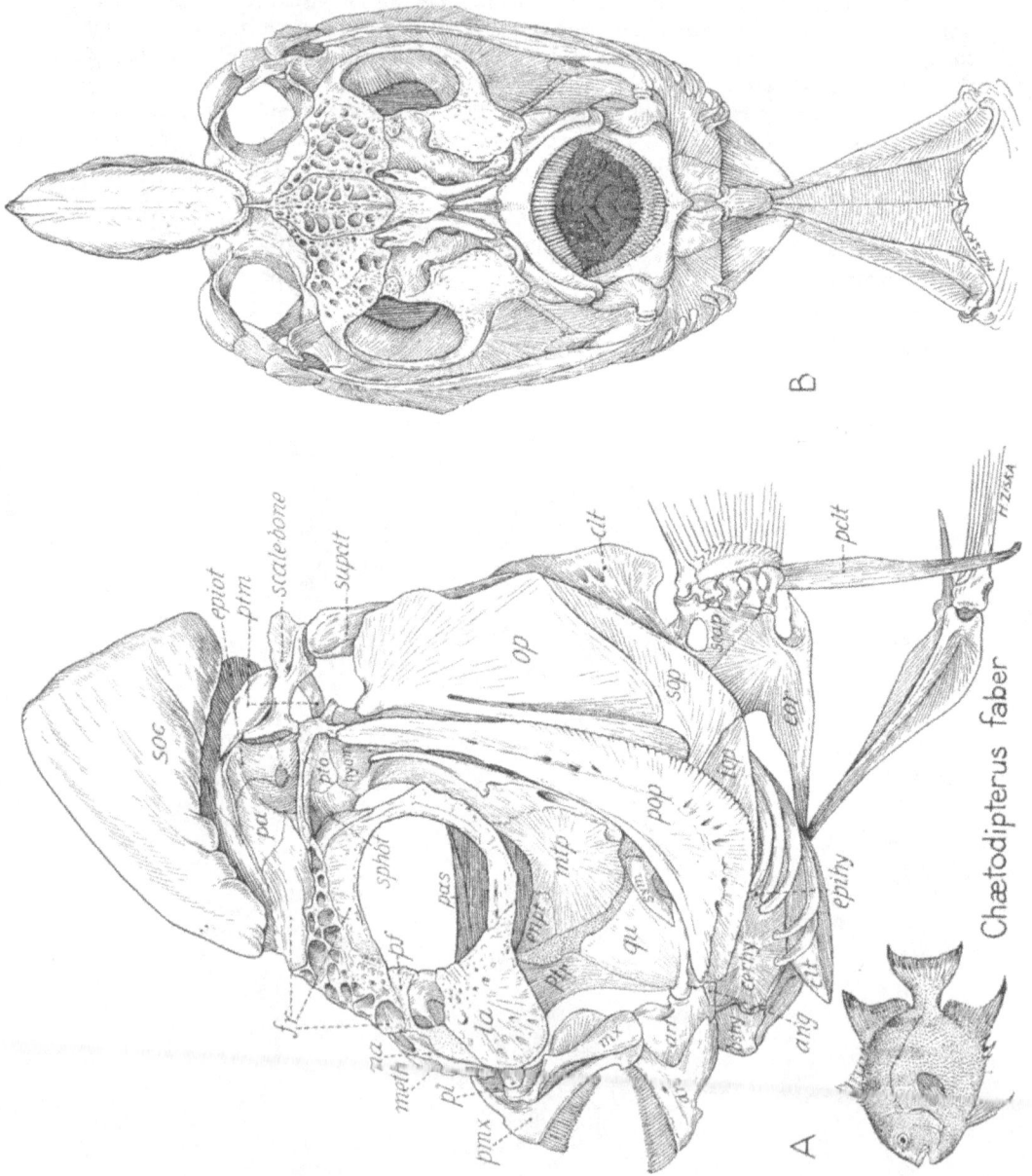

Fig. 152. *Chætodipterus faber.* A. Side view. B. Front view.

Chætodipterus faber

(3) The opercular is smooth, thin and covered with silver pigment; the preopercular has a smooth border.

(4) The subocular shelf is absent.

(5) The first vertebra is closely appressed to the occiput.

No special resemblances to the chætodonts were noted.

The otolith (sagitta) of *Psettus argenteus* of the family Scorpididæ, according to Frost (1927*b*, p. 304), resembles those of *Perca* and *Centropomus*, except in details. The sagitta of *Ephippus faber*, according to Frost (1927*b*, p. 304), is very peculiar, but resembles that of the young form of *Mullus* except in details. This resemblance is so striking that it would not be surprising if it gave a clue to the origin of the scorpidid-chætodont series from the perciform stock near the hæmulid-sparid-mullid branch.

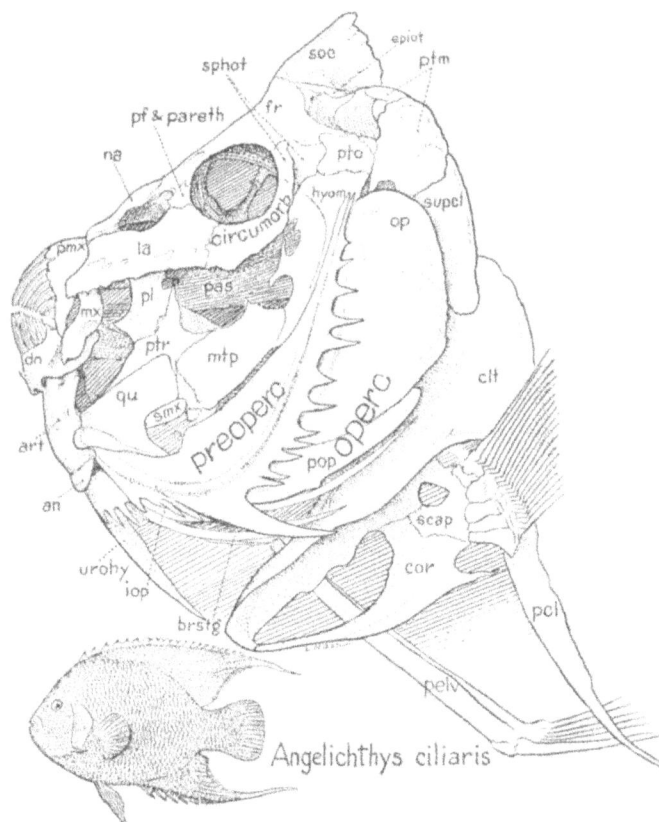

Fig. 153. *Angelichthys ciliaris.*

Platax.—This fish (No. 462, Brit. Mus. Nat. Hist.) of the family Ephippidæ is definitely advanced beyond that of *Scorpis* in the general direction of the angel-fishes and chætodonts. Its skull (Fig. 151) is an immediate structural ancestor to that of *Ephippus* (*Chætodipterus*).

Chætodipterus.—The skull (Fig. 152) is aberrantly specialized in the swelling of the ethmoid, frontals and supraoccipital, and in the shortening of the lower jaw. It approaches the scarids in the marked shortening of the dentary and articular, which brings the biting surface near the fulcrum and gives it great power.

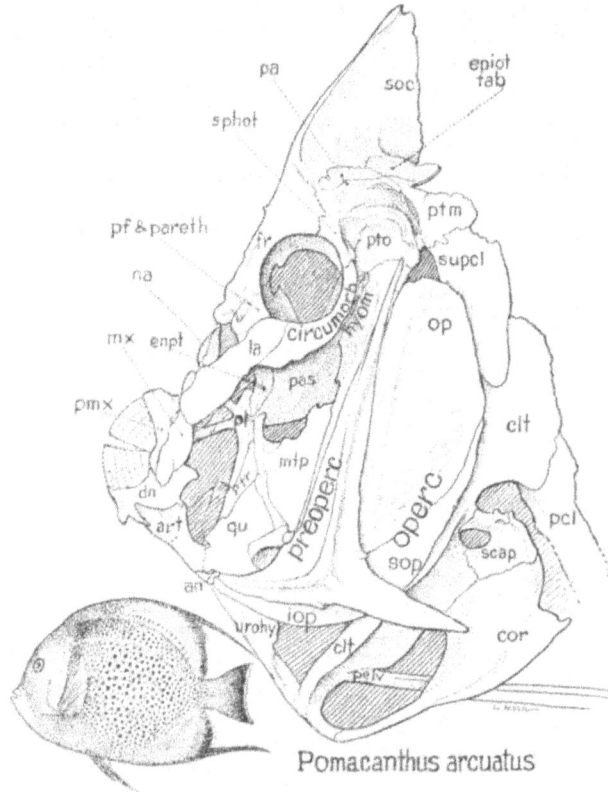

FIG. 154.   *Pomacanthus arcuatus.*

Angelichthys and Pomacanthus.—The deepening and shortening of the head is further accented (Figs. 153, 154). The eyes are smaller and displaced upward and backward beyond the level of the hyomandibular. Reminiscent of *Zeus*-like ancestors with highly protrusile mouths are the elongation of the naso-lacrymal region and the upward inclination of the mandible. The power of the bite has been greatly increased by swinging the insertion points of the adductor muscles backward nearly above the quadrate-articular fulcrum and (in *Angelichthys*) by the development of an incipient joint between the dentary and the articular. Meanwhile the palatoquadrate border has been swung backward past the vertical, so that the lower borders both of the mandible and of the palatopterygoid bar are nearly at right angles to their usual positions. The heavy spikes in the preopercular of these forms are also a sign of relatively high specialization.

Chætodon.—This fish (Fig. 155) has a minute protrusile mouth with bristle-like teeth, a long lower jaw and moderately ascending processes of the premaxillæ. The head and body are vertically deepened so that the quadrate-articular fulcrum is brought far below the mouth; thus its protrusility is increased. The opercular elements share in the general

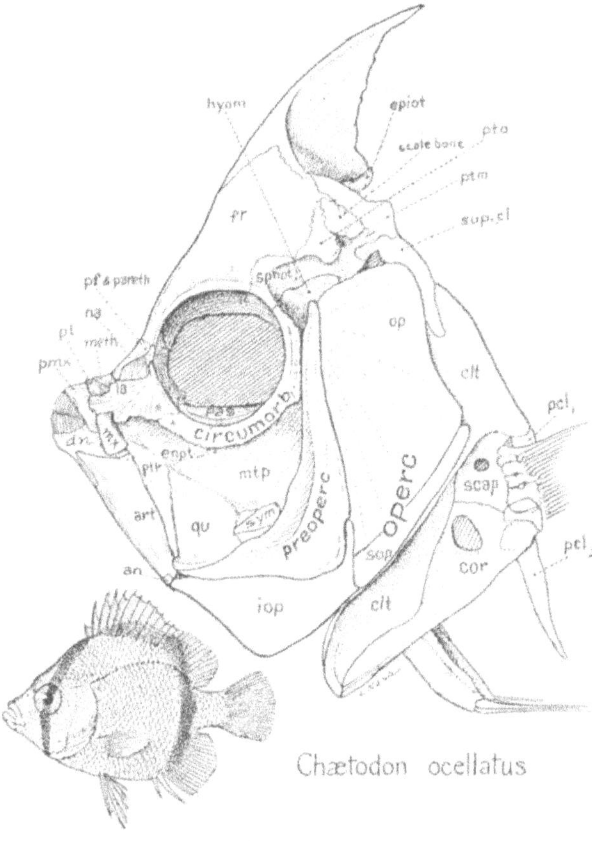

Fig. 155. *Chætodon ocellatus.*

deepening. The extremely high supraoccipital crest is concurrent with the crest on the pterotic. This feature is also conspicuous in other related families. The posttemporal is closely ankylosed with the skull and the supracleithrum is nearly vertical. The eye is large and the fish very alert, quick and even aggressive.

BALISTOIDEI (ACANTHURIDS, ZANCLIDS, SIGANIDS, TEUTHIDS, PLECTOGNATHS)

According to Boulenger (1910, p. 668), the surgeon-fishes (Acanthuridæ) (Figs. 156, 157) and their allies "form a connecting link between the Chætodontidæ and the Plectognathi." Starks (1926a, p. 277) from his studies of the ethmoid region judged that

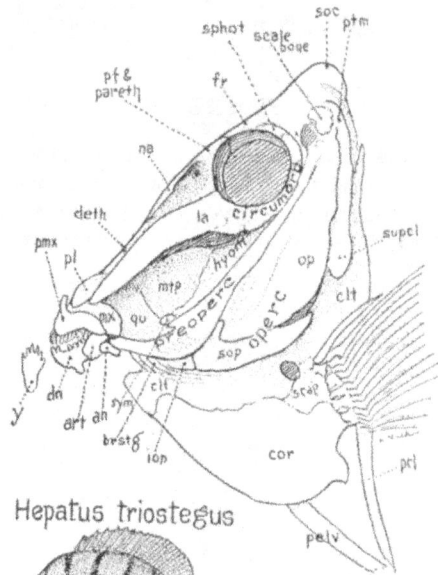

Hepatus triostegus

Hepatus sandvicensis

Fig. 156. *Hepatus triostegus.*

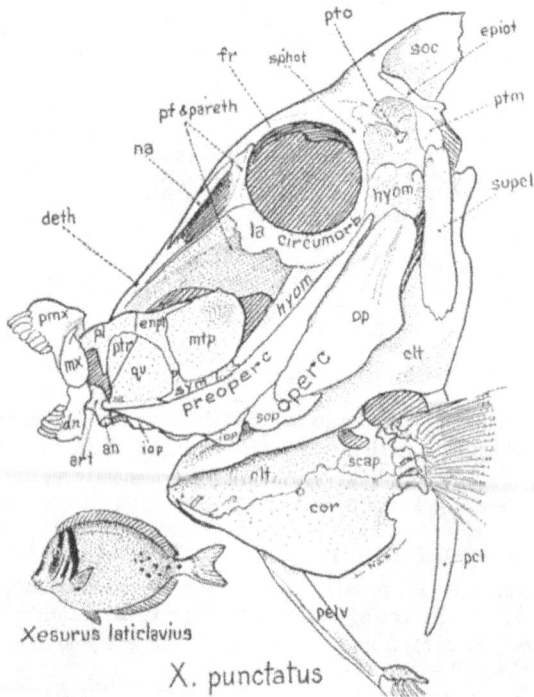

Xesurus laticlavius

X. punctatus

Fig. 157. *Xesurus punctatus.*

Zanclus (Fig. 158), which is usually classed as a chætodontoid, really belongs with *Acanthurus* (*Hepatus*) and its relatives. This seems at first sight to strengthen Boulenger's conclusion. But a comparative study of the skulls suggests that it is only the stem forms of the chætodonts, perhaps most nearly represented today by *Platax*, which can claim a

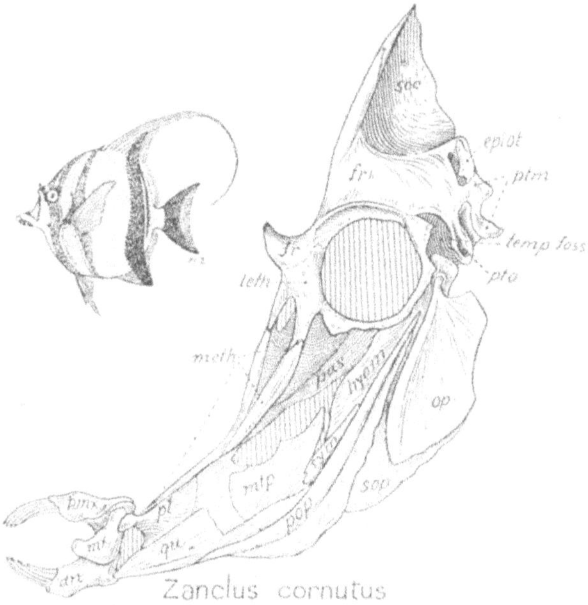

FIG. 158. *Zanclus cornutus.*

remote relationship with the acanthurids, zanclids and balistids, and that the cleft between the typical chætodonts and the acanthurids is much greater than that between the latter and the plectognath stem as represented by the balistids.

The chætodonts and their relatives are referred by Tate Regan (1913a, p. 112; 1929, pp. 320–323) to the perciform division of the suborder Percoidea of the order Percomorphi. The acanthurids and their allies form another suborder, Teuthidoidea, of the same order, while the balistids are classed with the puffers, etc., as a separate order, Plectognathi. The reference of the teuthidoids to the order Percomorphi rests chiefly upon their retention of several primitive percomorph characters in the fins and vertebræ, while the separation of the balistids is due to their loss of primitive characters. Although the plectognaths have acquired conspicuous specializations, which have masked their relationships with the teuthids, we may nevertheless assign them to a new group called the Balistoidei, including the zanclids, the teuthidoids and the plectognaths. In spite of their diverse specializations in the fins and body-form, the zanclids, teuthidoids and balistids possess a common heritage in their skull type, which is already curiously specialized even in the teuthids (Fig. 159), on the whole the least advanced members of the series.

We have seen in the chætodonts and many earlier small-mouthed forms how the suspensorium is inclined and produced forward; in the teuthidoids, zanclids and balistids the continuation of this process leads to some very strange results, for the hyomandibular now comes into direct contact with the circumorbital bones, which become fastened to it, while at the lower end the quadrate-articular joint is finally produced to a point far in front of

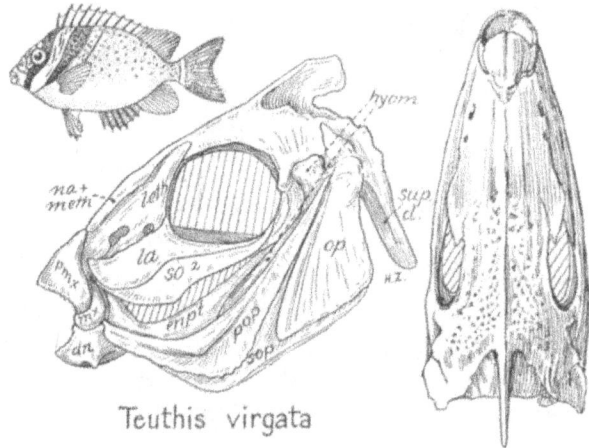

FIG. 159. *Teuthis virgata.*

the eye.   The angel-fish (ephippid) division of the chætodontoids also shows us how the lacrymal on the outside and the lateral ethmoid and mesethmoid on the inside became prolonged downward and forward; again in *Zanclus* (Fig. 158) this condition is emphasized to a hitherto unheard-of degree.   The result of this forward growth of the suspensorium

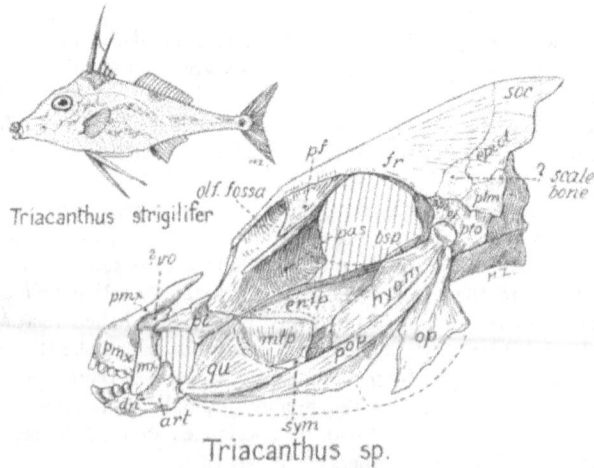

FIG. 160. *Triacanthus* sp.

below and of the ethmoid region above has been that in the zanclids, acanthurids and balistids the palatine, pterygoid, quadrate and metapterygoid found themselves still small and delicate and in harmony with the requirements of the minute mouth, although displaced so far forward that now even the metapterygoid and the adductor mandibulæ muscles lay

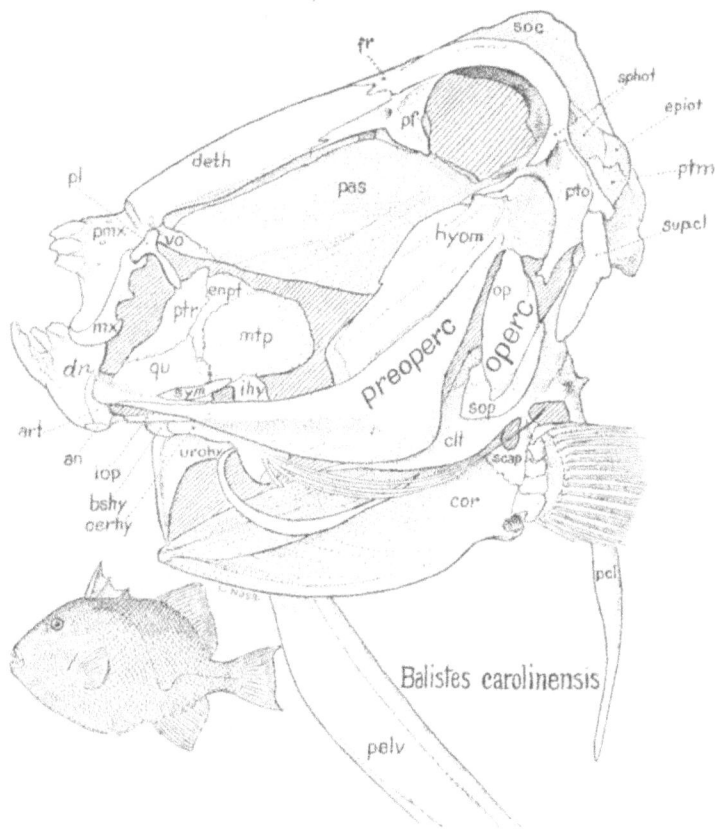

Fig. 161. *Balistes carolinensis.*

below and in front of the eye. Meanwhile in the zanclids the premaxilla, by reducing its ascending process, became able to turn upward, while in *Xesurus* (Fig. 157) the dentary developed a sliding connection with the articular so that it could turn downward at a sharp angle, thus enabling a small mouth to open very widely without loss of biting power.

In all these forms, including the balistids, the lateral ethmoids have remained with the orbits in a relatively posterior position while the mesethmoid has grown forward, along with the parasphenoid, for the support of the snout. This is well shown in the detailed topographic descriptions of the ethmoid region by Starks (1926a) who called attention to the extreme anterior position of the mesethmoid in these forms.

Balistids.—It may be an advantage for a nibbling fish to have its eyes as far back as possible from the tip of the pincer-like mouth. At any rate, in *Balistes* (Figs. 161, 162) the posterior border of the orbits is almost flush with the occiput and practically on top of the pterotic and hyomandibular socket. This fact, together with the extreme forward

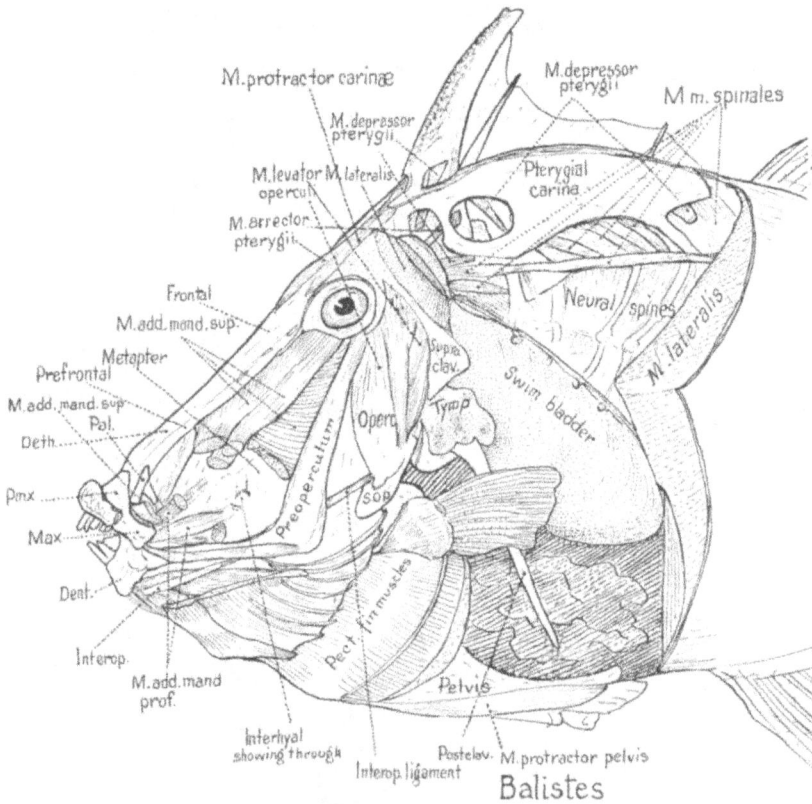

Fig. 162. *Balistes* sp.

prolongation of the suspensorium, results in some anomalous topographic reversals of primitive conditions, especially in *Aleutera* (Fig. 163): the posterior process of the pterotic now points downward and forward instead of backward; the pedicel for the opercular on the hyomandibular also points forward and downward and the hyomandibular itself points almost forward. Meanwhile the preorbital part of the parasphenoid has developed a great median keel which coöperates with the elongated and well-stiffened mesethmoid to form the base for the upper jaw. The opercular series, already restricted in *Xesurus*, has lost further territory in the balistids.

A dissection of *Balistes* (Fig. 162) reveals many interesting relations between the skeleton and the soft parts by which it is moulded. The adductor mandibulæ muscles occupy the long narrow area beneath and in front of the eye and are divided into several superficial and deep heads which are inserted on the upper as well as on the lower jaw. The interopercular is tracker-like and connects the back of the mandible with the opercular, thus enabling the opercular flap to move with the mandible rather than with the branchial arches. The point of attachment of the interopercular to the opercular is shifted dorsad, perhaps to lessen the movement of the opercular when the jaw opens.

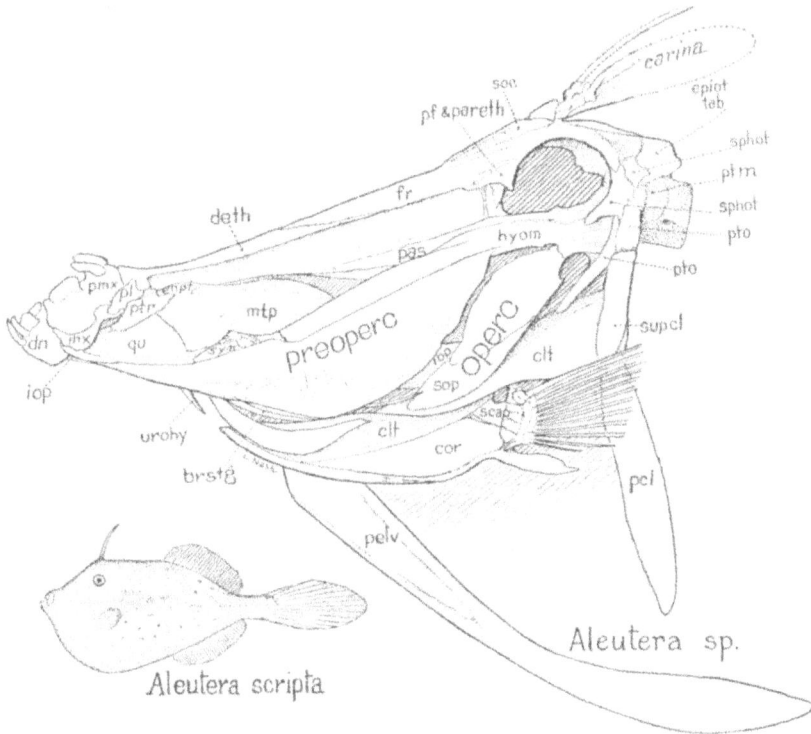

FIG. 163. *Aleutera scripta.*

Immediately behind the opercular is a toughened patch of skin, which I have called the "tympanum." This is connected with the swim-bladder, with the rod-like post-clavicle and the supraclavicle. It would not be surprising if this tympanum should serve to transmit either sound waves or pressure reactions along the supraclavicle to the back of the otic capsule.

The elaborate apparatus that includes the "trigger" or erectile spine is well shown in Figure 162. The erector muscles of the trigger run forward on the cranial roof and in *Aleutera* (Fig. 163) the entire apparatus is connected with the skull roof so that a new supraoccipital crest is rising.

The dissections also suggest why the occiput of *Balistes* should be so strongly braced by heavy ridges above and behind the eyes and on the occipital surface, for these regions serve for the insertion of two powerful muscles: first, the dorsolateral muscle, which, arising from the superficial parts of the myomeres, has been compounded into a single muscle inserting into the base of the tail; second, the deep vertebral muscles, which pass obliquely downward and backward from the back of the occiput toward the anal fin. The figure of this highly specialized fish thus particularly well illustrates something that is true of all vertebrates, namely, that the back part of the skull serves as the anchor or pivot for two sets of stresses, one coming from the locomotor apparatus behind and another from the jaws and branchial apparatus in front and below the occiput.

Anacanthus barbatus

FIG. 164. *Anacanthus barbatus.* After Day.

Anacanthus (Figs. 164, 165) carries the grotesque specializations of *Aleutera* to an amazing extreme. The tendency to secondary lengthening results in transforming an originally deep fish almost into a tube-fish.

Ostraciontids.—In the trunk-fish (*Lactophrys*) (Fig. 166) of the family Ostraciontidæ the external appearance is quite different from that of the balistid type, principally because of the replacement of scales by a rigid body armor mostly formed of hexagonal plates, which are to some extent foreshadowed in certain balistids. The spiny dorsal has been eliminated and in this genus horns are added. The eyes are enlarged and raised, drawing the frontals, sphenotics and pterotics up with them. The mouth has been brought down to the ventral border. But when the armor is stripped off, the skull (Fig. 166) is seen to be fundamentally identical with that of *Balistes* with but a few minor changes. The mouth is now tilted partly downward, which has required the extension of the mesethmoid past the quadrate-articular joint so as to form an abutment for the stout premaxillæ. The teeth are small and incisor-like, belonging to a number of successive sets; the maxillæ are firmly united with the premaxillæ, the palatine is apparently absent. This powerful nibbling mouth is supported above by the strong coalesced ethmoid and parasphenoid and below by the rather delicate quadrate. The hyomandibular, essentially similar to that of *Balistes*, is broadly expanded and meets the parasphenoid in a secondary sutural contact. The opercular and especially the preopercular are reduced. The interopercular retains its tracker-like character. The gill-chamber is very small.

Triodonts.—In *Triodon bursarius* of the Indian Ocean we have an important structural link between the balistid stock on the one hand and the puffers (tetraodonts) and porcupine-fish (diodonts) on the other. As noted by Tate Regan (1929, p. 325), *Triodon* has an air-sac or diverticulum of the œsophagus like the puffers and porcupine-fishes. But it retains the pelvis which these others have lost and it "resembles the Balistidæ in having the pelvis a long movable bone that dilates the air-sac. It differs, however, in having no spinous dorsal fin and in having the teeth represented by a beak." Here again it shows intermediate conditions: in the upper beak, the two halves are separated by a median suture as in the puffers; in the lower jaw the two halves are united into a single beak like that of the porcupine-fishes.

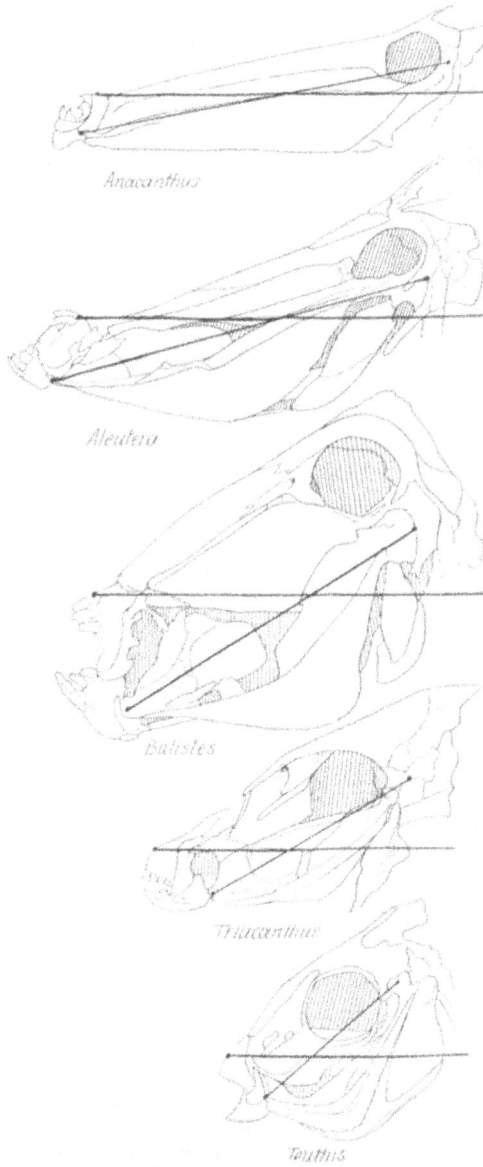

FIG. 165. Comparative series of balistoid skulls.

Tetraodonts.—It is noted above that in the balistids there are many sets of teeth; experience with other groups, such as the elasmobranchs, the scarids, oplegnathids, etc., shows that a beak may be produced in such cases by the multiplication of these teeth and the subsequent close appression or perhaps complete fusion. In the trunk-fishes (Ostraciontidæ), as above noted, the teeth are small and numerous and close-set but still distinct. In the tetraodonts or puffers (e.g., *Spheròides, Lagocephalus*, Figs. 167–169) the beak is complete except that the right and left halves remain quite distinct. Here the beak has more of a shearing character, while in the porcupine-fishes (Diodontidæ) the beak (Figs. 170, 171) is more massive and more adapted for crushing.

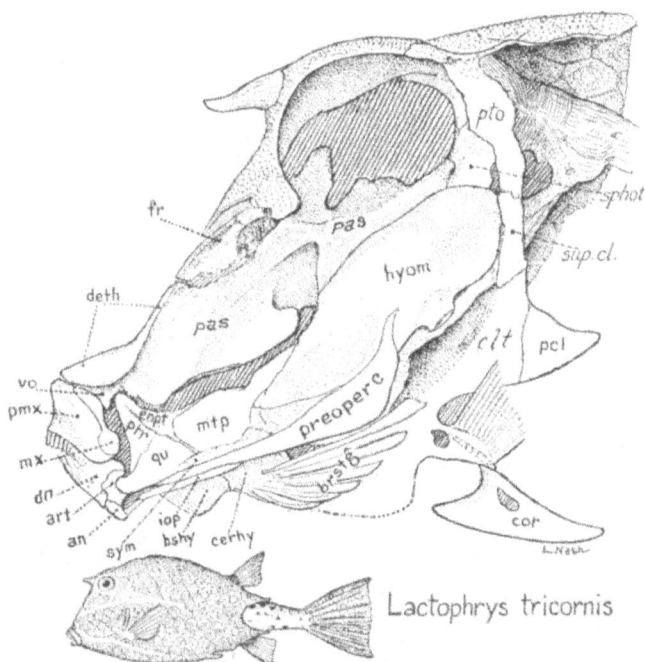

FIG. 166. *Lactophrys tricornis.*

With the acquisition of sharp-cutting edges on the four dental plates of the puffers there has probably been a rapid increase in the size of the mouth and a change perhaps from nibbling to shearing; also the form of the body when uninflated is of a swifter type than that of the other families, so that puffers ought to have a more omnivorous diet, including crustaceans. It is therefore not surprising to read in Bigelow and Welch (1925, p. 299) that puffers ". . . feed on small crustaceans of all sorts, especially crabs, shrimp, and amphipods, as well as on small mollusks, worms, barnacles, sea urchins, and other invertebrates, which they find on the bottom. Young fry of 7 to 10 mm., examined by Doctor Linton at Woods Hole, had eaten copepods and crustacean and molluscan larvæ."

The general hypothesis that I am defending, namely, that the whole plectognath stock was derived from very small-mouthed nibbling forms at least closely related to the balistids, requires then that, *pari passu* with the acquisition of a cutting beak in the ancestors of the puffers, the jaws increased greatly in breadth and to a less extent in the length.   But so large a cutting and crushing mechanism could not advantageously be operated at the distal end of so prolonged and narrow a rostrum as that in the trunk-fishes.   Hence, according to our hypothesis, there was a rapid period of transverse growth of the skull-top accompanied by a marked secondary shortening of the rostrum and by the enlargement of all the elements

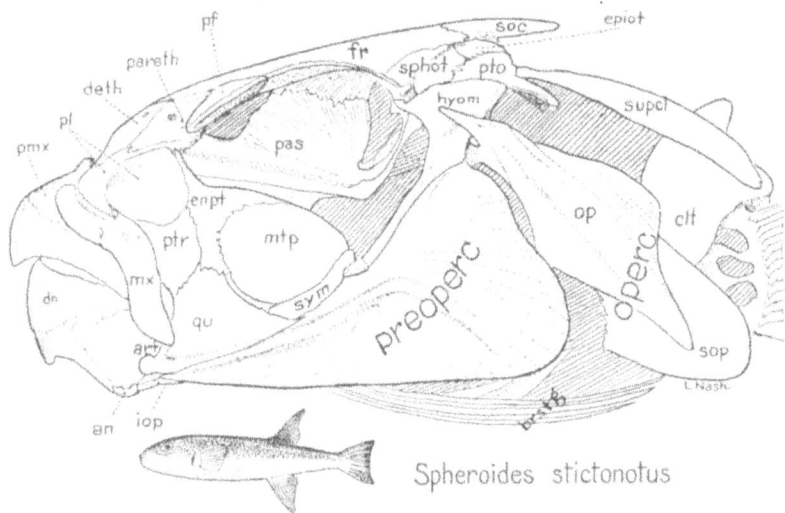

Fig. 167.   *Spheroides stictonotus.*

that brace the jaws posteriorly.   That the rostrum has been shortened is indicated, among other reasons, by the marked difference in the length of the mesethmoid in different species of puffers.   In *Lagocephalus* (Fig. 168) it is relatively narrow and much resembles that of the narrow-nosed balistids.   In *Spheroides* (Fig. 167) it has become very wide and short, due partly to the dorsal encroachment of the posterior processes of the premaxillæ.   That the palatine has received a marked secondary increase in size is evident by comparison with the minute palatine of the balistids, teuthids, chætodonts and other nibbling types.

In *Lactophrys*, of the family Ostraciontidæ, the entopterygoid and metapterygoid share this enlargement and still retain much of the arrangement shown in *Triacanthus* (Fig. 160) where they were placed near the lower end of the downwardly elongate face. But in *Spheroides* (Fig. 167), according to our hypothesis, a marked relative shortening of the face from the postorbital process to the quadrate-articular joint has shifted the entire palato-metapterygoid tract backward beneath the orbit.   However, even with this secondary backward shift, the upper part of the metapterygoid has not regained its primitive percoid connection with the upper part of the hyomandibular, from which it is still separated by a deep indenture.   This hypothesis of a secondary shortening of the face and its bony scaffolding, in the line leading to the puffers, does not in any true sense require an exception to the law of the irreversibility of evolution.   It is rather a special application of that law, which, as originally enunciated by Dollo, recognized clear evidence for changes in the direction of evolution.

In the sagittal section of the skull of the puffer *Lagocephalus* (Fig. 168) we notice that the tip of the upper beak is almost in horizontal line with the basis cranii and the parasphenoid is horizontal or even tilts slightly upward, whereas in the small-mouthed nibbling fishes such as the teuthids and balistids, which I am supposing to represent the remote ancestors of the puffers, the upper jaw is far beneath the level of the basis cranii and the

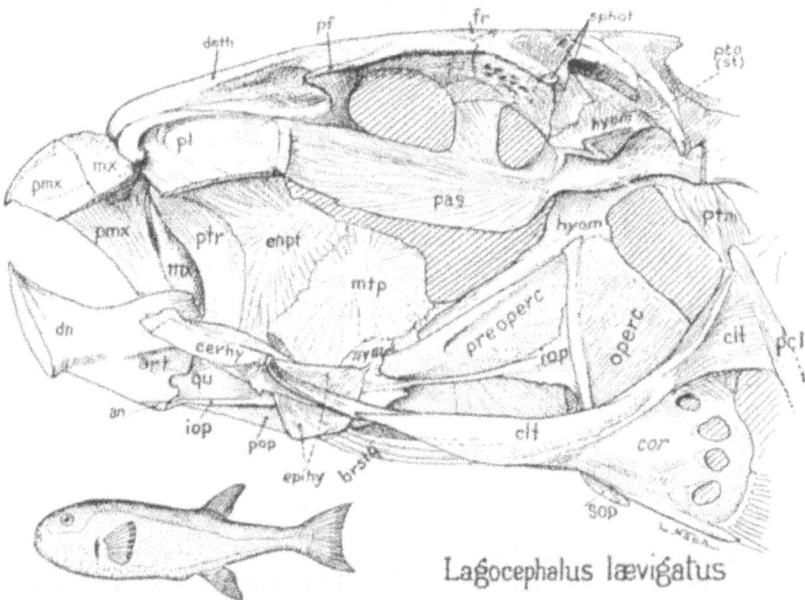

FIG. 168.   *Lagocephalus lævigatus.*   Right half of skull viewed from within.

parasphenoid is directed obliquely downward.   Four factors have probably taken a leading part in producing this result.   In the first place, as I have suggested above, there has been a marked shortening of the preorbital face length, this in itself bringing the dorsal border of the mouth up to a higher level.   Secondly, there has probably been a marked decrease in the total height from the summit of the back to the lower border of the throat when a primitive high-bodied fish was changed into a shallow-bodied and secondarily elongate type.   Thirdly and almost indissolubly connected with the last factor, there has been a great secondary increase in transverse diameters, finally producing the sub-spherical contours of the porcupine fishes.   This brings into view the fourth and perhaps leading factor of all, the modifying influence of the dominant "puffing" apparatus in both the puffers and the porcupines.

In the mechanism of inflation, as so clearly exposed by Parr (1927c), the broad obliquely-placed muscles of the body-wall force water or air into the œsophageal diver-

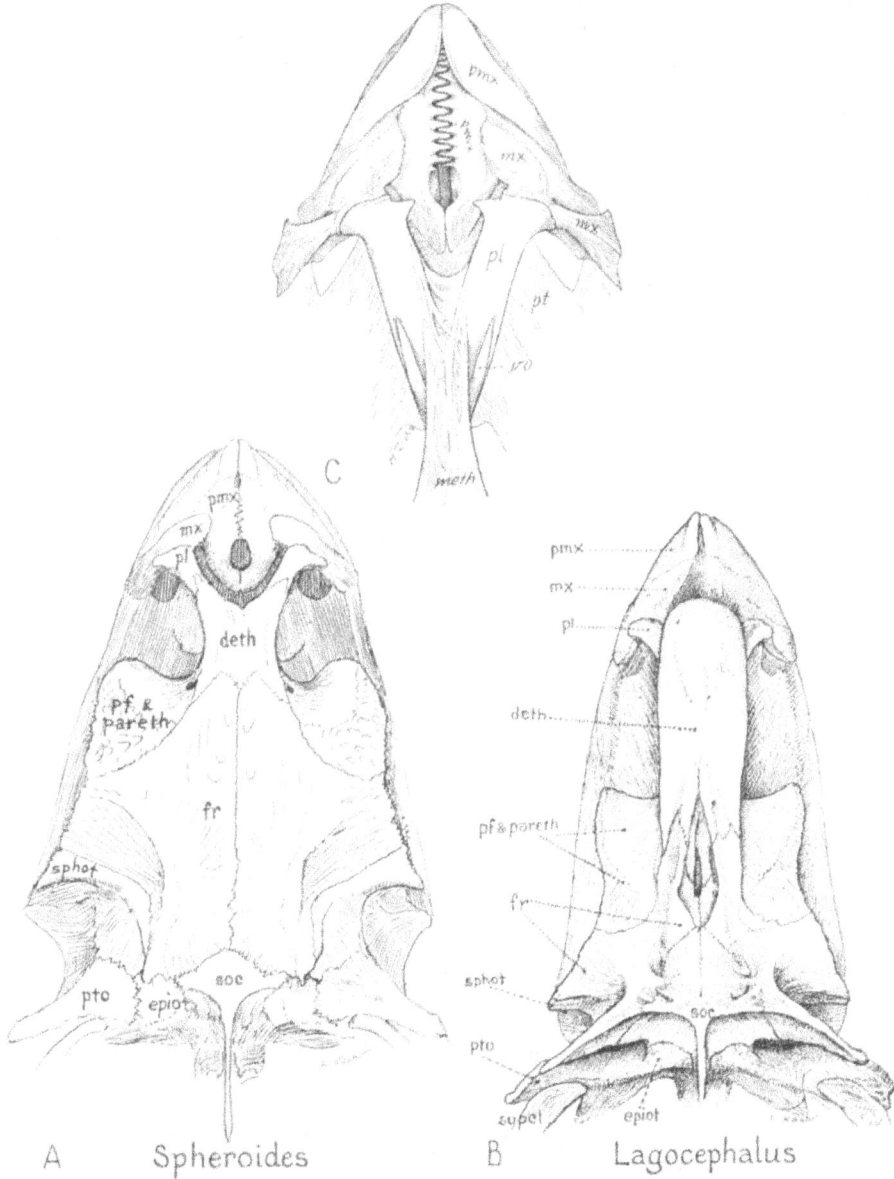

FIG. 169. A, C. *Spheroides*. B. *Lagocephalus*. Top view, showing dentate suture of rostrum.

15

ticulum, using the greatly elongated supracleithra and postclavicles as levers to work the bellows.   The supracleithra, which are quite short and vertically-placed elements in the teuthids and lower plectognaths, then became pulled out into long oblique or nearly horizontal rods, which are so conspicuous that they have been used as diagnostic characters for the different families; but that the elongation and sub-horizontality of the supracleithra are quite secondary is indicated not only by their association with the newly-evolved puffing apparatus but also by the parallel elongation and oblique position of the opercular and subopercular and by the pulled-out appearance of the lower corner of the preopercular.

This backward prolongation of the subopercular might have caused the disruption of the bony tracker of the interopercular but this important mechanism for tying the opercular series to the jaw was thereupon shifted from its normal contact with the

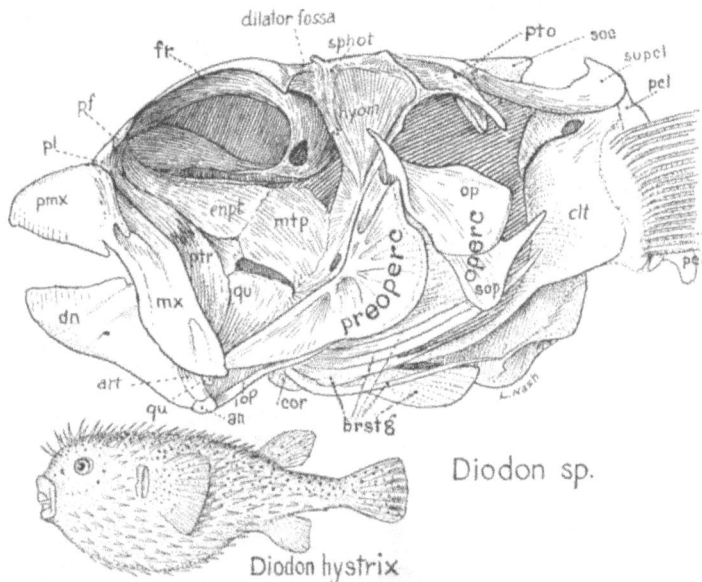

FIG. 170.   *Diodon hystrix.*

junction of the opercular and the subopercular to a position on a special inner ridge of the opercular (Fig. 168*B*).

The backward pulling of the opercular also caused the dorsal process of the opercular to overlap the outer side of the hyomandibular.   The same backward dragging movement also shifted the posterior process of the pterotic from its more vertical position in *Lactophrys* to its backward and downward direction in *Spheroides*.   In brief, all the signs point indubitably not only to a shortening of the muzzle but also to a secondary flattening down of the whole head in *Lagocephalus*, thus producing the horizontality of the parasphenoid and the terminal position of the now enlarged and widened mouth.

Diodonts.—In the diodonts or porcupine fishes the puffing apparatus is doubtless responsible for the elevation of the pectoral fin, for the sub-horizontality of the supracleithrum, for the backward expansion of the opercular and preopercular.

In this family (Fig. 170) the opposite halves of both beaks have fused together into an unusually broad, massive beak, which now serves the double function of cutting and crushing. Secondary shortening of the muzzle has now gone so far, however, that the mesethmoid is very short anteroposteriorly.

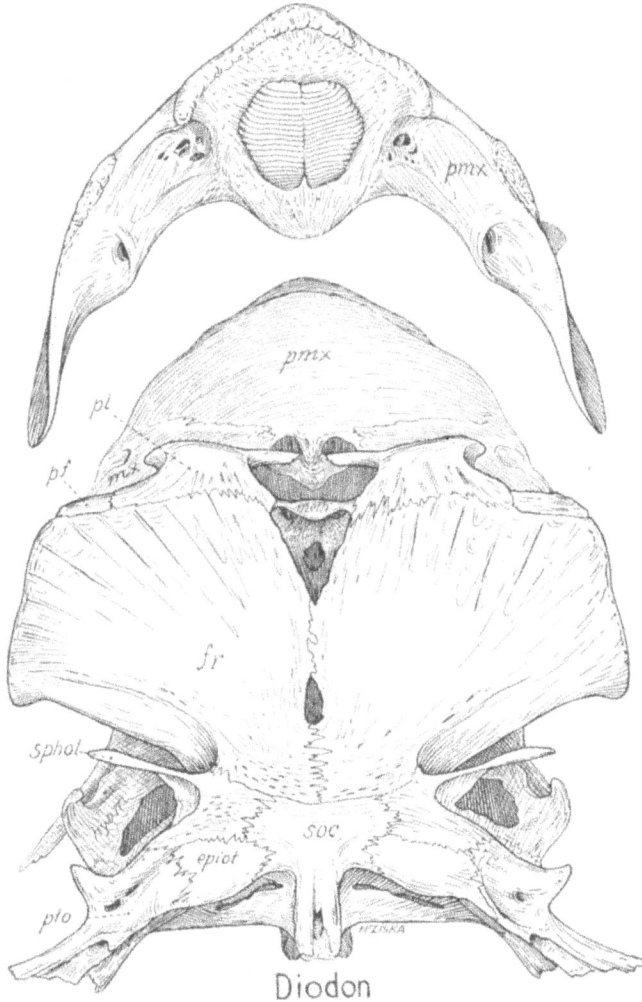

FIG. 171. *Diodon hystrix.* Top view.

Support for the massive beak is given by the greatly enlarged palatine, whose hammer usurps the place of the vestigial prefrontal. There is a marked secondary enlargement of the metapterygoid, quadrate and hyomandibular. In the top view (Fig. 171) the skull is extremely wide and short.

Molids.—Although excessively specialized in the form of the body, the Molidæ retain gills on all four branchial arches instead of on three only, as in both tetraodonts and diodonts (Tate Regan, 1929, p. 325); their beak is undivided as in the diodonts but in their osteology they are very similar to the Tetrodontidæ (Regan). Their larvæ also suggest close re-

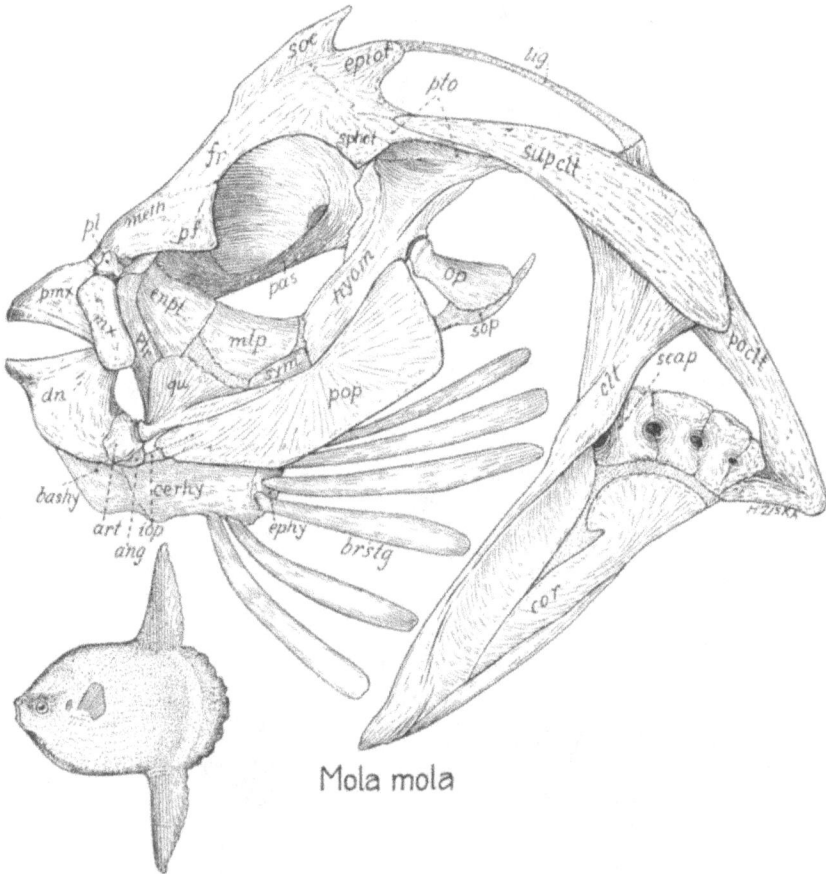

Fig. 172. *Mola mola.*

lationship with the Diodontidæ. Hence we may assume that they are an offshoot of the common stock of the tetrodonts and diodonts. The entire osteology of *Mola* has been figured by Cleland (1862) and Steenstrup and Lütken (1898). The skull bones of *Mola* are remarkably spongy and delicate, since their supporting function is largely taken over by the extremely thick hide, which envelopes the entire body like a coat of blubber. As the examination of dried skulls reveals only a sorry mess of distorted fragments, we present a figure from a dissection of a fresh specimen prepared by my colleague Mr. H. C. Raven.

The fairly large skull is buried in the orbicular contour of the very short deep body. In the front view the greatest diameter is above the relatively small eyes, which look outward and slightly forward. The adipose rostrum forms a prominent rounded bump above the very small thick-lipped mouth and thin, very sharp-edged, pointed beak. The adductor muscles of the jaws are thick and fleshy with few or no tendons, and fill the large space between the eye and the crescent-shaped suspensorium. The hyomandibular is prolonged above and behind with the long horn-like projection of the pterotic, which in turn is braced above by the large curved supracleithrum. Although the opercular is almost vestigial, it is still functional, operated by broad fleshy dilatator and levator operculi muscles. These muscles doubtless serve to expand the branchial chamber and draw in water past the large oral valves. The large opercular flap has a small crease near the posterior end, which enables this part to act as a hinge valve regulating the escape of water from the opercular chamber. The six branchiostegals are likewise buried deep beneath the thick hide but are still freely movable. The subopercular is vestigial and the interopercular reduced to a thread-like tracker that is attached in front to the angle of the mandible.

The viscera are extremely voluminous, especially the liver, but as the body is narrow transversely, the large kidneys have grown far forward and upward. The neurocranium, as described by Steenstrup and Lütken is largely cartilaginous.

Thus may be traced the progressive stages of specialization from generalized percoid fishes to deep-bodied nibblers, thence to long-faced balistoids, thence to secondarily short-faced, large-beaked and low-headed plectognaths. The key to this transformation is that the "habitus" of the remote ancestor becomes the "heritage" of the descendant after a change of function and a change in the direction of evolution.

In conclusion, when the advanced stage of specialization that is seen in *Xesurus* is reached, one might well doubt the ability of Nature to produce viable creatures of any greater degree of specialization. But Nature's limits are not so easily determined. Not satisfied, as it were, with *Xesurus*, she next evolved the triggerfish (*Balistes*), going on to *Aleutera*, which is a libel even on *Balistes*, and culminating in *Anacanthus*, which is almost a tube-fish in appearance. Returning to the pre-balistid model, she made some minor changes and brought out the trunk-fish. Rising then to still more daring improvizations, she invented the unique mechanisms of the puffers. But each new "invention" implies also a further sacrifice of the capital stock of well tried, normal fish arrangements of the earlier types, so that when *Mola* at last issues from Nature's experimental laboratory its grotesque body might appear to the inexperienced to be fearfully handicapped by intensive specialization. *Ranzania* is then brought forward, an elongated *Mola*, the latest but perhaps not the last word in the evolution of the plectognaths.

Postscript.—The otoliths of the order Plectognathi, as studied by G. Allen Frost (1930*b*, p. 621), are "curiously aberrant in form showing little affinity with those of other orders."

ALLOTRIOGNATHI (OPAH, OAR-FISH, ETC.)

In 1907 Tate Regan established the "Allotriognathi" as a suborder of the Teleostei, to include such highly different looking forms as the orbicular Moonfish (*Lampris luna*), and the ribbon-like King-of-the-Herrings (*Trachypterus*) and Oar-fishes or Sea-Serpents (*Regalecus*). I treat it here as one of the main divisions (suborders) of the Acanthopterygii even though its fin rays are mostly soft, since the skull characters rather definitely remove it from truly soft-rayed fishes and ally it with the berycoids. In *Velifer*, according to Frost (1927*a*, pp. 440, 444), the principal otolith is fairly generalized and resembles that of the berycoid *Polymixia*. The sagitta of *Trachypterus* may be a reduced and simplified derivative of the *Velifer* type. In *Lampris* extreme specialization of the sagitta has wiped out all resemblances to the other allotriognaths; the second otolith (asteriscus) is as high as the sagitta—a very unusual condition (pp. 439, 444).

The remarkable modifications of the mouth parts and of the cranium in the different families have been fully described by Tate Regan. In *Trachypterus* and *Velifer* (Fig. 173)

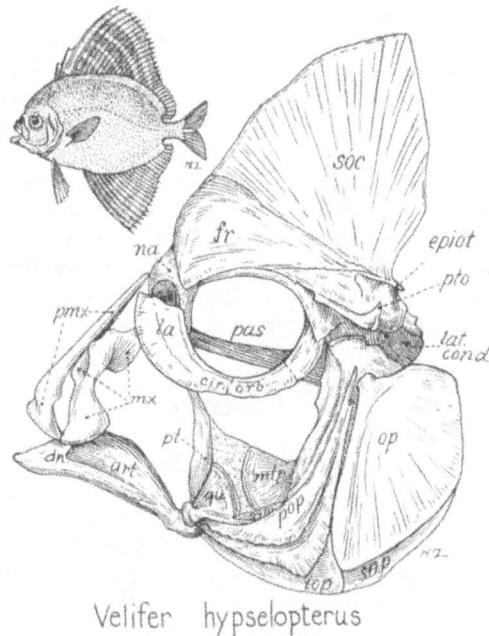

Velifer   hypselopterus

Fig. 173.   *Velifer hypselopterus.*

the protractile mouth has a peculiar type of maxillæ which can be protruded along with the premaxillæ; the lower forks of the maxillæ meet below the ascending processes of the premaxillæ and slide backward and forward on each side of a median keel on the vomer or on the preëthmoid cartilage (Regan, 1907*a*, p. 639). The protrusility is lost in the highly specialized lophotid *Eumetichthys*, in which the posterior ends of the premaxillary processes are attached to the anterior face of the vomer (p. 639).

The cranium of *Lampris luna* (Fig. 174) strongly resembles the primitive berycoid type as described by Starks (1904*b*) except that it bears a huge sagittal crest on the supra-occipital and frontals which extends forward above the prefrontals (Tate Regan, 1907*a*, p. 635). In *Velifer* (p. 636) the cranial elements are arranged according to the same plan, but the head is shorter and the sagittal crest much higher. *Trachypterus* (p. 637) retains the marked shortening of the skull, but the occipital crest has disappeared and the bones are very thin and light, almost papery. The skull of *Regalecus* (Fig. 175), which is de-

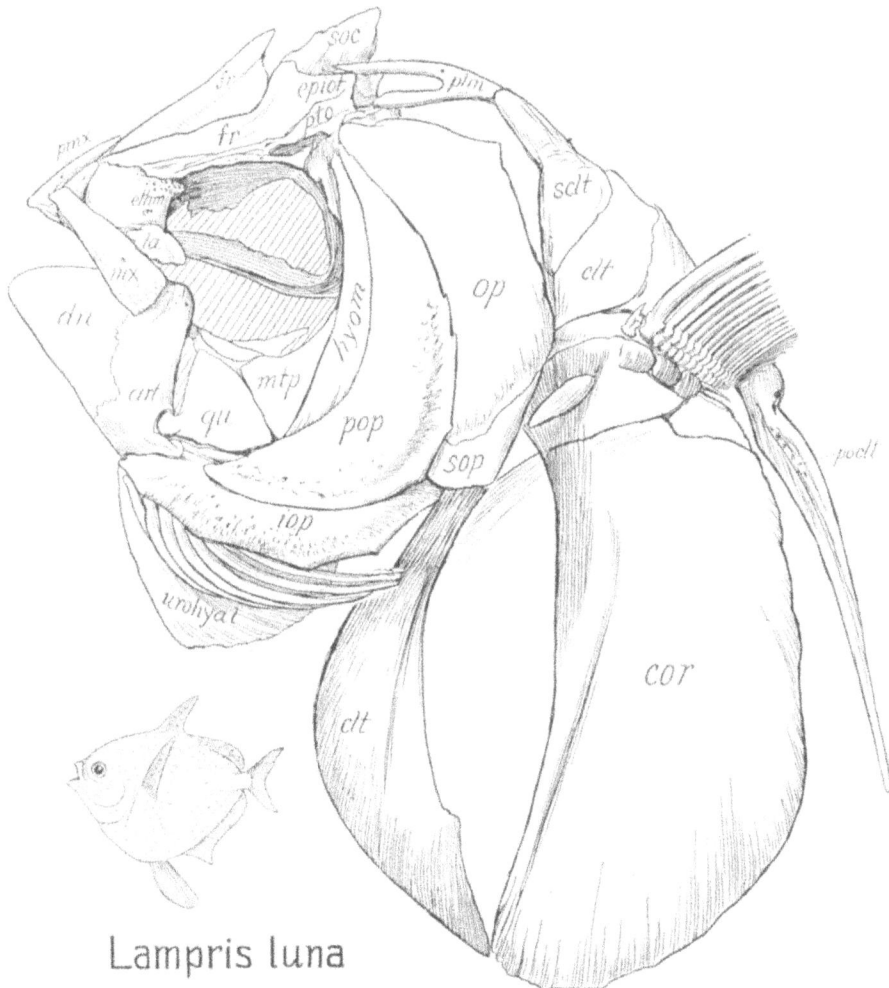

Lampris luna

FIG. 174. *Lampris luna.*

scribed in a beautifully illustrated memoir by T. J. Parker (1886),[1] is essentially similar to that of *Trachypterus*, except that it lacks the basisphenoid (Regan, 1907a, p. 638).   The opisthotic is absent, the bone so named by T. J. Parker and by Dunbar (1906) being part of the proötic.   In the Veliferidæ and the Trachypteridæ there is a large chamber on the

[1] See also Benham and Dunbar, 1906.

anterior part of the cranium, the walls formed by the frontals, the floor partly by the mesethmoid (Regan, 1929). This chamber is absent in the other families of the order.

Thanks to the kindness of Dr. Tate Regan and Mr. J. R. Norman of the British Museum (Natural History), I have had the privilege of studying excellently prepared skulls of *Velifer hypselopterus* (No. 527*A*, Brit. Mus. Nat. Hist. (Fig. 173)) and of *Lampris luna* (Fig. 174). The former appears to me to be much nearer than the latter to the origin of the group. Apparently the peculiar character of the maxillæ, noted by Tate Regan, permit nearly the entire protrusile portion of the upper jaw, including the articular processes of the maxillæ, to slide on the vomer and thus to be retracted into the chamber that lies above the mesethmoid and below the dorsal flanges of the frontal. In *Lampris* this arrangement appears to be in a degenerate condition and the median cavity between the

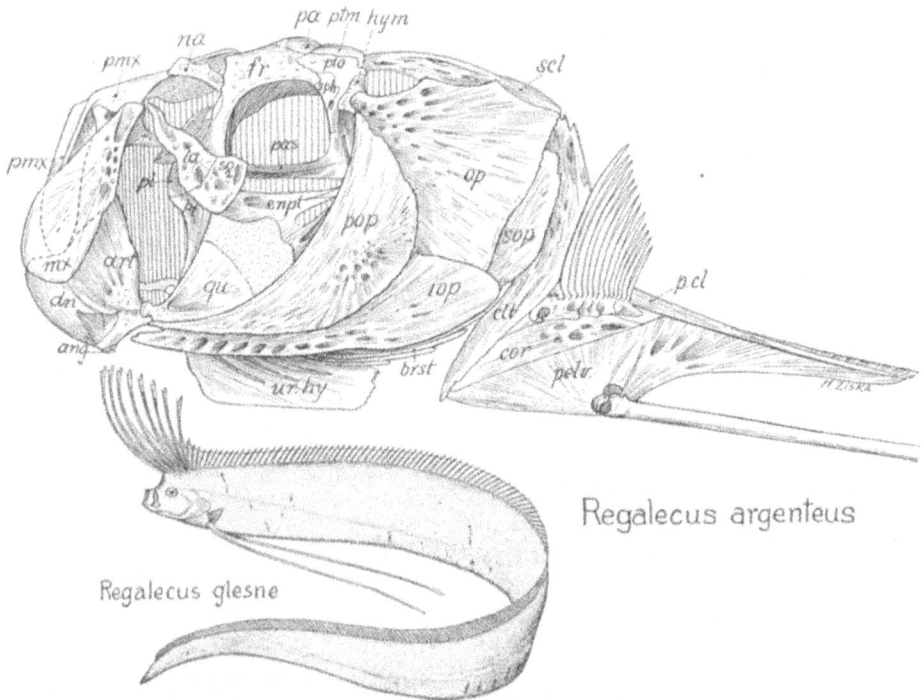

Fig. 175. *Regalecus argenteus*. After T. J. Parker.

frontal flanges has disappeared. It need hardly be said that the latero-dorsal surfaces of the frontal flanges serve as a base for forwardly-produced muscles of the nuchal crest.

Velifer also seems to be more primitive than *Lampris* in the entire opercular region. In the line leading to *Lampris*, apparently in correlation with the change in body form from very elongate to orbicular, the horizontal diameter of the fore part of the skull has been shortened in relation to its vertical diameter, that is, the mandible has become much shorter and thicker, presumably implying a change from more delicate to larger food, the upper jaw has been shoved against the lateral ethmoids, which have become very thick, even if cellular. Meanwhile with the subsidence of the nuchal fin crest, the crests of the supraoccipitals and frontals have greatly diminished. The very peculiar downward and backward turning of the cleithrum in *Lampris* has perhaps influenced the notable spreading of the preopercular and opercular in the same general direction.

On account of the retention of the orbitosphenoid bone and of the basic resemblance of the cranium of *Lampris* and *Velifer* to the berycoid type, Tate Regan (1907*a*, pp. 641, 642) infers that the order Allotriognathi has been derived from Cretaceous berycoids. Jordan (1923, p. 166), on the other hand, places the Lampridæ next to the Cretaceous *Semiophorus*, a carangoid fish with a high dorsal fin extending on the forehead,—probably a convergent resemblance.

One of the most curiously specialized members of the order Allotriognathi is the deep-water fish *Stylephorus chordatus*, the osteology of which has been described by Starks

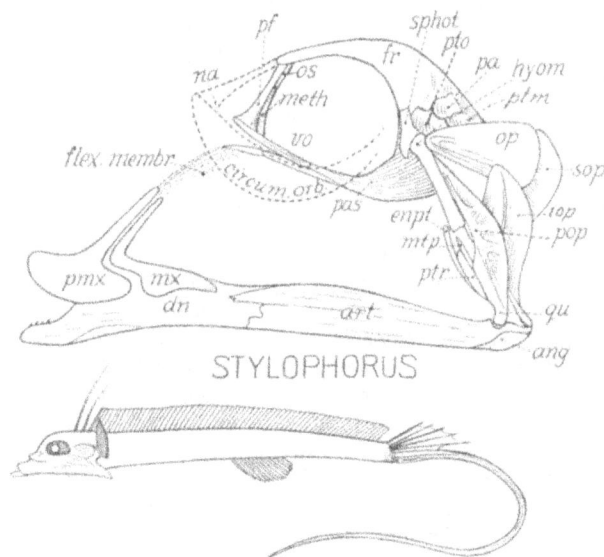

FIG. 176. *Stylophorus chordatus.* After Regan.

(1908*a*), and for which he established the suborder *Atelaxia*. The very long body (Fig. 176) is surmounted by a continuous dorsal fin and ends in an excessively long flagellum-like tail. The large telescopic eyes are directed forward. The head as a whole is long with the very long lower jaw projecting backward behind the gill region in a sharp elbow. One would therefore expect to find a huge mouth, but on the contrary the mouth is quite small. The relatively small premaxillæ and maxillæ both bear long dorsal processes. Tate Regan (1929, p. 319) has described the action of this curious apparatus as follows:

"By a downward movement of the lower jaw the upper is pulled right away from the skull, and the mouth appears at the end of a membranous pouch; the downward projection would not catch the prey sighted by the forwardly directed eyes, and this is remedied by the head being thrown up as the mouth is protruded."

The cranium of *Stylephorus*, as figured by Starks (1908*a*, Pls. 1, 2) bears a strange, spurious resemblance to the skull of a bow-head whale. The size of the enormous orbits determines the marked constriction and upward arching of the interorbital bridge and in part also the narrowing of the ethmoid rostrum. The occiput is inclined forward (as in whales), the supraoccipital being in broad contact with the frontals. The hyomandibular is a small rod, sharply inclined backward and with a single articular head; the preopercular is narrow and assists in the support of the simplified quadrate; the pterygoid arch is reduced to a vestigial sliver. The opercular is reduced to a narrow triangular plate which bears a tab-like subopercular. The subopercular is relatively large but has to be inclined backward to reach the angular region of the relatively huge mandible, which is composed of the usual elements. From the large size of the glossohyal and the ceratohyal it may be inferred that the muscles of the floor of the mouth are strongly developed and that the whole apparatus acts as a suction trap to drag in the prey. Notwithstanding these strange specializations, both Starks and Tate Regan consider that this fish is distantly related to the Tæniosomi (trachypterids and *Regalecus*), Tate Regan finally (1929) assigning it as a section of the Allotriognathi.

In conclusion, the skulls of the Allotriognathi are so very highly specialized that it seemed best to defer treatment of them until after the more central percomorph types had been described, even if the Allotriognathi an as order may have sprung off from the berycoid stem independently of the percoid groups.

SCOMBROIDEI (CREVALLES, MACKEREL, TUNNIES, ETC.)

Pomatomus.—The blue-fishes (*Pomatomus*) are generally recognized as connecting the Carangidæ with the Serranidæ (Tate Regan, 1909*b*, p. 68) or at least as affording intermediate characters, not only in the skull (Fig. 177) but in general body-form and in the caudal peduncle and caudal fin. However, it seems possible that while *Pomatomus* and *Scombrops* may now be on the way to becoming carangoids, the real ancestors of the latter may be some deep-bodied Cretaceous form, such as *Aipichthys*, which Smith Woodward refers, along with other Cretaceous genera, to the Carangidæ (1902, p. 3).

Carangids.—It has been customary to bracket the Carangidæ with the Scombridæ and related families in a single section of the percomorph order called by Jordan and Evermann (1896, p. 863) the Scombroidei and by Boulenger (1910, p. 675) the Scombriformes. But those recent American authors who have abandoned all superfamily groups simply list these families near each other, while Tate Regan (1909*b*, 1929, p. 321) refers the carangoids to the percoid suborder and the true scombriforms to a separate suborder, Scombroidei. Starks, however, as a result of his monographic studies (1909, 1910, 1911*a*) on the osteology and relationships of the scombroid families, concludes that while the carangids are closely related to the percoid fishes they are even more closely related to the true scombroids.

A comparison of the skulls of various carangids and scombroids with each other and with those of *Pomatomus* and other percoids, checked by reference to vertebral, rib and fin characters, inclines me to the opinion that the position of Boulenger and of Starks is essentially correct.

That the most primitive acanthopterygians had a short vertebral column of ten abdominal and fourteen caudal vertebræ was the suggestion of Boulenger, which finally provided the clue to many of the puzzling facts recorded by Gill, Jordan and others (see Jordan, 1891, pp. 202–216). Smith Woodward also records this number in several Cretaceous acanthopts, including the berycoids *Hoplopteryx* and *Aipichthys*, *Vomeropsis*, *Mene*, which he refers to the Carangidæ and which Tate Regan refers to the Berycoidea. Boulenger's family definitions of Carangidæ, Scombridæ and related families show the more primitive numbers (24–26) in Carangidæ, the higher numbers (30–50) in Scombridæ and the extreme (32–60) in the progressively long-bodied Gempylidæ and Trichiuridæ. Even

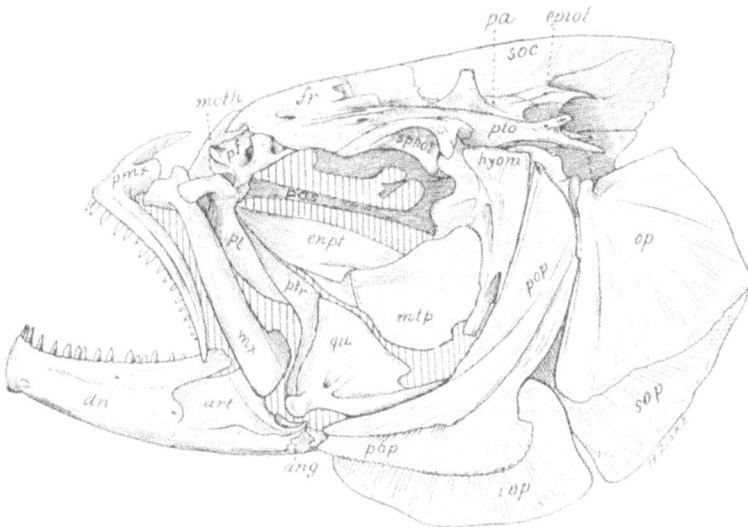

Pomatomus saltatrix

FIG. 177. *Pomatomus saltatrix.*

without material increase in number, the lengthening of individual vertebræ causes a progressive increase in body length in the various species of the genus *Seriola*. As regards the fins, the allied genus *Trachinotus* is the most nearly ovate to orbicular species. *Trachinotus falcatus* (Fig. 178) and *T. culveri* (see Meek and Hildebrand, 1925, Pls. XXXIII and XXXV, pp. 378, 381), have the obviously least specialized condition of the dorsal and anal fins, while in the related but longer-bodied genus *Oligoplites* (*Scomberoides*) the spinous dorsal and anal are either reduced or separated from the soft-rayed parts of the fin. In *Oligoplites refulgens* (Meek and Hildebrand, 1925, Pl. XXXIX) the body has become almost mackerel-like in its length and slenderness. Meanwhile the head has also become more elongated, the opercular broadening antero-posteriorly. Even in these elongate forms the lateral line retains the great curve above the pectoral fin which is characteristic of the ancestral ovate to orbicular body-form of the Carangidæ.

Again, if we study the vertebræ and ribs it will be seen that the carangoids are indubitably nearer to the primitive percomorph type than are the highly peculiar scombroids, also that *Coryphæna*, although clearly allied with the carangids in skull structure, has the vertebral and rib characters more or less transitional to the scombroid type.

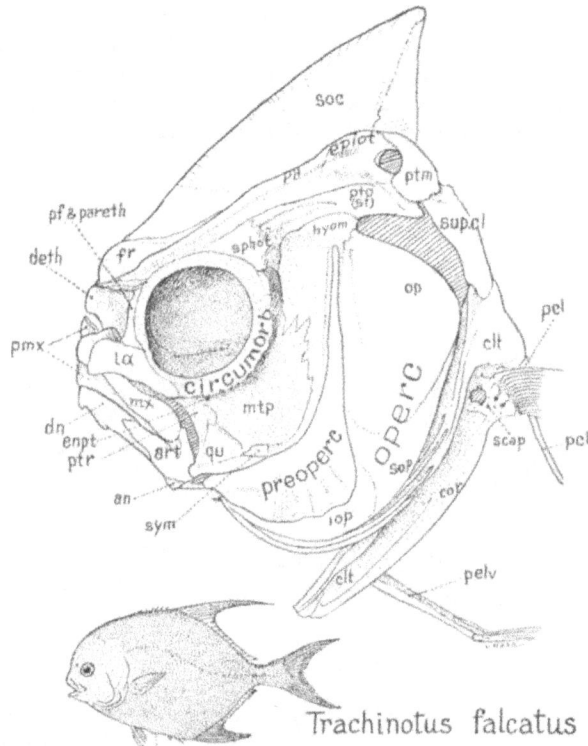

FIG. 178. *Trachinotus falcatus.*

Turning now to the skulls (Figs. 178–181), all carangids have a conspicuous extension of the sagittal crest from the supraoccipital to the frontal, which is higher and more steep in the more primitive compressed types and lower and more elongate antero-posteriorly in the derived long-bodied types. The variations in the relative antero-posterior width of the opercular flap are also correlated in part with body-length; very probably the primitive carangid had a relatively short, deep opercular. It also appears highly probable that in the primitive carangid the mouth was small, with rather long ascending processes of the protrusile premaxillæ, and that the quadrate-articular joint was moderately far forward, perhaps beneath the middle of the orbit; that subsequently larger mouths were acquired along with longer, swifter bodies and more predaceous habits, either by a moderate backward shifting of the quadrate-articular joint, as in *Coryphæna*, or by a forward growth of both snout and jaws, as in *Trichiurus*.

Among the numerous skull characters of the Carangidæ noted by Starks (1911a, p. 30)

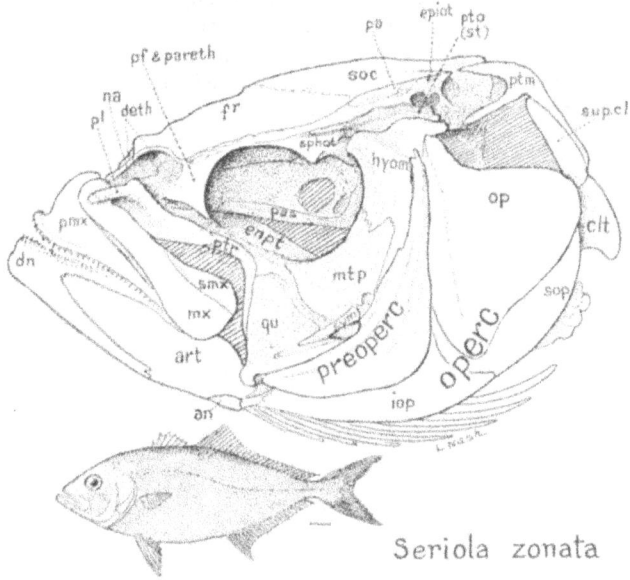

FIG. 179. *Seriola zonata.*

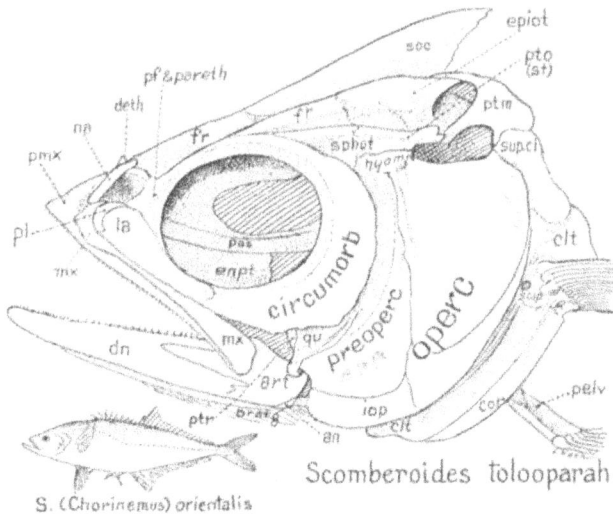

FIG. 180. *Scomberoides tolooparah.*

only a few contrast with those of the Scombridæ, especially the following: opisthotic never as much interposed between the exoccipital and pterotic as it is in the Scombridæ; eye with bony sclerotic case less well developed than in most of the Scombridæ; premaxillæ usually protractile.   All the other skull characters listed (on pp. 30, 31) are either fully shared by the Scombridæ or are differences of degree only.   Starks (1909, p. 573) notes

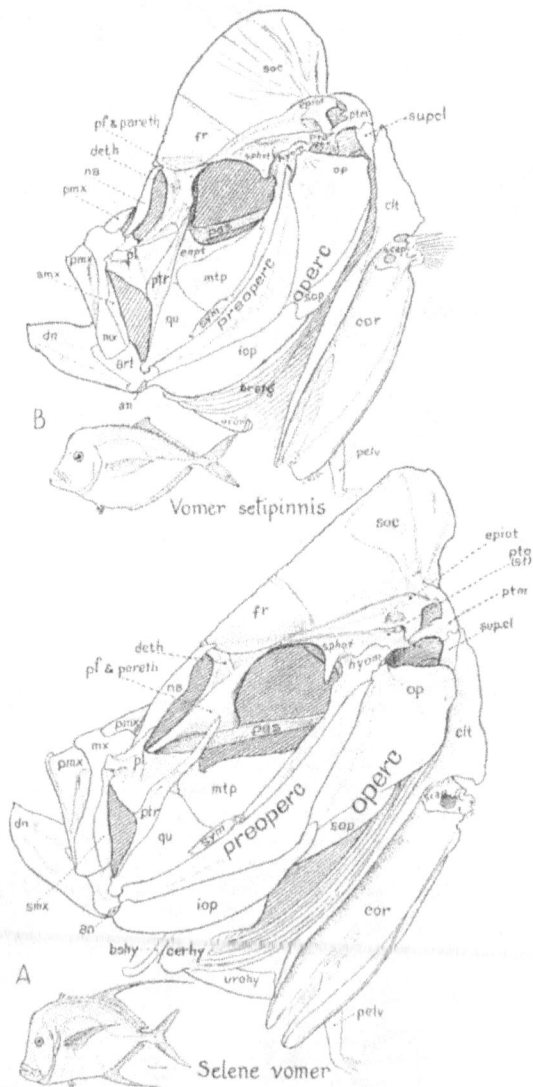

Fig. 181. *Selene vomer* (A). *Vomer setipinnis* (B).

that in Tate Regan's diagnosis of the suborder Scombroidei (1909b, p. 70), after exclusion of the carangoids, "the characters with a few minor exceptions are characters of the percoid fishes and spiny-rayed fishes in general, including the carangoids." After showing that even the exceptions are variable in the Carangidæ and Scombridæ, Starks notes that "it does not appear from Dr. Regan's paper why the other scombroids should not follow the family Carangidæ into the group of percoid fishes." He then cites important characters in common to the two groups and favors retaining the old group of Scombroidei.

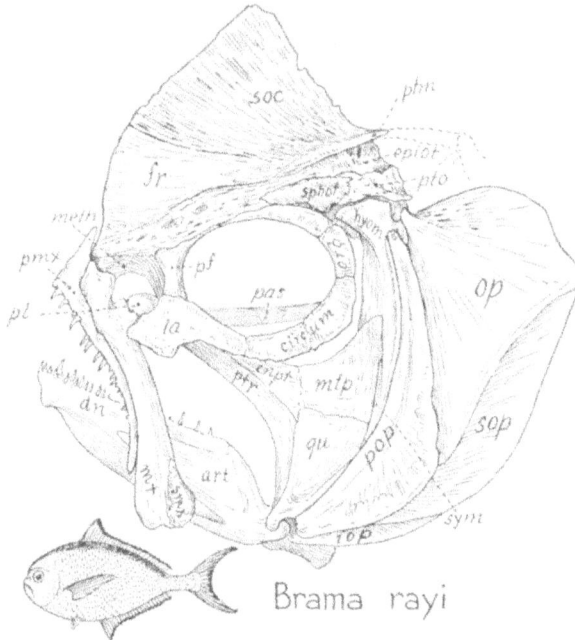

FIG. 182. *Brama rayi.*

Brama rayi is a moderately short-bodied fish of carangoid appearance. It would seem to be a good starting-point for *Coryphæna* especially in the skull (Fig. 182).

Coryphæna.—To continue the subject of the relationship with each other of the carangoids and scombroids, it has been noted above that as regards vertebral characters *Coryphæna* tends to divide the difference between the Carangidæ and the Scombridæ. Thus it clearly suggests how the peculiar vertebral and rib characters of the latter, as described by Boulenger (1910), have probably been derived from those of the former. The skull of *Coryphæna* (Fig. 183) retains the protrusile premaxillæ of the carangids, their distally developed sagittal crest, their stoutly developed parethmoid, their crested mesethmoid and large osseus narial cavity. Its opercular has shared in the anteroposterior elongation of the body but is not unlike those of the carangids *Scomberoides* (Fig. 180) and *Oligoplites.* Its opercular also differs from that of the true scombroids and resembles that of the carangoids in being truncate at top and more produced at the lower end. The preopercular agrees with that of the Carangidæ in being much less expanded posteriorly than that of typical scombroids. In brief, while *Coryphæna* appears to belong in the carangoid section rather than in the scombroid, its skull as well as its vertebræ, ribs, fins, caudal pedicle and tail, are all less specialized, more primitive, than those of the scombroids. Tate Regan (1909b, p. 69) notes that "the structure of the pectoral arch and of the caudal fin is as in the Carangidæ, to which family the Coryphænidæ may be related." He notes also that *Brama* and *Mene* appear to be related to each other and that the cranium of *Brama* (Fig. 182) is strikingly similar to that of *Coryphæna*.

Stromateoids.—Apparently allied with the carangoid stem are stromateoids, which have teeth in the gullet. In *Rhombus* (Fig. 184) the bony bases of the median fins are tied together in a continuous line which suggests the outline of *Luvarus*.

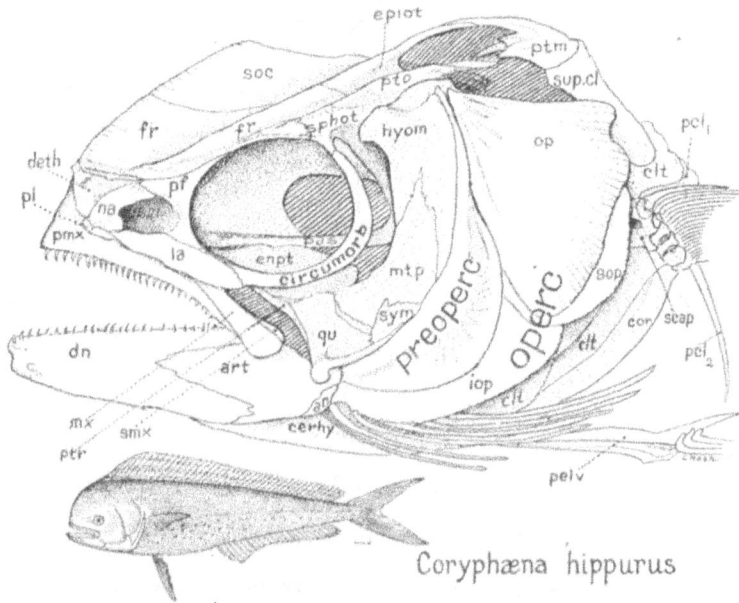

FIG. 183. *Coryphæna.*

Luvarus.—Apparently related to *Rhombus* is the rare *Luvarus*, the skeleton of which is described and figured by Waite (1902). The fish somewhat resembles a bonito in general outline, the upper and lower borders of the skeleton being defined by strong longitudinal arches formed by coalescence of the bases of the interneurals and interhæmals, respectively. The vertebræ recall those of *Coryphæna*. The skull roof (Fig. 185) is depressed to the plane of the vertebral column by the steeply developed epaxial muscles which extend forward to the ethmoid, forming a high crest. The large eye therefore lies well below the level of the backbone. The mouth is small, slightly upturned and apparently not protrusible. The suspensorium is curved far forward. The large preopercular and very large opercular are produced downward and backward.

Rachycentron.—This fish (Figs. 186, 187) is classed by Boulenger among the Scombriformes and its family is placed next to the Carangidæ and ahead of the Scombridæ in his system. But an examination of its skull and vertebral column suggests to me that *Rachycentron* is merely a somewhat mackerel-like offshoot of the true percoids, which has likewise been derived from a short-bodied type that has later become elongated. The side view of the skull in general appearance somewhat suggests the scombroid type, especially in the general contour, broad curved preopercular, rather small opercular. The top view (Fig. 187), however, affords a very wide contrast to the scombroid type in the small develop-

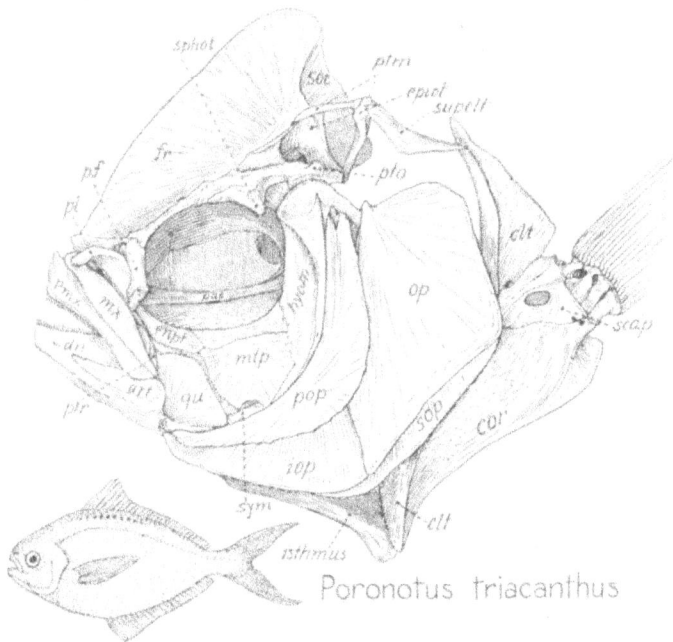

Fig. 184. *Poronotus.*

ment of the supraoccipital, very much flattened, crestless frontals, long epiotic "horns," absence of parietals, etc. In short, while it seems highly probable that the scombroids are much more nearly related to the carangids than to any other modern type, it does not seem that *Rachycentron* serves to connect the scombroids either with the percoids or with the carangoids.

Tate Regan (1909*b*, p. 68) notes that "*Rhachicentrum*" shares a number of primitive skull characters with both the normal Perciformes and the Carangidæ and that its pectoral arch, vertebral and rib characters conform to the perciform type.

In brief, the hypothesis that seems to accord best with the known facts is that the scombroids represent a highly progressive but now isolated derivative of some Cretaceous carangoid that has rapidly become long-bodied as its predaceous powers and swiftness increased.

16

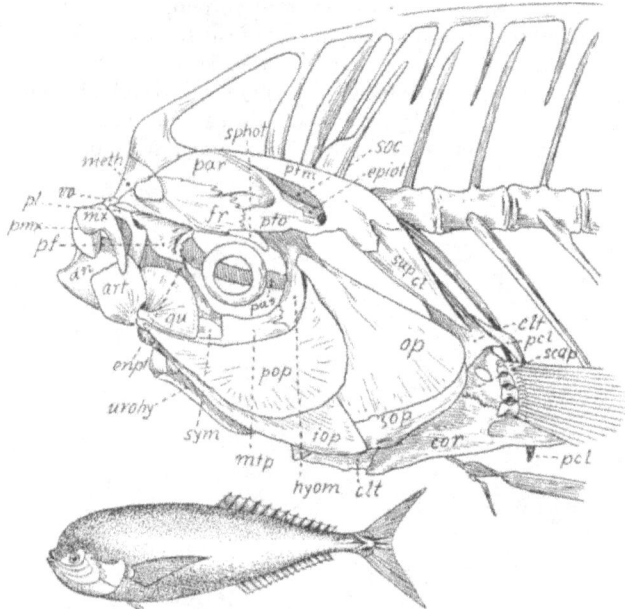

Fig. 185. *Luvarus.*

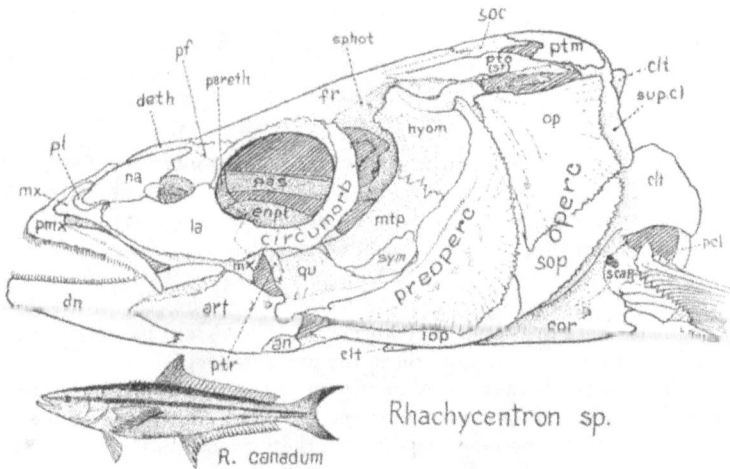

Fig. 186. *Rhachycentron.*

Mackerels, Tunnies, Bonitos.—The mackerels and their allies, as is well known, are among the fastest fish in the sea and their locomotor apparatus is excessively specialized, not only in the skeleton and musculature, but in the vascular and respiratory system as well, to such a degree that their body temperature is higher than that of ordinary fishes. Even the bones reflect the altered physiological properties of the blood-stream, as they are saturated with oil and difficult to degrease.   The surface bones are very thin, often smooth.

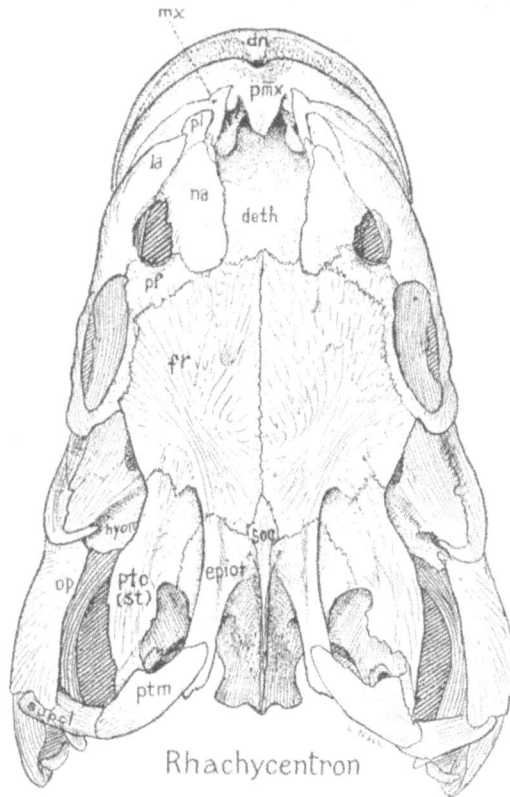

FIG. 187. *Rhachycentron.* Top view.

The marked anteroposterior elongation of their opercular and preopercular bones is probably conditioned also by the voluminous development of the branchial apparatus.   They ought to be able to bite hard, for the upper ends of their premaxillaries are coalesced into a sharp, short beak, which lacks distinct ascending processes, is not protrusile and abuts against the massive prevomer and ethmoid.

The skull of *Scomber scomber* (Fig. 188) is minutely described by Allis (1903) in a superbly illustrated monograph on the anatomy of this fish.   This work is a veritable treasury of knowledge of structural detail, but for the most part it deals only with *Scomber* and is not concerned with the relationships of that fish with other scombroid fishes. This subject is dealt with by Starks (1910) in a paper that is illustrated with excellent views (Fig. 189) of the dorsal aspect of the crania of *Scomber*, *Scomberomorus* and *Sarda*, while the osteology of these and related genera are described in the text.

In a diagram illustrating the inferred relationship of the various genera, as based on the data recorded in this paper, Starks puts the true mackerel (*Scomber*) on or near the trunk of the tree; the Spanish mackerel (*Scomberomorus*) and the tunnies (*Thunnus*, *Auxis* (Fig. 190), *Gymnosarda*) belong on opposite forks, while *Sarda*, which combines the cranial features of the two forks, stands between them. It will readily be seen that the arrangement of the crests on the cranial roof of *Scomber* (Fig. 189) is basically the same as in the normal

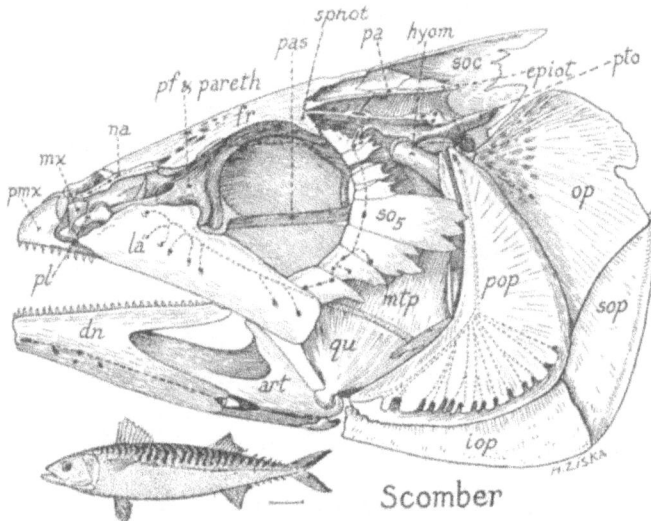

Fig. 188. *Scomber*. After Allis.

percoid genera, in which the skull is narrow in front; while that of *Scomberomorus* (Fig. 189B) differs in the broadening of the interorbital bridge and prefrontal area so that the skull top appears more or less oblong. At the same time the anterior end of the ethmoid has grown forward to form a concave facet for the blunt ascending process of the premaxillæ, the nasals are stout and tend to face laterally. The occipital crest, although low, extends forward over the frontals to the ethmoid, while parallel and continuous crests on either side run forward from the epiotic over the parietals on to the frontals. The skull top of *Sarda* (Fig. 189C) appears to me to have been derived directly from some ancestral form that much resembled that of *Scomberomorus*, by the widening and shortening of the whole skull, by the spreading laterally of the concave ethmoid facet for the premaxillæ, by the divergence of the epiotic parietal crests. The opisthotics widen transversely and the posterior pterotic processes become elongate.

In the tunny (*Thunnus*) the skull becomes very wide, in accordance with the robustness of the smooth rotund body. Figures 191–193 bring out well the curious features of this skull, which, while highly specialized, are clearly derivable from the less advanced conditions in *Sarda*. Some outstanding features are: the breadth across the lateral ethmoids and the massiveness of this region, the lessening of the diameter across the epiotics and the increase across the posterior wings of the pterotics, the development of paired fenestræ in

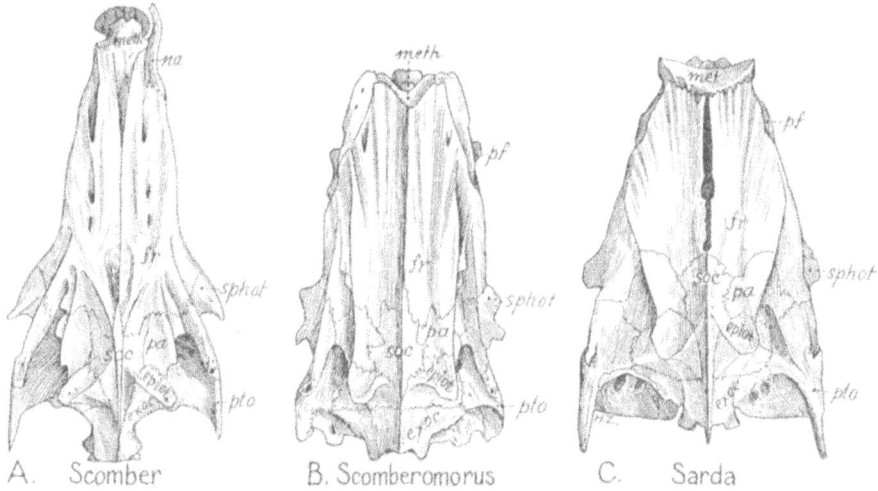

Fig. 189. Top views of skulls. *Scomber* (A), *Scomberomorus* (B), *Sarda* (C). After Starks.

the skull-roof leading downward on either side of the interorbital septum (Fig. 192), the excavation of the basis cranii as part of the myodome (Fig. 191).

From the viewpoint of a student of phylogeny one of the most constructive mono-

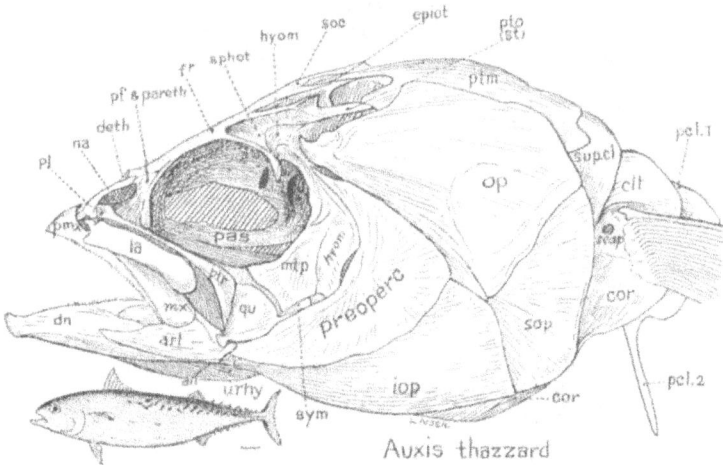

Fig. 190. *Auxis thazzard.*

graphs in the whole field of ichthyology is the "Contributions to the Comparative Study of the So-called Scombroid Fishes" by Professor Kamakichi Kishinouye of the Tokyo Imperial University. For here we find a comprehensive synthesis of prolonged investigations in field and laboratory on the external characters, skull, backbone, muscles, ligaments and tendons, vascular system, brain, biology, ecology, classification and phylogeny of the

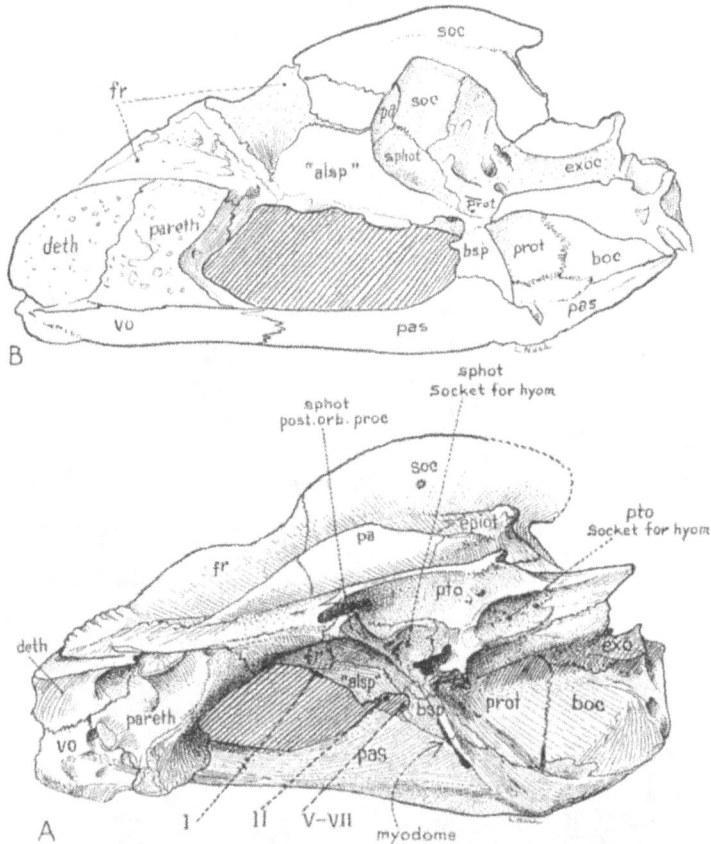

FIG. 191. *Thunnus.* A. Side view. B. Median sagittal section of skull.

"so-called scombroid fishes"; these the author classified under twenty-one species, thirteen genera, four families and two orders.

The feature of this work which is the least likely to be acceptable to ichthyologists is the proposal to separate the tunnies and bonitos from all other scombroid fishes and to create for them an "order" (Plecostei) equivalent in rank with all other Teleostei, the latter being regarded as an "order" of teleostomous fishes. In defense of this procedure the author cites his discovery that in the "Plecostei" (tunnies and bonitos) the vascular system is remarkably different from those of all other fishes; for not only do they have a greater quantity of blood, a much greater number of blood-vessels and a larger heart, but they also have a new and remarkably complex cutaneous vascular plexus, developed as sheets in the lateral muscle, which is the cause of the dark red or nearly black surface color of the muscles under the "corselet." They also have another highly developed local

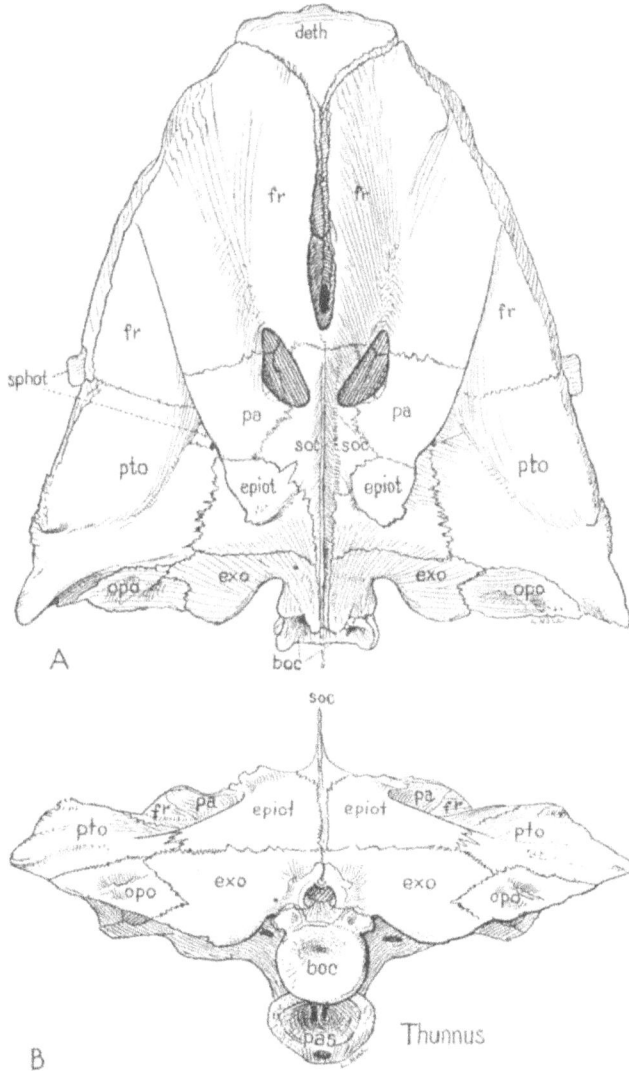

FIG. 192. *Thunnus.* Top (A) and occiput (B).

vascular plexus either within the liver or in the hæmal canal.   It is to this elaboration of the vascular system that the tunnies and bonitos owe their higher body temperature and great speed and activity.

While the beautiful color plates XIII, XIV, XV of Dr. Kishinouye's monograph clearly reveal the marked advances in the vascular system in the bonitos and tunnies as

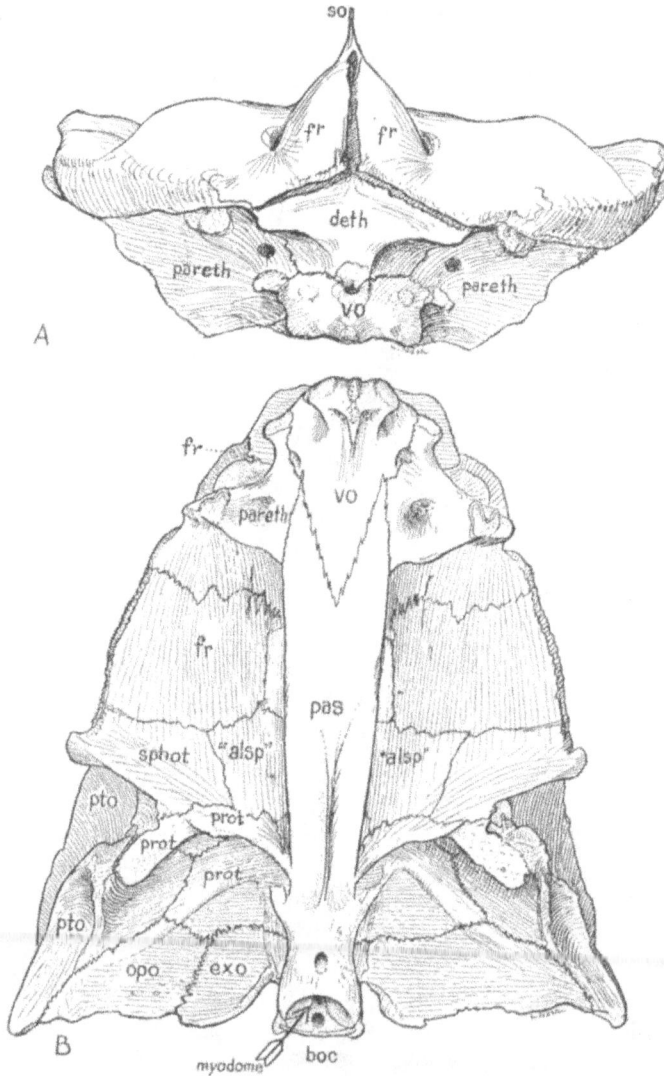

FIG. 193. *Thunnus.* Basal (B) and front views (A).

compared with the more primitive conditions in the ordinary mackerels, they also bring out the relatively close relationships in all other characters even between the extremes of the series, *Scomber* and *Auxis*, so that the separation into higher and lower "orders" seems to be an extreme application of the "horizontal" system of classification. Both the illustrations and the descriptions, however, support the author's conclusion that in a general way the "scombroid" group (as he limits it) exhibits four successive grades of evolution, typified respectively by the mackerels ("Scombridæ"), seer-fishes ("Cybiidæ"), tunnies ("Thun-

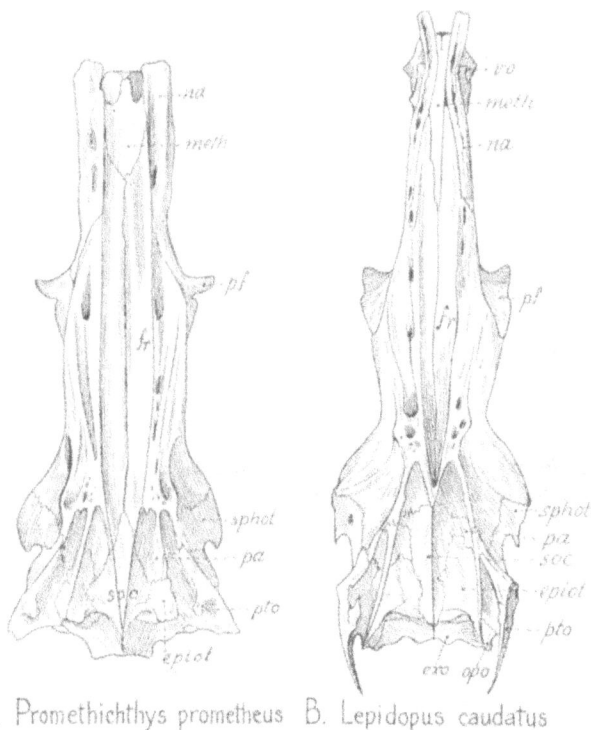

A. Promethichthys prometheus B. Lepidopus caudatus

Fig. 194. A. *Promethichthys prometheus.* B. *Lepidopus caudatus.* Top views. After Starks.

nidæ") and bonitos ("Katsuwonidæ"). It has been shown above that with regard to skull structure this sequence involves the transformation of the relatively primitive and narrow percoid skull top of *Scomber* into the specialized broad-snouted, broad-skulled type of the tunnies.

Gempylids.—This family, including the escolares, oilfishes, cutlass fishes and their allies, begins with *Ruvettus*, which is quite near to *Scomber*. Starks (1911a) figures the top view of the skull of *Promethichthys prometheus* (Fig. 194) of this family and we see at once that it is a modification of the *Scomber* type, masked by general prolongation, marked increase in size of the nasals and by the possession of a concave facet on the end of the vomer for articulation with the short ascending branch of the premaxillæ.

The skull top of *Lepidopus caudatus* as figured by Starks is even narrower and more elongate than that of *Promethichthys*, especially in the ethmo-nasal region; its crests seem easily derivable from the *Scomber* type.   The same is true of the skull of *Trichiurus* (Fig. 195), which well illustrates the extremely predaceous habits of this family.

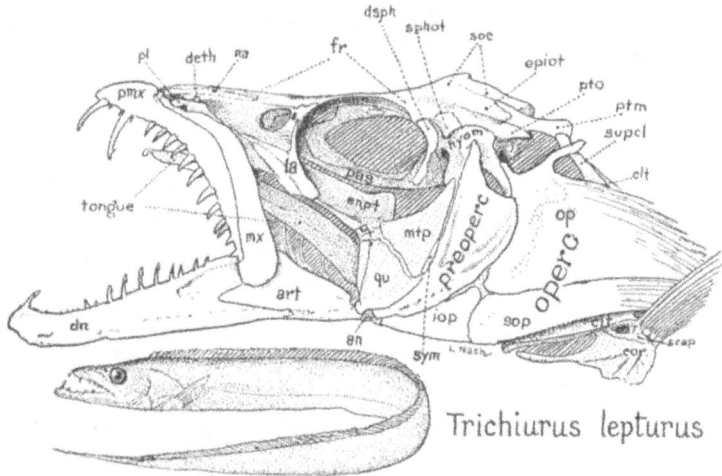

FIG. 195.   *Trichiurus lepturus.*

Sailfish, Marlin, etc.—In a quite different direction, the skull of the sailfish *Istiophorus* (Figs. 196, 197) is likewise derivable from a primitive scombroid type.   The outstanding feature is the great development of the rostrum from the premaxillæ and maxillæ.   The

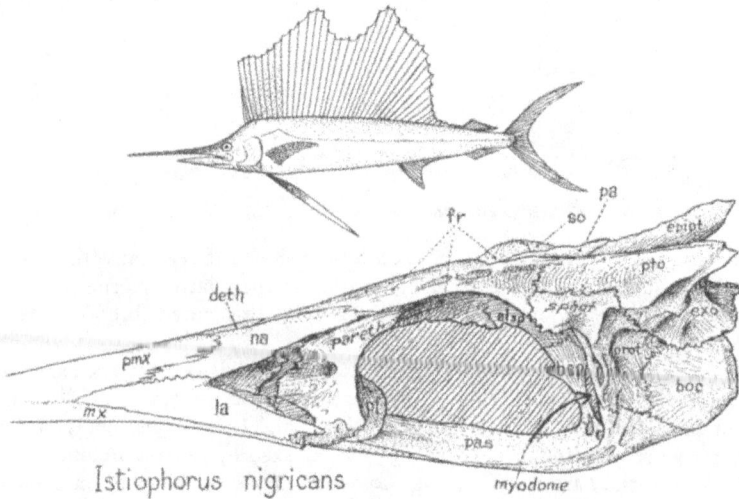

FIG. 196.   *Istiophorus* sp.   Side view.

rostrum (Fig. 197) is braced especially by the broad anterior end of the vomer and by the broad nasals. Tate Regan (1909b) has traced the evolution of the rostrum in Histiophoridæ, Xiphiidæ and related families from the beak-like premaxillæ of *Acanthocybium* of the Scombridæ, while Starks (1910, p. 81) states that it was long ago pointed out that *Acanthocybium*

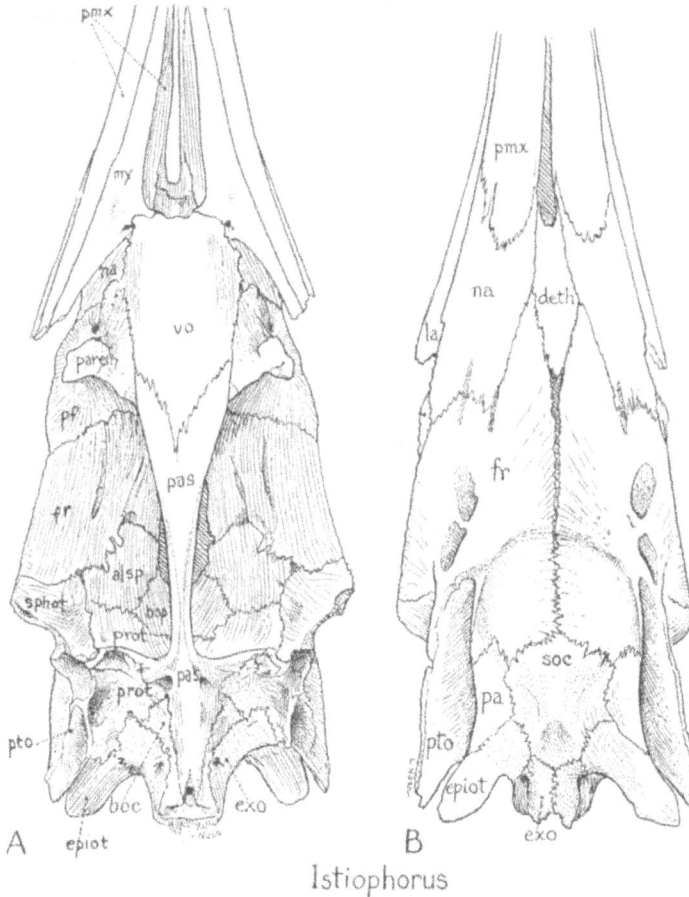

Istiophorus

FIG. 197. *Istiophorus.* A. Under side. B. Top view.

shows an apparent divergence toward the swordfishes. *Scomberomorus* shows the initial step. *Cybium chinese* of the Scombridæ as figured by Kishinouye would indeed appear to form an ideal structural ancestor for the marlins and sailfishes both in external characters and skull structure (Plate I).

Postscript.—The comparative study of the otoliths by G. Allan Frost (1927b, p. 302; 1928b, p. 328) shows that *Trachurus* of the carangids has a sagitta resembling that of *Perca*,

PLATE I

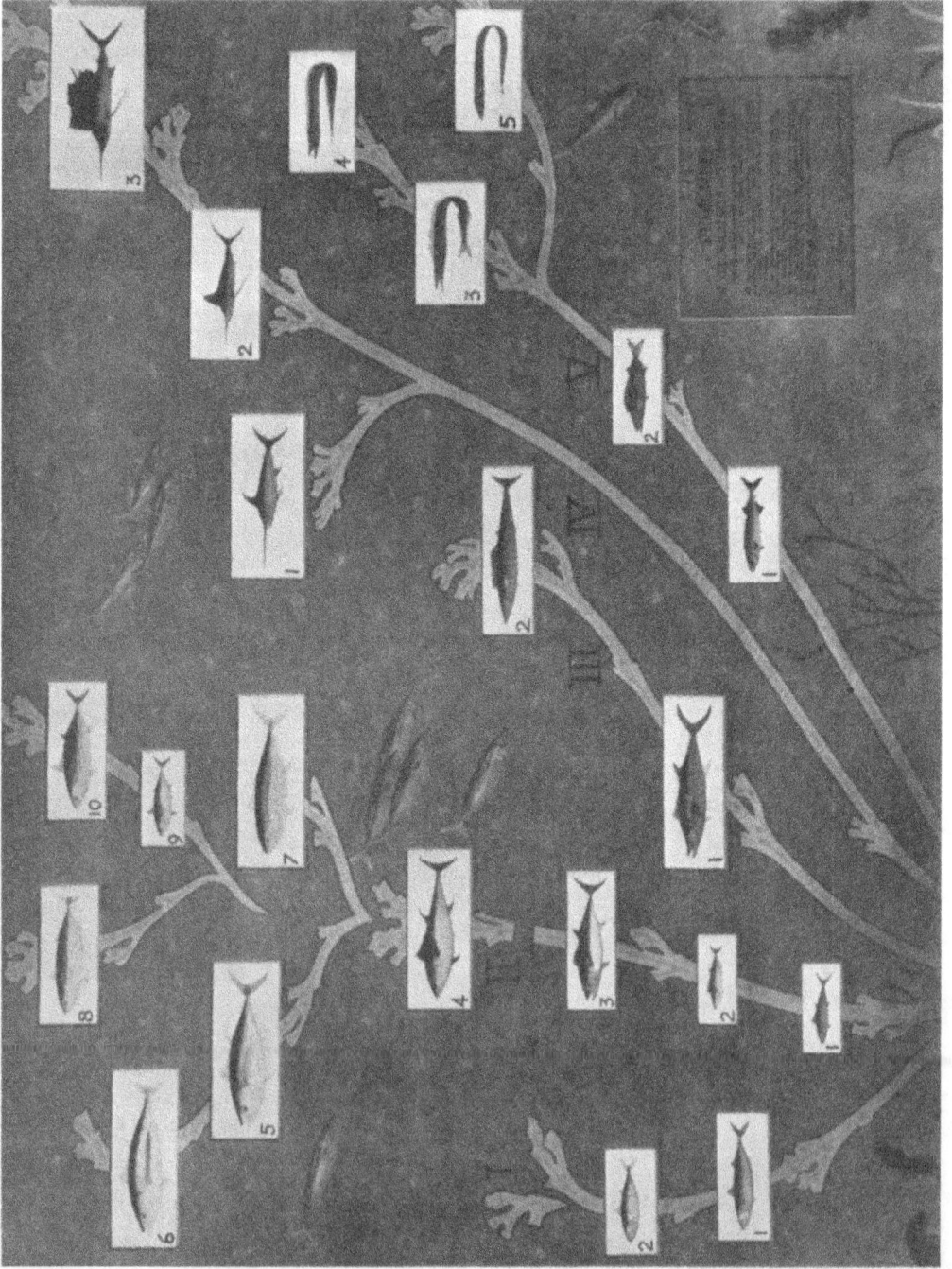

except in certain details in which it recalls the "Elopine" type; that the otoliths of the trichiuroids are peculiarly specialized; those of the true Scombriformes are small, frail, curved, and elongated, but "otherwise resemble the Elopine type." Since there is no reasonable probability that the Scombriformes in the restricted sense have been derived, independently of the carangids, from an elopine type, the evidence of the otoliths, in this, as in certain other cases, does not in itself reveal the remote heritage of the group.

The phylogenetic relations of the different branches of the scombriform fishes, as inferred from the data reviewed above, are illustrated in Plate I.

DISCOCEPHALI (SUCKING–DISC FISHES)

The amazing modification of an anterior spinous dorsal fin into a sucking-disc has made these fish the objects of much study. But their real relationships within the Percomorphi remain a mystery. Dr. E. W. Gudger (1926) has called attention to the remarkable resemblance between the young of certain echeneids and the young of *Rachycentron*, a curiously modified percomorph. But a comparison of the skull, vertebral column and fins of the adults lends no definite support to this view, as already noted by Boulenger (1910, p. 691). The top of the skull (Fig. 198) has been considerably modified for the support of the sucking-disc, notably by the widening of the mesethmoid and flattening of the cranial table. The presence of parietals and the wide separation of the epiotics by the supraoccipital and the lack of a "suborbital stay" (Fig. 199) indicate that this fish cannot be derived from such a highly advanced percomorph as *Scorpæna*, while the fairly normal characters of the pectoral pterygials exclude the blennies and other advanced groups of percomorphs. Comparison with the skull of the anacanth *Lota* reveals only general resemblances. The fairly normal predaceous mouth indicates a primitive percomorph ancestor.

On the whole, these comparisons strengthen the inference that the echeneids have been derived from relatively primitive percoids not dissimilar from the barrel-fish (*Palinurichthys*); this fish has the habit of lurking under floating logs and its arrector and depressor muscles of the short spinous dorsal are unreduced. Once the habit of pressing the spinous dorsal against the under side of the log was established as an indirect means of remaining in the vicinity of a bountiful food supply, Natural Selection might soon be concentrated on

---

PLATE I

Inferred phylogenetic connections of the main branches of the scombriform fishes. From an exhibit in the American Museum of Natural History. Drawings by Dudley Blakely chiefly from the monographs of Kishinouye.

KEY TO PEDIGREE OF MACKERELS, ETC.

I.	1. *Scomber japonicus*		III.	1. *Cybium chinese*	
	2. *Rastrelliger chrysosomus*			2. *Acanthocybium solandri*	
II.	1. *Grammatorcynus bilineatus*		IV.	1. *Xiphias gladius*	
	2. *Sarda orientalis*			2. *Tetrapturus imperator*	
	3. *Gymnosarda nuda*			3. *Istiophorus greyi*	
	4. *Thunnus thynnus*				
	5. *Parathunnus mebachi*		V.	1. *Ruvettus pretiosus*	
	6. *Germo germo*			2. *Epinnula magistralis*	
	7. *Neothunnus macropterus*			3. *Gempyla serpens*	
	8. *Auxis hira*			4. *Trichiurus lepturus*	
	9. *Euthynnus yaito*			5. *Lepidopus caudatus*	
	10. *Katsuwonus pelamis*				

the survival of those larval fishes in which the union of the right and left halves of the spinous dorsal was more and more delayed and in which the principal direction of growth became transverse rather than vertical.

An alternative hypothesis is that the ancestral echineid at first simply followed closely in the wake of the shark and then moved up under its ventral surface, gradually learning to

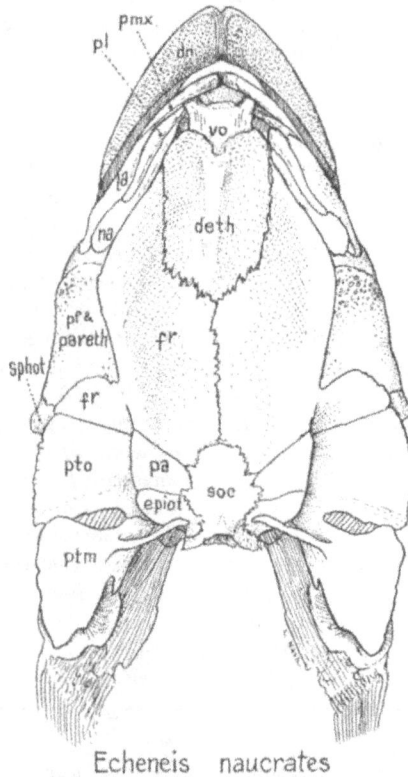

Echeneis   naucrates

FIG. 198. *Echeneis.* Top view.

"steal a ride" by pressing its short spinous dorsal against the underside of the host.   In this connection Dr. William Beebe (1932) has called attention to the presence of numerous suckers on the dorsal surface of the protruding lower lip of the larval *Remora remora* as probably serving for attachment to the host before the dorsal sucker was fully developed. It is quite possible however that this is a purely cænogenetic or larval adaptation without special phylogenetic significance.

In any case, the evolution of such an indubitably well adapted organ as the sucking-disc from a structure with quite different functions constitutes strong evidence for the potency of Natural Selection in controlling changes in the direction of evolution, by the selection of indiscriminate hereditary variations with reference to their utility in a given situation.

The setting apart of the echeneids as a distinct "order" Discocephali, solely on account of their possessing this specialized structure, the sucking-disc, assuredly contributes little to the question of the phylogenetic relationships and origin of the group. Finally, the echeneids afford a good example of a group which has acquired one marked regional specialization, with only minor modifications of the rest of the body, in contrast with such a form as *Mola*, in which all parts of the body have become profoundly specialized.

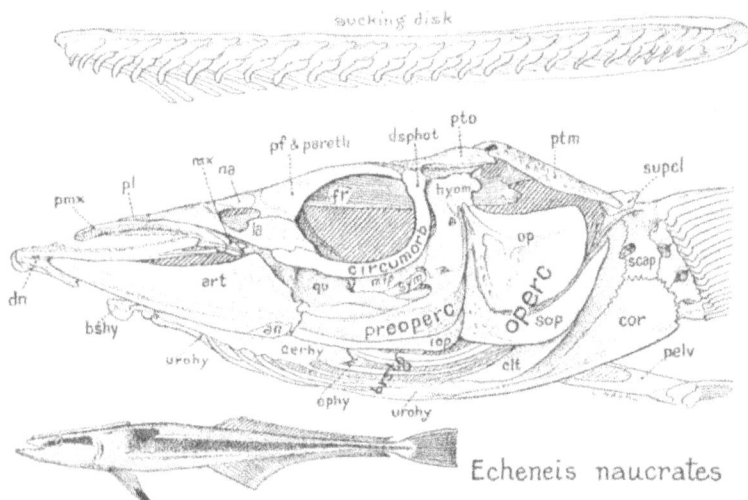

Fig. 199. *Echeneis*. Side view. Sucking-disc seen obliquely from below.

Postscript.—The recent studies of Mr. G. Allan Frost on the otoliths of Neopterygian fishes may possibly afford a clue to this long-standing puzzle. He states (1930a, p. 621) that the otolith of *Echeneis naucrates* is of the percid type and resembles that of *Chromis chromis* of the family Pomacentridæ, except in certain details; in one detail it resembles that of *Anableps tetrophthalmus* of the order Synentognathi.

Renewed examination of the skulls in question does not, however, reveal much evidence in support of the possibility of *Echeneis* being related either to the pomacentrids or to the Microcyprini. The entire skull, in top, side and bottom views, has been profoundly modified as a result of the presence of the modified dorsal fin so that the habitus has very largely concealed the superfamily heritage. This fact gives a good example of the plasticity of the fish skull and of the appearance of many coördinated, but pseudo-Lamarckian responses of its various parts to the evolution of an organ lying outside of it.

SCORPÆNOIDEI (ROCK-FISHES, COTTIDS, GURNARDS, ETC.)

It has long been recognized that the scorpænoid or mail-cheeked fishes are closely allied with the percomorphs, but they, being a large and diversified group with a very constant leading character (the "suborbital stay"), have often been set off as a distinct order or suborder, variously named Loricati, Pareiopliteæ, Scleroparei, Scorpæniformes.

The cranial anatomy of the mail-cheeked fishes is the subject of a classic monograph by Allis (1909), containing accurate descriptions of the cranial elements, the muscles, nerves and blood-vessels, with many superbly executed illustrations of the skulls of representatives of the principal families.   I have also studied the dried skulls of most of the genera mentioned below.

Scorpænids.—Among the least specialized of the living genera are *Sebastes* and *Sebastodes*, with numerous species.   In general appearance these fish are distinctly bass-like, except that the eye dominates the head and that the posterior border of the preopercular is armed with five conspicuous points.   In a dried skull of *Sebastes marinus* these points

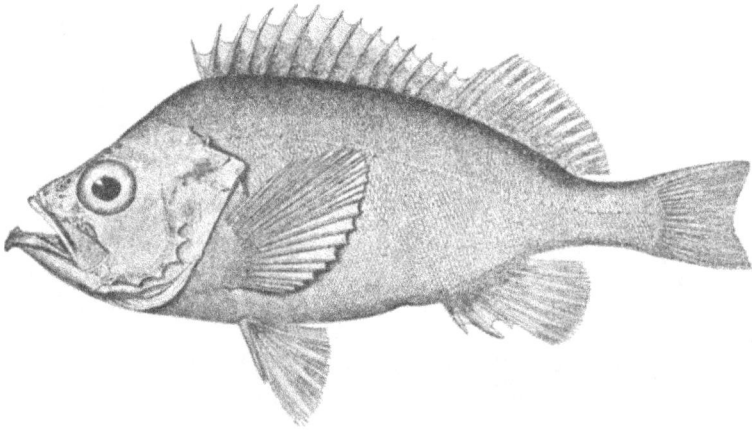

FIG. 200.   *Sebastes marinus.*   After Bigelow and Welsh.

alternate with the large openings of the preopercular branch of the lateral line canal.   The projections on the preopercular margin are conspicuous in many scorpænoids but subject to wide modification in detail.   Apparently they were characteristic of the stem scorpænid and may have originated in a pre-scorpænid or bass-like stage, since more than a beginning of this character is attained by the serranid genus *Lates.*

The skull of *Scorpæna* also shows two prominent spikes on the posterior border of the opercular (Fig. 201), projecting back over the opercular aperture.   Both spikes are large enough to inflict a wound by vigorous side strokes of the head and possibly the preopercular spikes may serve the same purpose.   There are also several pairs of small sharp spikes on the top of the skull and on the posterior borders of the posttemporal and supracleithral plates.   In *Scorpæna scrofa* (Fig. 203) and various other species of this family the whole head bristles with sharp spikes.

The two spikes on the opercular plate of our specimen of *Sebastes* (Fig. 202) have left a clear record of their growth in the form of delicate wavy parallel lines like tide-marks or folds.   Evidently the two spikes were, so to speak, the growing tips of the outer border; as they grew backward they also left another trail in the form of supporting ridges on the scale-like plate.   But why do these growing points have dense, sharply pointed tips like the spiny rays of the dorsal fin? Why are all the little spikes on the top of the skull (Fig. 203) similarly dense and horn-like? Very possibly because spikes may result from the dense crowding, near the center of growth, of rapidly proliferating tissue in which the

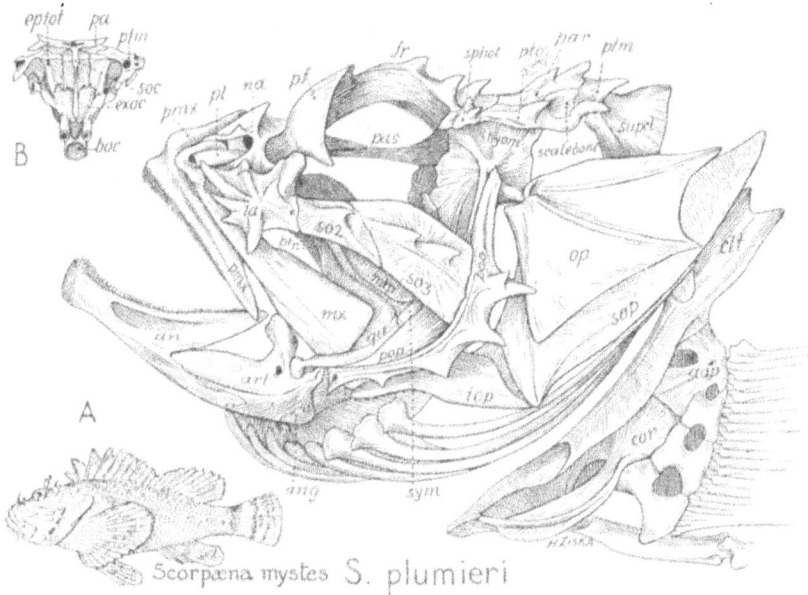

FIG. 201. *Scorpæna plumieri.* A. Side view. B. Occiput. After Allis.

inelastic outer layers tend to inhibit or restrict the expansion but not the multiplication of the inner layers. While these surface phenomena no doubt reflect obscure and deep-seated molecular activities, it is worthy of note that a horn-like texture of originally bony

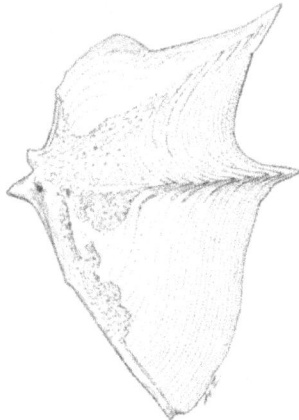

FIG. 202. *Sebastes marinus.* Left opercular bone enlarged.

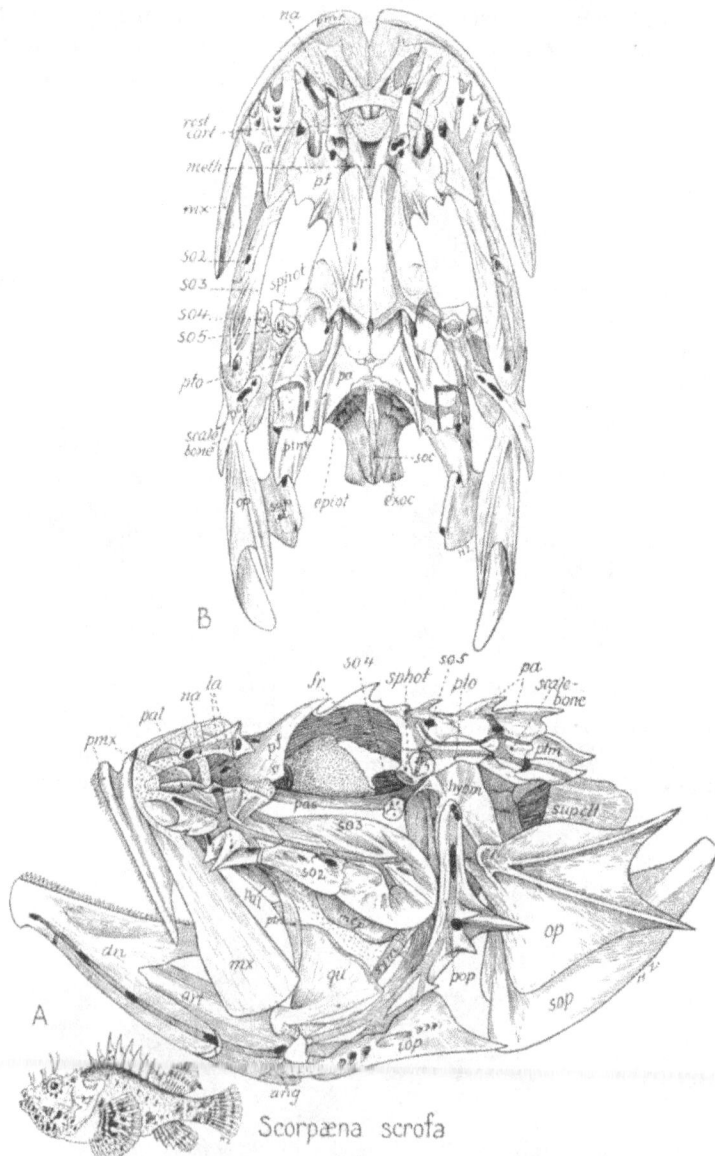

Fig. 203. *Scorpæna scrofa*. A. Side view. B. Top view. Both after Allis.

plates is a concomitant and perhaps a cause of spinescence in this family as well as in some others.

The second most conspicuous character of typical scorpænids is the presence of a "suborbital stay," consisting of three flat suborbital bones (Fig. 201) conjoined with each other and extending from the preorbital projection of the lacrymal backward to the anterior border of the preopercular, to which the posterior edge of the elongate third suborbital is more or less securely fastened. How did this contact between the third suborbital and the preopercular arise? Is it a souvenir of the remote epoch when the ancestral percomorph was short-bodied somewhat like the Cretaceous berycoids, when also the posterior end of the third suborbital was in contact with the preopercular? Did the subsequent antero-posterior lengthening of the body as a whole affect the suborbital bar without breaking its contact with the preopercular? The latter hypothesis now appears to me likely, in spite of the fact that in many respects even the most primitive known scorpænoid skull seems to have been derived from a form not earlier than the primitive bass type, by which stage the body presumably had already passed from the very short to the normal bass type in which the third postorbital was already moved away from the preopercular. The adhesion between the suborbital bar and the preopercular may easily have arisen at the time of the crowding of the suborbitals against the preopercular in embryonic and larval stages and through the subsequent retention of the adhesion in the adult. This early crowding, as well as the subsequent lengthening of the space between the orbit and the preopercular, is well shown in the larvæ and fry of the rosefish (*Sebastes marinus*) as figured by Bigelow and Welsh (see below, 326). The backward growth of the third suborbital bone also may have been favored by a tendency of the suborbital canal (Fig. 203) to send a horizontal branch backward toward the preopercular branch of the canal system, especially as the lateral line canals are so vigorously developed in the skull of scorpænids, where they form prominent channels and tubes of bone. In the percomorph *Cirrhitus* (Fig. 135) the posterior tip of the third suborbital is fastened to the upper bar of the preopercular, which is near to the orbit. Such a connection might later have moved downward to near the middle of the rim of the preopercular.

However it may have arisen, there seems little doubt that the stiff "suborbital stay" as a whole serves to protect the jaw muscles, the eyes, and to some extent the delicate palatopterygoid tract, while also bracing the spike-bearing preopercular and opercular arches.

As to the posterior and upper elements of the circumorbital plates, I could not find them in the fragments of a dried *Sebastes marinus* skull, but Allis (1909, p. 97) states that in *Sebastes dactylopterus* there are two postorbitals, which are delicate semi-cylindrical bones bounding the hind edge of the orbit and transmitting the main infraorbital canal from the second suborbital to the postfrontal (sphenotic). These are also seen in his figure of *Scorpæna scrofa* (*cf.* Fig. 203).

The premaxillæ and maxillæ of *Sebastes* and *Scorpæna* (Fig. 203) are completely percoid in fundamental plan. Between the ascending and articular processes of the premaxillæ is a large median cartilage, the rostral cartilage of Allis, which slides on the dorsally-keeled mesethmoid and is received between the diverging anterior horns of the mesethmoid. In *Scorpæna scrofa* Allis also figures the elaborate system of check-ligaments (Fig. 203) by which the premaxillæ are prevented from dorsal and lateral dislocation and which also ties

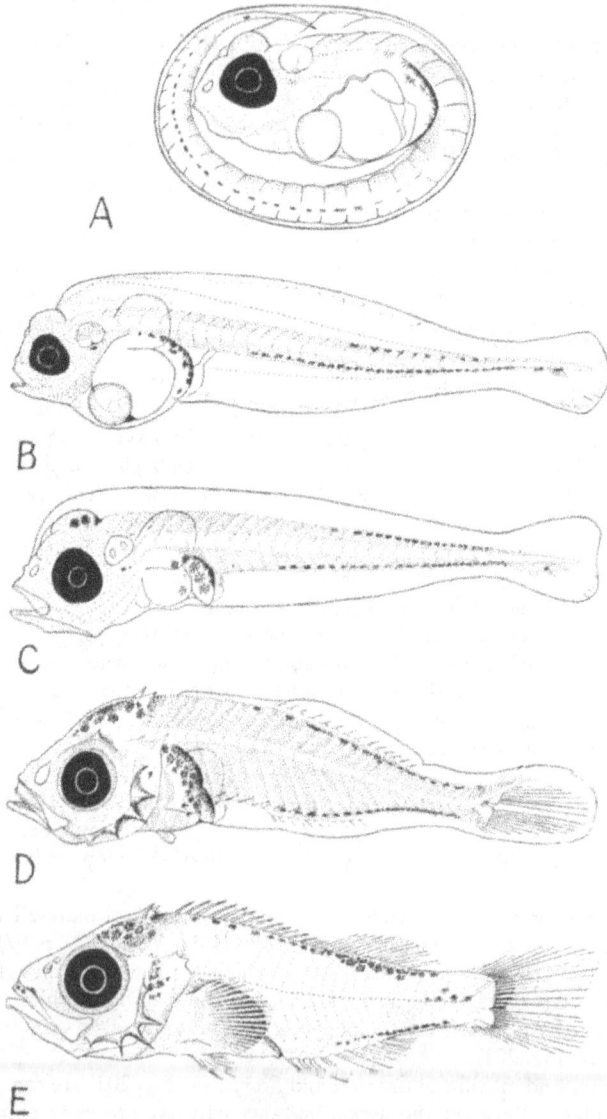

FIG. 204. Stages in the Development of the Rosefish (*Sebastes marinus*). After Bigelow and Welsh. A. Egg from the oviduct of a female. B. Larva, 6 millimeters. C. Larva, 9 mm. D. Larva, 12 mm. E. Fry, 20 mm.

them in with the maxillary processes of the palatines, the anterior horns of the mesethmoid and the ascending processes of the suborbital stays.

Six stages (Fig. 204) in the development of the rosefish (*Sebastes marinus*) are figured by Bigelow and Welsh (1925, pp. 311, 305). As is usual among teleosts, the eyes are very large in the late embryos, larvæ and fry, the brain swells dorsally, the mouth and jaws are at first very small. The body as a whole is elongate and slender, the myomeres forming a

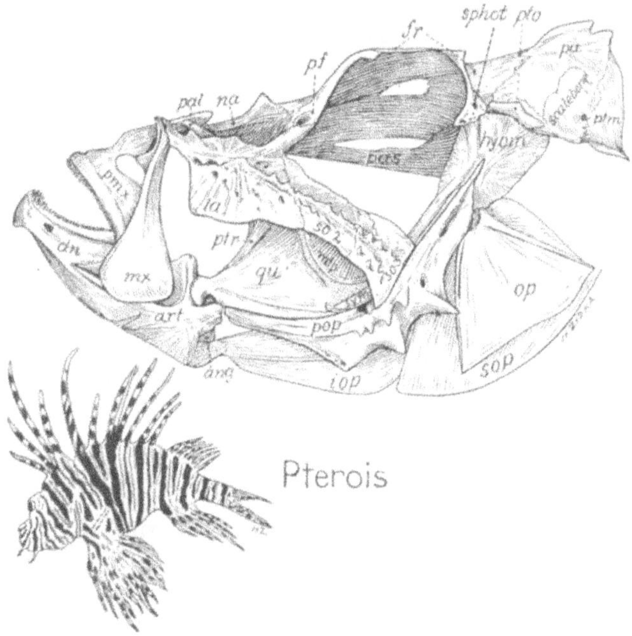

FIG. 205. *Pterois*. Fish from photograph of specimen in action, published by Breder.

narrow strip along the flanks. As the fry become larger the body deepens rapidly, carrying with it the posterior part of the head; the eye becomes relatively smaller, the jaws longer. Spikes appear on the preopercular rim in the late larval stage. Thus, as in other teleosts, the shape of the head changes profoundly during late larval and young stages. Crests and ridges do not appear until the related muscles are well developed; vertical growth of the occiput waits for the deepening of the back. The dependence of adult skull-form upon body-form could hardly be better illustrated.

The mail-cheeked fishes are such an extensive group that a detailed review of their cranial osteology would unduly expand the present paper. I shall try nevertheless to sketch some of their main specializations, referring the reader for details to the monograph by Allis, to Jordan and Evermann's "Fishes of North and Middle America" and similar sources. With regard to general features the bass-like *Sebastes* and *Sebastodes* seem to stand near the ancestral stock, which early divided into several main branches. First, within the family Scorpænidæ many variations on the main theme were played. For example, in

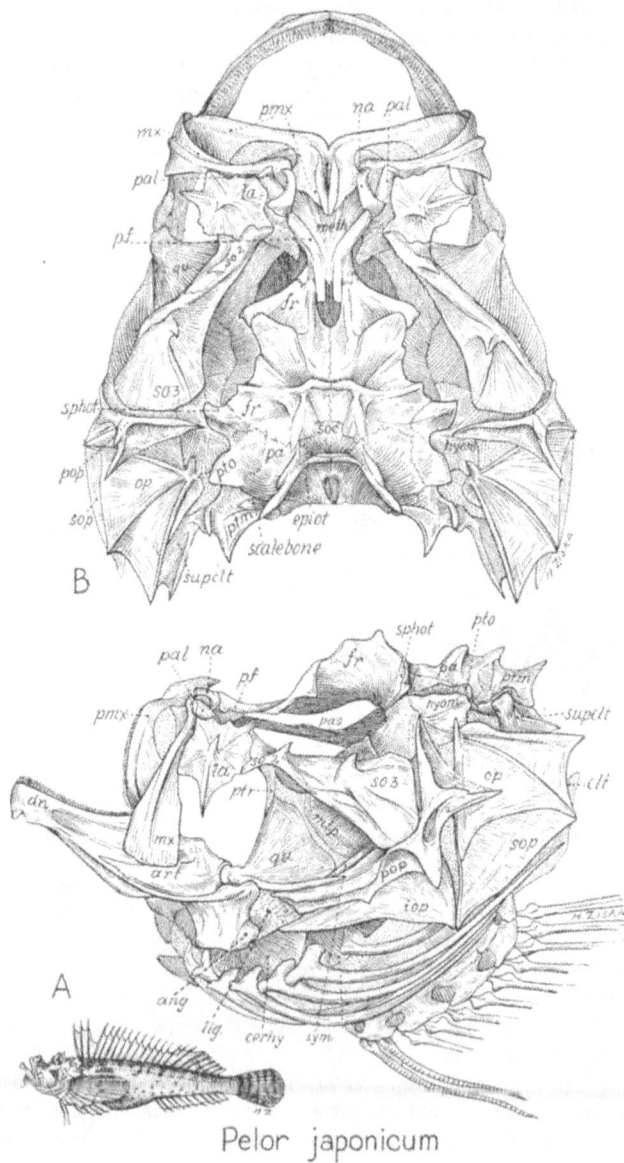

Fig. 206. *Pelor.* A. Side view. B. Top view.

some species of *Sebastolobus* the skull became long and narrow, the bones extremely thin and the mucous cavities on the suborbitals, preoperculars and frontals quite large; in another direction, certain species of *Scorpæna* became very broadheaded, the bones very dense, with smooth, ivory-like surfaces, the spikes numerous and sharp. In *Pterois* the dorsal fin spines are much prolonged and the color patterns are bizarre and supposedly terrifying. The fish has been seen to snap up small fish that actually drift toward it (Breder, 1932). The skull (Fig. 205) is a moderately specialized derivative of the scorpænid type, with thin spineless bones and depressed opercular.

In *Pelor japonicum* (Fig. 206) the broad short skull has the orbits on top, like those of a hippopotamus; the broad mouth opens upward; the upper lip being raised very high up, the parts of the suspensorium and jaws are correspondingly modified.

Synanceja (Fig. 207) is a highly peculiar genus which is even more specialized than *Pelor*. An accurate description of the skull has been given by Leighton Kesteven (1926, pp.

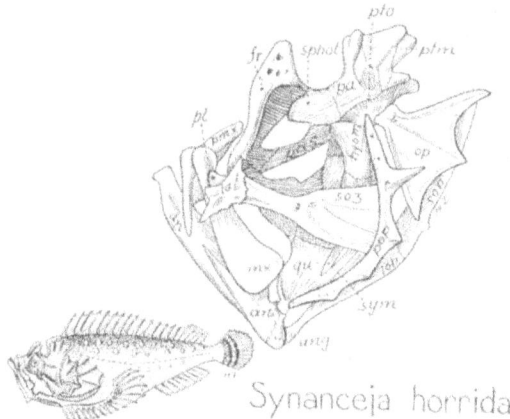

FIG. 207. *Synanceja.* Skull re-drawn from Leighton Kesteven's figure.

224–230). The top of the skull is like a "Turkish saddle" with high supraorbital protuberances on top of the frontals and a suddenly elevated occiput. The eyes are very small and so are the orbits. The mouth and jaws are directed sharply upward. Leighton Kesteven remarks that the synanceian skull presents several characteristics "which make its inclusion in the present company (Scleroparei) seem a *mésalliance;* much more do these characteristics appear in evidence against the inclusion of the genus in the Scorpænidæ." He accordingly makes it the type of a new family, the Synanceiidæ; but a careful reading of this description of the skull fails entirely to convince me that *Synanceja* is anything but a highly specialized derivative of the scorpænid type, much less modified away from that type than are the triglids and dactylopterids. In a word, its *habitus* is new and peculiar but its phylogenetic heritage is thoroughly scorpænid.

Patæcus (Fig. 208A) is another highly specialized scorpænoid; the enlarged dorsal fin has been prolonged forward above the swollen forehead and in front of the downwardly and almost backwardly developed face. The general effect is ludicrously like the head of a

Sioux Indian chief with a feather head-dress. *Aploactis* (Fig. 208*B*) supplies a structural starting-point.

Hexagrammids.—In these more or less cod-like fishes, the skull (Fig. 209) is entirely devoid of spines, but relationship with the scorpænoid group is definitely indicated by the presence of a complete suborbital stay. The family are also more specialized than the Scorpænidæ in the marked increase in number of the vertebræ and their relative de-differentiation. Externally the Hexagrammidæ differ widely from the other scorpænoids in their smooth, more or less cod-like appearance. Possibly they may be related to the ancestral Cottidæ through *Ophidion* (Fig. 210).

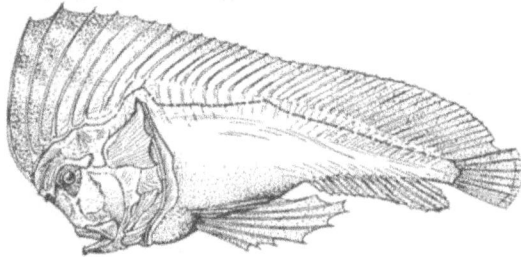

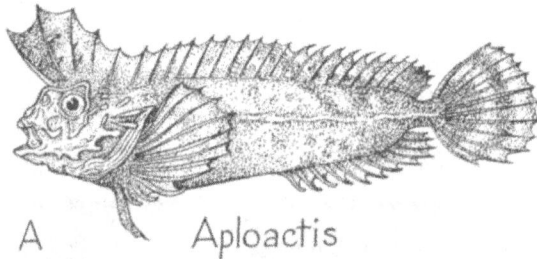

Fig. 208. A. *Aploactis milesii.* B. *Patæcus.* After McCulloch.

In the genus *Platycephalus*, which is the type of the family Platycephalidæ, the skull (Fig. 211) is elongate, more or less flattened; the eyes are directed partly upward.   Leighton Kesteven (1926, pp. 208–218) gives excellent figures of the skull of *Platycephalus marmoratus*, which seems to be a relatively primitive species, not very far from the central scorpænoid type. The suborbital stay extends to the backwardly inclined preopercular, which bears two small spikes apparently homologous with the two main ones in *Scorpæna*. The body-form as a whole (Day, 1878–1888, Pls. LIX, LX) suggests relationships with the Cottidæ and Agonidæ, near which Boulenger (1910, p. 699) locates this family.

Hemitripterus is classed among the Cottiformes but its skull is derived from the *Scorpæna* type; the skull top is broad with thin translucent bones and blunted spikes; on the preopercular only the upper pair of spikes persist (Fig. 214*B*).

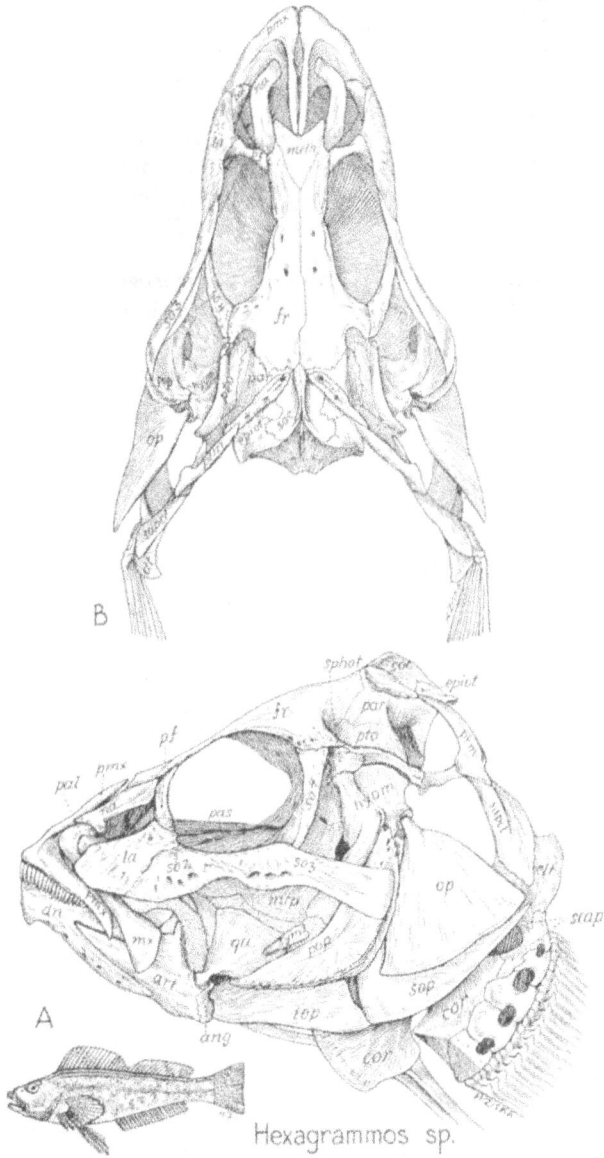

FIG. 209. *Hexagrammos.* A. Side view. B. Top view.

Cottids.—Another great subdivision of the scorpænoid series produced a highly diversified family, the Cottidæ, especially characteristic of northern waters. These are mostly long-bodied with very large pectoral fins, long and divided dorsal fin, large tails, big eyes and protruding lips. The spikes on the preopercular border vary greatly; they are often reduced to two conspicuous ones and of these the upper is much produced, sometimes forming a curved hook either with lateral accessory processes (*Icelinus*) or without them (*Artediellus*). A curved preopercular spike with two small upwardly directed accessory spikes similar to that of *Icelinus borealis* (Jordan and Evermann, 1896, Pl. CCLXXXIV) is found in *Callionymus rubrovinctus* Gilbert (Gilbert, 1905, p. 650) belonging to an entirely different family and suborder.

The *Cottus octodecimspinosus* (Fig. 212) of Allis's monograph appears to be identical with *Myoxocephalus octodecimspinosus*, several of which have been available for study.

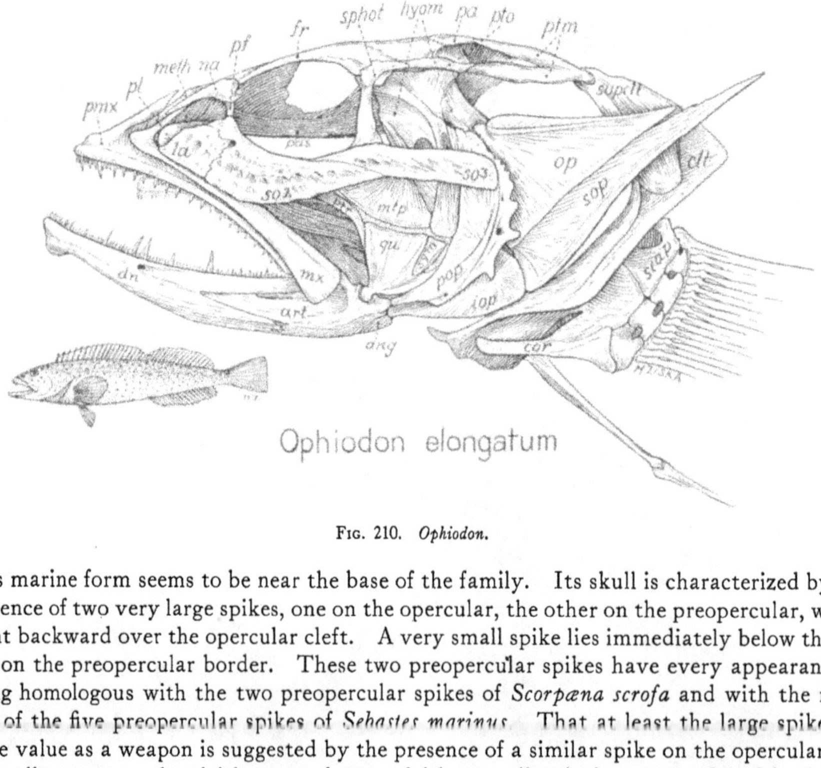

FIG. 210. *Ophiodon.*

This marine form seems to be near the base of the family. Its skull is characterized by the presence of two very large spikes, one on the opercular, the other on the preopercular, which point backward over the opercular cleft. A very small spike lies immediately below the big one on the preopercular border. These two preopercular spikes have every appearance of being homologous with the two preopercular spikes of *Scorpæna scrofa* and with the main two of the five preopercular spikes of *Sebastes marinus*. That at least the large spike has some value as a weapon is suggested by the presence of a similar spike on the opercular and of smaller ones on the cleithrum and supracleithrum, all pointing outward and backward and collectively forming a sort of *chevaux-de-frise*. The preopercular has lost its squamous expansion and now consists chiefly of a narrow bent rod, very densely built to support its huge spike and tunneled from top to bottom by the lateral line canal. Six openings to the latter on the posterior border mark the position of short side branches. The lower end of the preopercular is deeply forked, the posterior branch terminating in a curved spike pointing downward and forward. The bent preopercular rod braces and is wrapped over the quadrate and is supported by a strong short crest from the thickened hyomandibular.

Even in this species the skull as a whole is notably broader than that of *Sebastes*, but in *Myoxocephalus jaok* (Fig. 213) and allied species the breadth becomes extreme, while the eyes are directed chiefly upward, this apparently indicating a bottom-living habit. The skull bones in *M. octodecimspinosus* are thin and delicate and the spikes reduced or wanting.

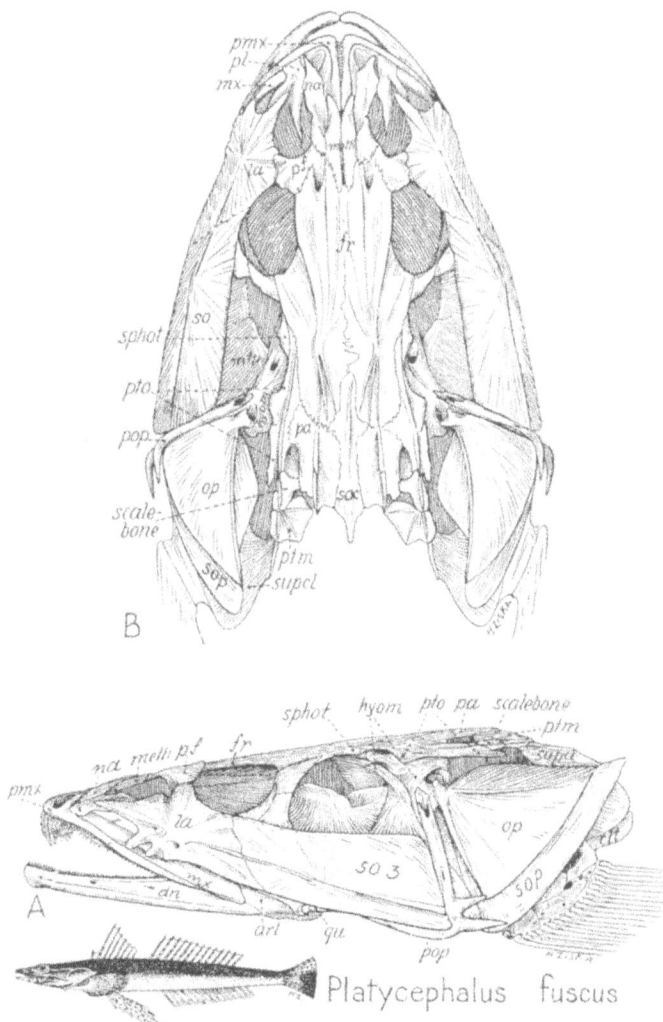

Fig. 211. *Platycephalus*. A. Side view. B. Top view.

This seems to suggest genetic instability with but little regard to the usefulness of the experiments.

These thin-skulled scorpænoids afford some interesting records of the development and

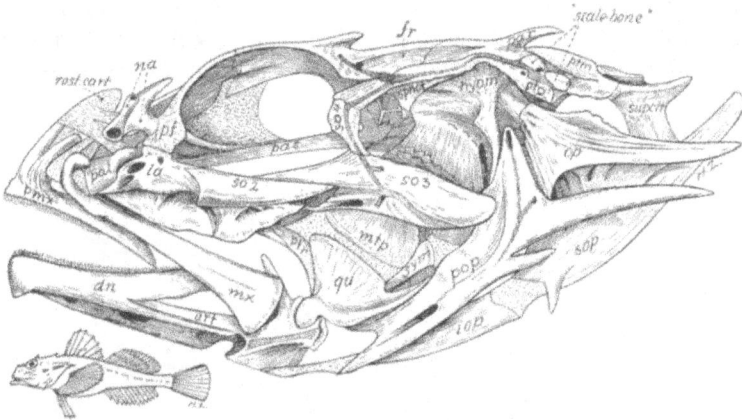

Fig. 212. *Cottus octodecimspinosus.* After Allis.

growth of the skull elements (Fig. 214). The hyomandibular, for example, has certain densely built tracts, including the two projections for articulation with the sphenotic and pterotic, the pedicle for the opercular, the crest for the preopercular and the lower limb

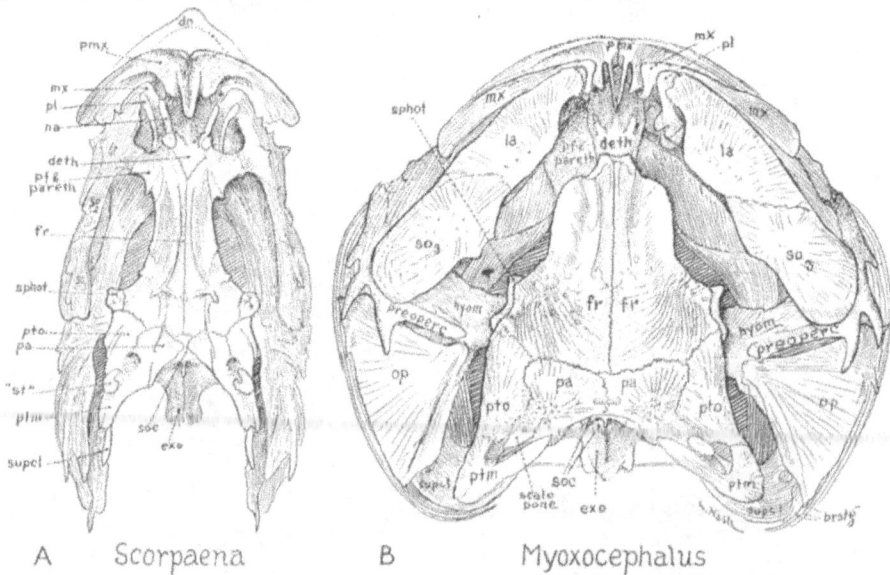

A   Scorpaena   B   Myoxocephalus

Fig. 213. A. *Scorpæna* sp. B. *Myoxocephalus jaok.* Top view.

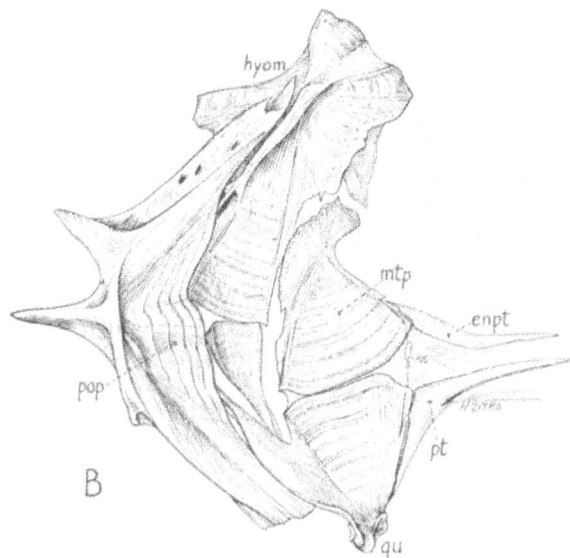

FIG. 214. *Hemitripterus.* B. Suspensorium. A. Lacrymal. Enlarged.

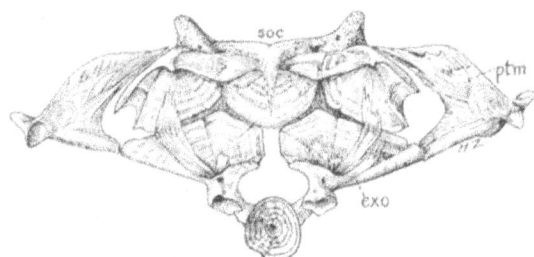

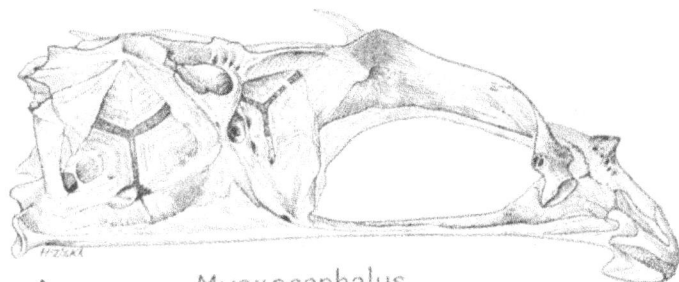

FIG. 215. Neurocranium of scorpænoids, showing growth zones and triradiate sutures.

for the symplectic.   In the central parts of these tracts, which are subject to great stresses, the trabeculæ are so crowded together that the bone is more or less opaque even to strong light; but these dense parts are connected by a web of bone of varying thickness, with zonal growth-bands traversed by trabecular ridges that radiate from the growth-center of the bone.   In the quadrate the growth-center appears to be located in the dense articular facet at the lower end; from this the bone has seemingly grown upward in a fan-like way, a very strong ridge near the hinder border carrying the growth lines obliquely upward while the web-like plate of the bone has grown by transverse zones crossed by upwardly-streaming rays.

The endochondral bones of the braincase (Fig. 215A) likewise consist of zonal and trabecular regions.   On the side of the braincase there is a prominent triradiate suture marking the adjacent boundaries of the pterotic, proötic and exoccipital bones; each of these exhibits zonal plates strengthened by trabecular ridges.   The triradiate suture is probably somehow due to uniform rates of growth from three equally distant centers.   Another triradiate suture separates the proötic, the alisphenoid and the sphenotic.

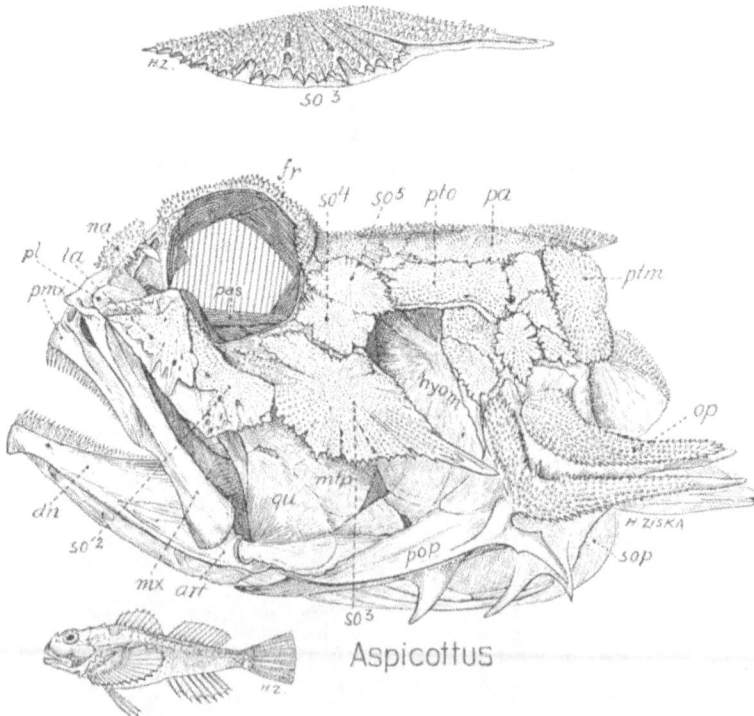

Fig. 216.   *Aspicottus (Enophrys)*, with oblique view of so³.

In a general way the ectosteal or derm bones resemble the endochondral bones in being composed of zonal and trabecular elements.   The same principles of growth are clearly

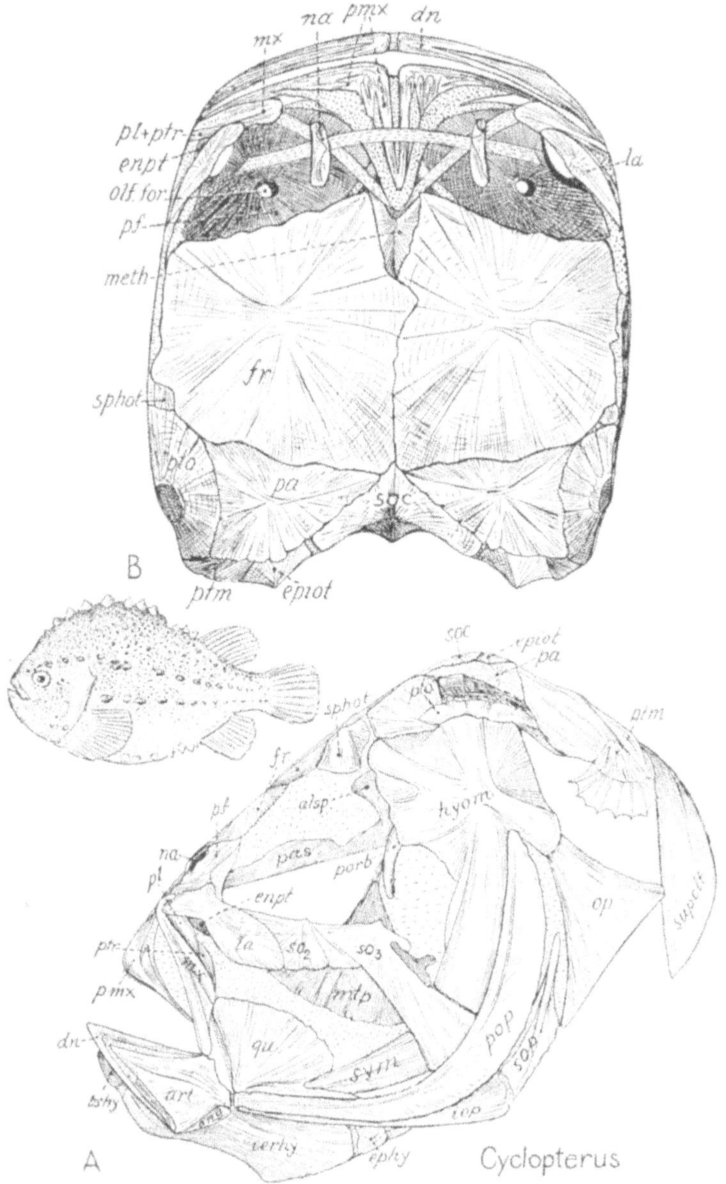

FIG. 217. *Cyclopterus*. After Uhlmann.

illustrated in a large skull of *Hemitripterus americanus*. Here (Fig. 215*B*) all the bones of the skull roof, braincase, jaws, etc., are so thin that the growth zones and trabeculæ may readily be seen with the naked eye. The cranial spikes are seen to have definite relations

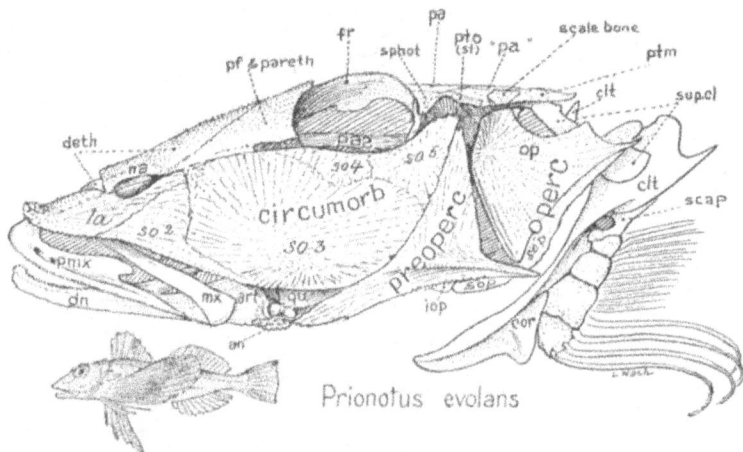

Fig. 218. *Prionotus.*

with the growth-centers. In general, radiating fibres predominate in the ectosteal bones, while zonal growth predominates in the endochondral bones of the braincase, suspensorium and primary upper jaw. Three beautiful triradiate sutures are seen on the occipital surface;

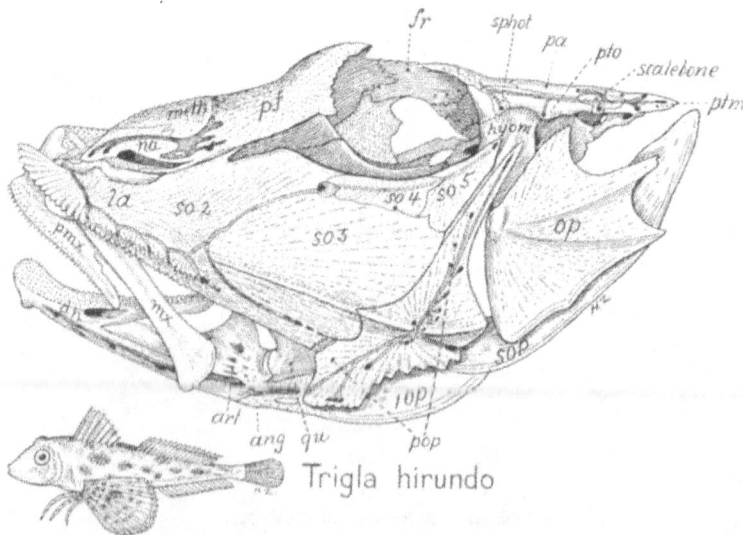

Fig. 219. *Trigla.* The epiotic is seen only in the occipital view. After Allis.

one between the supraoccipitals and the other two between the supraoccipitals and the epiotics. The scorpænoid heritage of this skull is seen in many characters, as of the lacrymal and suborbital stay, small preopercular spikes, etc.

In a very wide-skulled species of *Myoxocephalus* (Fig. 214) from Alaska (presumably *M. jaok*) the strongly-raised, fine trabeculæ of the skull roof show a tendency to give off very numerous small points on ridges radiating from the center of the bone. A continuation of this tendency might give rise to the finely-papillate or villous surface of the cranial armor of the gurnards (Triglidæ). In another *Myoxocephalus* skull the radiating ridges are equally evident but are not subdivided into papillæ; in a third species the trabeculæ do not project above the surface and there are no villi. Here then we have in allied species interesting differences in bone texture which imply corresponding differences in the under side of the skin.

A strange modification of the cottid type is seen in the skull of *Enophrys* (*Aspicottus*) *bison* (Fig. 216). Here the ectosteal bones of the face and head are thickened and their

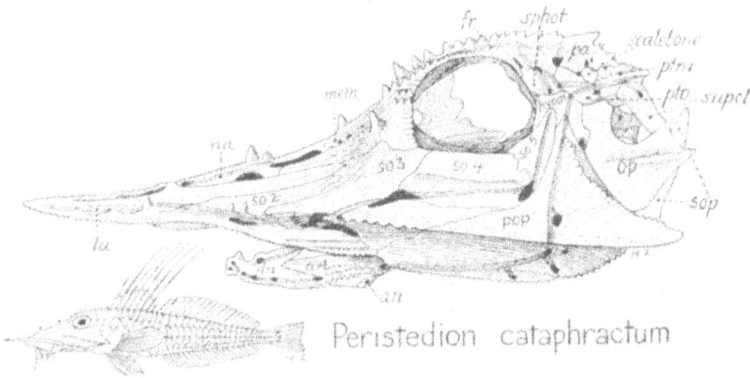

FIG. 220. *Peristedion.* After Allis.

surface is studded with a carpet of minute horny pustules or denticles arranged on radiating ridges from the growth-centers and covering even the two large spikes on the preopercular and opercular. These pustules or villi are much more numerous and uniform in size than those described above as occurring in *Myoxocephalus jaok*. They are usually at right angles to the direction of the ridges or trabeculæ, like beads of water on a thin wire.

The third suborbital is pointed posteriorly and in the dried skull diverges widely from the preopercular. The endochondral quadrate and metapterygoid bones show the usual zonal structure but are very thin and translucent with fine, delicate, radiating fibres.

Cyclopterids.—Nothing could well be more unlike the bristling, aggressive scorpænoids than the obese lump-fishes, Cyclopteridæ, yet the possession of a well developed suborbital stay (Fig. 217) seems to offer a reliable indication of their taxonomic position. Indeed Boulenger (1910, p. 698) states that the Cyclopteridæ are very closely related to the Cottidæ, with which they are connected through *Psychrolutes*, and that "it is even doubtful whether they deserve to be separated from them."

The peculiar genus *Caracanthus* (*Micropus*), which is widely distributed among the

18

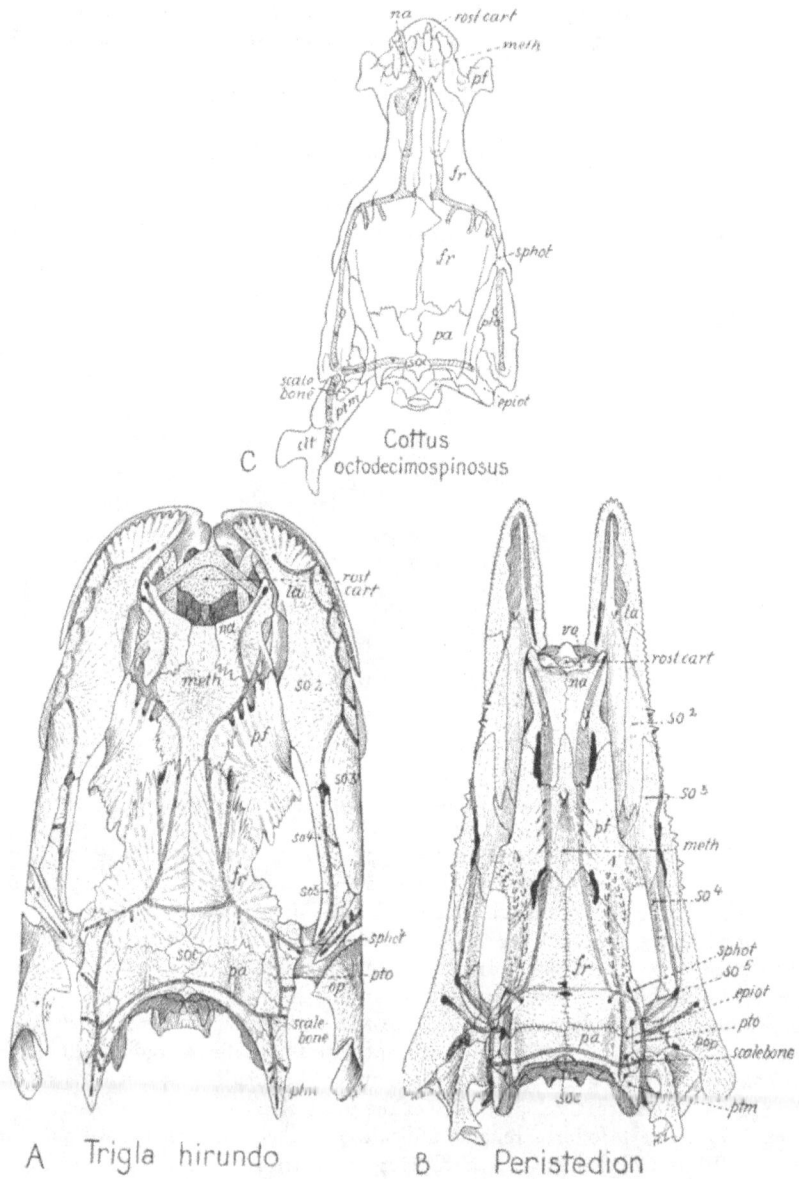

FIG. 221. A. *Trigla*. B. *Peristedion*. C. *Cottus*. All after Allis.

islands of the tropical Pacific, is said by Jordan and Evermann (Hawaiian Fishes, 1905, p. 453) to be closely related to the Scorpænidæ but differs in the compressed, deep body and vestigial pelvic fins. Several bones of the head are strongly armed, the preopercular and interopercular bear strong spines directed downward.

Triglids.—In the gurnards (Triglidæ) and their allies the surface bones of the cheek (Figs. 218–221), especially the suborbitals, have become enlarged into two great flat armor plates, one on each side; these together with the heavily armored skull-roof form a strong shield for the whole head. Although the orbits are widely open in the dried skull, the eyes are perhaps protected by a tough sclerotic coat, cornea and surrounding skin; but, as remarked above, the eyes seem to have a dominant role in the scorpænoid group and apparently the fish must depend upon its visual alertness to avoid injury to its eyes.

The outer surface of all the bones of the cheek-plate is covered with sharply raised ridges that radiate from the ossific centers. In *Prionotus* and other genera the edge of

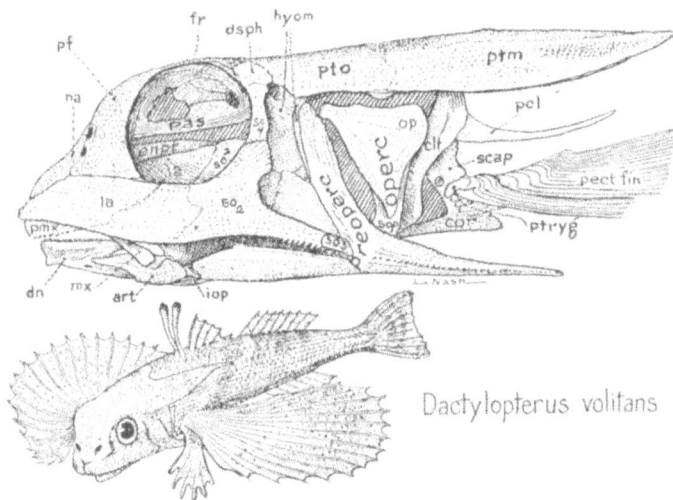

FIG. 222. *Dactylopterus* (*Cephalacanthus*).

each ridge bears a single row of more or less sharp, thorn-like projections, which are so small as to be seen as individuals only with a pocket-lens. As their tips are all in nearly continuous planes these denticles impart a smooth, velvet-like appearance and sandpaper-like feel to the surface. They are enlarged in some places, notably at the front edge of the lacrymal, into a short-toothed, comb-like edge. Besides the minute denticles there is a fairly large spike on the posterior border of the preopercular, two smaller ones on the opercular and one above the orbit, others on the skull-top behind the orbit. By comparison with other scorpænoids it seems evident that the raised radiating ridges of the cheek-plates represent an extreme emphasis of the radiating trabeculæ of less specialized forms and that the minute denticles on the surface are comparable with the villous outgrowths on the surface of the trabecular ridges in the peculiar cottoid *Enophrys* (*Aspicottus*) (see p. 339). Detailed and repeated comparisons of the surface ridges and denticles of the triglids with those of

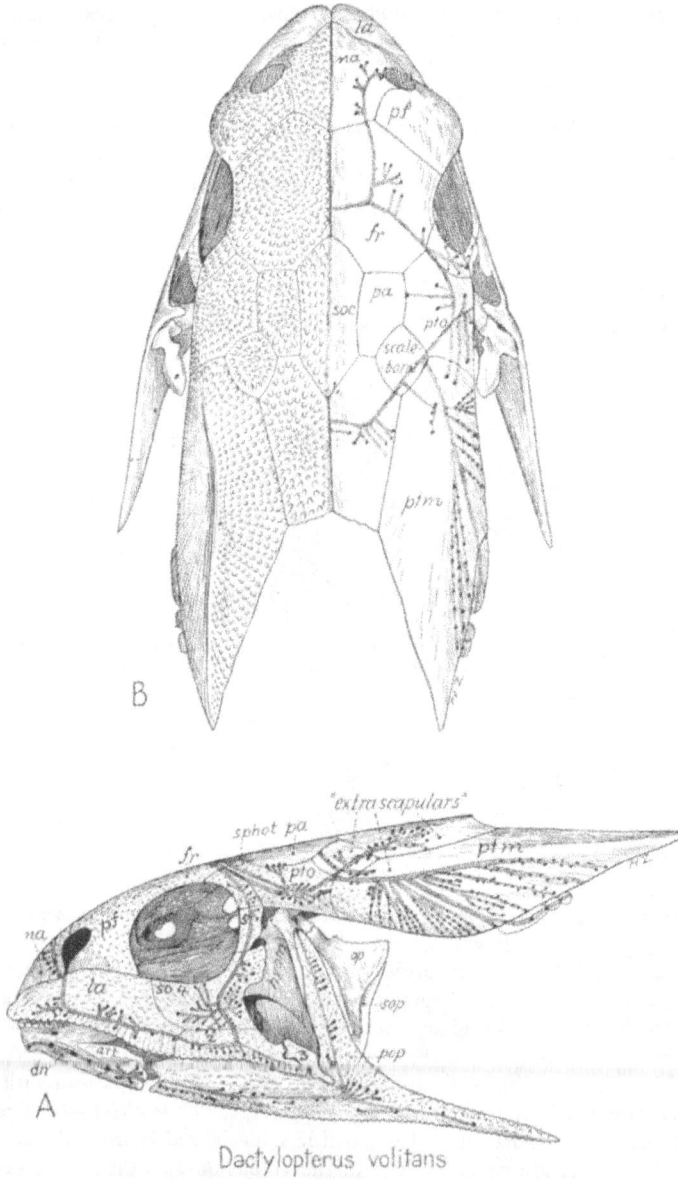

FIG. 223. *Dactylopterus*. A. Side view. B. Top view. Both after Allis.

Myoxocephalus jaok and other cottoids suggest that in more primitive conditions the bone was porous with minute tubes opening on the exterior. Then the radiating trabeculæ began to assert themselves, the pores became open channels between the trabeculæ, and with further increase in the growth pressure the trabeculæ began to sprout into irregularly cuspidate edges as in *Myoxocephalus jaok*. Finally the almost uniform distribution of the villi or denticles was attained, as in the triglids. In *Peristedion* (Fig. 220) the growth momentum of the lacrymal bones has resulted in a pair of long processes which project far in front of the nasals. This, together with the reduction of the jaws, produces a somewhat sturgeon-like appearance of the head and is doubtless associated with the habit of resting on the bottom.

In spite of these and other specializations, the distribution of the lateral line canals on the cheeks and skull roof assists us in homologizing the several elements with those of normal scorpænoids (*cf.* Allis, 1909, Taf. VI).

On the whole, the skull of *Prionotus*, as well as those of other triglids, is remarkably heavy, especially in its outer parts, while it is equally delicate and thin in the well-covered parts, including the jaws. In many, but not in all *Prionotus* skulls, there is a pair of peculiar dense swellings on the lower part of the prefrontals. The base of the cranium is smooth and compactly built, with three wholly distinct and well-rounded occipital condyles, essentially as in Cottidæ and Scorpænidæ.

The flying-gurnard *Dactylopterus volitans* is the type of a very peculiar family whose connection with the scorpænoid group is indicated by the presence of a suborbital stay, by the forward extension of the lacrymal, and perhaps other characters. But the skull (Figs. 222, 223) has become so highly specialized that the more precise relationships of this family to the other scorpænoids is uncertain. Comparison of the skull-top of *Dactylopterus volitans* and of *Trigla hirundo* as figured by Allis (1909, Taf. VIII) reveals immense differences in the pattern and in the arrangement of the individual elements. In *Dactylopterus* the plates above the shoulder-girdle have been expanded into a huge neck shield, which is conjoined by immovable sutures with the expanded cranial roof. The differences in relations of all the bony elements and the course of the lateral line canals between *Dactylopterus* and *Trigla* seem in fact to be irreconcilable with the idea of a near relationship between the two families. This negative conclusion is strengthened by the recent paper of Starks on the shoulder-girdle of the teleost fishes (1930) in which it is shown (p. 71) that the pectoral girdle of *Dactyloptena orientalis* is extremely different from those of the cottoid and triglid fishes, so that Starks notes its typical percoid characters and says that it is "difficult to understand why this evidently aberrant trigloid form should possess the typical percoid shoulder-girdle of the main line of descent, rather than that of its immediate relatives of the family Triglidæ, which have the cottoid shoulder-girdle."

The otoliths of the scorpænoids, as studied by G. Allan Frost (1929b, pp. 257–263), "show a strong affinity with those of the suborder Percoidea, and, although in some cases considerably modified, resemble either the Percid or the Labrid type. . . . In the family Triglidæ, the otoliths are high in shape and present certain Labrid features; they may usually be distinguished by the contorted appearance of the sulcus, due to the uneven position of the upper and lower angles, and to the elevation of the cauda. In the families Scorpænidæ, Cottidæ, and Dactylopteridæ, the otoliths resemble the percid type." With some exceptions the otoliths vary from the primitive, slightly elongate percid type of

Scorpæna dactyloptera either toward the high type of *Trigla* or toward the elongate low type of *Platycephalus*.

HETEROSOMATA (FLATFISHES, SOLES)

The outstanding feature of the skull of flatfishes is undoubtedly the transference of both eyes to one side of the skull. Tate Regan (1929, p. 324) summarizes the embryological studies of Williams on this topic as follows:

"Williams (Bull. Mus. Comp. Zoöl., 1902) has studied the migration of the eye; in the cartilaginous skull of the larva two bars above the eye connect the lateral ethmoid cartilages with the otic capsules; preparatory to the migration of one eye the bar above it is resorbed and becomes reduced to projections of the lateral ethmoid and otic capsule with a gap

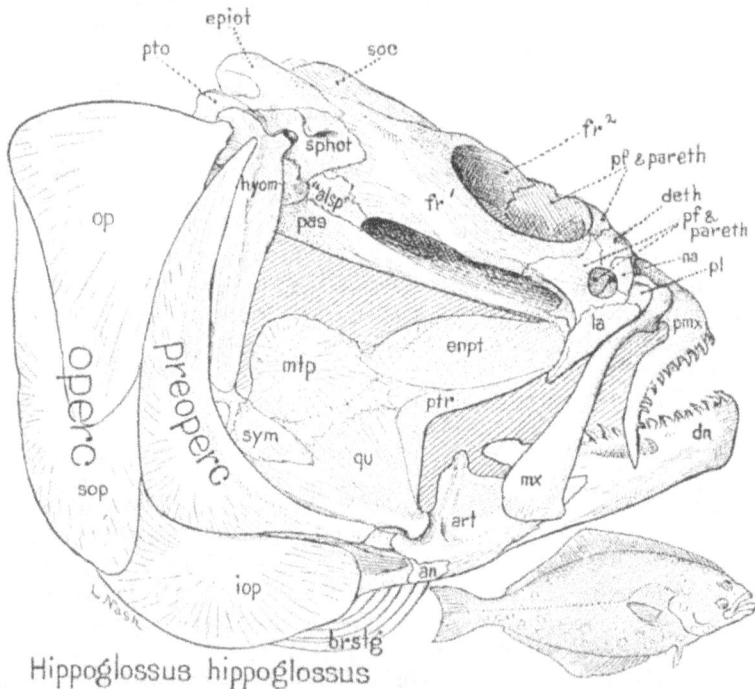

FIG. 224. *Hippoglossus.* Eyed side with twisted interorbital bar (fr¹).

between them. Through this gap the eye migrates until it reaches the other supraorbital bar, when both eyes move to their final position, causing a torsion of the bar between them which also affects the ethmoid region; when the shifting is complete ossification takes place, and the main part of the frontal bone of the blind side forms on the wrong side of its eye. Thus the essential feature of the skull of the flatfishes is that the interorbital bar is formed mainly by the frontal of the eyed side and that the frontal of the blind side extends forward to the ethmoid region outside the upper eye."

In Figure 224 is shown the twisted interorbital bar of the frontal (*f*1) and the secondary bar formed by the junction of the postorbital part of the frontal (*f*2) with the lateral ethmoid.

In Figure 225, the left side of the same skull, we see the blind side, with the secondary bar (*f*2). The remaining elements of the skull are readily identifiable.

According to Boulenger, the flatfishes have been derived from symmetrical deep-bodied fishes with a short body cavity, represented by the Eocene *Amphistium*. Tate Regan, however, has shown (1929, pp. 214, 324) that *Psettodes*, the most primitive member of the group, points strongly toward the derivation of the flatfishes from normal Percomorphi. "Except for its asymmetry and the long dorsal and anal fins *Psettodes* is a typical perch and

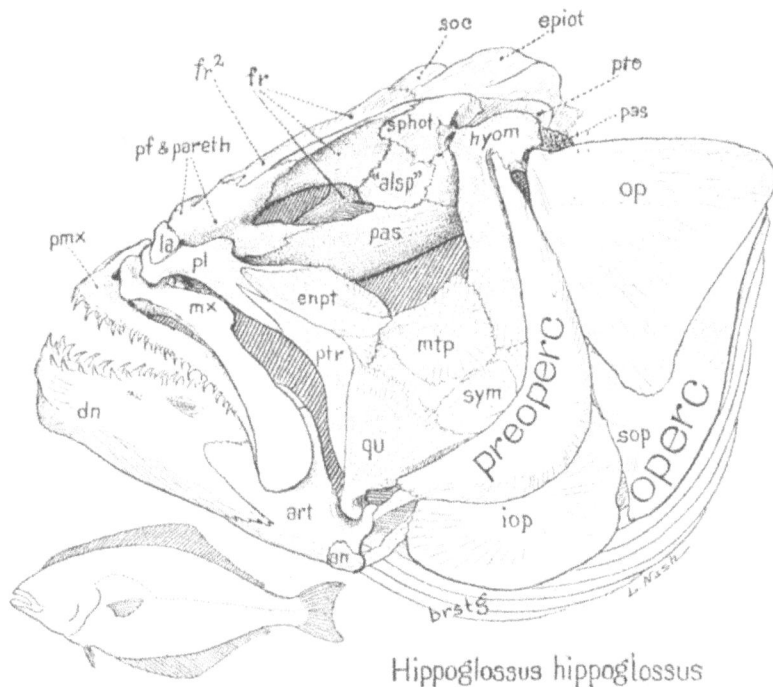

Hippoglossus hippoglossus

Fig. 225. *Hippoglossus.* Blind side showing secondary union of prefrontal and frontal (fr²).

might almost be placed in the Serranidæ. . . . It may have retained so many percoid features because it has not adopted progression along the bottom by undulatory movements of the body and marginal fins to the same extent as other flat-fishes. . . ."

From a comparative study of the otoliths G. Allan Frost (1930*a*, pp. 231–239) supports Tate Regan in deriving the Heterosomata from a percoid stem. "The otolith of *Psettodes erumei*," he writes (p. 232), "of the family Psettodidæ, resembles in every feature those of the suborder Percoidea, the only modification being the presence of a slight crest on the anterior rim in front of the sulcus. The shape resembles that of *Perca* and the sulcus reproduces that of *Centropomus*, both of the suborder Percoidea; the sulcus has an upper and a lower angle, and a curved and pointed cauda.   It differs in the presence of a narrow crest which encloses the front of the ostium, and in the dentations of the ventral rim.   Except for these slight modifications, it is essentially the otolith of a Percoid Fish.   It is rather more curved in its length than those of the remainder of the Flat-fishes."

GOBIOIDEI (GOBIES)

Most of the gobies are quick-darting little fishes but *Eleotris* (Fig. 226) is a sizable fish of aggressive, even ferocious looks.   In general the gobies are close to the typical percoids

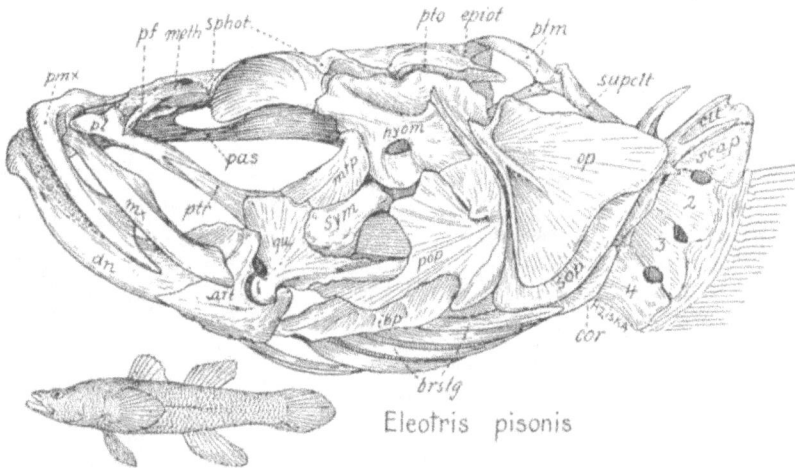

FIG. 226. *Eleotris.*

but more specialized in many features, including the close appression of the opposite pelvic fins, which often serve as suckers.   The jaws are of the percoid type, with more or less protrusile premaxillæ and toothless maxillæ which are excluded from the gape.   The suspensorium is prolonged to a point beneath the small orbits, but as the depth beneath the orbits is somewhat less than usual, the lower border of the mandible is directed only moderately upward, even though the snout is short.   The small eyes look partly upward.   They are guarded posteriorly by a large flange of the dermosphenotic.   Due to the subdorsal position of the orbits, the interorbital skull-roof is narrow, while the postorbital roof is wide and flat. The hyomandibular is elongate anteroposteriorly but short vertically, the symplectic stout. Behind it is a fenestra which has doubtless been formed subsequently to the thinning-out of the bone at this point and the strengthening of the surrounding elements under the stresses of the powerful adductor muscles.

The opercular and subopercular are fairly normal, the preopercular is abbreviated dorsally, the subopercular extends upward and backward so as nearly to exclude the opercular from the border of the opercular slit, as in the gadoids and ophidioids.   For further details of skull structure see Tate Regan, 1911c, pages 729–733.

In the famous mud-skipping goby *Periophthalmus* (Fig. 228) we view some extraordinary specializations of the goby type of skull. The very large orbits are lifted above the rest of the skull and the eyes are erectile (Tate Regan, 1911c, p. 733). The quadrate-articular joint is moved forward in front of the anterior orbital rim; the jaws are of a modi-

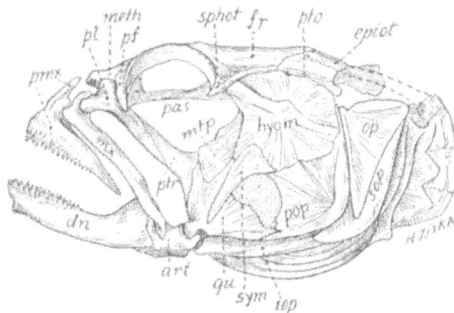

Fig. 227. Unidentified Goby from Bahamas.

fied nipping type with remarkably stout protrusile premaxillæ, reduced maxillæ and inter-locking maxillary processes of the palatines.

The hyomandibular is remarkably wide anteroposteriorly; on the lower end of this bone the preopercular appears as an almost vestigial patch. The interopercular forms the tracker that is characteristic of advanced derivatives of the percomorph stock. The bran-chial chamber is small and with it the opercular and subopercular. The use of the pectoral

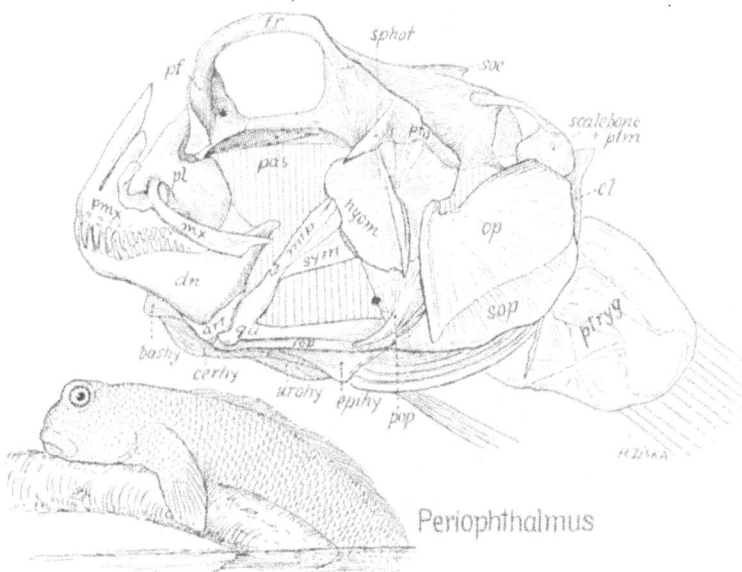

Fig. 228. *Periophthalmus.*

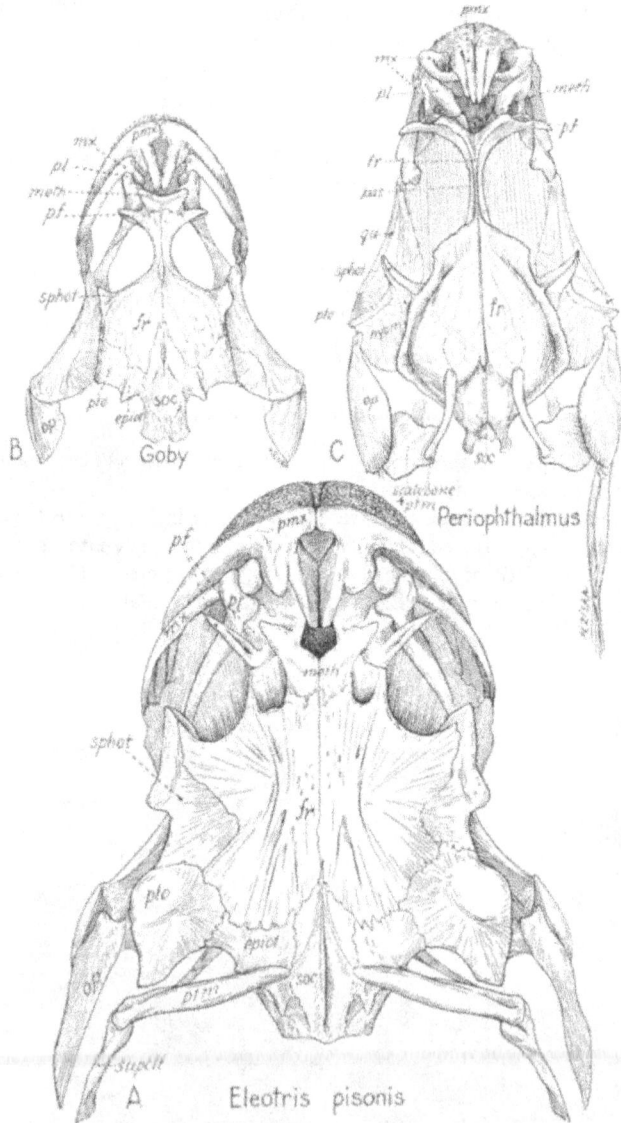

Fig. 229. A. *Eleotris*. B. Unidentified goby. C. *Periophthalmus*. Top views.

limb for climbing and hopping is reflected in the large size of the pterygials. The post-temporal, however, is not remarkably large and has but a small contact with the stout occiput.

As to the relationships of the gobies, Dareste (quoted by Emery) and Emery (1880, p. 21) stressed the resemblance of the goby skull to those of the Ophidiidæ among the blennioid group. Emery also noted that the general disposition of the bones in the gobies,

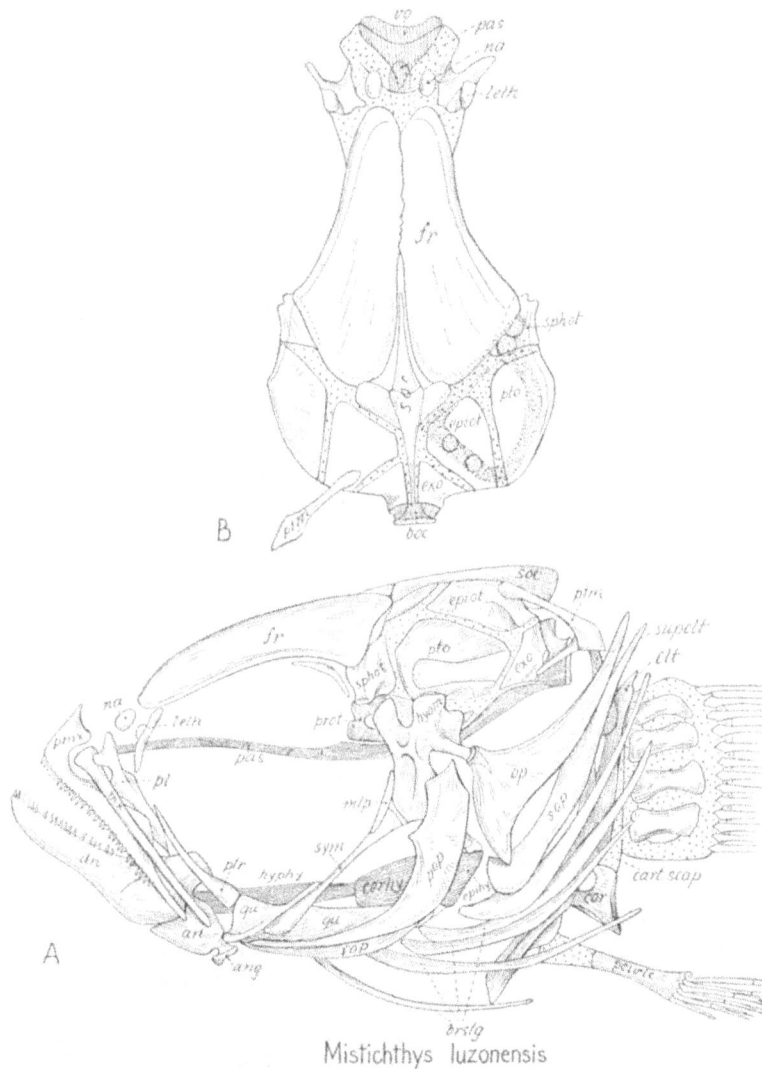

Mistichthys luzonensis

FIG. 230. *Mistichthys.* A. Side view. B. Top view. Loaned by Miss Lois Te Winkel.

gadoids and ophidiïds shows a certain conformity of structure, through which these fishes seem to constitute a group with depressed skulls, in contrast with the compressed skulls of the greater number of acanthopterygians. Starks (1911b, pp. 747–748) suggested that the gobioids may be an offshoot of the scorpænoids; but Regan (1911c, p. 731) is unable to accept this suggestion and refers the gobies to the Percomorphi, defining for them the suborder Gobioidea.

In a valuable study of the skull of *Mistichthys* (Fig. 230) the smallest known fish, Miss Lois Te Winkel, who has courteously given me permission to reproduce the accompanying figure, finds that the skull retains many features that are usually found in young stages of teleosts.

The otoliths of the Gobioidea, as described by Frost (1929a, pp. 126–129), are highly specialized; more or less oblong with an enclosed ovate or biovate sulcus and other peculiarities. This "Gobiïd" form is "unique and unmistakable, and has persisted since Eocene times, large numbers occurring in the Barton Clay of Hampshire, which closely resemble the otoliths of recent fishes" (p. 126). Starks (1930, pp. 218, 219) gives abundant data for the present conclusion that the details of the scapula, coracoid and actinosts in the gobies, go to make up a distinctive gobioid type, which is convergent in some features toward the cottoid type but must have been derived independently from a more primitive percoid girdle. We seem therefore to be compelled to regard the gobioids as one of the early offshoots of the primitive percoid group.

SYMBRANCHII

In the fresh and brackish waters of India and Burma occurs a greatly elongate eel-like fish, the *Amphipnous cuchia*, which, according to Tate Regan (1929, p. 327), "spends the greater part of its life out of the water, wriggling along the banks, in which it burrows during the dry season. It visits the water in search of food, worms, crustaceans and small molluscs." Boulenger (1910, p. 599) says that "this amphibious fish, when in the water, constantly rises to the surface for the purpose of respiration, and it is often found lying in the grassy sides of ponds after the manner of snakes." Thus it is one of several fishes of different orders which have been able to come up out of the water or burrow in the mud by virtue of possessing a respiratory air-sac of some sort. Regan (1929, p. 327) states that in the *cuchia*, the air-breathing sacs are a pair of "diverticula of the pharynx which lie on each side of the back-bone above the gills. . . ." According to Boulenger (1910, p. 598), "Of the three branchial arches the second alone possesses gill-filaments; the third supports, in their place, a thick and semi-transparent tissue; the principal organs of respiration are two small bladders, resembling the posterior portions of the lungs of snakes, which the animal has the power of filling with air immediately derived from the atmosphere. . . ."

The skull of *Amphipnous* is stated by Tate Regan (1912d, p. 390) to differ from that of the Symbranchidæ (see below) only in features which are connected with the presence of these respiratory sacs; these have pushed away the pectoral arch from the skull, so that the posttemporal is reduced or absent . . . on the outside the sacs are covered by the operculum and suboperculum, "which are enlarged and form thin, almost membranous laminæ," (1929, p. 327).

In the fresh or brackish waters of Central and South America, West Africa, India, Southeastern and New Guinea, are found the representatives of the family Symbranchidæ, with one marine form (*Macrotrema*), from Penang and Singapore (Tate Regan, 1929, p. 327).

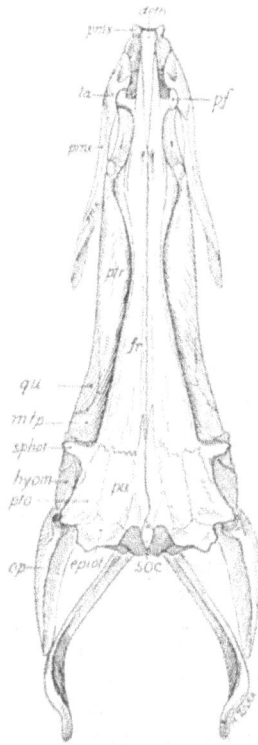

Symbranchus

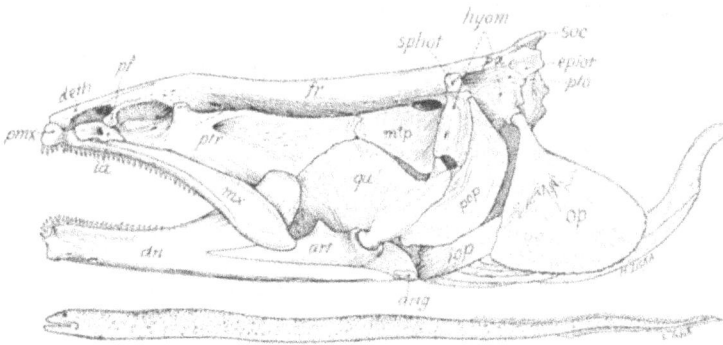

Symbranchus marmoratus

FIG. 231. *Symbranchus.*

The skull of one of these symbranchids, *Monopterus javanensis*, has been described and figured both by Boulenger (1910, pp. 597–598) and Tate Regan (1912c, pp. 388–390). Our figure (Fig. 231), however, represents a skull of the allied genus *Symbranchus*. This type of skull is very long and narrow, mostly in the interorbital region. In some ways it is strongly suggestive of the skull of the so-called electric eel (*Electrophorus*), an interesting example of convergence. The cranial table consists largely of two wide flat parietals which meet in the mid-line above the minute supraoccipital. This arrangement, as in so many other cases, appears to be correlated with a long flat-topped skull in which the axial muscles do not extend on top of the occiput.

All the vertical diameters are reduced so that the braincase appears like a long, gently tapering tube in side view. The middle part of this tube is formed by a long descending flange of the frontal which has a long contact with the stout parasphenoid. On the whole, this part of the cranium rather suggests modern amphibian skulls with elongate spheneth-moid bones,—another case of convergence.

The jaws in this fish are fairly long, although the suspensorium is inclined forward. The premaxillæ, thickened in front, are elongate and slender posteriorly; they are appressed to the maxilla, which is rod-like anteriorly and wider posteriorly. The premaxillæ bear very short, recurved, conical teeth; the dentary, very delicate, minute, recurved teeth. The rear part of the mandible is disproportionately large with strong ascending or coronoid process. There is a crescentic band or series of small teeth on the palatines and pterygoids. The palatines, according to Tate Regan (1912c, p. 389), meet below the vomer and lack maxillary processes—an unusual arrangement. The mesopterygoid is absent and the ectopterygoid enlarged. The opercular is reduced to a rather small, downwardly-directed spatulate piece bearing a tab-like subopercular. The interopercular is a broad, short, curved slat. The curved preopercular covers the forwardly-directed suspensorium in the usual manner. Several elements of the normal fish skull have disappeared, including the maxillary process of the palatine, the mesopterygoid, the basisphenoid, alisphenoid, orbito-sphenoid, opisthotic, suborbitals (Regan, 1912c, pp. 388, 389). Thus this skull abounds in convergent resemblances to skulls of characin-eels and true eels, but evidently represents a quite different stock.

On the coasts of Australia and Tasmania are found small, eel-like fishes known as "shore-eels" (*Alabes* or *Chilobranchus*) but which Boulenger and Tate Regan refer to the order Symbranchii. Their external similarity to certain eel-like blennioids (Scytalinæ and Zoarcidæ) is so strong that Vaillant, quoted by Tate Regan (1912c, p. 388), expressed the opinion that they are indeed related with the blennioids rather than with the Symbranchidæ, but, according to Tate Regan, *Alabes* "in its osteology differs widely from the Blennioids, and although it also differs sufficiently from the Symbranchoids to be made the type of a separate suborder, its relationship to them is quite clear." The skull differs from that of *Monopterus* in the extreme shortening of the snout; this makes the skull in top view very short and relatively wide. The premaxillæ have a very stout, fairly long ascending process and a very short tooth-bearing branch. The maxilla is a small rod behind the large pre-maxilla; the dentary is short and thick with small erect teeth, the articular short with very strong short coronoid process. All this indicates a "small-mouthed nibbling" habit. The telescoping of the snout has widened the braincase and thrust apart the flat parietal plates, exposing the flattened supraoccipital plate in contact with the frontals. The short para-sphenoid is widely separated from the frontals by the expanded proötics and exoccipitals.

In conclusion the generally eel-like appearance of the Symbranchii, and the fact that the parietals meet in the mid-line on top of the skull, together with the lack of fin spines, have led most authors to place the order Symbranchii among the soft-rayed groups not far

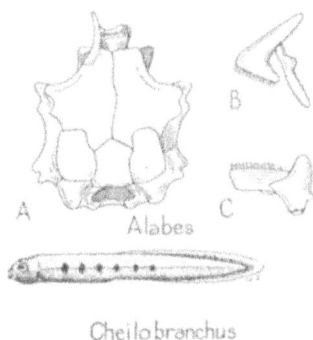

FIG. 232. Alabes. A. Cranium, top view. B. Left premaxilla and maxilla. C. Mandible. After Tate Regan.

from the true eels. But Tate Regan after a more thorough consideration of the osteological and other characters concludes (1912c, p. 387) that the "resemblances to the true eels are not due to relationship" and that *Alabes* gives a clue to the derivation from some group of acanthopterous physoclists, since *Alabes* possesses long ascending or articular processes of the premaxillæ, supraoccipital in broad contact with the frontals, triple occipital condyles, jugular pelvic fins, etc.

In *Monopterus javanensis* the otolith, according to Frost (1929a, p. 128), resembles that of *Cepola rubescens* of the suborder Percoidea in general shape and in the sulcus. It differs in certain details. The sulcus also resembles in its relative proportions those occurring in *Neobythites* and *Genypterus* of the division Ophiiformes (suborder Blennioidea). In *Symbranchus marmoratus* the sagitta is ovate and apparently more specialized than that of *Monopterus* (p. 128).

OPISTHOMI (MASTACEMBELIDS)

These "spiny-finned eels," or more properly eel-like spiny-fins, of Southern Asia and Tropical Africa, show convergent resemblances to the eel-like dipnoan *Protopterus*, since the Indian *Rhynchobdella aculeata*, "conceals itself in the mud and becomes drowned in water if unable to reach the surface, as it apparently requires to respire air directly" (Day, quoted by Boulenger, 1910, p. 717). The dorsal fin is greatly elongate and is continuous around the tail with the less elongate anal. The front half of the dorsal is represented by a series of short detached spines. The ventral fins are absent. The head is very long and narrow, the snout produced in front into a fleshy median tentacle bordered by the produced tubular anterior nostrils, which lead into a tube that opens in front of the eye. The fishes are carnivorous, with small eyes and hypertrophied olfactory chambers.

The osteology of *Mastacembelus* has been described by Tate Regan (1912d, pp. 217–219), who figures the cranium. Our figure, however, is from the specimen. The cranium is very elongate, more tapering in front, with very large long nasals covering the large nasal chambers, long narrow frontals and short parietals, which are well separated by the supraoccipital.   There is no opisthotic, orbitosphenoid nor basisphenoid; the proötic separates the parasphenoid from the alisphenoid; the proötics are elongate, flanking the narrow parasphenoid; the palatine is a long narrow lamina which is firmly united to the elongate vomer and

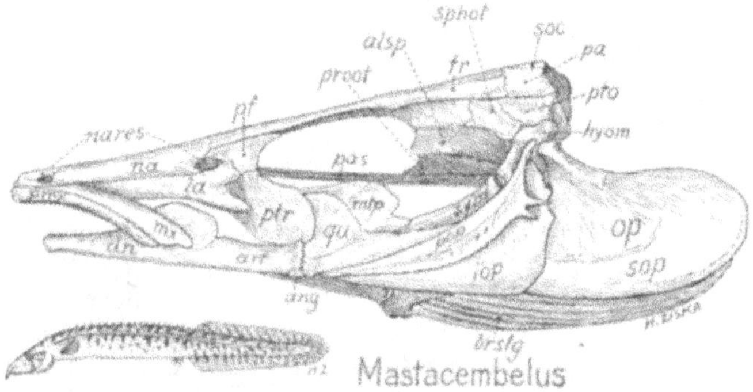

FIG. 233. *Mastacembelus.*

is wedged in between the opposite lateral ethmoids.   The pterygoid is movably articulated with the lateral ethmoid external to the palatine.   The mouth is small and subterminal, teeth villiform on the premaxillæ and dentaries, maxillæ firmly attached to the nonprotractile premaxillæ.   The hyopalatine and opercular bones are all present (Regan).   Gill-cleft inferior (Boulenger), as in the symbranchoids; branchial arches percoid (Regan); posttemporal absent, the pectoral arch suspended from the skull far behind the skull (whence the name Opisthomi).

Boulenger suggested (1910, p. 716) that this single family is possibly derived from the Blenniidæ.   Tate Regan, however, concludes (1912d, p. 218) that these fishes "are related to but more specialized than the Percomorphi, but they show no particular affinity to any group of Percomorphous fishes."

In *Mastacembelus armatus*, according to Frost (1930b, p. 625), the otolith (sagitta) is of the percid type, except in certain details.   No special resemblances to the blennioid types figured by Frost (1929a, Pl. I, Figs. 9–19) were noted.

AMMODYTOIDEI (SAND-LANCES)

Another offshoot of the percomorph group which is otherwise *incertæ sedis* is the family of the sand-lances, Ammodytidæ.   These small sagittiform fishes dart into the sand; perhaps their acutely pointed mandible serves as the point of the lance.   The long narrow skull (Fig. 234) presents to my scrutiny no hint of its derivation.   There are several very curious features; the end of the mandible bears a unique elevation, which may support a pad of skin in life.   The long premaxillæ are very slender, much more so than the long maxillæ.   The proximal end of each of the latter is expanded.   The long narrow bars which Thilo (1920) mistook for the ascending processes of the premaxillæ are in reality the extremely narrow anterior extensions of the mesethmoid, the true ascending processes of the premaxillæ being very small.   Hence the fish cannot have a protractile mouth and Thilo's description of its mechanism must be erroneous. At most the upper border of the mouth could be everted slightly. The delicate opercular bones, while peculiar, have not helped to solve the problems of relationship.

The otolith of *Ammodytes tobianus* described by Frost (1928a, p. 454), is amygdaloidal and biconvex. It gives no clue to near relationship but serves to emphasize the relative isolation of the group.

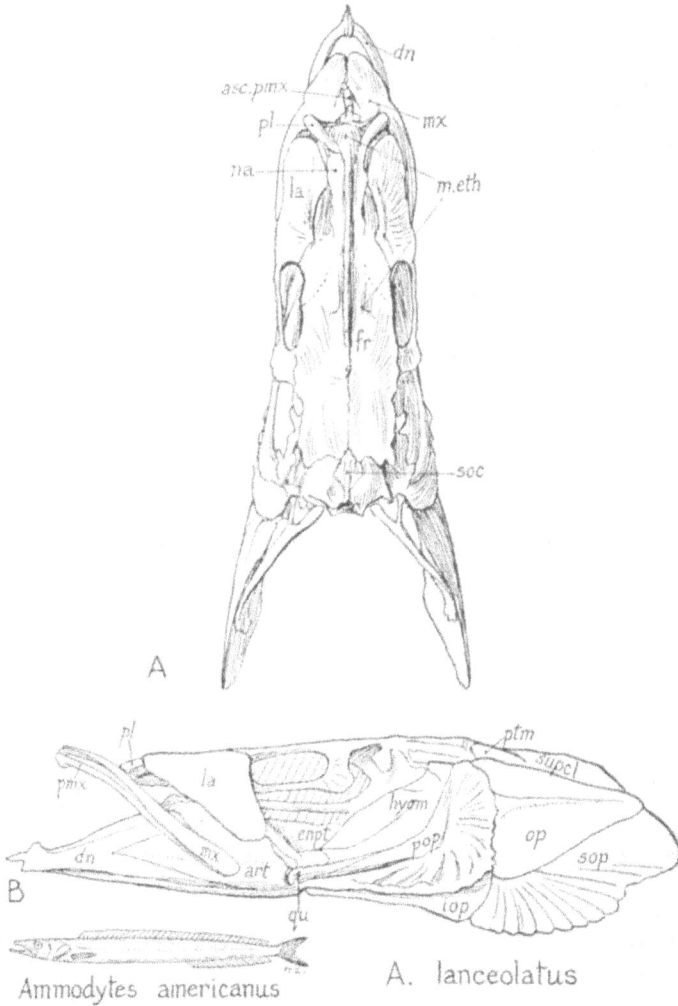

Fig. 234. *Ammodytes.*

TRACHINOIDEI (NOTOTHENES, WEAVERS, STAR-GAZERS, DRAGONETS)

If hard and fast definitions were to be sacrosanct, this group would not have any more standing than the old "Jugulares." Nevertheless the fishes in it show rather strong affinities with each other, now in certain characters and again in others.

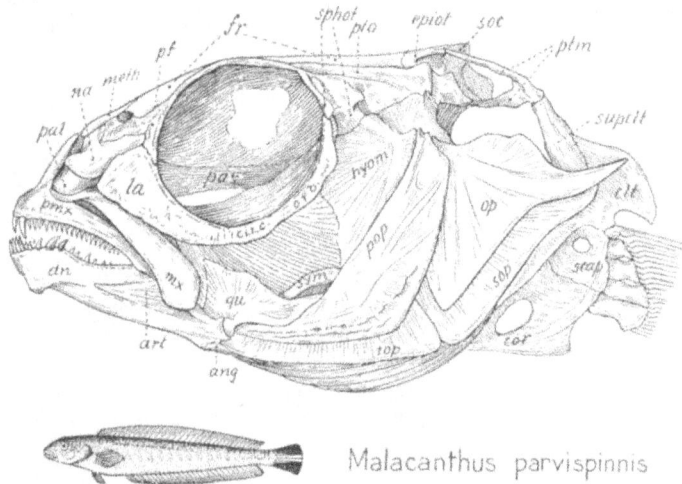

FIG. 235. *Malacanthus.*

Malacanthus.—Of this family, Malacanthidæ, Jordan (1923, p. 202) says that "it bears strong resemblance to the trachiniform fishes, differing mainly in the thoracic insertion of the ventrals." The skull (Fig. 235), however, hardly affords convincing support for this view, as the presence of a spike on the opercular may be a mere parallel both to the cottids and trachinoids.

The skull of the related genus *Lopholatilus* is quite percoid in general features and so is the shoulder-girdle. Nevertheless it is rather evident from comparison with the subsequent figures that *Malacanthus* indicates the structural path leading to the notothenes and trachinoids.

Pinguipes.—A skull (Fig. 236A) of *Pinguipes chilensis* (No. 286, Brit. Mus. Nat. Hist.) conforms in general to the percoid type and the family Pinguipedidæ, consisting of *Pinguipes*, *Neopercis* and *Parapercis*, is assigned by Tate Regan (1913a, pp. 112, 139) to the perciform division of the Percoidea. Nevertheless, in certain features it suggests several of the higher teleosts, especially the notothenioids and the batrachoids. The mouth has large well-toothed premaxillæ with stout ascending processes, thin rod-like maxillæ not unlike those of the batrachoids and clinids. In the top view the long straight interorbital bridge and large orbits also rather suggest the batrachoids, and the same is true of the hyomandibulars, which flare somewhat laterally and bear lateral oblique ridges. The small occipital crest, indistinct parietal crests and expansion of the skull behind the orbits are suggestive of such primitive notothenioids as *Cottoperca*. The pectoral girdle (Fig. 237) and actinosts, al-

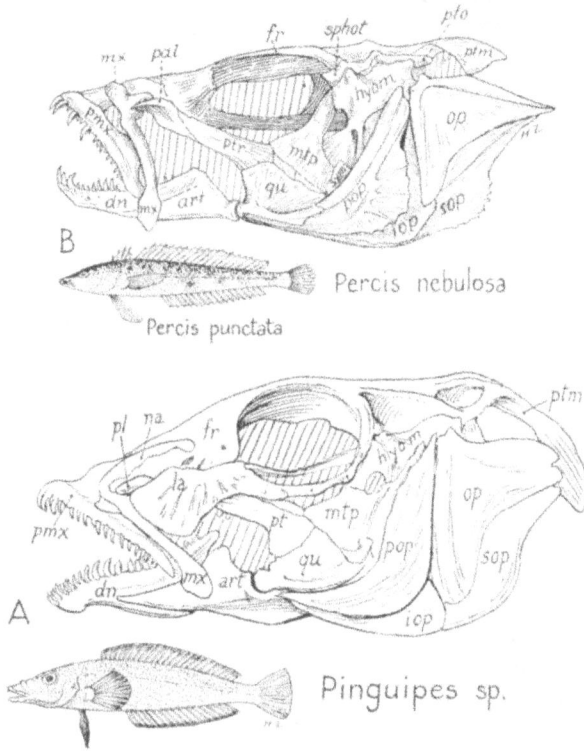

Fig. 236. A. *Pinguipes*. B. *Percis*.

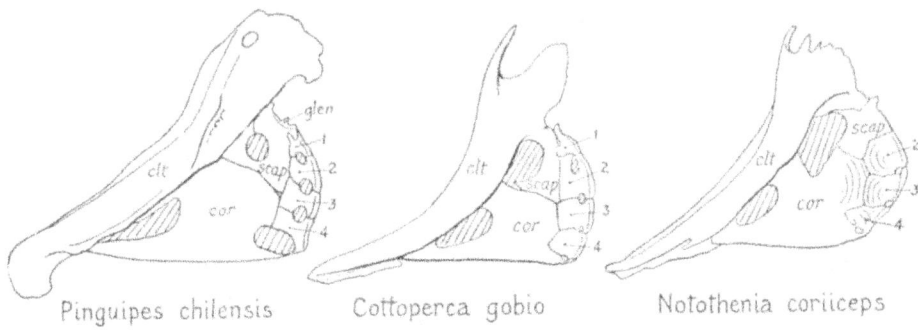

Fig. 237. Pectoral girdles of *Pinguipes, Cottoperca, Notothenia.*

though falling by definition with the percoid series would, I think, make an ideal starting-point for those of *Cottoperca*, chiefly by the enlargement of the scapular foramen and the fusion of the first actinost with the glenoid process of the scapula.

Percis nebulosa.—This skull (No. 277, Brit. Mus. Nat. Hist.) (Fig. 236*B*) agrees in essentials with that of *Pinguipes* but is longer.

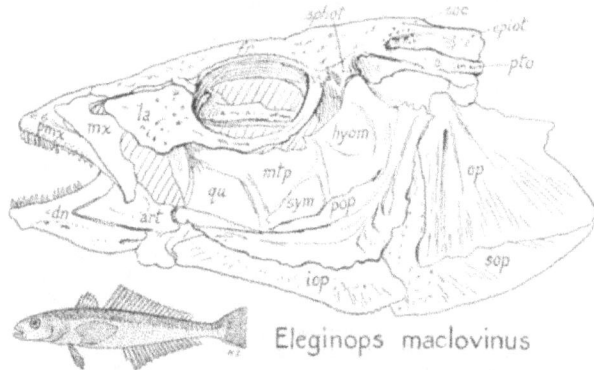

FIG. 238. *Eleginops.*

Eleginops.—This form (Fig. 238) differs from the typical notothenes in its small strong jaws.   The opercular series is remarkably large but at the same time more or less resembling both *Pinguipes* and *Notothenia*.

Bathymaster signatus.—This skull (No. 287, Brit. Mus. Nat. Hist.) looks like a large-mouthed pinguipid, especially in the palatal aspect.   The premaxillæ have normal ascending processes but rather short alveolar borders, exposing the narrow maxilla in the lower half of the gape.   The suborbitals are thin, the lacrymal small and delicate, the nasal fossa large.   The interorbital bridge is narrow, the eyes being large and directed upward; cranial roof smooth, without crests.   The opercular is small and percoid, the subopercular large and forked, as in *Leptoscopus*.   The preopercular is very narrow and reduced, but folded around the lateral line canal.   The pectoral girdle, including the pterygials, is of percoid type, and this fish is referred to the Percoidea by Tate Regan.

Notothenioids.—The most primitive genera of this diversified Antarctic group, according to Tate Regan (1914), are the two genera of the Bovichthyidæ, *Cottoperca* and *Bovichthys*. Regan (1913*a*, p. 112) treats the group as a whole as a division, Nototheniiformes, of the Percomorphi.

Cottoperca and Bovichthys.—Skulls of these interesting forms were examined in the British Museum (Natural History).   In *Cottoperca gobis* (No. 278) the skull (Fig. 239*A*) in essential features strongly resembles the *Pinguipes* type described above (p. 356).   The *Bovichthys* skull (No. 285) is very close to that of *Cottoperca*, with the important exception that the opercular (Fig. 240*C*) has acquired a very large spine and a broad process from its dorsal edge, articulating with both the pterotic and the lateral surface of the stout posttemporal. The lower part of the opercular is reduced to a point which fits into a notch in the subopercular.   The very close resemblance in other features of *Bovichthys* to *Cottoperca* indicates

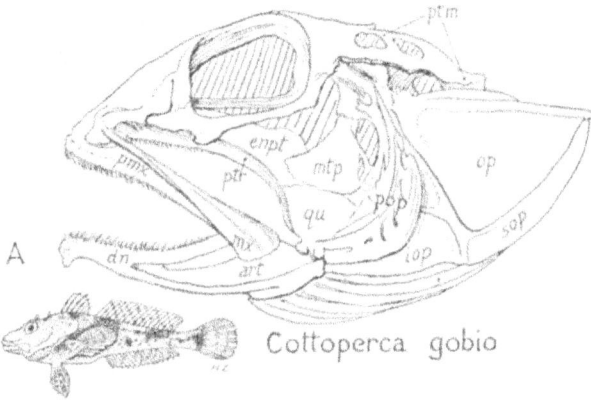

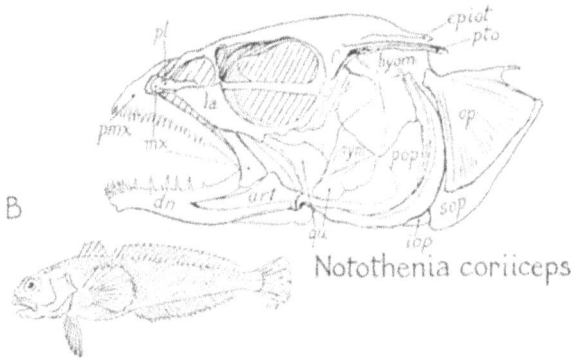

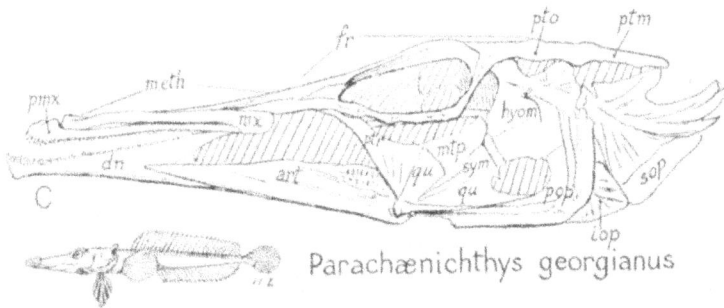

FIG. 239. *Cottoperca* (A), *Notothenia* (B), and *Parachænichthys* (C).

that the highly peculiar opercular of the former is a neomorph, developed within narrow taxonomic limits.

At the other extreme of the notothenioid series stands the deep-sea family of the Chænichthyidæ which have depressed, somewhat duck-like, non-protrusile snouts (Fig. 239C). The palatopterygoid arch is reduced to a thread; the opercular bears several

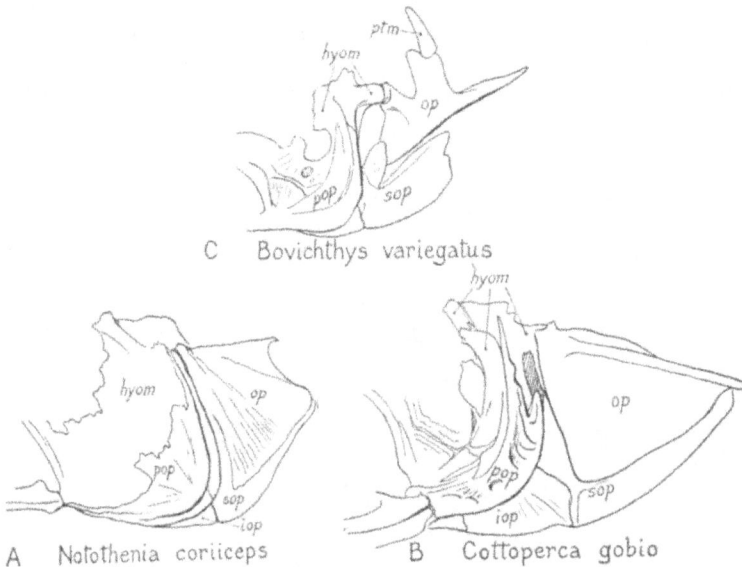

Fig. 240. Opercular region of *Notothenia* (A), *Cottoperca* (B), *Bovichthys* (C).

posterior spines, of which the most dorsal one appears to correspond with the dorsal process of the opercular of *Bovichthys*.

As seen from above, the skulls (Fig. 241) of notothenioids vary from fairly broad to long and narrow, as shown in the following figures:

	Width across pterotics	Length pmx. to mid-occiput	Index of width
Notothenia macrocephala, No. 294, Brit. Mus. Nat. Hist............	29	41	70.7
Cottoperca gobio, No. 278, Brit. Mus. Nat. Hist....................	26	51	50.7
Parachænichthys georgianus, No. 294A, Brit. Mus. Nat. Hist........	42	123	34

In the otoliths of the more primitive Nototheniiformes, according to Frost (1928a, p. 454), the sagitta is of the labrid type resembling that of *Scarus abildgaardii* in general features. This may possibly mean that the notothenioid group came off from the percoid stock near the labrid branch, or to put it another way, that the labrids, notothenioids and trachinoids may have diverged from such a relatively primitive percoid type as the *Pinguipes* or its ally *Parapercis*.

Callionymus lyra.—This is one of the most curiously specialized skulls (No. 496, Brit. Mus. Nat. Hist.) among all the hosts of teleosts. The ascending processes of the premaxillæ are enormous. They are received posteriorly into a deep fossa formed chiefly by the lateral ethmoids; the mesethmoid, affected by this as well as by the dorsal shifting of the huge orbits, has retreated behind the lateral ethmoids.

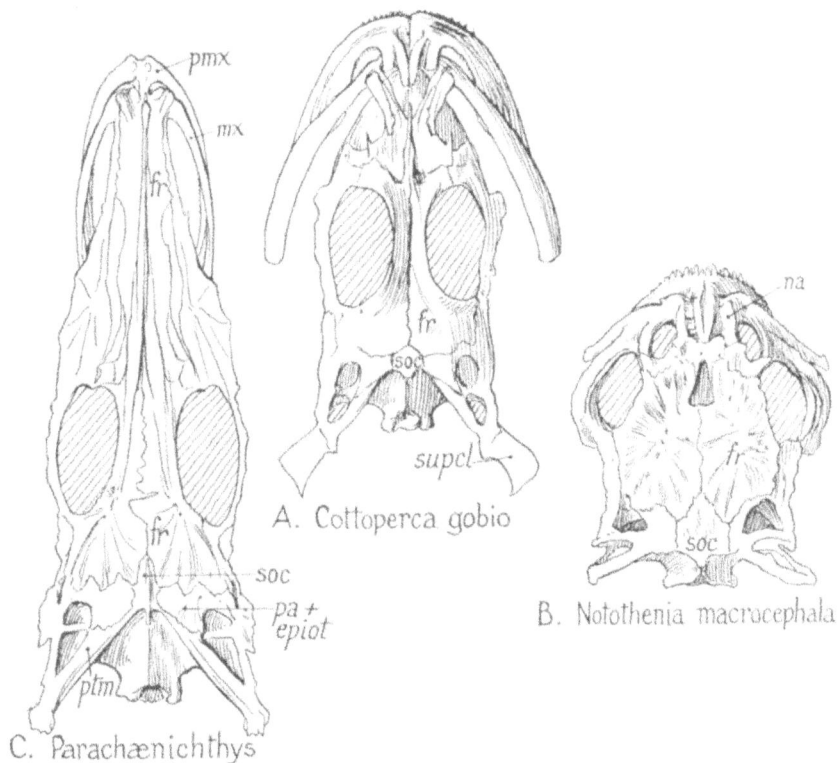

FIG. 241. A. *Cottoperca.* B. *Notothenia.* C. *Parachænichthys.* Top views.

Professor Starks (1923, pp. 267, 268) who drew attention to this unusual position of the mesethmoid, which here simulates an orbitosphenoid, seemed inclined to place a taxonomic value on this peculiar character, but it seems to be a result of the above noted changes in the orbits and rostral fossa.

The opercular apparatus is dragged backward under the influence of the pelvic suckingdisc. The opercular itself is reduced almost to a vestige, its place being largely usurped by the enlarged subopercular. The well-spiked preopercular is dragged backward as a ventral brace for this region.

The otolith (sagitta) of *Callionymus lyra* is described by Frost (1928a, p. 455), as resembling that of *Labrus* except in certain details.

The clue to the relationships of this strange fish has, I believe, been supplied by Professor Starks, who in his recent work (1930) on "The Primary Shoulder Girdle of the Bony Fishes" figures without special reference the strange pectoral girdle and actinosts of *Callionymus* on page 222 facing the figure on page 223 of the corresponding parts in *Notothenia*.

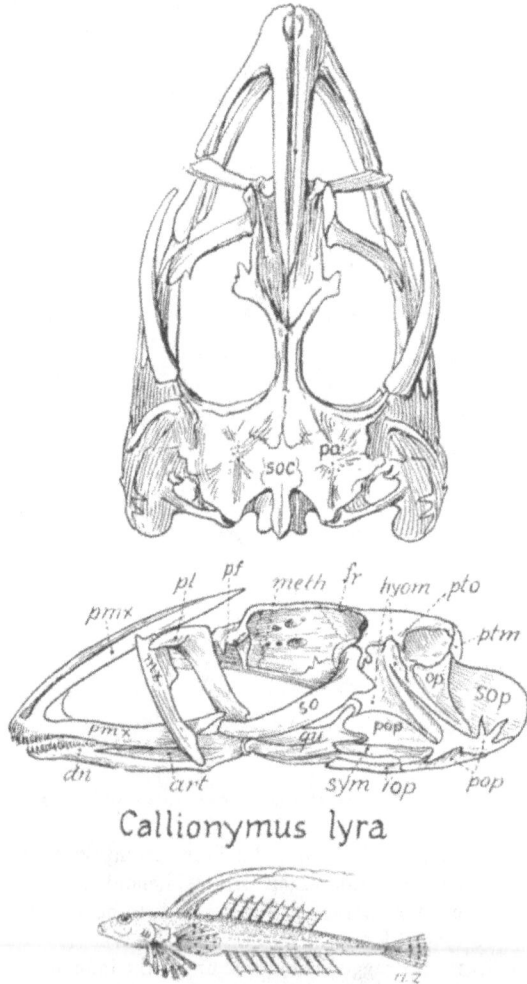

Callionymus lyra

Fig. 242. *Callionymus.*

Nowhere else, so far as known, than in these two families does the scapula form such a bridge between the three expanded actinosts. Evidently in its shoulder-girdle *Callionymus* is merely a highly specialized notothenioid, and a comparison of the skulls seems to me to strengthen this hypothesis.

Trachinus.—The outstanding feature of the skull (Fig. 243) (Nos. 273, 274, Brit. Mus. Nat. Hist.) is the very large spine which projects straight backward from the dorsal part of the opercular. This spine is said to be poisonous. The skull top (Fig. 245) is rounded, with no crests, the surface being covered with a close-set pattern of many tubules and openings, possibly derived by branching of the lateral line canals. There is a row of small denticulate projections above the orbit and three of them on the lateral ethmoid.

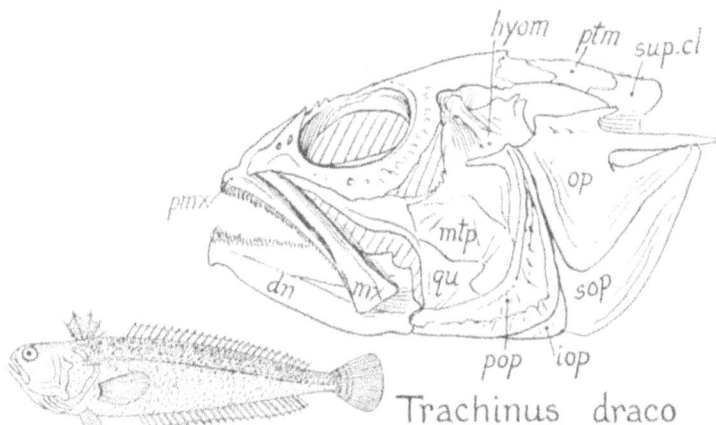

FIG. 243. *Trachinus.*

The general appearance is that of a small percoid which has begun to parallel the scorpænoids in a few features. The hyomandibular shows the beginning of a lateral process starting from the upper end of its preopercular crest, which process is greatly emphasized in the uranoscopoids. The preopercular offers nothing conspicuous; it differs widely from that of *Trichodon* (see p. 370), as does the lacrymal. The mouth is directed somewhat upward. Taken as a whole the skull could well be the starting-point for the specializations of the Uranoscopidæ but has not acquired their diagnostic characters. It is surely widely removed from *Trichodon*. The pectoral girdle, including the actinosts, also differs widely from that of *Trichodon*. On the whole *Trachinus* is nearer to *Uranoscopus* than to *Notothenia*. Its otolith, as described by Frost (1928a, p. 454), could be conceivably derived from the *Parapercis* type and give rise to the *Uranoscopus* type.

Percophis.—A skull of *Percophis brasilianus* (No. 283, Brit. Mus. Nat. Hist.) represents a fairly large-mouthed predaceous type, long and low with a projecting mandibular tip, the mouth opening obliquely upward. The jaws are moderately protractile. Tate Regan (1913, p. 140), who refers *Percophis* to the Percophiidæ, says that the skull is "much as in *Trachinus*, but more depressed, with the exoccipital united behind the supraoccipital, forming a roof for the foramen magnum." The pectoral girdle, including the actinosts, is of modified percoid type, in some features similar to that of *Pinguipes* (Fig. 237).

Dactylagnus.—In this genus of the family Dactyloscopidæ the skull (Fig. 244) as described and figured by Starks (1923) shows an early stage in the line leading toward *Uranoscopus* but also retains traces of more primitive percomorph characters. Thus the last traces of the myodome are retained as well as the basisphenoid (Starks, 1923, p. 282). The skull top lacks the extreme specializations of either *Uranoscopus* or *Astroscopus* and apparently could be ancestral to either. It shares with the former the contact of the sphenotic with the parietal. The supraoccipital widely separates the opposite parietals,

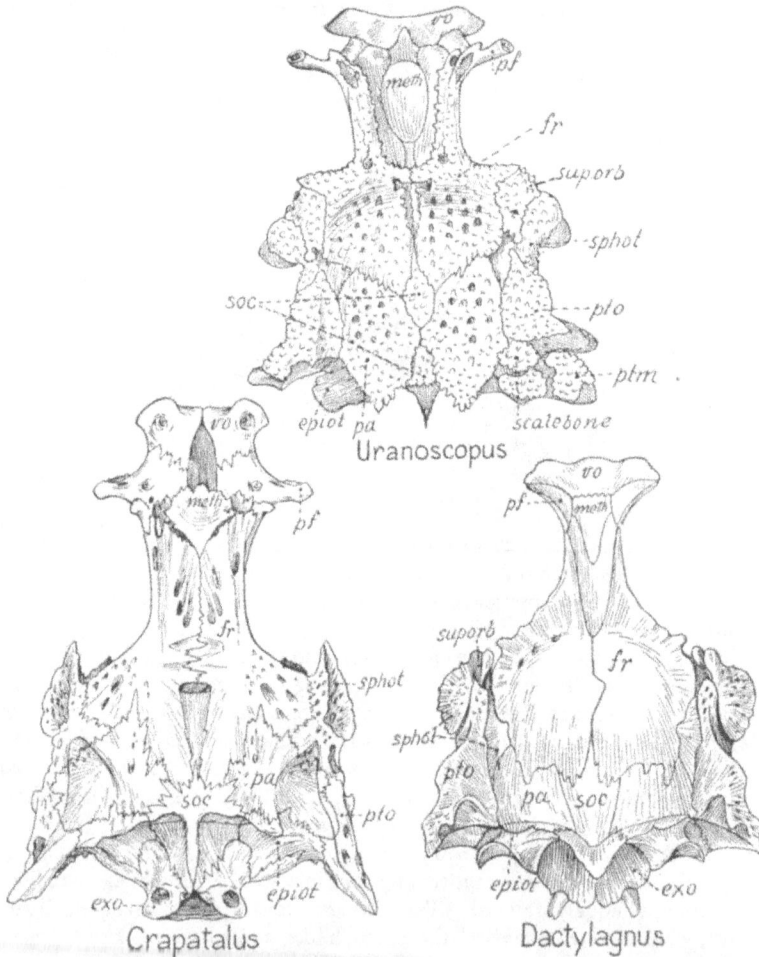

FIG. 244. *Dactylagnus, Crapatalus, Uranoscopus.* Top views. After Starks.

which have barely begun to overlap it. The very large orbits have raised orbital rims, which leave between them a median fossa for the reception of the ascending processes of the premaxillæ. The condition here suggested is that the ancestral uranoscopoid was a fish with broad, depressed head, dorsally-displaced large eyes and upturned, backwardly-pushed mouth. Thus would be initiated the wide overlap of the ascending processes of the premaxillæ above the mesethmoid, the widening of the vomer and the beginning of the median supra-ethmoid fossa.

Crapatalus.—In *Crapatalus* of the Leptoscopidæ, as figured and described by Starks (1923, p. 286 and Pl. III), a further step toward *Uranoscopus* is likewise shown in some features; but this skull (Fig. 244*A*) is much more primitive than that of *Uranoscopus* (*cf*. Fig. 246) in such characters as the following: the broadening of the braincase is more moderate, the eyes do not so greatly constrict the interorbital skull-roof and there is little or no median frontal-ethmoid fossa. Also the ectethmoids, while projecting widely laterally, are not nearly so much reduced and crowded as in the more advanced genera. Evidently also the nasal chamber was less reduced. The back part of the skull is not so much crowded anteroposteriorly; the pterotics project backward in a more normal percomorph way; the sphenotics project forward; the supraorbitals are not interposed between the frontals and the sphenotics; the large supraoccipital extends forward under the frontals and widely separates the broadened parietals.

There are also marked reductions, losses and specializations in this skull. The bones in the undried skeleton Starks tells us (1923, p. 280) were very much thickened with cartilage but in drying became thin and paper-like. The alisphenoid, basisphenoid and the myodome are all absent. The hyomandibular is longer longitudinally than vertically. The symplectic is extraordinarily large. Thus, as Starks notes (1923, p. 265), this fish is not closely related either to *Dactylagnus* or to *Uranoscopus*. Nevertheless in his recent paper on the shoulder-girdle of the teleost fishes Starks (1930, p. 226) notes that "In some respects the shoulder girdle of this form [*Crapatalus*] is intermediate between *Uranoscopus* and *Dactylagnus*." This is an important piece of evidence for a common origin of the three families.

Leptoscopus.—A skull of *Leptoscopus macropygus* (No. 267, Brit. Mus. Nat. Hist.) gives a strong impression of being closely related to *Uranoscopus*. The skull roof, however, is not roughened and pitted (perhaps it has sunk too far beneath the skin), the preopercular as well as the opercular is much less extended ventrally. The process on the lateral surface of the pterotic division of the hyomandibular turns forward towards the postorbital process but does not reach it. The dentary has fine teeth instead of large ones. The depression of the articular angular beneath the level of the dentary is pronounced.

Kathetostoma.—This skull (Fig. 245) is closely related to that of *Uranoscopus* but is even more specialized in its extreme width.

Uranoscopus.—In this skull (Fig. 246) the eyes are displaced forward and directed partly upward, but as there is a complete absence of the expanded postorbital fenestræ of *Astroscopus* (Fig. 247), we may safely assume that those eye-muscles which give rise to the electric organ of that genus had not yet in *Uranoscopus* become greatly hypertrophied or otherwise modified. Nevertheless the general configuration of the entire facial region of *Uranoscopus* and *Kathetostoma* is such that the very extraordinary specializations of *Astroscopus* might readily be derived therefrom. In other words, as in so many other cases, the "habitus" of the ancestor has become the "heritage" of its descendant, which has gone on to acquire a still later habitus.

Similarly the interocular channel for the reception of the ascending processes of the premaxillæ is far less specialized in *Uranoscopus* than in *Astroscopus* since it is widely open and not constricted, there being no electric organs to encroach upon it.

This broadly U-shaped fossa in *Uranoscopus* and *Kathetostoma* accommodates the broadly oval articulating processes and the long but delicate ascending processes of the premaxillæ. The unusual width across the vomer, which lies immediately beneath these broad articular processes of the premaxillæ, is apparently conditioned not so much by the

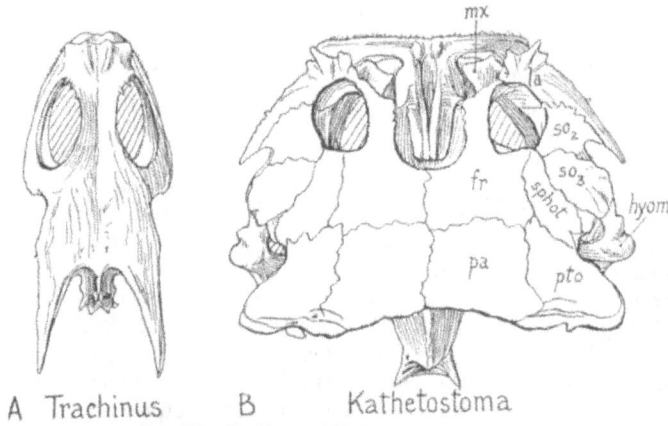

Fig. 245. *Trachinus* and *Kathetostoma*. Top views.

breadth of the latter as by the general breadth of the mouth cavity, of which the vomer forms the roof. Also the lower part of the hyo-branchial complex is unusually broad and this broad part occludes against the widened vomer. Finally the opposite lacrymals and prefrontals spread wide transversely, further broadening the roof of the mouth. The mesethmoid is small, partly because it is squeezed down between the ascending and

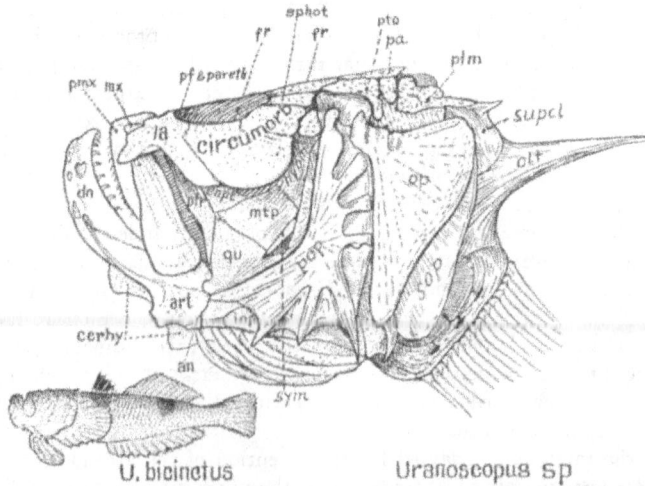

Fig. 246. *Uranoscopus*.

articular processes of the premaxillæ above and the widened vomer below. Thus the consideration of these adaptive or growth changes tends to make intelligible the details of the ethmoid region, which appear so dry and meaningless when considered as detached from their surrounding parts. The derivation of this mesethmoid-prefrontal fossa itself becomes intelligible if we compare the skull of *Uranoscopus* with that of *Ophiocephalus*. In the latter fish this fossa is seen in its initial phase as a gentle concavity between the anterior forks of the frontals, its function being to accommodate the ascending processes of the premaxillæ.

The skull roof of *Uranoscopus scaber* has been carefully studied by Starks (1923), who shows that its broad parietals are beginning to overlap the supraoccipital. In my specimen of *Uranoscopus* sp. from Japan this process has gone much further, so that the parietals meet in the mid-line above and appear to separate the supraoccipital widely from the frontals (see also Fig. 245B). A better case of an apparent but not real exception to Dollo's "Law of the Irreversibility of Evolution" could hardly be wished for, since the parietals have not regained their original contact along the mid-line but only a secondary one.

As a whole, the skull of *Uranoscopus* closely resembles that of *Astroscopus* as described below, except that it is somewhat less specialized and less simplified secondarily. Thus the bones of the posterior part of the skull-roof maintain their sutural contacts, which may be seen plainly on the sculptured upper surface, while in *Astroscopus* sutural lines are at best but vaguely suggested. Among additional striking differences from *Astroscopus* are the following:

The posttemporal is a very short thick bone with a sculptured dorsal plate. It lacks the long slender inner fork seen in *Astroscopus*. The supracleithrum bears a long sharp spike directed upward and backward. The preopercular is a broad curved plate with a high and flat, vertically concave anterior border, which bears in all eight radially diverging processes, not extending beyond the web-like outer border but standing in high relief above it; the lowermost three processes end in sharp, horny, downwardly-directed thorns. Obscure traces of this complex armature may be detected in the somewhat degenerate preopercular of *Astroscopus* (Fig. 247).

A similar downwardly-directed thorn is borne on the lower end of the subopercular. This spiniferous tendency is also expressed in the presence of pairs of thorns elsewhere, i.e. on the pelvic bones just in front of the pelvic fins, on the postero-superior angles of the small supracleithra, on the antero-inferior projection of the lacrymal.

Several features in the skull of *Uranoscopus* are rather suggestive of relationship with the scorpænoid fishes; the third suborbital is produced backward and downward (although it fails to reach the preopercular border); the relations of the radiating ridges and spikes on the preopercular recall the conditions in the scorpænoids, and so does the presence of spikes on the supracleithra, the peculiar beaded surface of the skull, the radiating trabeculæ and zonal growth lines. The pectoral pterygials are more specialized than those of the more primitive scorpænoids but do not appear to exclude derivation from the scorpænoid stem. Nevertheless the derivation of the uranoscopoid series from relatives of *Trachinus* and *Notothenia* appears far more probable.

Astroscopus.—An advanced stage of specialization of the family Uranoscopidæ is represented by the skull of *Astroscopus y-græcum*. In this very strange fish (Figs. 247, 248) the eyes are shifted to the dorsal surface; parts of the eye-muscles are greatly hypertrophied and modified into powerful electric organs (White, 1918).   These newly-evolved organs· are evidently responsible for the presence of the very large postorbital vacuities with which the orbits are confluent.   Their presence has also caused the extreme constriction of the frontals and the approximation of the side walls of the deep groove above the ethmoid, the fore part of which receives the ascending processes of the premaxillæ.   The electric organs evidently press posteriorly against the braincase, which is excessively short and wide.

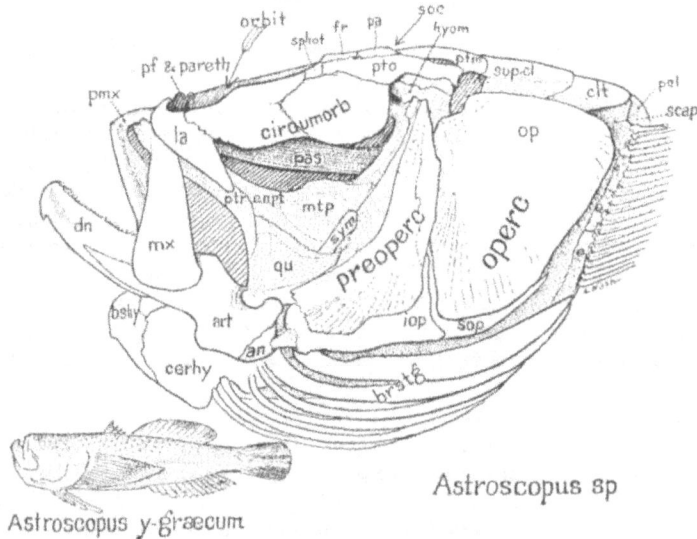

Fig. 247.   *Astroscopus.*

The front walls of the braincase are formed from ascending flanges of the parasphenoid as they meet descending flanges from the frontals.   With the dorsal shifting of the eye-muscles and their derivatives, the primitive eye-muscle canal beneath the floor of the brain-case has been abandoned and closed up, so that the systematists report "no myodome" as characteristic of this superfamily.   The lack of a myodome in turn probably conditions the fact that the parasphenoid lacks the normal posterior extension beneath the braincase.

The electric organs extend so far posteriorly that only an antero-posteriorly short and transversely very wide and flat cranial roof is left, consisting of conjoined frontals, followed by small flat parietals which seem to have joined secondarily above the small supra-occipital.   The occipital border of the skull roof is formed by a transverse row of small "neuromastic" or lateral line bones, apparently representing respectively the "scale bones" (extrascapulars), the posttemporals and the supracleithra.   The suborbital plates, along with the dorsally displaced eyes and related parts, have likewise moved upward to form the outer border of the upward-looking face.   Together they form a stiff armor for the cheek and one of them (the third) is supported by a special process of the hyomandibular, just in front of the top of the preopercular.   This third suborbital is excluded from the border of the orbit and, being also braced posteriorly by the preopercular (through the special process mentioned above), the arrangement is suggestive of certain of the mail-cheeked fishes.

The premaxillæ have long ascending processes, which, as already stated, are received by a diverging fossa above the mesethmoid; the upper jaw must therefore be to some extent protrusile. The opposite ascending processes of the premaxillæ are closely appressed and united by connective tissue. The articular processes of the premaxillæ are expanded ovals, much as in *Uranoscopus*. The maxillæ have the proximal fork and rounded articular

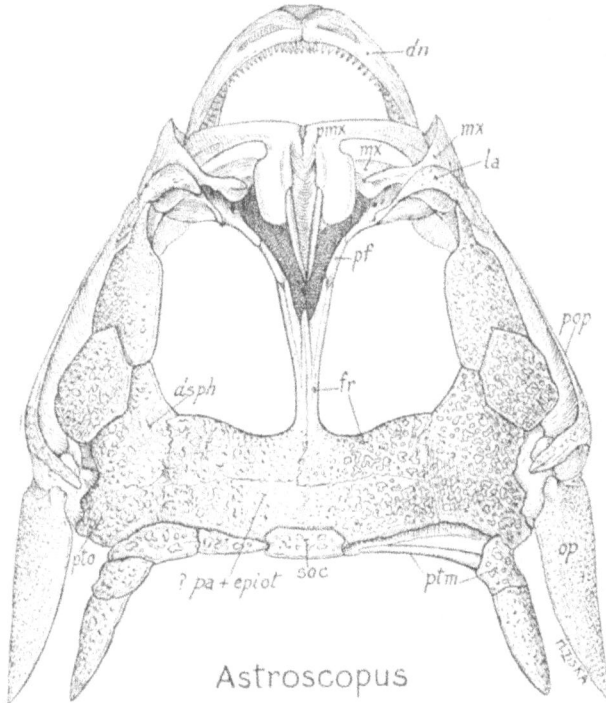

Fig. 248. *Astroscopus*. Top view.

processes, as in other derivatives of normal percoids. The olfactory chambers are very small, squeezed in between the articular processes of the maxillæ and the forwardly-displaced flanges of the lateral ethmoids. The constriction of the olfactory chamber is primarily due to the small size of the enclosed sensory capsule but the extreme forward displacement of the small orbits has probably also contributed to it. The dorsal rim of the mouth has been raised until it is level with the cranial table, while the premaxillæ have, as it were, been pushed back into the face. At the same time the quadrate-articular joint has been drawn forward so as to be only a little way behind the strongly upturned mouth. As a result of all this the mandible is inclined sharply upward, the palatopterygoid arcade is steeply inclined.

The head as a whole is very wide. Hence the opposite pterygo-metapterygoid tracts are removed far from the mid-line and there are no mesopterygoids. The hyomandibulars are very thick and strongly built, forming the main lateral braces of the head and being supported in turn by wide transverse extensions of the braincase, formed distally at least chiefly by the stout sphenotics and pterotics. The broad flat occiput evidently afforded insertion for thick axial muscles. It is braced posteriorly by coalescence with the expanded neural arch of the first vertebra. The basioccipital condyle is tightly appressed to the first centrum, while the exoccipitals meet the anterior zygapophyses of the first vertebra in dentate sutures, which have replaced the lateral occipital condyles.

The much expanded pectoral girdle is tied to the rigidly braced occiput by the chain of small bones already mentioned and by a long underlying rod, which may be either an ossified tendon or the medial process of the posttemporal or both. In this feature *Astroscopus* is closer to *Crapatulus* of the Leptoscopidæ than to *Uranoscopus* (*cf.* Starks, 1923, p. 289). At its antero-inferior end the pectoral girdle is tied by ligament to the greatly expanded lower end of the hyoid arch, namely, to the enlarged ceratohyal and basihyal.

The opercular series has remained fairly normal in external appearance. The rigid preopercular braces the quadrate posteriorly. The interopercular shares the enlargement of its associates, the cerato- and epihyal; it is fastened as usual by ligament to the small angular bone of the mandible. The dermarticular is suddenly expanded downward, below the quadrate-articular joint; it probably afforded insertion just in front of this expansion to the transverse muscles of the mandible. The most remarkable feature of the entire opercular apparatus is the reduction of its mobility. In the first place, the normal laterally-expansive movement of the hyomandibular, which carries the opercular plates, appears to be definitely stopped by a rigid transverse bar of bone connecting the hyomandibular with the proötic and formed by both elements. Secondly, the hyomandibular articulation with the sphenotic and pterotic is extended transversely as well as antero-posteriorly and would appear to permit little if any motion. Thirdly, the opercular process of the hyomandibular is directed downward and outward, while its cartilaginous tip extends inward and is received into a deep funnel on the inner side of the opercular in such a way as to allow at most only a small outward and inward swing of the posterior part of the opercular. Again, the opercular has two contacts with the preopercular: one, a vertically extended, bevelled facet just above the middle of the anterior edge of the opercular; the second, above and in front of the first, comprising a short, forwardly-directed process with a concave inferior contact just behind the prominent vertical ridge of the preopercular. This second contact is shown on the right side only in the single specimen studied. The left opercular lacks this process and may have been movable on its hyomandibular pedicle. Along with all these indications of greatly restricted mobility a small but distinct fossa for the dilatator operculi muscle remains beneath the dermal roof of the sphenotic and pterotic and even in the dried skull it contains a number of tendinous bands which are apparently remnants of this muscle; so it would not be surprising if in life the cartilaginous covering of the main joint between the pedicle of the hyomandibular and the opercular permitted a minimal movement of the opercular even in this exceptionally rigid skull. One may suspect that the inhalation of water into the branchial chamber is effected chiefly by rhythmic movements of the hyoid and branchial arches, since the enlarged hyoid arch retains its movable articulation with the hyomandibular.

Trichodon.—This is an isolated and puzzling form. The small skull (No. 285, Brit. Mus. Nat. Hist.) shows but few definite indications of its relationships. In the dried condition it is shriveled through the loss of the cartilage. Starks (1926a, p. 298) notes that the bones of the ethmoid region are very thin surface ossifications, filled with cartilage and overlying masses of cartilage. "There appears to be no endosteal bone in the ethmoid region whatever. . . . A dried cranium becomes very much distorted." The flattish skull roof is crowded with irregular thin-walled cells, and is without crests. The fairly large orbits are directed laterally. The skull top is blunt anteriorly and widens posteriorly toward the attached posttemporal horns. The ethmovomer block forms a small prominence in the top view, but in the side view is very shallow, due to the reduction of the mesethmoid which, according to Starks (1926a) is a thin simple plate lying below the frontals between the upper end of the projecting prefrontals. This reduction may well be correlated with the marked reduction of the ascending processes of the premaxillæ. The mandibular teeth are small. The lower anterior border of the lacrymal bears a small downwardly directed spike followed after an interval by a much larger one. The opercular is small and without spikes. There is no "suborbital stay" connected with the preopercular. The broadly crescentic, large preopercular bears six radially directed points on its raised outer surface, which are faintly suggestive of the eight radially diverging processes on the preopercular of *Uranoscopus*, but the opercular region differs widely from that of *Trachinus*.

Tate Regan (1913a, p. 136) puts the Trichodontidæ as "Division 7, Trichodontiformes" of the suborder Percoidea, "differing from the Perciformes in the pectoral fin skeleton." Jordan (1923, p. 203) holds that the trachinoid fishes follow the Trichodontidæ in natural sequence, "a fact not to be shown in a linear series." Starks (1923, p. 265), in considering the relationships of the uranoscopoid fishes, wrote that *Trichodon* (as compared with *Parapercis* and *Bathymaster*) "with its much reduced mesethmoid, its sphenotic extending inwards to the parietal and separating the frontal from the pterotic, its widely separated ectethmoids and its large opisthotic, shows a somewhat closer relationship with the Uranoscopoids, but not nearly close enough to be admitted into the group." But in his recent work on "The Primary Shoulder Girdle of the Bony Fishes" (1930, p. 75) he notes that "the shoulder girdle [of *Trichodon*] is strikingly like that of some cottoid fishes," which, as he shows, are widely different from those of the uranoscopoid and notothenioid groups. Consequently it is at present doubtful whether the skull characters which *Trichodon* shares with the uranoscopids are enough to indicate a real relationship.

The otolith of *Trichodon trichodon* is described by Frost (1928a, p. 455) as resembling in shape that of *Iniistius*, an aberrant member of the division Labriformes of Tate Regan's classification of the Percoidea. It differs widely from those of the Trachiniformes.

XENOPTERYGII (CLING–FISHES)

Some very unusual specializations from the percomorph ground-plan are embodied in these little fishes, which have the pelvic fins modified into a sucking-disc but are certainly not related to the gobies and other forms with similar pelvic sucking-discs.

The skull (Figs. 249, 250) is extraordinarily specialized in many directions. The short stoutly-built jaws bear forwardly inclined incisors resembling somewhat those of man and evidently adapted for nipping. Replacing teeth lie in the alveoli. The short dentary forks over the massive articular, which bears a transverse hinge-like joint with the quadrate. The ascending process of the articular is vertical and very stout, indicating powerful adductor muscles. These were braced by a strong backwardly directed process of the

20

quadrate, by the broadened hyomandibular and backwardly produced preopercular. Starks (1905, p. 295) shows that in *Caularchus*, a member of this family, the pterygoid is reduced to a vestige adhering to the front edge of the quadrate, while the meso- and meta-pterygoids are absent. The palatine is a narrow rod, connected with the vestigial pterygoid

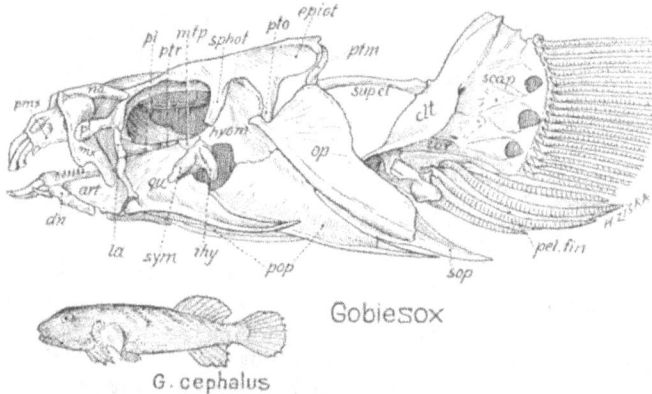

Fig. 249. *Gobiesox*. Side View.

by a ligament but hooking over the maxilla in the normal way in front. The only bony contact between the hyomandibular and the quadrate is by means of the small but strong symplectic which is received into a notch in the middle of the quadrate, whereas in most fishes this notch is in the back part of the quadrate. A prominent fenestra lies between the preopercular and the quadrate, recalling the conditions in the goby *Eleotris*. Possibly the presence of the great pelvic sucking-disc may have somehow caused the strange backward growth of the opercular and subopercular and of the parts attached to them, which together form a sharp lower rim projecting backward on either side of the disc. They may serve for the origin of the disc muscles. The skull is broad and low, with long lateral processes from the prefrontal (lateral ethmoid) and sphenotic. In *Caularchus* the vomerine region is broadly notched in front (doubtless to receive the enlarged premaxillæ), and the flat, long ascending rami of the premaxillæ lie in a broad depression on the upper surface of the rostrum and between the orbits. These features are more emphasized in *Caularchus* than in *Gobiesox* (Starks, 1905, pp. 292, 293). The parasphenoid has broad lateral wings in front. Three occipital condyles are present, one on the basioccipital, the other two on the exoccipitals, so that the three parts of the condyle are in a horizontal line (Starks).

Many other strange cranial characters are recorded by Starks, who also made a painstaking comparison of skeletons of cottoid, blennioid and gobioid fishes in the effort to determine the affinities of the Gobiesocidæ, but with "small results." The families Batrachididæ and Callionymidæ offer some slight indications of relationship to the Gobiesocidæ, and the weight of evidence is thrown toward the former family by the young of some or all of them having a ventral sucking-disc just behind the base of the pectorals. The family Batrachididæ further resembles the Gobiesocidæ in having the suborbital ring reduced to a small preorbital bone, only very small parapophyses present posteriorly, no myodome, and a single superior pharyngeal present on each side. As opposing the idea of relationship, the Batrachididæ have five long actinosts, the posttemporal forms an integral part of the cranium, the palatine is normally joined to the pterygoid, and the mesopterygoid, metapterygoid, alisphenoid, and basibranchials are present.

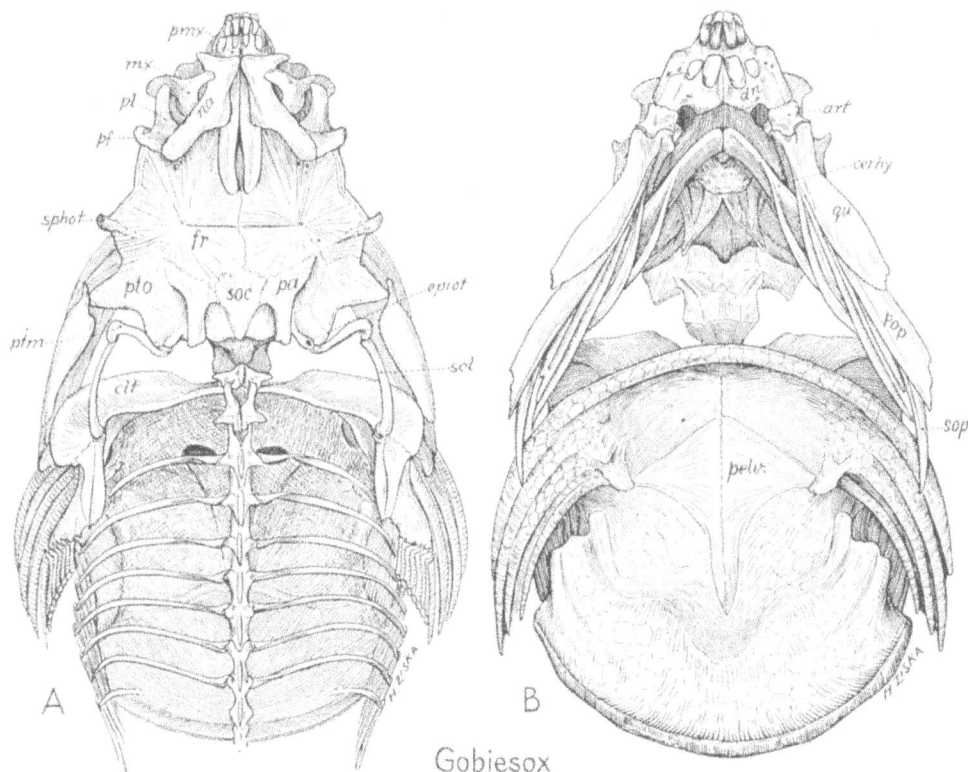

FIG. 250. *Gobiesox.* A. Top view. B. Underside view.

As to the possible affinities with the Callionymidæ, (Fig. 242), Starks writes as follows: "The family Callionymidæ resembles the Gobiesocidæ in having no mesopterygoid or metapterygoid, thus leaving the symplectic to form part of the anterior border of the cheek bones, in having no myodome or suborbitals, in the ventrals being widely separated, as well as in the general form of the body. The Callionymidæ, however, possess some important and well marked characters not possessed by the Gobiesocidæ, and these probably more than counterbalance the characters held in common. These characters are briefly: a spinous dorsal present; the ethmoid extending back and forming a bony interocular septum; the frontals reduced and occupying little more than the interorbital space; the posttemporal forming an integral part of the cranium; the actinosts all abutting against the hypocoracoid; the hypercoracoid foramen between the coracoid elements cutting an equal notch from each;

the palatoquadrate arch normal; three superior pharyngeals present on each side; basi-branchials present; the neuropophyses and hæmopophyses ending each in two spines between which the interspinous elements fit."

Tate Regan (1929, p. 326) agrees with Gill and Starks that there is nothing to do but to put this strange fish in an order by itself (Xenopterygii) allied to the Percomorphi.

Apparently not much aid in solving the problem is given by the otoliths. According to Frost (1930a, p. 623), the sagitta of *Lepidogaster gouanii* of the family Gobiesocidæ is of the percid type; it resembles that of *Gerres rhombeus* of the suborder Percoidea in some features and that of *Smaris australis* (also of the Percoidea) in others, but its genetic relations with either of these forms can hardly be very close.

BLENNIOIDEI (BLENNIES, BROTULIDS, ETC.)

Notwithstanding the advanced position of the blennies, the skull is relatively little modified from the percomorph type. The premaxillæ of the most generalized form (*Clinus*, No. 517, Brit. Mus. Nat. Hist.) are in essentials of pure percoid type (Fig. 251). The

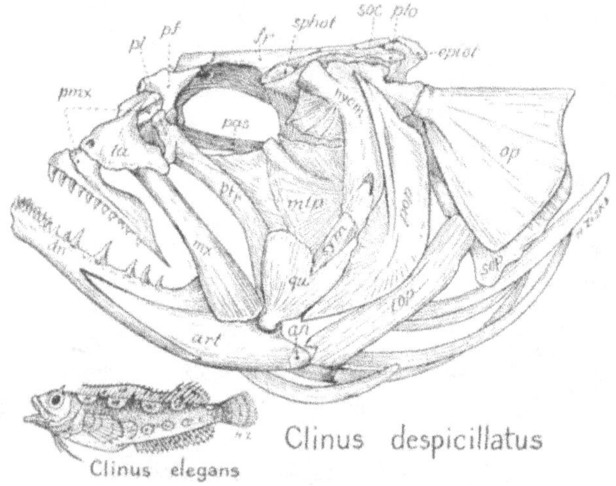

FIG. 251. *Clinus.*

maxilla has a lump-like process for articulation with the vomer and a small proximal fork around the articular process of the premaxillæ. In *Blennius*, however, the skull (Fig. 252) is much more specialized, due to the development of a battery of nipping teeth which collectively are curiously suggestive of those of the sauropod dinosaur *Diplodocus*. In order to operate this nipping dentition the jaw muscles are evidently stout, as indicated by the strong sagittal and lambdoid crests. The large strongly-rimmed orbits project upward. Thus with regard to adaptive skull characters there is a greater difference between *Blennius* and *Clinus* than there is between the latter and *Pinguipes*, which is classed by Regan as a percomorph. The osteology of the blennies has been reviewed by Tage Regan (1912e).

Even in the highly specialized *Zoarces anguillaris* (Fig. 253) the skull retains most of the familiar landmarks. Apparently this is a predaceous, eel-like form with narrow skull and fairly large biting teeth in the front of the jaws. The hyomandibular is much enlarged and forms a firm pivot for the jaws. The preopercular is closely appressed to the hyoman-

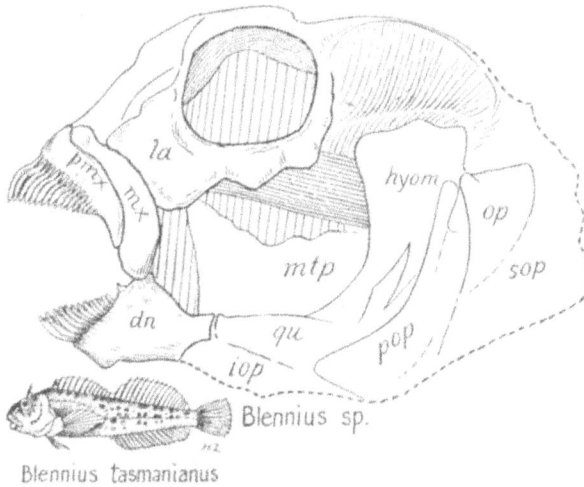

Fig. 252. *Blennius.*

dibular and the back of the quadrate. The upper part of the small triangular opercular is sharply truncated, perhaps to make room for muscles running obliquely above it.

In the top view of the skull (Fig. 254) we note the relative length and narrowness of the braincase, the large orbits and much constricted interorbital bridge, the relatively heavy muzzle formed by the stout prefrontals (parethmoids) and elongate mesethmoid,

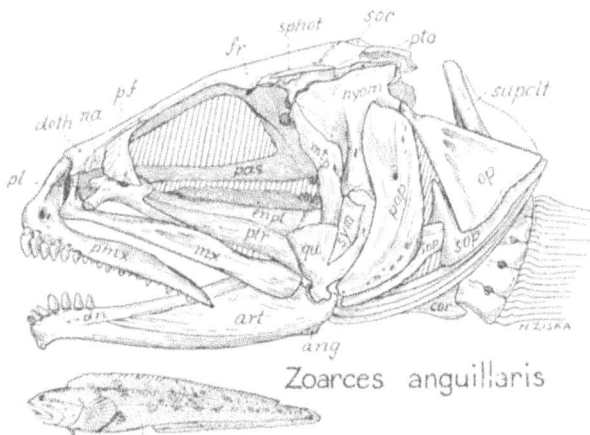

Fig. 253. *Zoarces.*

which widens in front, bearing a socket for the relatively strong ascending processes of the premaxillæ. The more technical characters of this skull, as given by Tate Regan (1912e, p. 275 (Blennioids)) tend to ally it with the Blenniidæ.

The skull (Fig. 255) of the wolf-eel (*Anarhichas*) likewise shows relationships to those of *Blennius* and *Zoarces* but is modified for the support of the massive caniniform front

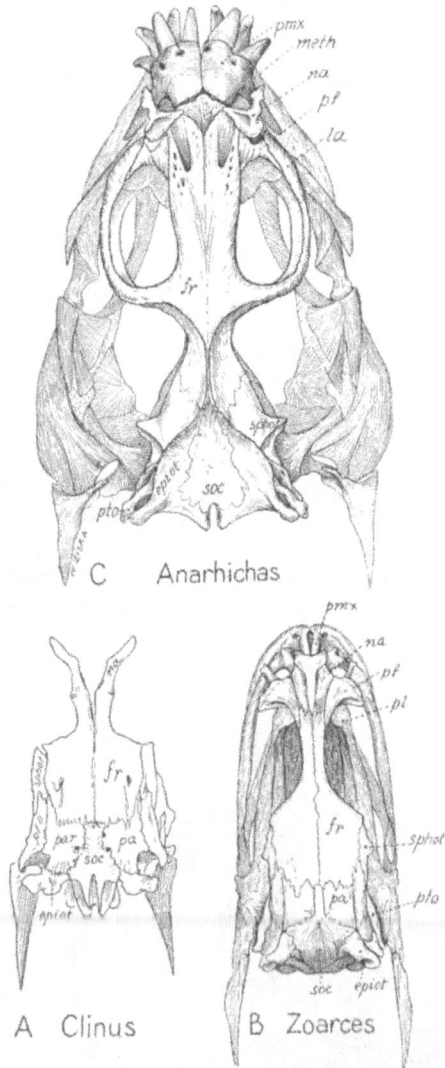

Fig. 254. A. *Clinus*. B. *Zoarces*. C. *Anarhichas*. Top views.

teeth and broad rounded molars, which enable the fish to crush sea-urchins and sand-dollars (Adams, 1908, p. 332). The braincase in the top view (Fig. 254C) is narrow, in part at least due to the powerful development of the jaw muscles. The orbits are far forward and widely removed from the sphenotics. The ethmoid and prefrontals are necessarily massive to withstand the thrust of the heavy premaxillæ. The ascending processes of the latter are extraordinarily massive; they abut against the deepened ethmoid. The preopercular

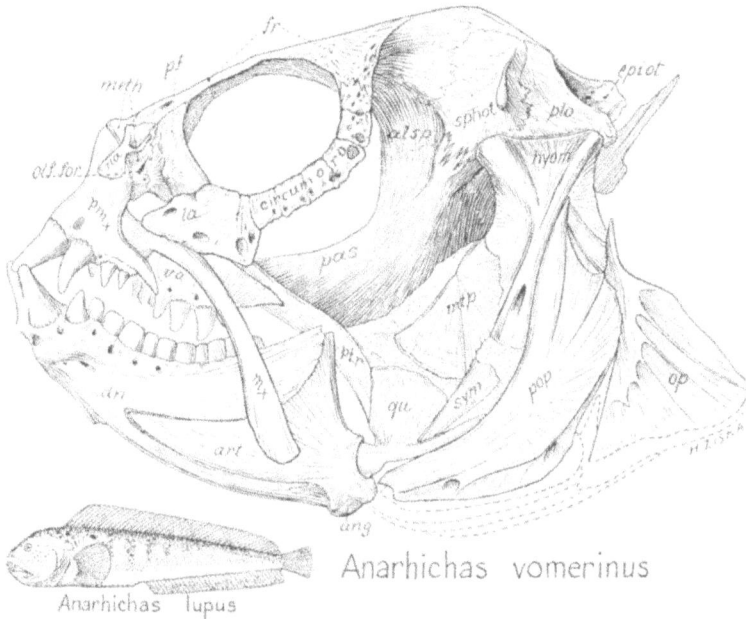

FIG. 255. *Anarhichas.*

appears as a heavy vertical bar, which reinforces the hyomandibular and projects downward behind the huge mandible.

According to Tate Regan (1912e, p. 277) the brotulids (Fig. 256) and ophidiids belong in the blennioid group. The skull of a small ophidiid suggests that of *Zoarces* but is in some features less specialized, e.g. a small supramaxilla is retained and the teeth are less specialized.

The remarkable genus *Fierasfer* (Fig. 257) is an eel-like fish that lives in sea-cucumbers. According to Tate Regan (1910a, p. 15) the skeleton agrees in many important features with that of *Brotula* of the ophidioid group. Skulls of two species of *Fierasfer* are carefully figured in the monograph by Emery (1880, Tav. III) along with comparative figures of *Ophidium, Pteridium, Motella, Gobius*, etc. Emery also referred *Fierasfer* to the Ophidiidæ. Boulenger (1910, p. 622) placed it with the Heteromi, but its relationship with the ophidioids seems well established.

According to Mr. G. Allan Frost (1929a, p. 123) the very small otolith of *Labrisomus*

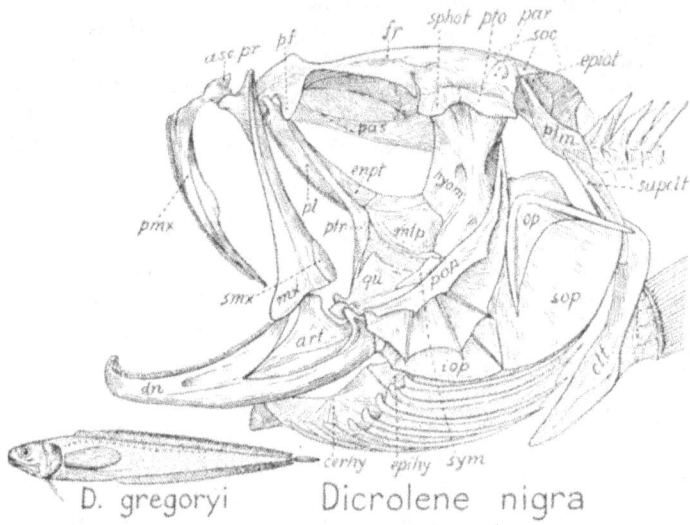

FIG. 256. *Dicrolene.*

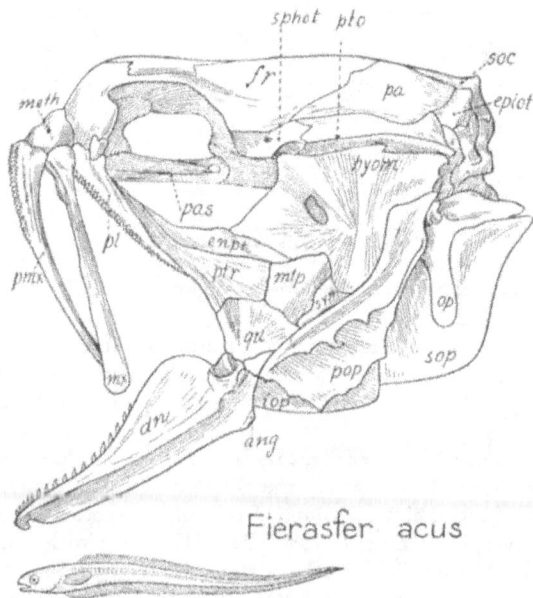

FIG. 257. *Fierasfer.* After Emery.

nuchipinnis of the family Clinidæ is of the Labrid type, and resembles very closely that of *Cheilinus fasciatus* of the Labridæ, except in minor details. In the Ophidiiform division of the Blennioidea, however, the otoliths are very large relative to the size of the fish and give rise to the so-called "Brotulid" type, which is frequent among fossil forms occurring in Tertiary formations (p. 125). The otoliths confirm the placing of *Fierasfer* in the division Ophidiiformes. Fossil otoliths of *Fierasfer* are frequent in the Upper Eocene (p. 126).

ANACANTHINI (CODS, ETC.)

According to Garman (1899), the families Zoarcidæ, Ophidiidæ and Brotulidæ are closely related to the Gadidæ and Macruridæ and belong in the order Anacanthini. Cockerell (1916) finds that the scales of the Brotulidæ resemble those of the Ophidiidæ and certain genera of the Gadidæ, while Miss E. S. Trotter (1926) records several interesting physio-

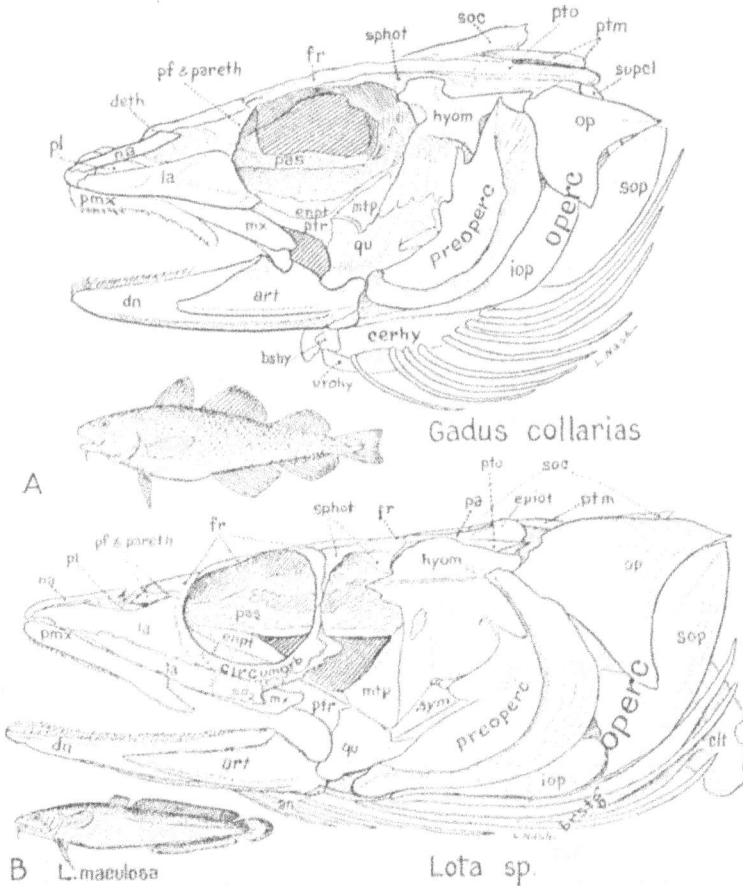

FIG. 258. A. *Gadus.* B. *Lota.*

logical resemblances between brotulids and macrurids. The skull of a small ophidiid is strongly suggestive of the gadoid type. It is rather hard to believe that all this is mere parallelism. Nevertheless, according to Tate Regan (1903, p. 460) and Boulenger (1910, pp. 646, 702, 703), the codfishes and their allies have not been derived from degraded blennies allied to the Ophidiidæ, and the striking resemblances between certain members of the two groups is probably to be ascribed to convergence, brought about in terminal lines of the descendants of some very early group.

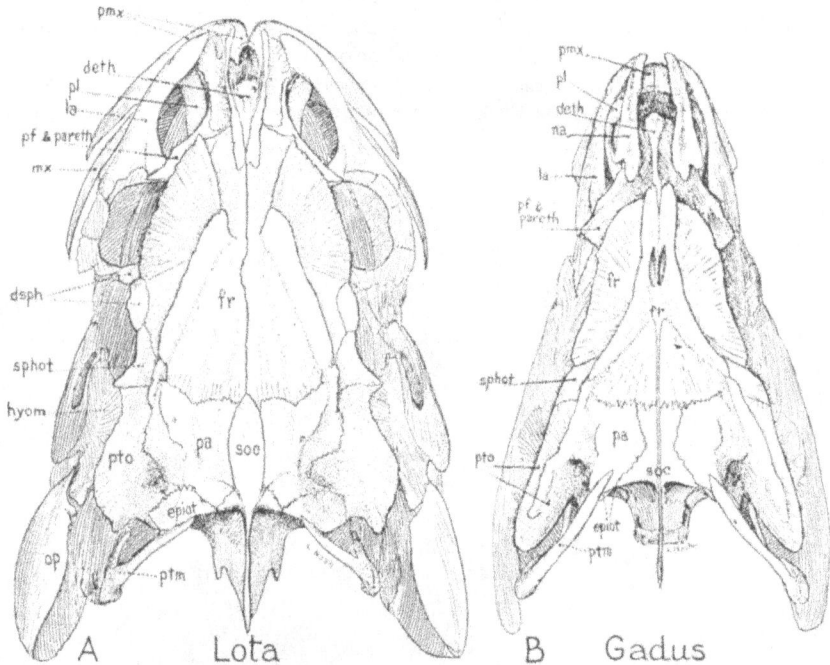

FIG. 259. A. *Lota*. B. *Gadus*. Top views.

Regan considers (1910a, p. 11) that "the absence of spinous fin-rays, the large number of rays in the pelvic fins and the indirect attachment of the pelvic bones to the clavicles are evidences that the Anacanthini are much more generalized than the ophidioids, near which they have been placed by some authors. They are perhaps derived from generalized scopeloids, such as the Aulopidæ." But the skull pattern of *Aulopus*, so far as I can see, does not bear any special resemblance to the anacanthine types. On the other hand, many features of the latter suggest relationship to various percomorphs. The opercular region of *Gadus* and *Lota* (Fig. 258) differs most widely from the *Aulopus* type. The antero-posteriorly wide hyomandibular recalls those of many higher percomorphs such as the scombroids, gobies, jugulares, batrachoids; the texture of the skull suggests that of *Lophius*, the enlarged opisthotic suggests gobioids, scorpænoids, ophidioids. There are also note-worthy resemblances in the skull to those of sciænids.

As to the loss of spines in the fin and the loss of the homocercal caudal skeleton, these characters have also been attained by some of the blennioids, while the presence of numerous rays in the pelvic fins is not necessarily a sign of primitive descent, since soft rays in the other fins have assuredly increased in number.

The otoliths of some Macruridæ, according to Frost (1926d, p. 489) resemble those of *Umbra* of the order Haplomi; but if the figures of these otoliths are correct the alleged resemblance hardly carries definite evidence of relationship, especially in view of the im-

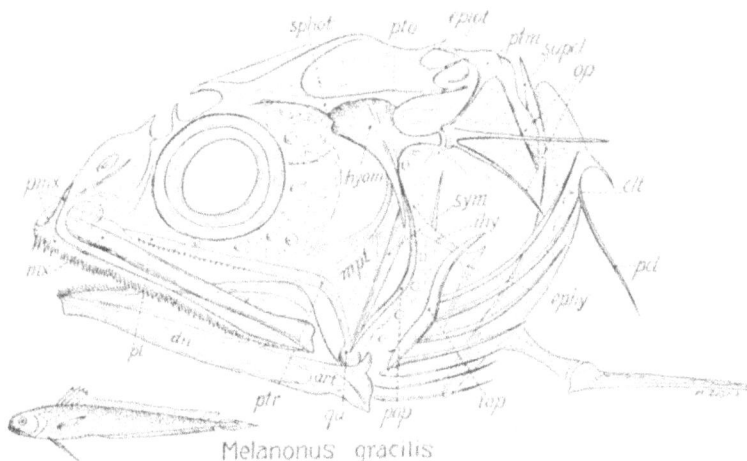

Fig. 260. *Melanonus* fry. Sketch of skull from specimen prepared by Miss Gloria Hollister for Dr. William Beebe.

mense differences between the skulls of macrurids and *Umbra*. Equally unconvincing is the suggestion (*op. cit.*, p. 483) of resemblance between some otoliths of the Macruridæ to the more specialized forms of the order Apodes, since the skulls of macrurids and all Apodes are extremely unlike in basic features.

For the present then, it seems that the evidence is insufficient to warrant setting aside the many suggestions of relationships between the anacanths and the blennies and ophidiids.

The typical anacanth skulls shown in Figs. 258, 259 call for but little comment. The skulls are relatively long and low, the hyomandibular and preopercular being exceptionally elongate anteroposteriorly, as in *Ophiocephalus*. The opercular is small and practically excluded from the margin of the opercular flap by the enlarged subopercular. The opercular has a more or less concave postero-ventral border lying between two sharp processes, an arrangement that is seen also in some pediculates. Some of the branchiostegals are also very long, as in pediculates. The lacrymal is extended nearly to the end of the snout; it broadly overlaps the very weak maxilla. The bones are thin and tend to have membranous borders. The jaws are feeble and the ascending processes of the premaxillæ are weak or absent. The ethmoid of *Gadus*, however, still bears a median keel, as if for the support of protrusile premaxillæ. The opposite frontals of *Gadus* are fused into a median plate, which bears crests and creases probably caused by enlarged lateral line organs. The parietals are enlarged, perhaps secondarily, but do not meet above the well developed supraoccipital, which retains its percoid contact with the frontals.

Melanonus (Fig. 260) and *Bregmaceros* (Fig. 261) are highly specialized deep-water anacanths.

The macrurid skull (Fig. 262) is remarkable for its parchment-like texture, its many enlarged fossæ for mucous sacs, its enlarged ethmoid, forming a more or less projecting

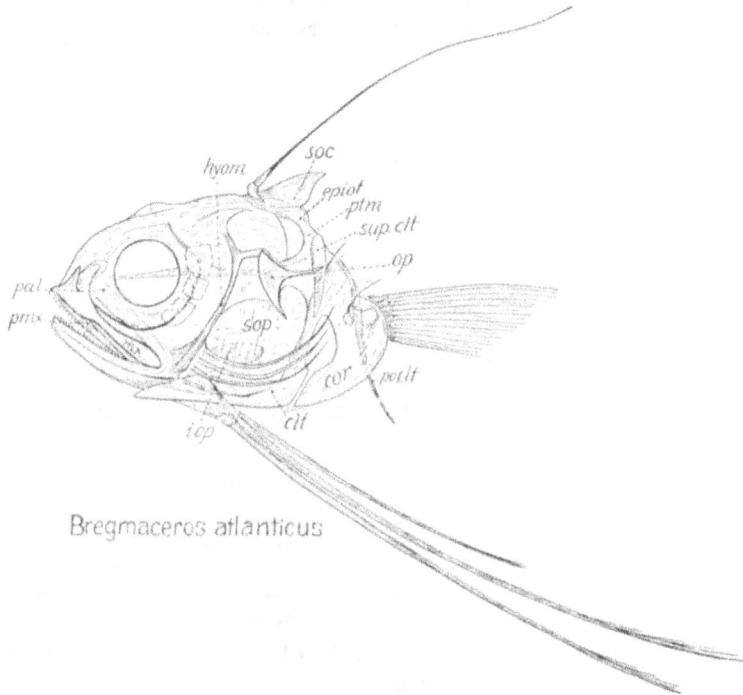

FIG. 261. *Bregmaceros atlanticus.* Sketch of skull from specimen prepared by Miss Gloria Hollister for Dr. William Beebe.

rostrum often ending in three burrs. The premaxillæ have retained large ascending processes. This skull to me retains but little that is reminiscent of a scopeloid type, but seems to be more suggestive of some secondarily spikeless derivative of the percoid group.

HAPLODOCI (TOAD–FISHES)

The skull of *Opsanus tau* (Fig. 263), representing the batrachoid fishes, presents extraordinarily interesting patterns to the student of animal mechanisms. The forces of growth and of evolution have favored the palato-pterygo-quadrate bars with their short conical teeth, especially those in front, and the corresponding teeth on the dentary, with all their supporting parts. The premaxillæ are reduced practically to thin protrusile lips, bearing little clusters of small pointed teeth at their front ends. The palato-pterygo-quadrate bars as seen from above form a pair of stout, widely diverging legs, starting in front from the broad, strongly braced vomer and abutting laterally and posteriorly on the laterally projecting suspensoria, comprising especially the hyomandibulars and preoperculars. Beyond the diverging palato-pterygo-quadrate legs are the widely arching halves of the mandible. This abuts posteriorly on either side against the firmly fixed quadrate, while its thickened distal ends are braced against each other at the symphysis. The thick sur-

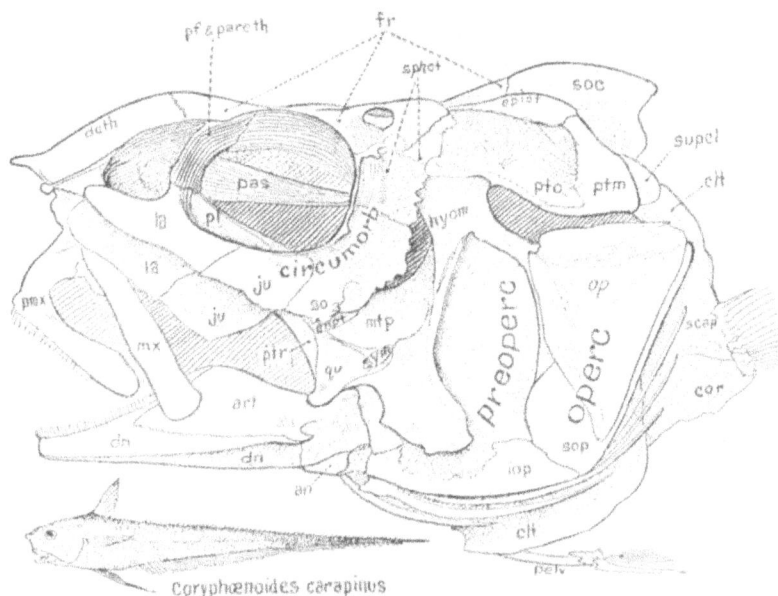

FIG. 262. *Coryphænoides.*

angular process of the articular bone, together with the stout ossified Meckelian region of the articular, afford insertion for the powerful adductor muscles, which stretch downward and forward from the oblique antero-lateral faces of the suspensoria. The relatively stout, partly inwardly directed teeth of the dentary do not occlude directly with the teeth on the palatopterygoid and vomer, but when the mouth is closed by the strong adductor muscles the mandibular teeth pass well to the outer side of the upper teeth. Hence the action here is not so much a shearing as a squeezing and breaking action. The much smaller teeth on the upper and lower pharyngeal bones may serve not so much for the trituration of small pieces of food as for manipulating the food by differential movements of the various parts of the mechanism, so as to facilitate the act of deglutition.

From the under-side view the wide, inverted V of the mandible is followed by the still more divergent inverted V of the lower hyoid arch, consisting of the thick basihyals, forming the keystone, and the stout cerato- and epi-hyal bars, to which are attached anteriorly the powerful geniohyoid muscles, and posteriorly the thick muscles that run backward to the cleithra; besides this, the basihyal and glossohyal afford support to the rather small branchial apparatus.

The bracing of the mandible, of the diverging palatopterygo-quadrate legs and of the suspensoria of the mandible, together with the necessity for protecting the brain and supporting the eyes and their associated muscles, have doubtless all contributed to the moulding of the entire skull roof (Fig. 263B), which appears to be stiffened in accordance with the truss principle so well known to civil engineers.   As seen from above, the frontal plate bears a

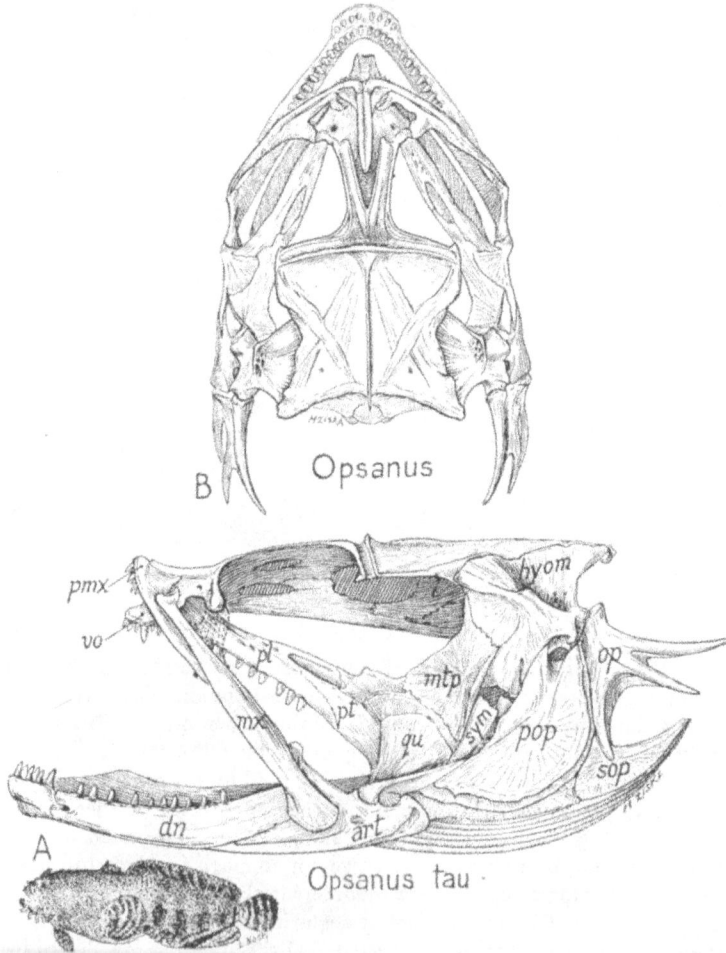

Fig. 263. *Opsanus*. A. Side view. B. Top view.

conspicuous T, of which the stem is formed by the sagittal crest of the supraoccipital, continued forward in the fused opposite frontals, while the cross-bar is formed by a stiff transverse crest borne by the frontals.   In front of the cross-bar, which is slightly convex, the raised postorbital rims afford further stiffening of the broad cranial table.   From the outer ends of the cross-bars, i.e., from the postorbital processes (presumably of the frontals) starts a broad, backwardly-pointing V, the limbs of which are low ridges that are more or less distinctly outlined in the cranial roof; the apex of this V is the posterior end of the occipital crest. Another V, pointing in the opposite direction, starts from the backwardly-projecting tips of the epiotics and runs forward, crossing the first V and having its tip at the junction of the cross-bar and the stem of the T. Still another V, with less divergent but very prominent limbs, lies on the surface of the interorbital constriction, its apex meeting the cross-bar of the T in the mid-line. The dorso-lateral surface of each of the widely-projecting hyomandibulars also bears a well defined but irregularly V-shaped crest; the apex of this lies above the stout pedicle for the opercular, the anterior limb runs up to the anterior or sphenotic head of the hyomandibular, the posterior limb of the V surmounts the posterior or pterotic head of the hyomandibular. So too in the inferior aspect of the skull one can readily distinguish two large and oppositely-pointed V's: the backwardly-pointed one has its limbs running obliquely across the proötic, its apex slightly truncated at the occipital condyle. The forwardly-pointing V has its apex under the parasphenoid; its sides are the diverging proötic borders of the braincase. In the occipital view of the skull the occiput itself is distinctly V-shaped, although its downwardly-pointing apex is rounded. From the limbs of this large inverted V diverge two small, more widely open triangles, whose legs touch the pterotics and supraoccipital respectively, and whose bases are formed by the flat skull.

This plethora of geometric and even Masonic designs in the skull of a fish which might well have been called the triangle-fish, besides affording a capital example of "Unnatural History Resemblances," to add to those cited by Dean (1908), may well excite our wonder even though we believe that these geometric figures are but the latest and most refined product of the selectively eliminative effect of Nature operating upon that well known property of bony tissue that gives it the power to thicken itself and develop trabeculæ along the lines of greatest stress. But to put the proposition the other way around, in order that the lines of greatest stress should conform to a system of opposing V's, it was necessary that the forces of growth, muscular contraction, water pressure, etc., should themselves be so orientated, at such positions and angles with reference to the main axes, as to produce the observed results.

No doubt the special patterns of a particular fish skull have grown out of the general symmetries and basic patterns of all fish skulls, e.g., the invariable antero-posterior sequence of the three main sense-capsules, the fundamental vertebrate relations of antero-posterior polarity, of bilateral symmetry and of the dorso-ventral sequence of neuron, notochord and enteron. But in the special case under consideration the dominant factor in the formation of the whole complex of interrelated V-like ridges in the skull of the common toad-fish appears to be the spreading apart of the opposite quadrate-articular joints and the consequent increase of transverse wrenching stresses and strains due to the increased transverse leverage of the powerful adductor mandibulæ muscles, especially since the thrusts of the inwardly directed teeth of the mandible have strong transverse components, which would tend to dislocate a weakly braced palatopterygoid arch. In fact, inspection shows that nearly all the V-shaped crests have plainly discernible relations to the wrenching strains from the mandible, which are relatively much greater than in fishes in which the mandible has a simple chopping movement. Hence it may not be improbable that many of the above described characters of the skull of *Opsanus* are no older than the inward-turning of the teeth of the mandible.   In other words, such "habitus" features may have been acquired in a relatively short (geologic) time and hence the relationships with other fishes with very different skull patterns may be less remote than their appearance would indicate.   This may be one of the reasons why some systematists like Boulenger have found skull characters relatively unstable and have had to seek for evidences of more remote genetic relationships in the relations of the ribs to the vertebral column and in the form and relations of the parts of the bones of the pectoral girdle and fins.

There is no reasonable doubt, for example, that the toad-fishes are related to the pediculates but they differ much in the construction of the ethmoid region (*cf.* Starks, 1926*a*, p. 306).   Another fact that tends to disguise relationship between fishes of different habitus is that a given bone may change its size or one of its contacts with surrounding bone either through what seems to be a growth or reduction of its own or through a more widely spread growth or reduction that affects whole regions rather than particular bones.   For instance, in the toad-fish the marked transverse growth and depression of the head as a whole apparently caused the hyomandibulars to become widely divergent, broadened transversely and shortened vertically.   The preoperculars shared the same distortion but the operculars immediately behind them have become quite small and now serve chiefly as a support for two aggressively large spikes, which not improbably are able to inflict severe wounds.

Again, the whole region of the primary upper jaw and its supporting parts in the toad-fish were enlarged, while at the same time the premaxillæ and maxillæ dwindled to slender bars.   This perhaps exemplifies the old principle of "compensatory growth."

In any case the fact remains that among the forms available for examination no one appears to give the clue as to the precise point of origin of the batrachoid-pediculate stock from any of the percomorph fishes.   Tate Regan (1912*f*, p. 279, Pediculati) even suggests the possibility that this group may have come off from a pre-percomorph stock in which the hypurals had the relations to the centra which are exemplified in the Salmopercæ. Probably this means at most that the batrachoid-pediculate branch (*Lophius* being already present in the Upper Eocene of Monte Bolca) may have begun to diverge from the acanthopt stock before the beginning of the modern families of percomorphs.

That the batrachoids are related to the pediculates is well attested by their reasonably close approach to that peculiar group in the characters of the pectoral fins, vertebræ, hypurals, as well as in many significant skull characters.   While in general more primitive than the pediculates, they are more specialized in the loss of the mesethmoids and of the epiotics, or possibly in the fusion of the latter with the parietals (Tate Regan, 1912*f*, pp. 279–280).

Mr. G. Allan Frost (1930*b*, p. 623) states that the otoliths of the order Haplodoci show no resemblance to those of the Pediculati but in their general features they resemble in a striking manner the otoliths of the family Macruridæ of the order Anacanthini, differing, however, in minor details.

Pediculati (Anglers, Sea-mice, Sea-bats)

Before discussing the evolution of the skull of the pediculates it is well to consider briefly the evolution of their general body-form.   The name Pediculati is given of course in reference to the peculiar construction of the pectoral fin, which can be turned downward, backward or forward, like an arm. Among existing pediculates the least specialized stage of the pectoral fins occurs in the toad-fishes (Batrachoidea). In *Batrachoides didactylus*, according to Regan (1912*f*, p. 279, Fig. 1), the broad pectoral fin is supported by four elongate pterygial bones, of which the lowermost is large and distally broadened; the uppermost is somewhat broadened distally, the others are slender. This construction, which is not yet completely pediculate, would appear to be well adapted for the mode of life of toad-fishes, which hide among rocks in shallow water. In the complete pediculate stage of *Lophius* the pterygials are reduced to two long rods, the lower one resembling a flattened ulna. *Lophius*, which is the giant of the order, has a broad flattened body for resting on sandy bottoms and huge jaws and throat. The sea-mice, or Antennariidæ, retain three pectoral pterygials; the pectoral extremities acquire amazing facility as the fish crawls among the sargassum weed. The benthonic sea-bats, Ogcocephalidæ, have more or less flattened and rounded bodies; their pectoral fins somewhat resemble the hind paddles of frogs or of seals and are doubtless able to propel the body forward by rapid strokes of their obliquely-placed surfaces. In the bag-like ceratioids the pectoral fins, while close to those of the antennariids in ground-plan, are more or less reduced and finally almost vestigial, while the pelvic fins are absent.

The question then is, which type of body and fins among the living forms may be considered to be nearer the starting-point for the group? The bat-fishes and all the cera-

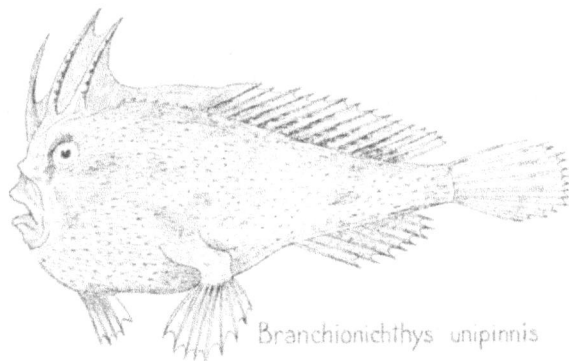

Fig. 264. *"Chironectes" (Branchionichthys) unipennis.* After Cuvier.

tioids may be at once eliminated, on account of their obviously extreme and diverse specializations. *Lophius* is specialized in its great size, in certain skull characters and in the reduction of the pectoral pterygials to two. Even *Chirolophius naresii* (Günther, 1880, Pl. XXV), in which the "illicium" or fishing-rod is still obviously only the first ray of the spinous dorsal fin, is relatively specialized in the huge size and extreme depression of the head. As to the antennariids, many also appear to be highly specialized in external appearance but, as will presently be shown, the South Australian antennariid *Brachionichthys* is much less specialized and in fact seems to give several clues to the origin of the entire order.

In 1817 Georges Cuvier, in a remarkable memoir entitled "Sur le Genre *Chironectes* Cuv. (*Antennarius* Commers.)," noted that the "fishing-rod" of *Lophius* was merely the

21

first ray of the dorsal fin and that in all their essential features the lophiids and their allies agree with the acanthopterygian fishes; that the "callionymes" show a reduction of the branchial orifice; that some of the blennies have the first dorsal on the head; that one of the gobies (the "*Cottus macrocephalus*" of Pallas) has the head as much depressed as that of *Lophius piscatorius;* that *Periophthalmus* has similarly elongate pectorals, which are also used for running about on the mud flats "*comme les chironectes.*" In Cuvier's figure of "*Chironectes (Branchionichthys) unipennis*" (Pl. XVIII, Fig. 3) the illicium differs but little from the two long rays behind it, while all three cephalic rays are connected by a web of skin with the long-based dorsal fin. This is the most primitive condition among the typical pediculates. In "*Chironectes*" (*Branchionichthys*) *punctatus* (Pl. XVIII, Fig. 5) the skeleton is relatively very primitive in appearance save that the stout pectoral pterygials are reduced to two and the pectoral fin is truly pediculate. This enables it to be turned below, behind or above the swollen throat and abdomen.

Here we have touched upon what is perhaps the primary adaptation of the pediculates, namely, the great enlargement of the throat and abdomen, which apparently permits them to devour either relatively large prey or a great quantity of small prey at one time. The "pediculate" portion of the pectoral is simply the remnant of a once very large and continuous pectoral fin which was spread around the side of the enlarged throat as it is in the batrachoids. The branchiostegal rays have shared in the enlargement of. the lower part of the opercular flap; in the antennariids this flap finally overlapped the shoulder-girdle and by adhering to the skin on its surface has closed off all the upper part of the normal post-opercular slit. In this way the exhalent respiratory current, instead of escaping in front of the pectoral girdle in the ordinary way, is led around through a special tunnel in the skin that opens above and behind the pectoral fin. According to Cuvier, Renard and Valentyn reported that the "chironectes" (antennariids) can virtually go on all fours and that they thus pursue their prey among the seaweed and on the mud. The small size of their branchial opening, he thinks, makes it very probable that they can live for some time in the air; he even approves the epithet of "amphibian" that Commerson had applied to the chironectes. He also notes that Margrave, Commerson and others testify that the chironectes have the power of inflating the belly like a balloon, and that anatomical inquiry shows that they could do this only by swallowing air and filling their great stomachs with it, as do the tetrodons. Aquarium specimens can also squirt water forcefully from their branchial orifice.

It seems possible that this undoubted power of inflation, which is conditioned by the closure of the normal branchial slit as well as by the enlargement of the branchiostegal flap, may partly compensate for the small size of the gills and their reduction in number to two half gills and two entire gills. It seems also that *Pterophryne* and other antennariids that are commonly found in sargassum weed hundreds of miles out at sea have simply stayed in the seaweed or been hatched in it after it has drifted far away from the shore where it originated. While on the New York Zoölogical Society expedition to the Sargasso Sea, we had the opportunity of studying the movements of these little fishes in our aquaria. In the seaweed the *Pterophryne* uses its arm-like, widely-webbed, almost hand-like pectorals and its foot-like pelvics in climbing about on the fronds of the sargassum weed, showing an amazing range of movements of these appendages. They often reach upward with one pectoral "flipper" and downward with the opposite one, at the same time reaching forward with the pelvic flippers. Even the dorsal and anal fin tend to press against or cling to the seaweed. If separated from its seaweed, a *Pterophryne* can develop a surprising burst of speed, wriggling its body and using all its fins to overtake its floating home.

As their body has tended to assume the globose form of the puffer or porcupine-fish type, the spinous dorsal has entirely lost its locomotor functions, while the soft dorsal tends to be paired with the anal in a wig-wagging motion.   Their highly cryptic coloration among the sargassum weed seems to enable them to stalk successfully the small crustaceans and other small animals and protects them from larger fishes that hunt in the weed.

From all this we draw the conclusion that the marked peculiarities of the skull and body-form in the higher pediculates were conditioned by the following primary adaptations:

(1) the development of an expanded oro-pharyngeal chamber for engulfing a large prey or a large amount of food;

(2) the posterior spreading of the branchiostegal membrane, the closure of the normal opercular slit and the migration of the lower part of this slit to a point behind and above the pectoral fin—all correlated with the reduction of the gills, the development of the power of inflating the throat and stomach with air or water, as well as the power of ingesting large prey;

(3) the elongation and specialization of two of the pectoral pterygials so as to enable the pectoral paddles to reach around below or above the swollen throat and abdomen;

(4) the change of function of the enlarged first ray of the spinous dorsal into a lure. Probably the initial step was the habit of living among seaweed, the acquisition of dappled color of skin and the development of excrescences or tags of skin on the body and anterior dorsal fin.   More or less directly progressive stages in this transformation may be seen in *Antennarius* ("*Chironectes*") *unipennis*, *A. scaber*, *A. lophotes*, *A. mummifer*.

(5) As to adaptive radiation into different habitats, I infer that the primitive habitat was in the kelp along rocky shores; that some of the antennariids clung to the Gulf-weed when it broke loose from the shore and was carried far out into the Sargasso Sea; that, as this weed eventually sinks, some of them became pelagic and free-swimming and thus gave rise to the various families of ceratioids; that in another direction, some of the early antennariids, becoming benthonic and with ever greater mouths, gave rise to the lophiids, while others through *Chaunax*-like forms passed into the sea-bats (Onchocephalidæ).

After the foregoing summary we are perhaps in a more favorable position to attempt an evolutionary interpretation of the architecture of the skull of pediculates.   The following sources and material have been used:

(1) various dried skeletons of *Lophius*, *Antennarius*, *Pterophryne* and *Ogcocephalus*, in the collections of this Museum;

(2) the fundamental article by Tate Regan on the classification of the order Pediculati (1912*f*) and his memoir on the Ceratoidea (1926);

(3) the ceratioids in the collections made by Dr. William Beebe for the New York Zoölogical Society.   These will be more fully described by him in subsequent publications, but while I was enjoying the hospitality of his laboratory in Bermuda he very generously invited me to study and sketch any of the deep-sea material and to use as much of it as I cared to for the present paper.

Antennariids.—The skulls of *Antennarius* and its allies (*Histrio*, *Pterophryne*, etc.) are on the whole perhaps the most central and least specialized of the pediculates above the grade of the batrachoids and well reflect the influence of the primary adaptations discussed above.   The syncranium (Figs. 265, 266) is large in proportion to the size of the short body, but it is extended downward rather than transversely.   The large mouth is inclined sharply upward, showing that the prey is engulfed from below, and the eyes are also directed partly upward.   The sharp upward inclination of the mouth is brought about by emphasis of the

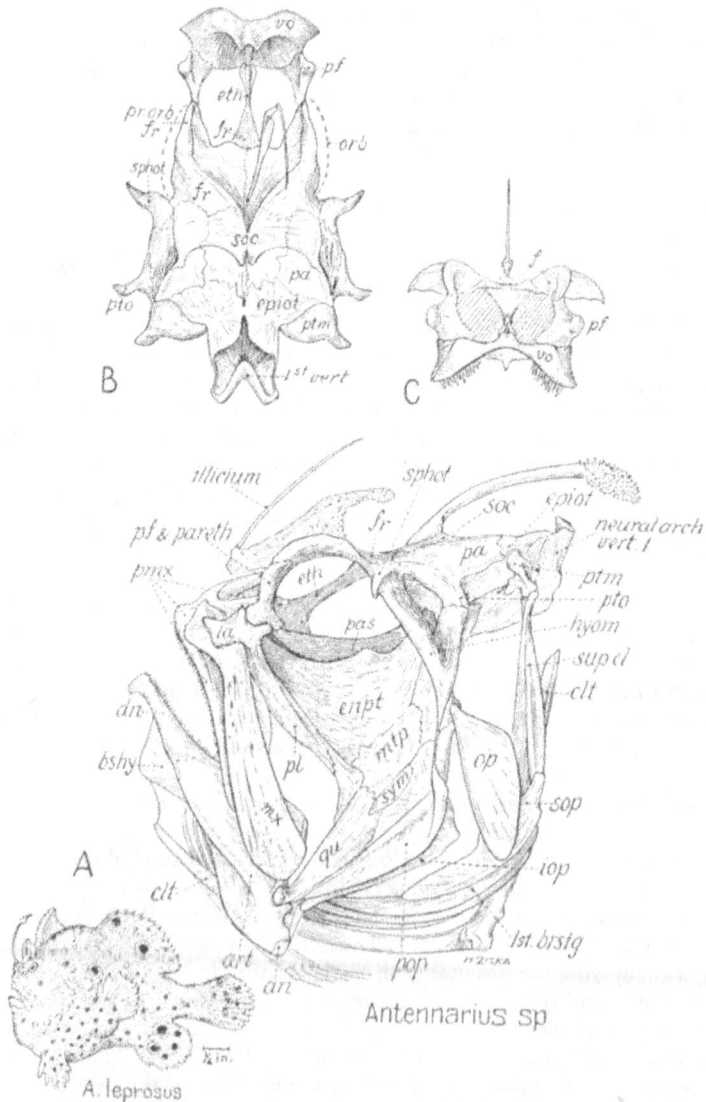

FIG. 265. *Antennarius.* A. Side view. B. Top view. C. Front view.

following factors: (1) marked depression of the quadrate-articular joint; (2) the retraction of the upper border of the mouth by shortening of the preorbital region.

The dorsal position of the mouth and the unusually large size of the ascending and articular processes of the premaxillæ, together with other factors to be noted below, have caused a deep V-shaped notching of the anterior median portion of the frontals and a sharp depression of the mesethmoid, which in the lateral aspect forms a narrow curved bar with an expanded lower end resting on the parasphenoid and vomer; the main bar lies behind and below the level of the everted orbital wings of the frontal; it runs backward and upward to the depressed median portion of the frontals. This V-shaped dorsal fossa is bridged in front by the transversely extended vomer, which forms the keystone of the large premaxillo-maxillary arch. Thus the strange characters of the forepart of the skull, as figured by Starks (1926a, pp. 320, 321) and by Regan (1912f) are fully explicable as part of the primary adaptive habitus of the pediculate mouth and jaws.

As the orbits are pushed backward so that they impinge upon the suspensorium, it is natural that the roof of the cranial vault should be short and wide. The supraoccipital has retained its acanthopterygian contact with the frontals; the middle of it coöperates with them to form a flat-bottomed valley that widens in front; toward the deepest part of this depression point the anterior end of the bony rod that supports the third dorsal ray and the posterior end of the bony rod of the second dorsal fin-ray. (The illicial ray and its short bony support rest in the skin above the ascending processes of the premaxillæ.) Various parts of the skull roof, in other cases, have readily moved forward and expanded laterally under the influence of cephalic extensions of the dorsal fin (e.g., in the echeneids, in *Coryphæna, Luvarus*). Thus the enlargement of the basal muscles of the second and third rays of the spinous dorsal may have initiated the enlargement of the supraoccipital, which even here is already the dominant median element of the cranial vault. As the supraoccipital has moved forward the epiotics have followed it, drawing in toward the mid-line but not quite meeting there except in front. The parietals have retained their primitive percoid positions, being separated by the supraoccipital. In my specimens they are represented by thin plates in front of the lunate epiotics and lateral to the supraoccipital; laterally they overlap the sphenotics and the pterotics. These two stoutly-braced elements have sharply projecting postorbital and posthyomandibular processes separated by a deep notch. This bears beneath its dorsal rim the usual facets for the anterior and posterior heads of the hyomandibular. All this region that receives the upward thrust of the suspensorium is strengthened by various trabecular tracts and crests on the lateral surface of the proötic, pterotic and exoccipital. A triradiate suture connects the sphenotic, proötic and pterotic, while another triradiate suture marks the contact of exoccipital, pterotic and epiotic. The posttemporal is a much shortened and broadened, more or less triangular bone which suspends the long pectoral girdle and is closely tied to the back of the epiotic, pterotic and exoccipital.

Inasmuch as the very long suspensorium and the long shoulder-girdle apply powerful forces to the back part of the cranial vault, the occiput is reinforced by the enlarged first vertebra, which has become almost immovably attached to it. However, the triple occipital condyle, which is characteristic of percomorph fishes, still marks the posterior limits of the skull proper. The stoutly-built centrum of the first vertebra articulates with the basioccipital by the usual subcircular cotylus but its posterior face forms a much widened, almost hinge-like articulation for the second vertebra, which, at least in *Histrio*, is bent downward at a sharp angle, as the column curves down and around the expanded body cavity.   The neural arches of the first vertebra are expanded anteroposteriorly and stiffened posteriorly. At the base they are excavated to receive the lateral exoccipital condyles.   They diverge gently at the top but are connected above across the top by a thin chevron, which may represent a modified epineural bone.

The base of the cranial vault is formed in the rear by the ventrally stiffened basioccipital, which narrows into a half-tube posteriorly but broadens anteriorly to hold the otoliths and to make contact with the posteriorly flaring parasphenoid.   The keel of the latter is narrow but greatly stiffened to receive the thrusts from the large transversely widened vomer and from the prefrontals.   The lateral ascending processes of the parasphenoids are short; they are appressed to the proötic but, according to Tate Regan (1912*f*, p. 282) they do not reach upward to the frontals as they do in *Lophius*.

Thus as a whole the neurocranium of *Antennarius* has a broad, rather stoutly-braced cranial vault and a large stiff parasphenoid keel supporting the very wide vomer; the slender interorbital region has raised orbital rims and a deep median V-shaped depression containing the large ascending and articular processes of the premaxillæ.   The mesethmoid, which forms the bottom of this deep depression, is reduced in the side view to a curved rod with an expanded lower end and in the top view to a V with a long anterior tip.   The median depression formed by the supraoccipital and the frontals is more or less filled by the basal rods of the second and third dorsal rays and their attached muscles.

It has already been shown how variously the neurocranium has been affected by the cavernous development of the jaws, mouth and throat; but the detailed interrelations of the various parts of the branchiocranium remain to be described.   The excessive depression of the quadrate-articular joint and the enlargement of the gullet have involved the marked vertical elongation of the hyomandibular, of the quadrate, preopercular, inter- and subopercular, cleithrum, supracleithrum and branchiostegals.   The space between the hyomandibular and the shoulder-girdle being very narrow in proportion to its height, the opercular is correspondingly narrow; it is also thin and membranous, the opaque bony parts being reduced to two forking streaks, including a much larger anterior branch, which is nearly vertical, and a much more delicate short posterior branch, the two meeting above and overlapping the large opercular pedicle of the hyomandibular.   The anterior pointed lower tip of the opercular is received between a small fork of the subopercular; this fork has a short anterior and a much longer and broader posterior branch; from the notch the subopercular continues downward and forward as a somewhat dagger-like blade with a curved posterior border; it is closely appressed to the lateral face of the longest and largest branchiostegal.   In the rear view the opercular is bowed outward beyond the straight cleithrum so that the posterior border of the opercular is no longer flush with the lateral surface of the cleithrum, as it is in the majority of fishes; thus in the dried skeleton the gap between the opercular and the cleithrum remains permanently open.   In life, however, this gap is covered laterally by the skin that forms the tube leading to the functional spiracle, behind and above the pectoral fin.

The preopercular is a thin vertically deep bone fastened tightly to the lateral crest of the hyomandibular and to the posterior ridge of the quadrate; it stiffens the long suspensorium.   Posteriorly it covers the long sliver-like interopercular, which as usual is attached by a stout ligament to the angular. In consequence of the marked forward displacement and downward depression of the quadrate-articular joint, there is a marked bend between the anterior rim of the quadrate and the pterygo-palatine arch. The latter meets the flaring lateral ethmoid and braces the lateral extension of the vomer; but the finger-like

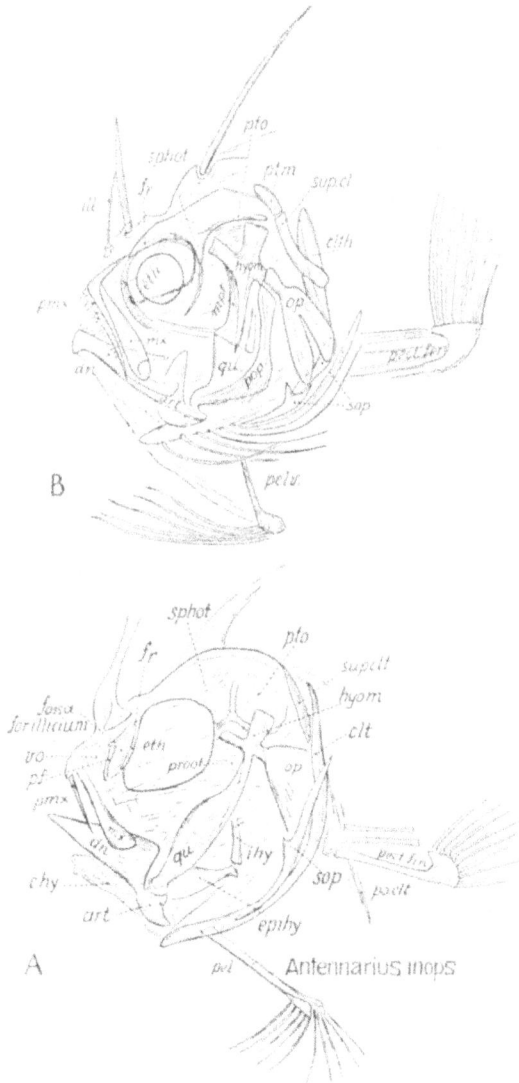

FIG. 266. *Antennarius.* A. Younger stage. B. Older stage. Specimens stained by Miss Gloria Hollister for Dr. William Beebe.

process which as usual overlaps the maxilla is fused with the small mallet-like lacrymal. The maxilla ends proximally in a small sculptured plate that overlaps the premaxilla, and in a short stout process that articulates on its posterior surface with the vomer. The alveolar branches of the premaxillæ are remarkably slender in comparison with the large size of the ascending and articular processes, which are received into the anterior facial fossa described above.

Passing to the branchial apparatus, we note that the lower segments of the hyoid arch are remarkably large, especially the basi-, cerato- and epi-hyals. Vomerine and upper pharyngeal teeth are numerous but not large, the latter forming two convex clusters of which the second is much larger than the first; the lower pharyngeal teeth are numerous and sharp but not large.

The adaptive radiation of the antennariid type, as already noted, appears to have led in one direction to the lophiids, in a second to the onchocephalids and in a third, to the ceratioids. But even within the antennariiform division, as recognized by Tate Regan (1912f, p. 282), there are considerable differences in general habitus. Thus, as noted above, the South Australian *Branchionichthys* shows primitive conditions of the illicium, which is still a simple dermal ray connected by a web of skin with the primitive second and third rays. In *Antennarius lophotes*, on the other hand, the illicium has a feather-like, many-branched tip; in certain antennariids the third ray is greatly enlarged and covered with tough skin, the surface studded with denticles, while the illicium itself is very small. A very peculiar side branch of the antennarioids is known only from a single species, *Tetrabrachium ocellatum* (Günther, 1880, p. 45, Pl. XIX, Fig. C) from the ocean south of New Guinea. Here the small projecting eyes are directed upward and so is the very small transverse mouth. The illicium is vestigial, the second ray is feather-like, the third very small; the spreading pectoral fin is divided into a large lower part and a small upwardly-directed part; the body is much more elongate than in typical antennariids.

Lophius.—The skull of *Lophius* (Fig. 267) is much more specialized than that of *Antennarius* in many details connected with the marked benthonic habitus, but the heritage is evidently antennariid. I thought at first that *Lophius* stood nearer to the starting-point of the higher pediculates than did *Antennarius* but, as noted above, further study has convinced me that the opposite is the case.

In the lophiids the fishing habits of the group attain their typical development. The successful fisherman is one who knows how to sit still and wait, while keeping his eye steadily on the bait, and for this congenial task the lophiids are eminently well adapted. In the first place, their enormously wide heads are flattened beneath so that they can rest comfortably on the sand, while the powerful pediculate pectorals and advantageously placed pelvic fins doubtless enable the fish to spring suddenly upward at the critical moment. As in the antennariids, the exhalent current instead of escaping in front of the pectoral girdle in the ordinary way is led around through a special tunnel in the skin, which in the lophiids opens in the lower axil, just behind the "fore-arm" of the pectoral fin. Meanwhile the strong development of the pulsing opercular flap, together with the immense deepening and widening of the mouth and throat, has caused the extension and marked narrowing of the opercular apparatus into a tracker-like interopercular, a slender subopercular with a forked posterior end and a narrow opercular. The branchiostegals, sharing the excessive expansion of the throat, have become very long and slender, while the supracleithrum is pulled out into a narrow rod.

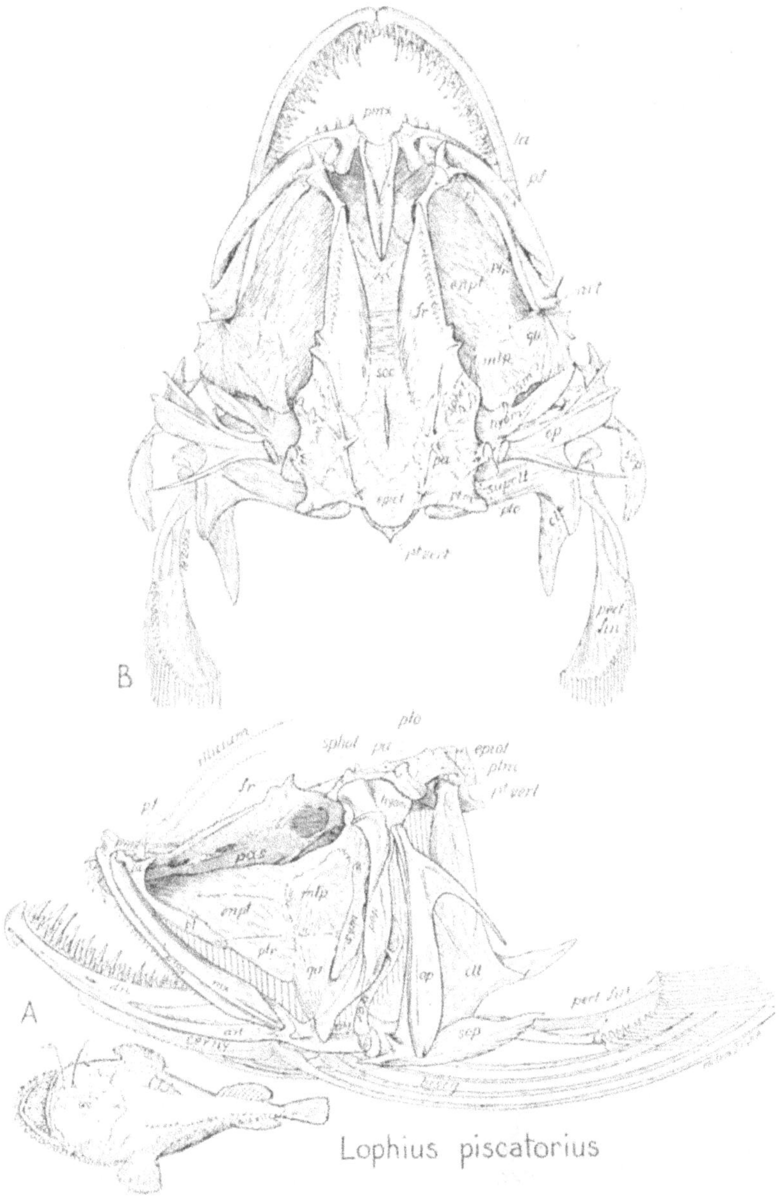

Fig. 267. *Lophius*. A. Side view. B. Top view.

The neurocranium of *Lophius* seems to me to be more specialized than that of *Antennarius* in the following features:

(1) it is much widened and flattened, in connection with the increase in the transverse diameter of the mouth;

(2) while the eyes retain their position immediately in front of the hyomandibular,

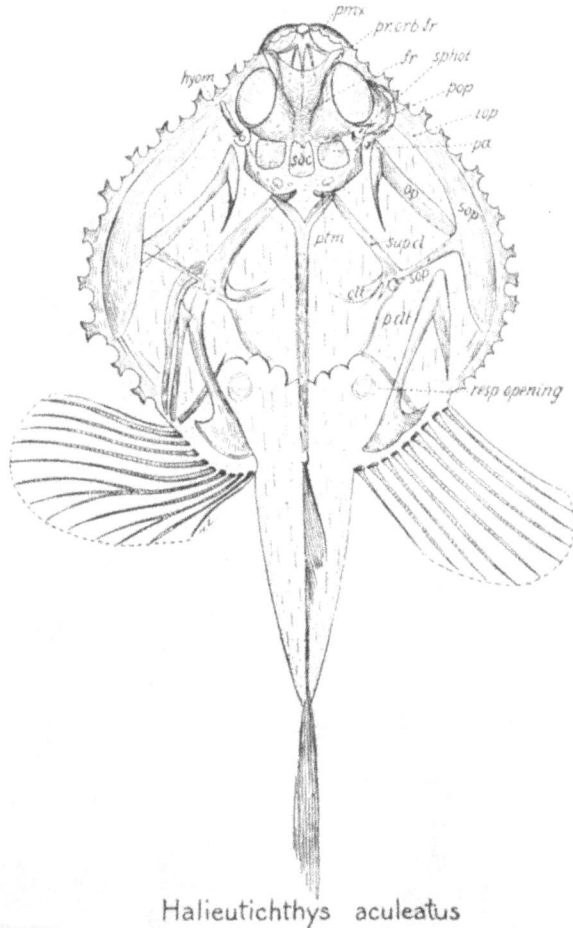

Halieutichthys aculeatus

FIG. 268. *Halieutichthys*. Young specimen stained by Miss Gloria Hollister for Dr. William Beebe.

the preorbital portion of the neurocranium has been lengthened and widened into a shallow trough, floored chiefly by the anterior wings of the frontals, which are secondarily widened;

(3) perhaps in consequence of the flattening of the head the parasphenoid has established a broad contact with the frontals in front of the orbits, which braces the enlarged interorbital bridge;

(4) thorn-like processes have been developed on many points on the surface of the skull.

As *Lophius* dates back to the Upper Eocene, the time of the supposed origin of the lophiid from the antennariid stock must probably be not later than Basal Eocene or Upper Cretaceous.

Onchocephalids.—In the most highly specialized members of the group, which are typically benthonic, the body (Fig. 268) is depressed, subcircular and almost lense-shaped, studded all over the dorsal surface with spicules and thorns. The greatly enlarged pectoral girdle is included in the disc and free portions of the pectoral fins protrude at the side, like the flippers of a sea-lion. The pelvic fins lie beneath the body disc and, spreading apart widely, serve as hind limbs. The skull in these round-disced forms protrudes but little

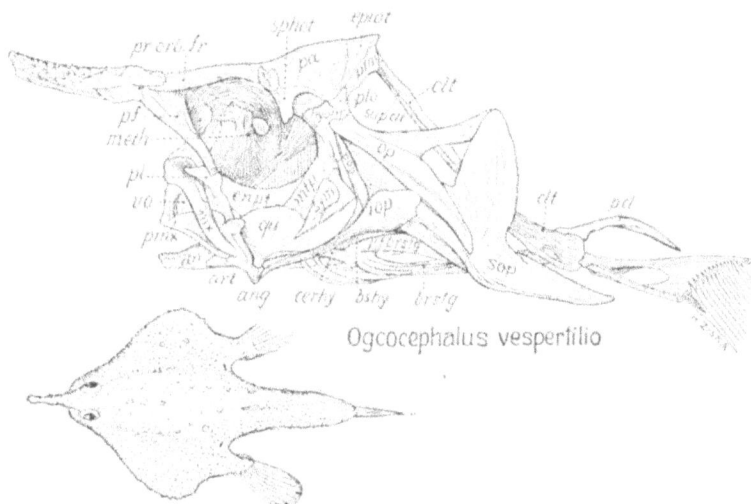

FIG. 269. *Ogcocephalus.* Side view.

above the disc. In others, however, such as *Ogcocephalus vespertilio*, the skull (Fig. 269) projects well above the level of the disc and its fundamental characters agree with those of the *Antennarius* type, as follows:

(1) the median interorbital and supra-ethmoid fossa is present and bordered by raised orbital flanges of the frontals, as in *Antennarius*, but in consequence of the shrinking of the illicium the function of this fossa is less apparent;

(2) the supraoccipital occupies the same median position on the occipital roof but its dorsal surface has now sunk down into a median groove between the raised epiotics;

(3) the preorbital fork of the frontals and prefrontal is bridged transversely by the vomer and the mesethmoid is depressed below the illicial fossa;

(4) the neural arches of the first vertebra are enlarged and appressed against the occiput;

(5) the relations of the opercular, subopercular, branchiostegals, supracleithrum, cleithrum and pediculate pectorals are fundamentally the same as in *Antennarius;*

(6) the respiratory pore lies behind the enlarged second pterygial in the position that is normal for antennariids.

We may safely conceive the onchocephalid skull to have been derived from the antennariid type by the following modifications:

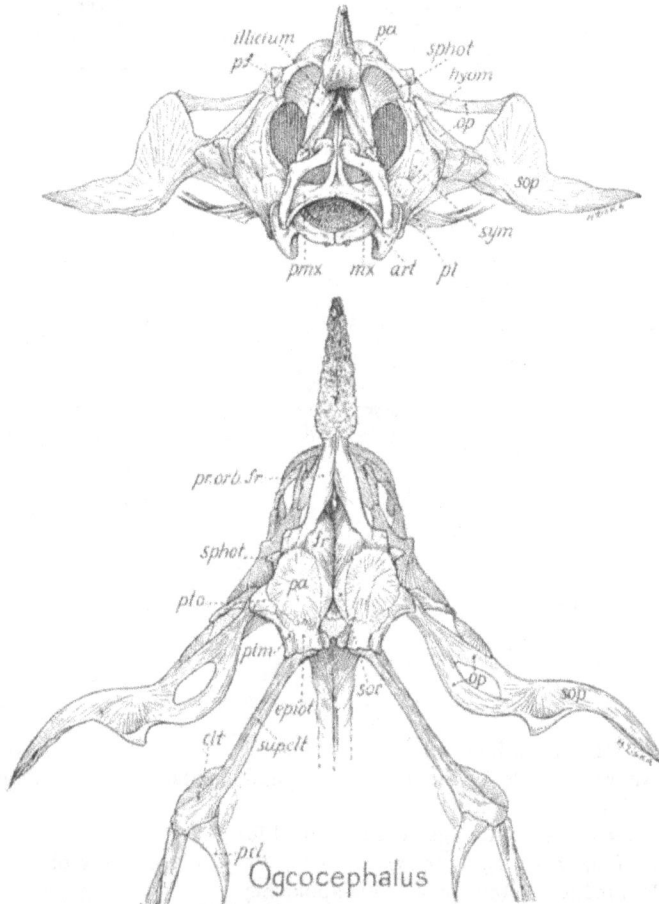

Fig. 270. *Ogcocephalus*. Top and front views.

(1) the great increase in size of the subopercular and its associated branchiostegal, as it came to form the lateral margin of the disc;

(2) the reduction of the illicium and the forward displacement of the illicial fossa beneath the newly-formed rostrum;

(3) in *Ogcocephalus vespertilio* the "rostrum" appears to represent a great tower of dermal bone, which has grown forward and upward to support excrescences of the skin;

these have perhaps enveloped or replaced the great horn-like spike of *Antennarius*, which represents the second dorsal fin-ray. In this species the forward growth of the pseudo-rostrum has necessitated the bracing and readjustment of all the supporting bone beneath it. Thus the orbital rims of the frontals have been stiffened and the lateral ethmoid ridges have been greatly strengthened and rotated forward;

(4) possibly the forms with a short rostrum have been derived from those with a long rostrum, such as *Ogcocephalus vespertilio;*

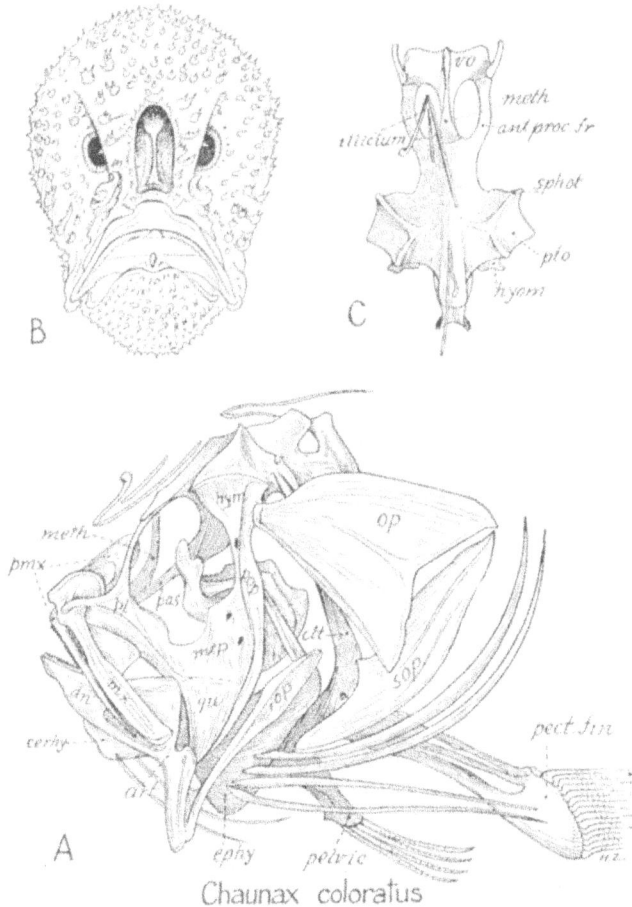

FIG. 271. *Chaunax.* After Garman. A. Side view of skull. B. Front view of fish. C. Top view of skull.

(5) the mouth parts of the sea-bats are relatively small and apparently have been derived by reduction from a more normal condition. Nevertheless the premaxillæ retain long ascending processes, which pass upward between the maxillæ in the normal way;

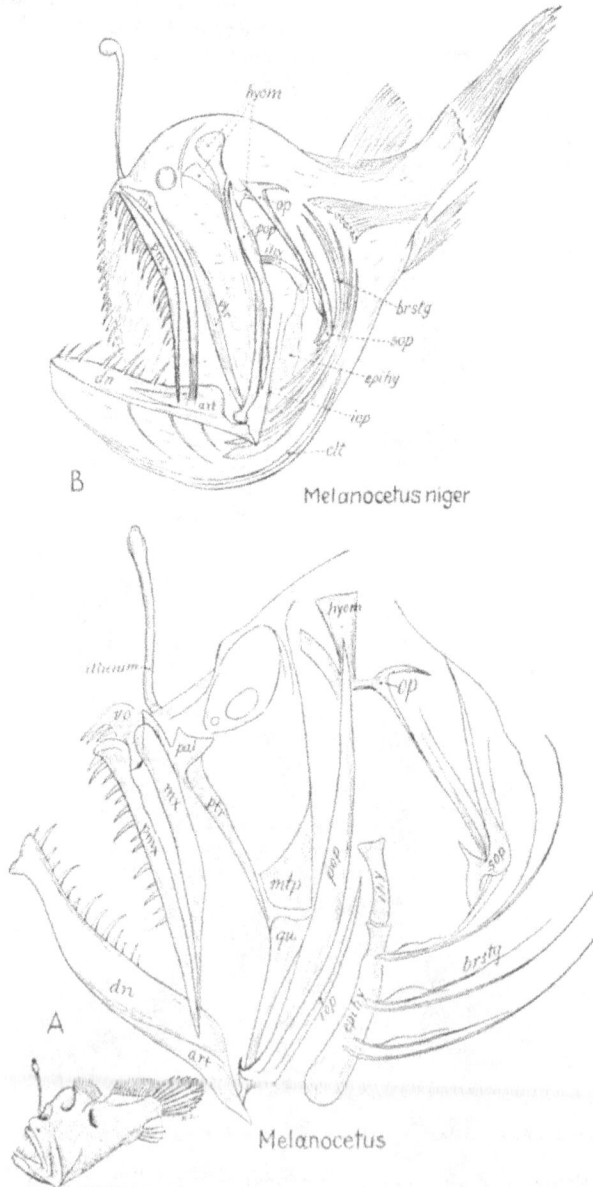

FIG. 272. *Melanocetus.* Young specimen (A) and younger specimen (B); both from stained preparations by Miss Gloria Hollister for Dr. William Beebe.

(6) although the mandible is delicate, the palato-quadrate arch is unusually massive, apparently a secondary enlargement to support the massive neurocranium which rests upon it;

(7) while in *Antennarius* and *Lophius* the opercular is exceptionally small, in *Ogcocephalus vespertilio* it has enjoyed a secondary enlargement, so that it has grown backward, retaining its contacts with the still more enlarged subopercular.

The skeleton of *Chaunax coloratus* as figured by Garman (1899, Pls. XVI, XVII) would seem to afford an ideal intermediate stage leading to the sea-bats from a primitive *Antennarius*-like type. The illicium (Fig. 271) is received into a fossa in the ethmoid region,

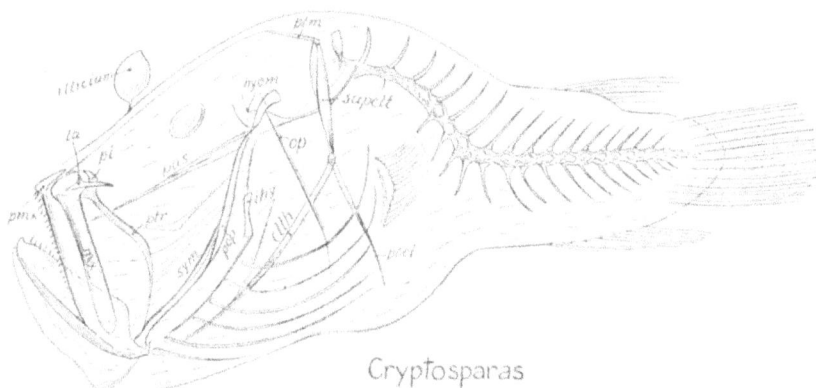

FIG. 273. *Cryptosparas.* Sketch of very young specimen from stained preparation by Miss Gloria Hollister for Dr. William Beebe.

the opercular and subopercular are enlarged and foreshadow the relations of these elements in the sea-bats; but the mouth is less reduced in size and the skeleton abounds in deep-seated resemblances to the antennariids. Hence it is not surprising that Tate Regan (1912f, p. 283) puts the Chaunacidæ between the Antennariidæ and the Onchocephalidæ and in the division Antennariiformes.

The Ceratioids.—The oceanic anglers ordinarily live at depths from about 500 to 1500 metres below the surface (Tate Regan, 1926). Accordingly they are usually black but occasionally have a pale translucent skin. The body is typically short, more or less globose, with a great upturned mouth bristling with long sharp teeth. The stomach, as in other pediculates, is distensible. The illicium usually bears a glowing bulb at the tip, but it may be vestigial or specialized in various ways. The eyes are typically small and are doubtless directed toward the prey in front of the illicium. The small backwardly directed, soft dorsal fin usually lies well behind the swollen abdomen and is frequently paired with the small anal. The caudal peduncle is usually wide and the tail has more or less webbed-rays and is convex posteriorly. Probably the fish moves up quietly toward its prey by the undulations of these three fins. The pectorals are small but retain their pediculate type and are often directed upward and backward. The ventral fins are absent, but Parr (1930, pp. 11, 13) reports the presence of pelvic bones in *Rhynchoceratias longipinnis.*

Although the skull of ceratioids is highly and diversely specialized, there is a curious constancy in the general features of the opercular, hyoid and branchiostegal series, as far as I could ascertain on stained preparations (Figs. 272–278) of various ceratioids in Doctor Beebe's collection (including especially *Cryptosparas, Melanocetus niger, Oneirodes, Lophodolus, Lasiognathus, Haplophryne hudsonius*). The two heads of the hyomandibular are very pronounced, as is also the pedicle for the opercular; the latter is thin and membranous but is stiffened by a short upper and a long lower bony fork; the opercular stands out laterally

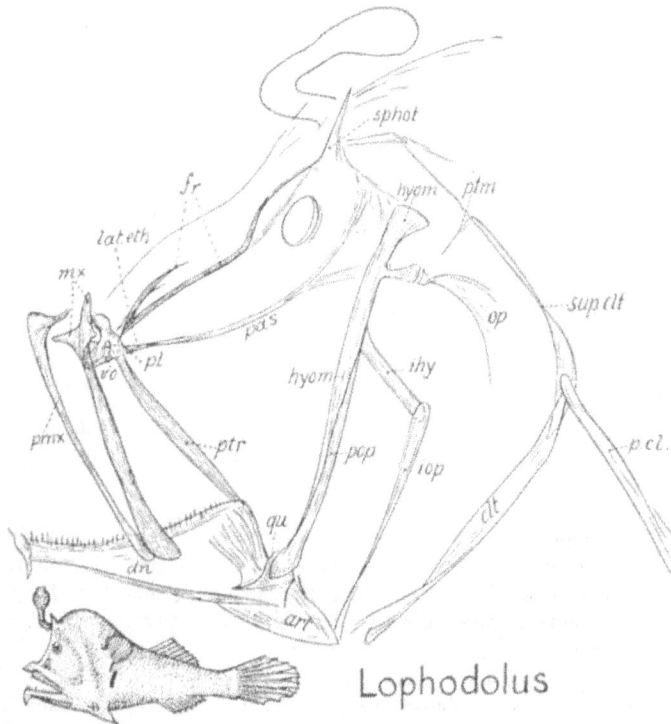

Fig. 274. *Lophodolus.* From stained preparation by Miss Gloria Hollister for Dr. William Beebe.

and supports the membrane, which continues backward over the excessively long branchiostegals; these overlap the base of the pectoral fin. The preopercular is curved and very slender; the cerato- and epihyals are usually large and connected with the hyomandibular by a slender, rather short interhyal. The interopercular is a long delicate tracker connected as usual with a ligament fastened to the posterior angle of the mandible. The latter is frequently stout, especially at the posterior end. The palatine usually has a normal process projecting over the maxilla and fused with the reduced lacrymal,—all these features being easily derivable from the *Antennarius* type as described above.

Even in antennariids the supraoccipital had already worked its way to the middle of the roof of the cranial vault, where it lay beneath two of the enlarged dorsal fin spines. In many of the higher ceratioids the supraoccipital becomes dominant as a large round plate, while the parietals disappear and the epiotics become greatly expanded. Occasionally, however, as in *Haplophryne hudsonius* Beebe (Fig. 280) and in *Rhynchoceratias longipinnis* (Parr, 1930c), good-sized but thin parietals are retained. In these cases we may well suspect a sporadic reappearance of a variable or dormant character, rather than an independent derivation from very ancient ceratioid types, as suggested by Parr (1930c, p. 5.)

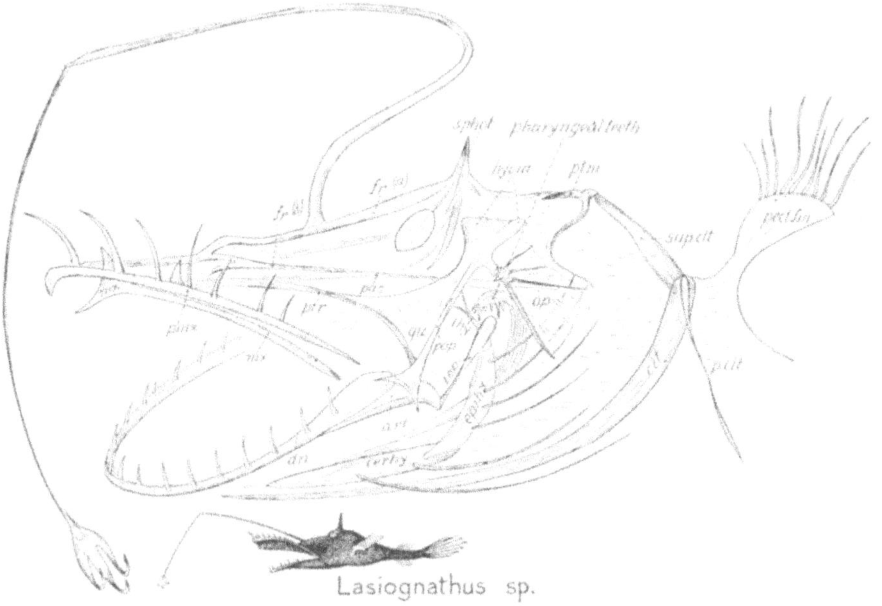

Fig. 275. *Lasiognathus.* From stained preparation by Miss Gloria Hollister for Dr. William Beebe.

The illicial trough, already pronounced in *Antennarius* (Fig. 265B) forms a prominent feature in ceratioids. It represents a caving-in of the skull-roof due to the presence of the enlarged basal bone of the illicium and of the usually strong muscles, arranged in three pairs, which are attached to it. Tate Regan has shown (1926) that in more advanced ceratioids, such as the Ceratiidæ, this trough forms a deep groove extending back on the roof of the occiput and lying between the raised epiotics.

Regan's figures of the skull-tops of ceratioids show great contrasts in proportion between the very wide skull of *Melanocetus johnsoni* and the very elongate skull of *Cryptosparas*. In the relatively primitive skull of *Borophryne apogon* (*op. cit.*, Fig. 8) the arrangements of the lateral ethmoid, mesethmoid, interfrontal fenestra and of the bones on the roof of the cranial vault all seem readily derivable from the antennariid type. The vomer, however, has become excessively small, while in *Melanocetus* it is very wide. The mesethmoid is very small in *Melanocetus*, very long and large in *Gigantactis*. Thus the bones reflect the great differences in the adjacent soft parts.

22

All the bones of the preorbital face in the side view also differ widely in proportions, sometimes even in rather closely related genera, as in the short-faced *Lophodolus* (Fig. 274) and the very long-faced *Lasiognathus* (Figs. 275, 276).

Of the family Oneirodidæ, in which, according to Regan, the illicial trough extends the whole length of the upper surface of the skull, the most primitive-appearing forms (Fig. 277) are *Dolopichthys luetkeni* and *D. danæ*. The former seems to be well fitted to give rise to the Melanocetidæ, the latter to forms that culminate respectively in *Lasiognathus*, with a very long skull, and *Lophodolus*, with a short deep one. The skulls (Figs. 274–276) of these

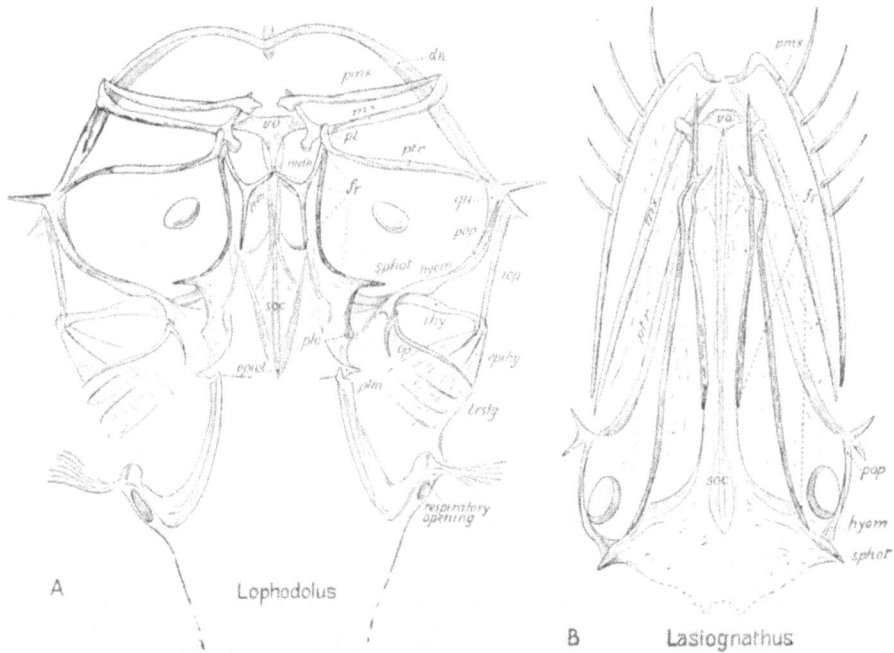

Fig. 276. A. *Lophodolus*. B. *Lasiognathus*. Top views. From stained preparations by Miss Gloria Hollister for Dr. William Beebe.

forms (which I have had the privilege of studying in Doctor Beebe's laboratory) conform in all respects to the family characters of the Oneirodidæ as defined by Regan, and show a remarkably close agreement in the relations of all parts of the opercular, branchiostegal, hyoid and other series.

The Ceratiidæ, according to Regan, are related in skull structure to the Oneirodidæ but the different species differ from *Dolopichthys* in retaining either a well developed or a vestigial second dorsal ray. The Gigantactinidæ, according to Regan, are related to the Ceratiidæ; their enormously elongated illicium is supported by enlarged mesethmoid and lateral ethmoids.

The Melanocetidæ, according to Regan, are also "evidently related to the Oneirodidæ."

The skull is very broad, with widened vomers and triradiate frontals. Parietals are retained as in the Oneirodidæ. The elongate, many-rayed dorsal of *Melanocetus polyactis* (Fig. 272) might at first sight be deemed more primitive than the short six-rayed dorsal of *Dolopichthys luetkeni;* but it seems far more probable that the increase in the number of the dorsal fin-

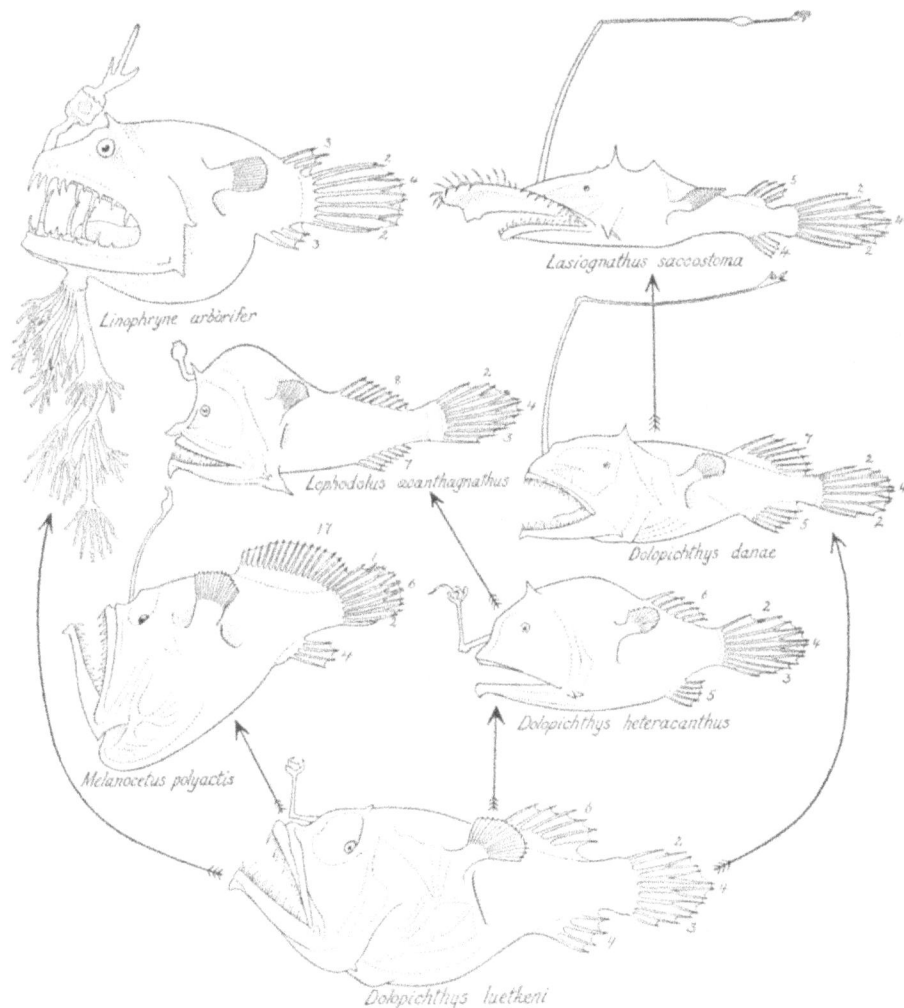

FIG. 277. Divergent Evolution in Ceratioids. Figures from Tate Regan.

rays is quite secondary and that by creating a downward and forward thrust on the column this long fin serves to redress the balance of the fish, which, with its enormously deep jaws and upwardly slanting column, would otherwise find it difficult to move forward rather than downward.

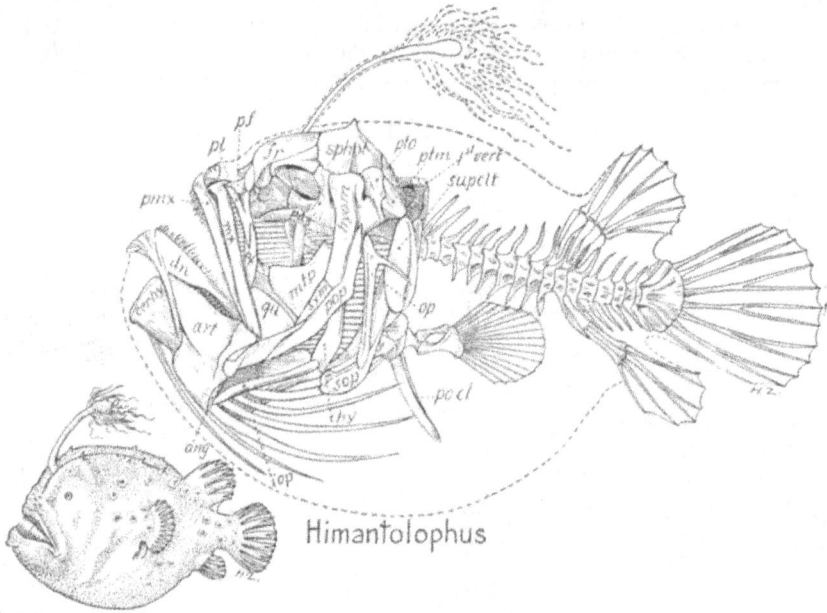

FIG. 278. *Himantolophus*. After Lütken.

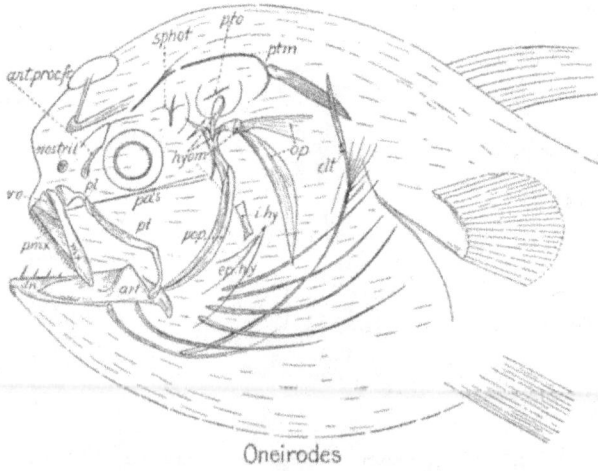

FIG. 279. *Oneirodes*. Very small specimen prepared by Miss Gloria Hollister for Dr. William Beebe.

Himantolophus and its ally *Diceratias* would appear to be almost immediately derivable from *Oneirodes* by the loss of the parietals and the dominance of the supraoccipital and epiotics (Regan, 1926, p. 21). The interfrontal fontanelle is nearly closed. *Himantolophus groenlandicus* Luetken has an oval body (Fig. 278) and a thick skin with large scattered bony plates. The teeth are depressible in 3–5 series, the inner largest. The illicium in this species bears several long branching tentacles. In *Diceratias* of the same family the mouth is upturned. The skull (*op. cit.*, p. 20) has the family characters. *Oneirodes eschrichtii* of the family Oneirodidæ is a rather large fish of oval, rather deep form; it is more specialized than *Dolopichthys luetkeni* in general appearance, especially in its small eyes, very small pectoral fin and thick rays of the median fins. In a very small fish referred to Oneirodes (Fig. 279) the body is globose and the endocranium but little ossified. The short illicium and its basal rod occupy a deep fossa.

The "Aceratiidæ" are extraordinarily specialized and on the whole have strayed the farthest of all ceratioids away from the ceratioid norm. Parr (1930a) has shown that at least some of the "Aceratiidæ" are the dwarfed males of gigantic female ceratioids. In "*Aceratias macrorhinus*" Brauer the body is almost cylindrical in shape, with vestigial dorsal and anal fins. The large eyes are telescopic and directed forward. The illicium has disappeared, as such, but according to Parr it is represented by three denticles at the tip of the vestigial rostrum. On either side, in front of the bulging eyes, projects a great olfactory capsule with two nostrils, of which the posterior is larger and opens backward. The moderate-sized mouth is horizontal.

The top view (Fig. 280) of "*Haplophryne hudsonius*" Beebe (1929) shows that "*Haplophryne*" is already well on the way toward "*Aceratias*" of Brauer. For its eyes are not small but large, they do not look outward but outward and forward, they already protrude widely from their sockets, the orbits are guarded by a sharp sphenotic spine pointing outward and slightly backward. The olfactory sacs with seven olfactory laminæ, while not as well developed as those of *Aceratias*, are precisely in the right position. They have two openings, of which the posterior one faces obliquely backward toward the eye (Beebe, 1929, p. 36) fundamentally as in "*Aceratias indicus*" (*cf.* Regan, 1926, p. 45, Fig. 26). The rostral projection bears three denticles, one median and two lateral, much as in "*Aceratias indicus*." In the side view we note that in "*Haplophryne hudsonius*" the rather small mouth is nearly horizontal and very low down on the head, much as in "*Aceratias macrorhinus*." The general form of the body is such as could readily give rise to that of "*Aceratias*." There are nine caudal rays, as in the latter.

"*Haplophryne*" is, however, more primitive than "*Aceratias*" in certain features besides those noted above: thus it assuredly retains large parietal bones, as does "*Rhynchoceratias*," according to Parr (1930c, p. 9), in spite of the fact that according to Regan (1926, p. 42) these elements are absent in the family as a whole.

"*Rhynchoceratias*" (Fig. 281) has a large rostral bone, the mechanism of which has been figured and described by Doctor Parr (1930c, pp. 8–11). He has shown that this bone is operated by the basal bone of the illicium, with three pairs of muscles, so that it rocks forward and backward along the hinge-like vomer and rests laterally on the anterior symphysial processes of the opposite premaxillæ; thus its denticles oppose the anterior denticles of the lower jaw; he holds also that this bone is not a mesethmoid and that it occupies the position that a greatly enlarged and ossified illicium would occupy. If, however, the rostral denticles represent the illicium itself, it is difficult to account for the appearance of the pre-dentary denticles in the mandible, as already implied by Beebe (1929, p. 27).

That the illicium itself has disappeared in *Rhynchoceratias* is suggested by the fact that in the related *Haplophryne* the illicium, while still present, is minute. Its basal bone,

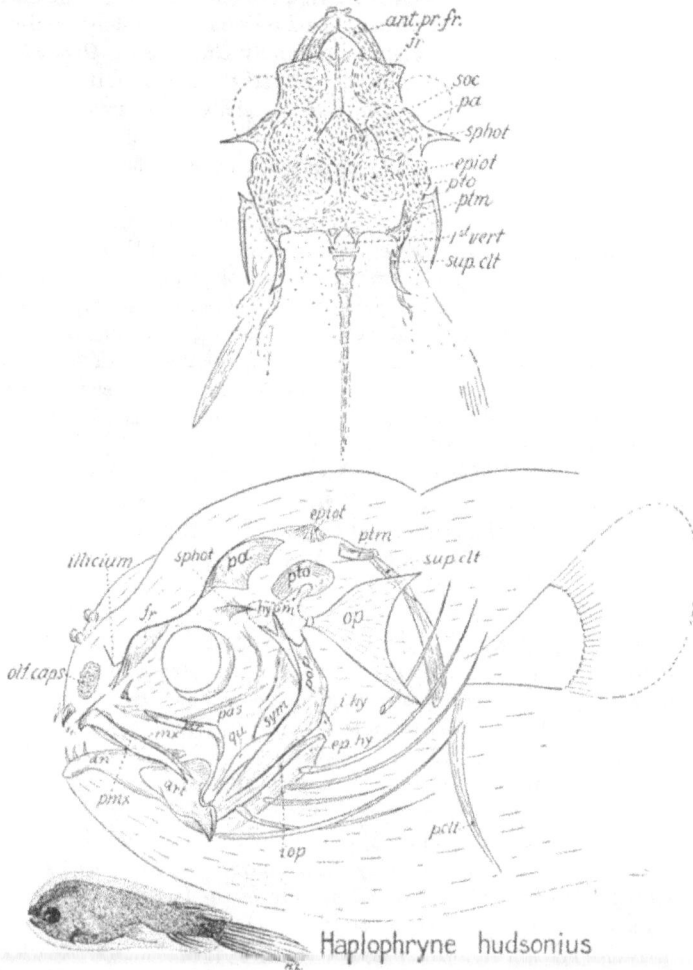

FIG. 280. *Haplophryne hudsonius.* A. Sketch of type. B. Sketch from photograph of type published by Beebe.

however, is present and gives attachment to the three pairs of illicial muscles (Beebe, 1929, p. 35). Doctor Parr endeavors (1930c, p. 15) to dispose of this difficulty by two assumptions: (1) that the so-called illicium of *Haplophryne hudsonius* might just as well represent the second and third tentacles as the first. But as to this, I have seen this little illicium in the type of *Haplophryne hudsonius*, the basal bone of which arises from the interfrontal depression precisely as does that of the true illicium of a young *Oneirodes;* assuredly the assumption that it is not the real illicium in *Haplophryne* but one of the postillicial rods which has moved forward to take its place, looks forced; (2) Doctor Parr further assumes

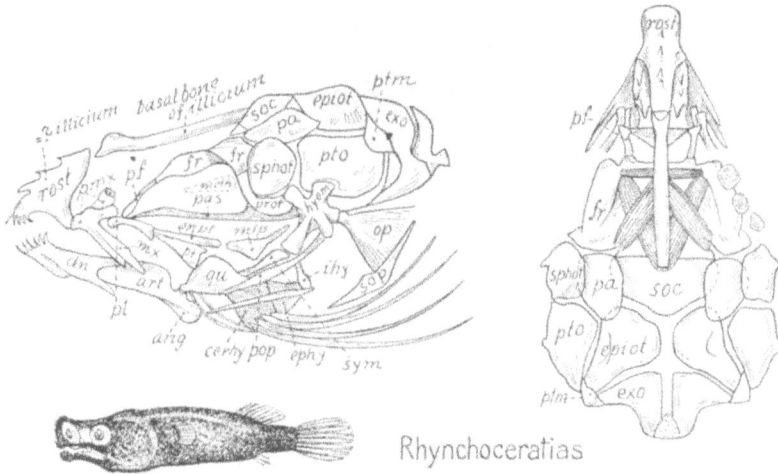

FIG. 281. *Rhynchoceratias.* After Parr.

(p. 17) that the rostral bone has secondarily disappeared in *Haplophryne*. The contrary assumption that *Haplophryne* retains the primitive unossified condition of the rostrum, along with the more primitive rostral denticles, the unexpanded nasal sacs and more normal premaxillæ, makes less demand upon our imagination.

"*Lævoceratias*" Parr (1930c, p. 19) is one of the most remarkable (Fig. 282) of all known pediculates, one which has departed very far from the typical pediculate type. Belonging

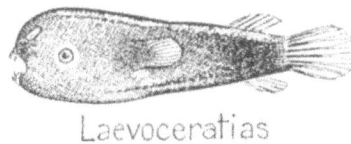

FIG. 282. *Lævoceratias.* After Parr.

to a predaceous group (the ceratioids) in which the mouth is typically cavernous, with long piercing teeth, its mouth has become minute and provided with feeble nipping denticles derived from the skin; belonging to a group whose members usually rely upon a highly developed illicium, it has lost all traces of the illicium except its slender basal rod and possibly the rostral denticles; belonging to a group whose primitive members have small olfactory sacs, the ancestors of *Lævoceratias* evidently first developed such sacs and then reduced at least their external openings.

In brief, whatever the taxonomic status of the several dwarf males referred to as "Aceratiidæ" may prove to be, it seems that this "family" must have diverged from *Haplophryne*-like ancestors into three lines, as follows:

(1) Development of a prominent rostrum and strong rostral bone moved by the former basal illicial muscles; development of varied rostral denticles, opposing similar denticles on tip of forwardly-prolonged mandible, thus producing a nipping type with small diastema; eyes directed outward...*"Rhynchoceratias."*

(2) Elongation of the body, great growth of the nasal sacs, protrusion and forward turning of the large eyes; rostral denticles dwindling; mouth remaining moderate.

"Aceratias."

(3) Marked reduction in size of mouth, which acquires nipping edges; illicial muscles weak, rostrum not ossified, olfactory sacs reduced; head becoming very broad, nearly circular in vertical view, with much shortened rostrum; eyes directed outward and partly upward; general appearance converging toward swell-fish type................*"Lævoceratias."*

PLATE II

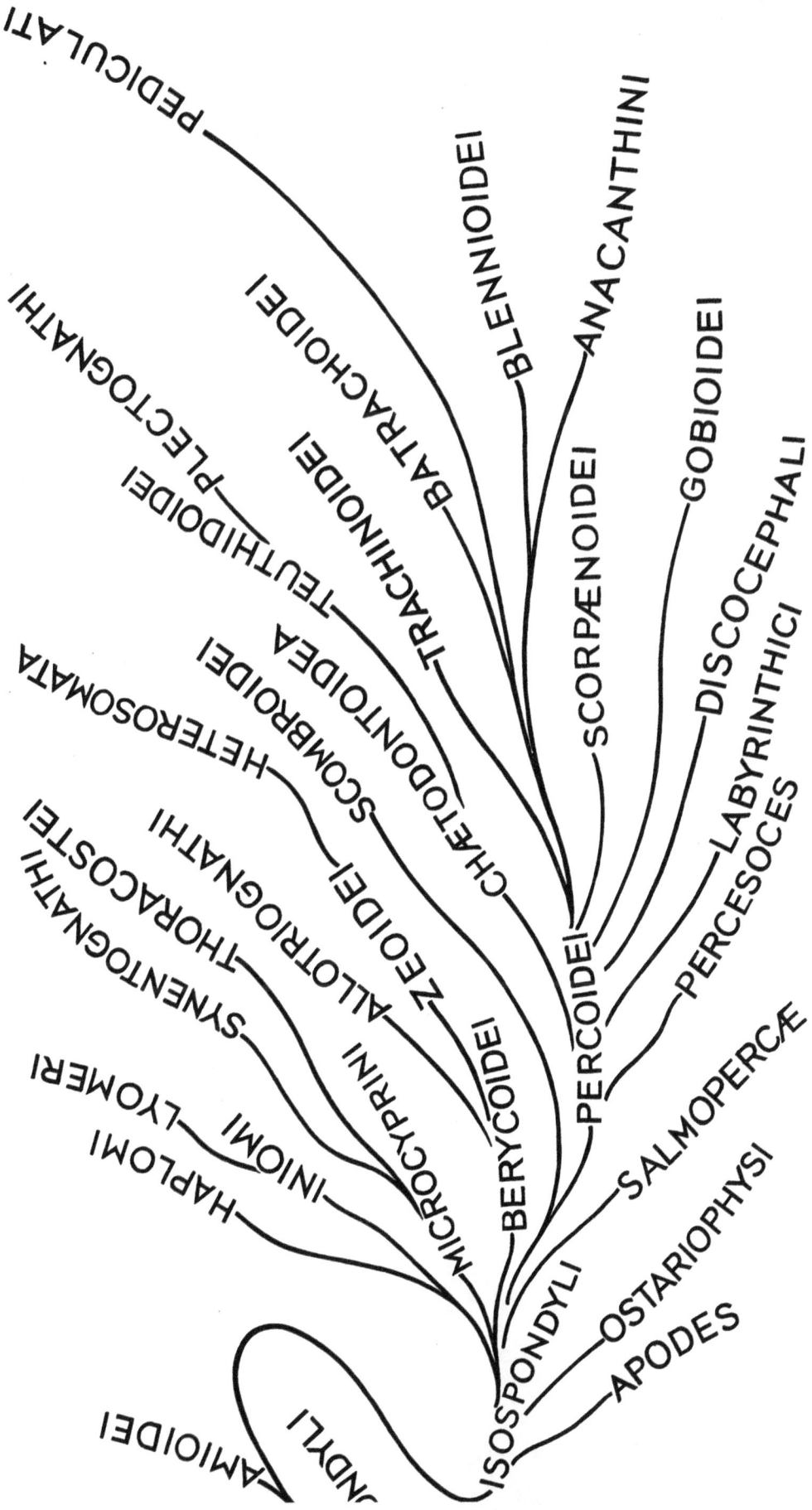

PEDICULATI

BLENNIOIDEI

ANACANTHINI

PLECTOGNATHI

BATRACHOIDEI

TEUTHIDOIDEI

GOBIOIDEI

SCORPÆNOIDEI

TRACHINOIDEI

DISCOCEPHALI

CHÆTODONTOIDEA

SCOMBROIDEI

LABYRINTHICI

HETEROSOMATA

PERCESOCES

THORACOSTEI

ZEOIDEI

ALLOTRIOGNATHI

PERCOIDEI

SYNENTOGNATHI

MICROCYPRINI

BERYCOIDEI

SALMOPERCÆ

HAPLOMI

LYOMERI

INIOMI

OSTARIOPHYSI

ISOSPONDYLI

APODES

AMIOIDEI

...ONDYLI

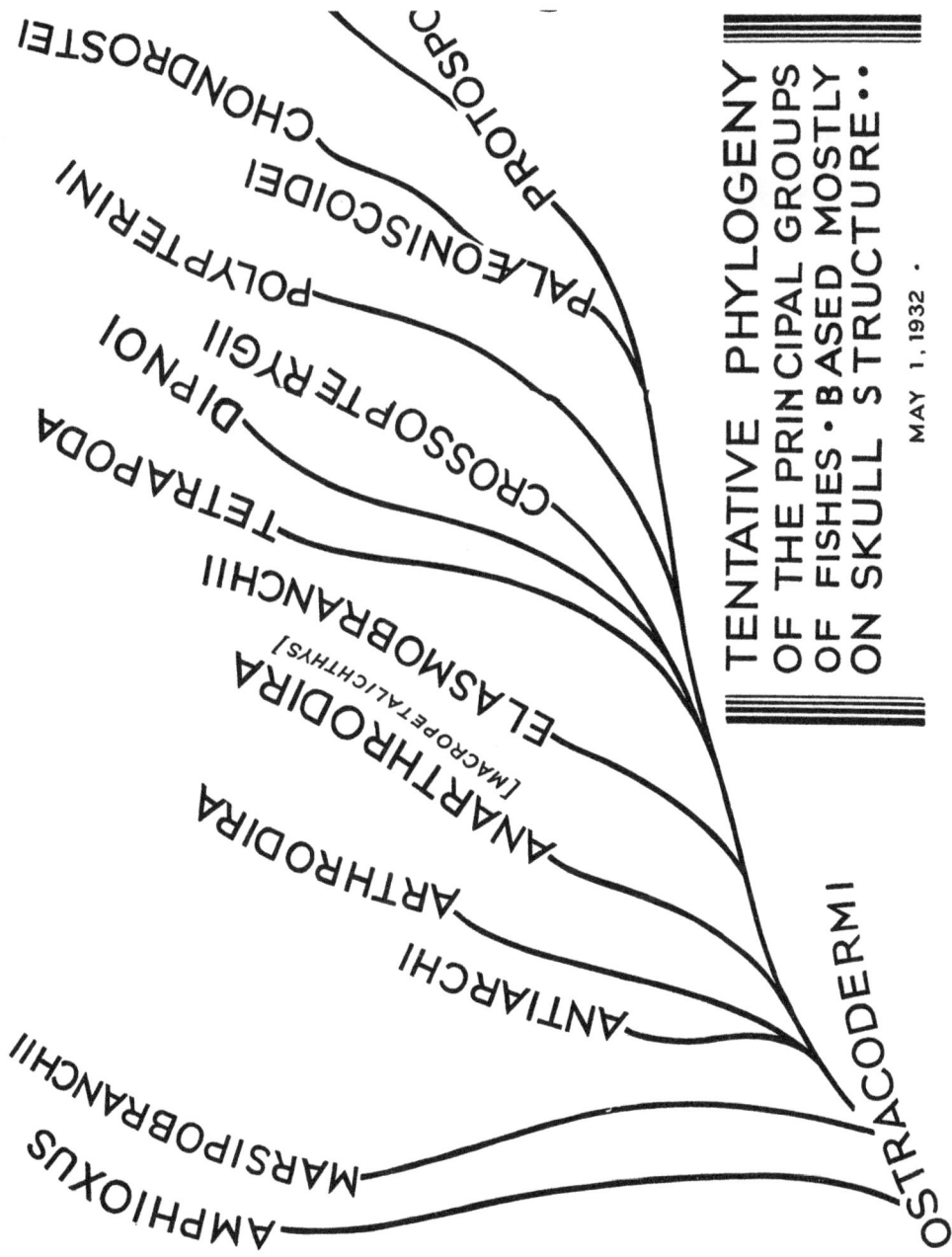

TENTATIVE PHYLOGENY
OF THE PRINCIPAL GROUPS
OF FISHES · BASED MOSTLY
ON SKULL STRUCTURE··

MAY 1, 1932 ·

CHONDROSTEI

PALEONISCOIDEI

POLYPTERINI

PROTOSP.

CROSSOPTERYGII

DIPNOI

TETRAPODA

ELASMOBRANCHII

ANARTHRODIRA
[MACROPETALICHTHYS]

ARTHRODIRA

ANTIARCHI

MARSIPOBRANCHII

AMPHIOXUS

OSTRACODERMI

OSTRACO

TENTATIVE PHYLOGENY OF THE PRINCIPAL SKULL TYPES

IN THE preceding sections the relationships of the group in question have been discussed and in this way the probable steps by which each particular skull type has arisen have been at least implicitly noted. The whole field, so far as I can interpret the data, is summarized in the subjoined diagram (Pl. II).

The groups omitted from this diagram are either *incertæ sedis* or have not been available for the present study.

THE FISH SKULL AS A NATURAL MECHANISM

THE characteristics of the skulls of fishes are mentioned in thousands of taxonomic papers solely because they are valuable as marks for distinguishing one kind of fish from another. A few modern monographs, chiefly by Allis, deal very thoroughly with several types of teleost skulls in the interest of descriptive morphology. Numerous papers or books, by Smith Woodward, Stensiö, Watson and others, give the palæontological data concerning the evolution of particular skull types, while still others, including some by Ridewood, Starks, Boulenger and Tate Regan, record the detailed cranial osteology either of particular groups or of special parts of the skull. Up to the present time, however, but few attempts have been made to study the fish skull as a "natural mechanism" in order to show the functional connections of its various parts.

A fish skull may fairly be called a natural mechanism because it is a product of nature and because it has the properties of mechanisms in general plus certain properties not found in human mechanisms. It is also part of a complete natural machine, the fish itself.

This machine captures stored-up solar energy from the environment and later utilizes part of this energy to operate its mechanism for the capture of more energy. It also normally steers away from danger and into favorable locations and uses another part of the energy in preparing the raw material of the next generation.

Form and Evolution of the Branchiocranium

The branchiocranium, including the jaws and branchial arches, obviously plays a vital part in the capture and turn-over of solar energy. It comprises the so-called visceral arches and all their osseous and cartilaginous appendages. Seen from below (Fig. 5), these arches form a series of inverted V's with the apices pointing forward. The successive apices are connected by median basal pieces. The first V is the mandible, those behind it are the lower segments of the hyoid and branchial arches. Seen from the medial aspect (Fig. 1), the branchial arches form a series of V's with the apices directed backward. The origin of this arrangement is discussed above (p. 83).

It is well known that there are two sorts of jaws, outer and inner. The outer upper jaws are bony dentigerous tracts, the premaxillæ and maxillæ, which in the oldest ganoids form part of the bony facial mask. They were at first fastened tightly to the inner or primary upper jaw (the palato-pterygo-quadrate arch). In the later ganoids and teleosts the premaxillæ and maxillæ become movably pivoted on the ethmo-vomer block. After this step was achieved these elements enjoyed a wide adaptive radiation in the teleosts (Fig. 283).

The inner or primary upper jaw seen from below forms an inverted V with the apex pointing forward. Its cartilaginous core, which in the shark is represented by the palato-pterygo-quadrate, gives rise in the teleosts to the paired quadrate, metapterygoid, mesopterygoid, "pterygoid" (ectopterygoid) and the palatine bones, the latter bearing dentigerous plates.

The mandible consists of an outer dermal shell, the dentary bone; and of an inner core,

412

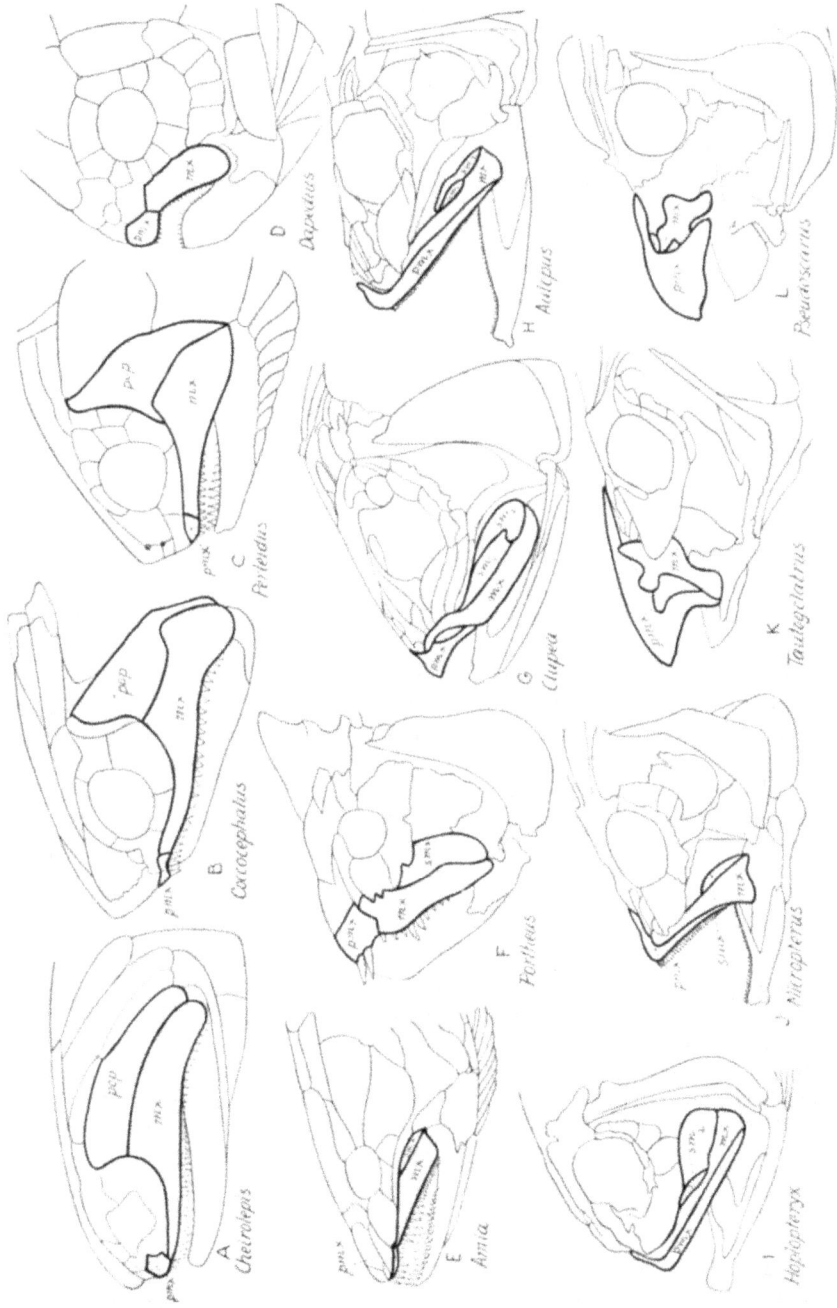

Fig. 283.   Adaptive Radiation of the Premaxilla and Maxilla.

homologous with the Meckelian bar of the mandible of the shark and represented by the endosteal portion of the articular bone. Each half of the mandible forms a simple lever of the third class, pivoted on the quadrate bone by a hinge-like joint (Fig. 284). The main

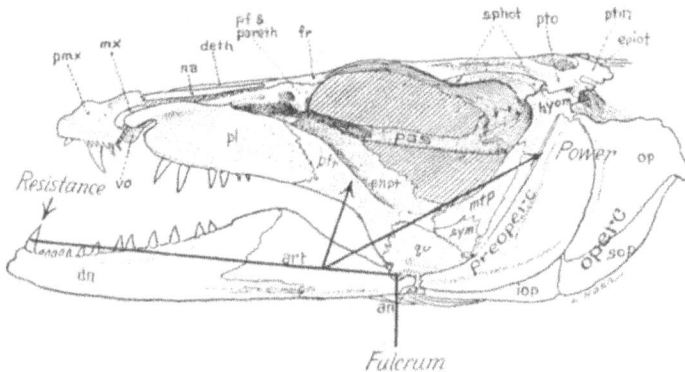

FIG. 284.   The Mandible as a Lever of the Third Class.   Skull of *Sphyræna* after removal of part of the premaxilla and maxilla

tendons of the jaw muscles are inserted in front of the pivot and behind the teeth, through which the power is applied to the resistence (Fig. 285). The jaw muscles are briefly noted below (p. 425).

The inner upper jaw and the mandible are attached to the cranium (Fig. 284) first, by a joint between the ethmo-vomer block and the palatine bone, secondly, through the hyomandibular, the metapterygoid and the symplectic bones, which support the quadrate. The head of the hyomandibular (Fig. 119) is movably attached to the upper outer rim of the otic capsule by two sockets, the anterior one borne jointly by the sphenotic and the proötic, the posterior by the pterotic. The lower segment of the hyomandibular usually tapers down to form the symplectic, which arises from a separate center of ossification; the symplectic is wedged into a groove on the posterior end of the quadrate. It appears to be only a process of the hyomandibular. Apparently the wedging of the symplectic into the quadrate stiffens the latter against the thrusts and pulls of the mandible, while the unossified joint between the hyomandibular and the symplectic permits both to grow in length. Just above the symplectic, on the medial surface, there is a movable ball-and-socket joint with the interhyal or stylohyal, which suspends the lower parts of the hyoid arch from the hyomandibular.

Hence the hyomandibular suspends the upper and lower jaws in front and the lower segments of its own arch below. Collectively, the palato-pterygo-quadrate series and the hyomandibular-symplectic act like a simple V-truss in resisting thrusts of the mandible (Fig. 284). Lateral warping and vertical stretching of the hyomandibular-symplectic-quadrate series is prevented by the lunate preopercular, which is wrapped tightly around the back of the quadrate-symplectic and hyomandibular and is often stiffened with prominent ribs.

The metapterygoid, although often thin, connects the upper part of the palato-pterygo-quadrate arch with the hyomandibular. It helps to strengthen the whole arch and to give insertion to part of the adductor muscles. In the oldest ganoids (pp. 112) there was a vertical wall above the palatine and pterygoid which must have greatly strengthened the palato-quadrate arch and protected the neurocranium.

Adaptive Radiation of the Mouth and its Parts.—The leading modifications of the bony parts of the mouth and throat have conditioned the size, the position, and the direction of

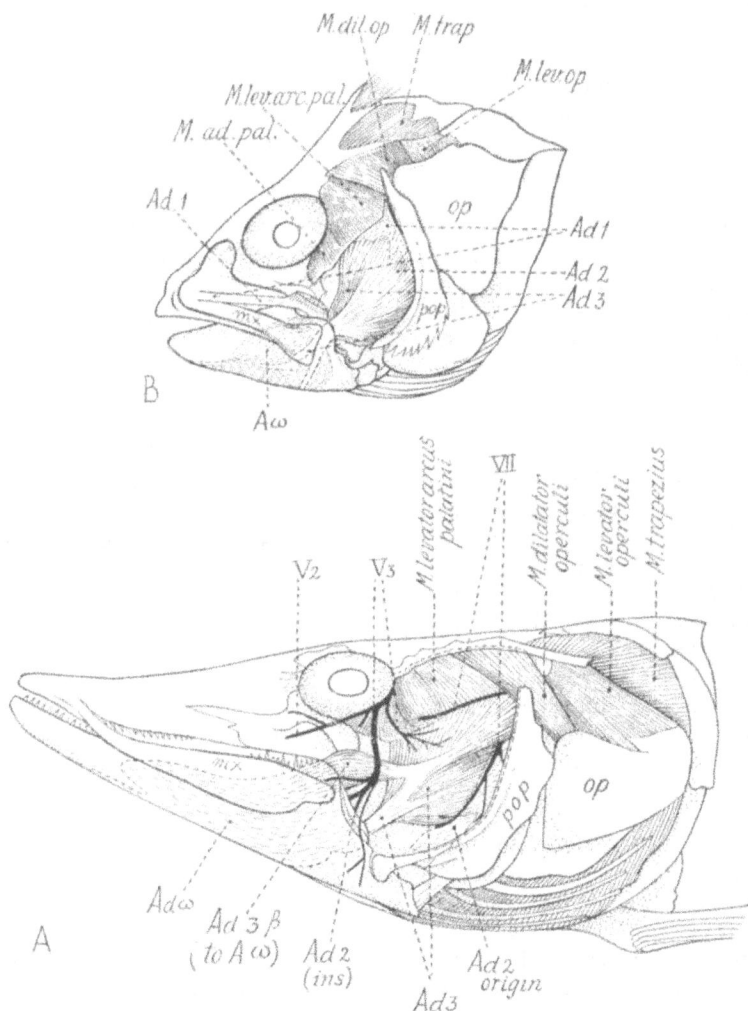

FIG. 285. Relations of the Jaw Muscles and their Tendons. After Vetter. A. Pike (*Esox lucius*) after removal of superficial layers A₁, A₂. B. Typical percoid (*Perca fluviatilis*) after removal of superficial portion (A₁) of the adductor mandibulæ.

The main tendon of A₁ inserts on the inner side of the maxilla. The tendon of A₂ inserts on the inner side of the coronoid process of the mandible. The tendon of A₃ inserts on the inner side of the mandible. Aω nearly fills the inner side of the mandible; its tendon joins that of A₁.

the mouth itself. When we say that the mouth has become larger or smaller or has shifted its position, what is really meant is that if we compare adult stages of successive geologic ages we shall find that the parts around the mouth have expanded or contracted in such a way as to cause the mouth itself to increase or decrease or to shift its direction or position. Similarly, in the individual development changes in size or position of the mouth, as between earlier and later stages, are doubtless due solely to the differential growth of cells surrounding the mouth.

(1) *Size.*—If the mouth increases in size, it is because the jaw-bones, which lie on its periphery, have been correspondingly lengthened. Here we note the existence of many gradations, from the normal predaceous mouth of the salmon (Fig. 45) to the immense cavernous opening of *Gastrostomus* (Fig. 94). In this case the jaws and hyomandibular have become extremely long and slender hoops, by means of which the highly distensible mouth, pharynx and stomach may be drawn gradually over the prey somewhat after the fashion of the anaconda. In these forms, as well as in *Cyclothone* (Fig. 54), *Chiasmodon* (Fig. 137) and others, the jaws and throat also act like a folding scoop-net. A very elaborate apparatus of this sort (Fig. 176) combined with protrusile mouth, is found in the rare *Stylophorus* (Starks, Tate Regan). As the mouth and pharynx increase in size it takes relatively greater power to push them through the water at high speed. Hence the advantage of angulated orobranchial arches, which after the distension in the act of swallowing, permit the folding-up of the apparatus, with consequent reduction in bulk and in resistence.

In *Lophius* (Fig. 267), which is an animated fish-trap, the enormous jaws bristle with pointed teeth. Very long jaws are operated at a mechanical disadvantage and are good chiefly for a sudden rush or a quick snap. Thus their teeth are usually long and narrow. Long teeth may be dagger-like and capable of inflicting severe gashes, as in *Sphyræna* (Fig. 141) and *Omosudis* (Fig. 89), or very long and needle-like, in order to penetrate easily to vital parts of the living prey, as in *Chauliodus* (Fig. 55).

Small mouths, with accompanying reduction in the dimensions of the jaws and related parts, seem always to have arisen by progressive reduction of normal jaws of a predaceous type. Thus, for example, the small mouths of the *Fundulus* group (Fig. 96) appear to have been derived from less minute mouths of the type seen in the Umbridæ, which type in turn is connected with the normal predaceous mouths of the typical Esocidæ, Enchodontidæ and Iniomi. In the actinopterygian fishes as a whole there are many gradations (Fig. 283) from the primitive predaceous type downward to the small mouths of sparids, hoplegnathids, pomacentrids, cichlids, labrids, scarids, chætodonts, teuthids, acanthurids, plectognaths, etc.

(2) The *position* and *direction* of the mouth are likewise determined by the form and arrangement of the jaw parts. The primitive mouth of the palæoniscoid *Cheirolepis* (Fig. 12) is almost terminal and directed slightly upward. From this developed, independently, the peculiar mouths of the sturgeons (Fig. 19) and of *Gonorhynchus* (Fig. 65), which are inferior and directed downward. A downwardly turned oval sucking-disc is developed in some of the cyprinids, loaches and in certain catfishes (Fig. 80). The opposite specialization in which the mouth is superior and directed more or less directly upward is seen in many groups, including the Microcyprini (Fig. 96), the extinct ichthyodectids (Figs. 35, 36), the deep-sea *Argyropelecus* (Fig. 52), *Chauliodus* (Fig. 55), the ceratioids (Figs. 272, 273), and in many bottom-living fishes such as *Lophius* (Fig. 267), the batrachoids (Fig. 263),

certain scorpænoids (Figs. 206, 207), the star-gazers (Fig. 246), etc. Finally, in the flat-fishes (Fig. 224) the mouth is twisted during development so as to face more on the upturned side of the body.

Tube-like prolongations of the preorbital part of the face terminating in small nipping or nibbling mouths are likewise developed independently in different groups. In certain mormyrids the extremely minute nibbling mouth is found at the end of a long decurved tube far beneath the eyes (Fig. 64). Among the tube-fishes (Fig. 105) and their allies the prolongation of the region between the eyes and the mouth becomes amazingly great. Appar-

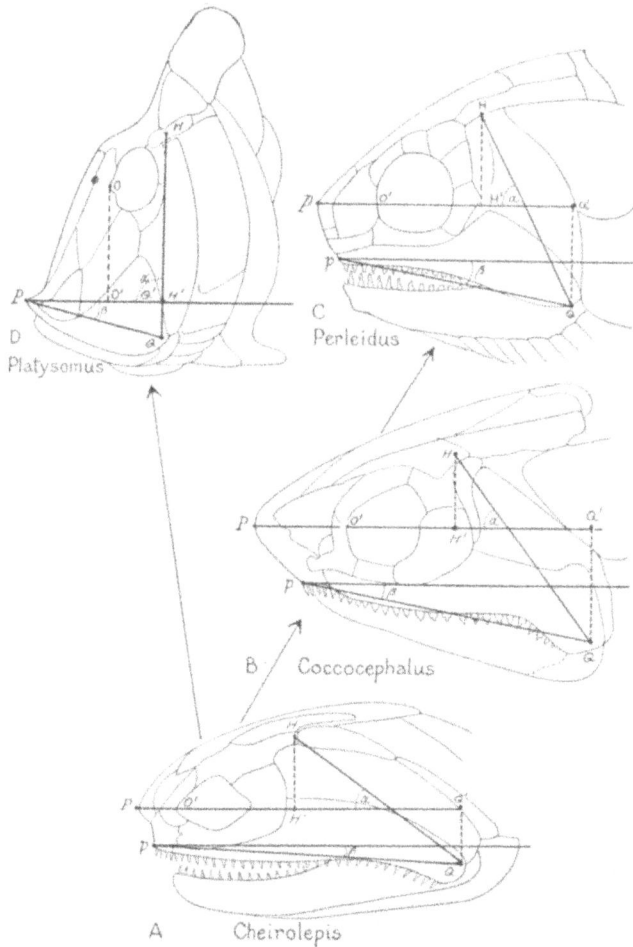

Fig. 286. Progressive reduction of jaw-length correlated with forward swing of the suspensorial line (HQ) in palæoniscoids. A. *Cheirolepis*. After D. M. S. Watson. B. *Coccocephalus*. After Watson. C. *Perleidus*. After Stensiö. D. *Platysomus*. After Watson.

ently such tubes may serve as a probe for poking into seaweed or into crannies, as among the sessile organisms of a coral reef, in search of minute food.

In every case the attainment of the specialized position and direction of the mouth has involved the differential growth of all the bones of the jaws and suspensorium. An increase in the length of the gape, involving a lengthening of the jaw, may be effected either

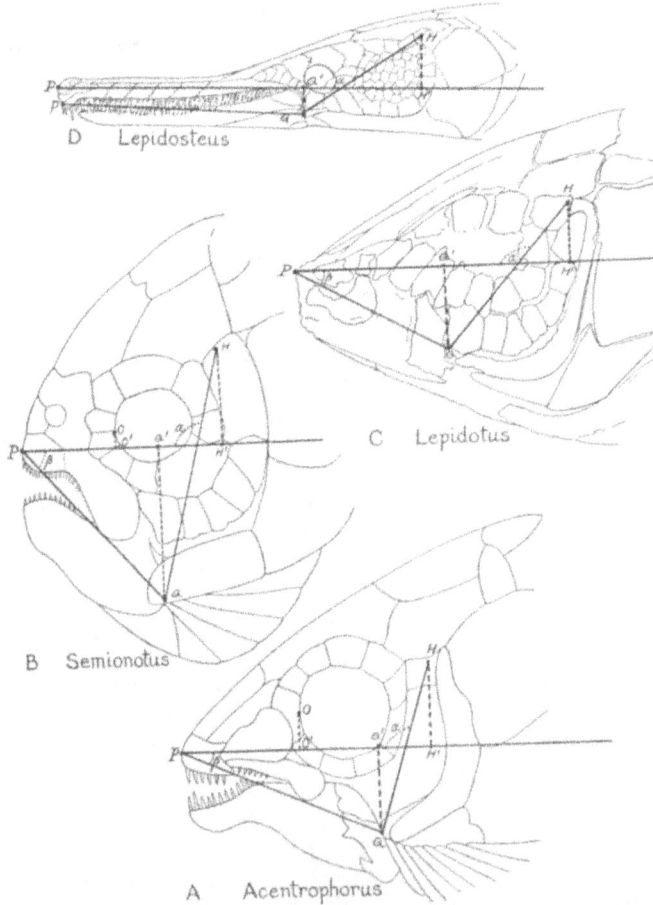

FIG. 287.   Forward swing of the suspensorial line (HQ) in holosteans. A. *Acentrophorus*. After E. L .Gill. B. *Semionotus*. After A. S. Woodward.   C. *Lepidotus*.   After A. S. Woodward.   D. *Lepidosteus*.

by moving the quadrate-articular facet (*Q*) downward and backward (Figs. 40, 94), at the same time rotating the suspensorium backward, or, if the suspensorium is already directed forward, the jaw may be lengthened if its anterior tip grows forward (Figs. 286*D*, 289). Similarly a reduction in the size of the mouth (Fig. 286) is very apt to be conditioned by a forward displacement of the quadrate-articular joint.

Dimensional Factors in the Mouth Region.—In order to attain a more graphic and precise formulation of the chief dimensional factors in the varying lengths, positions and directions of the mouth-parts, I have prepared the following definitions and the accompanying diagrams (Figs. 286–289).

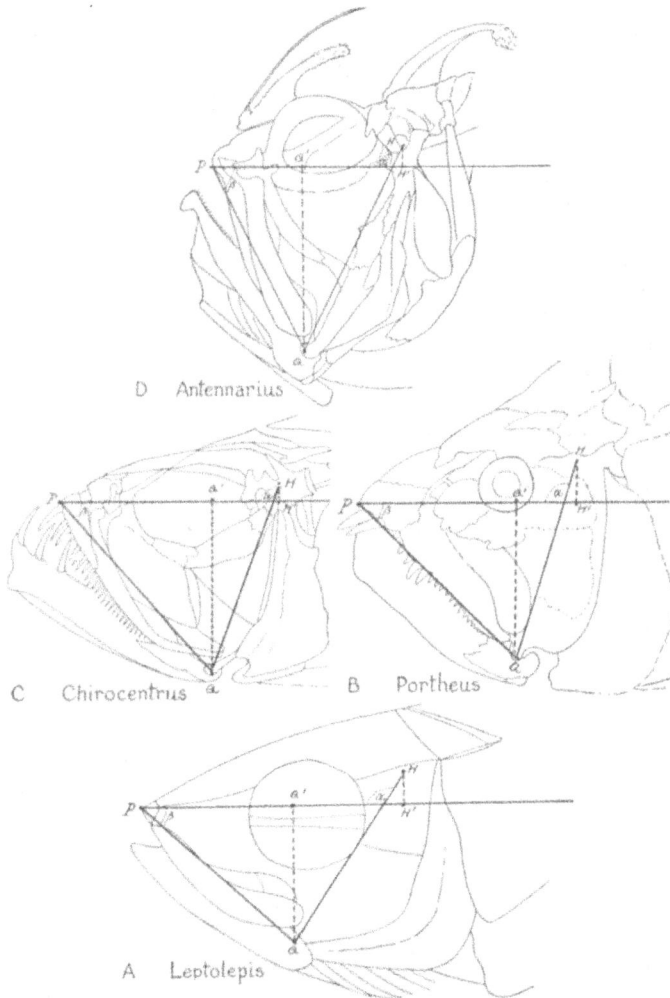

FIG. 288. Upturning of mouth correlated with depression of quadrate-articular fulcrum (Q).

The *prosthion* (P) is the most anterior point of the snout (Fig. 286).
The *menton* (M) is the most anterior point of the mandible.
The *pygidion* is the mid-point of the narrowest part of the caudal peduncle.

23

The *horizontal* is assumed as the line joining the prosthion and the pygidion.

The *suspensorial point* (*H*) lies midway between the centers of the anterior and posterior facets for the hyomandibular.

The *quadrate pivot* (*Q*) is the mid-point of the articular hinge of the quadrate.

The *suspensorial line* (*HQ*) joins the suspensorial point with the quadrate pivot.

The *suspensorial angle* (α) is the inclination of the suspensorial line to the horizontal.

The *premaxillon* (*p*), or *premaxillary tip*, is the most anterior point of the premaxillary in *norma lateralis* (often *p* coincides with *P*, as in Fig. 286).

The general *line of the gape* is usually the line *pQ* or *PQ*.

The *angle of the gape* (β) is the inclination of the line of the gape either to the horizontal or to a line parallel with the horizontal.

The *maxillary angle* (β) of the closed mouth is the inclination of the alveolar border of the upper jaw (projected in a parasagittal plane) to the horizontal axis of the body.

The *level of the quadrate-articular center* (*Q*) is its distance (*QQ'*) below the horizontal.

The *anteroposterior placement of the quadrate-articular center* (*Q*) is its distance *Q'H'* in front (+) or behind (−) the projection of the mid-point (*H*) of the hyomandibular socket upon the horizontal.

The *orbiton* (*O*) is the most anterior point of the orbit.

The *level of the orbit* is the height (*OO'*) of the orbiton above the horizontal.

The *height of the hyomandibular socket* is the height (*HH'*) of its mid-point (*H*) above the horizontal.

The *effective rostral length* (*PO'*) is the distance from the prosthion or tip of the snout to the projection (*O'*) of the orbiton (*O*) upon the horizontal.

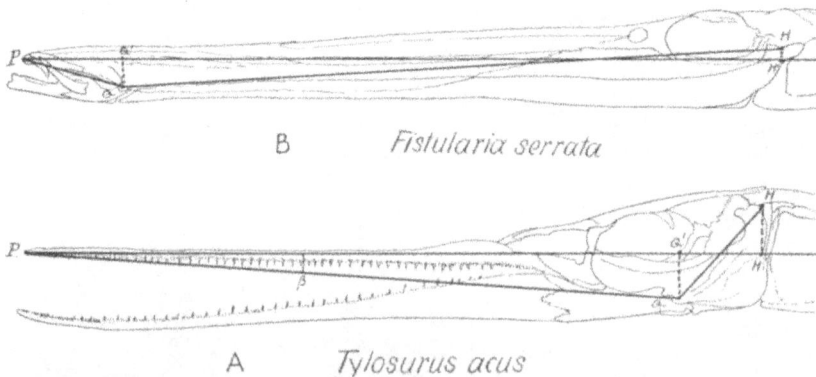

B   *Fistularia serrata*

A   *Tylosurus acus*

Fɪɢ. 289. Contrasting conditions of *Tylosurus* and *Fistularia*. (A) Short suspensorial line (HQ) combined with extreme lengthening of PQ'. (B) Extreme lengthening of suspensorial line (HQ) combined with very short PQ'.

In Figures 286–289 the points *Q*, *H*, *O* are projected upon the horizontal as *Q'*, *H'*, *O'*. Taking the point *H'* as zero, a positive displacement of *Q'* will lengthen the gape, as in Fig. 285*A*; a negative displacement of *Q'* will shorten the gape, as in Fig. 288.

A lengthening of *PQ'* will decrease the angle of the gape (β), as in Fig. 289, while a shortening of *PQ'* will increase the angle of the gape, as in Fig. 287*C*.

A raising of the point Q (with shortening of QQ') will decrease the angle of the gape, as in Figs. 287D, 286D; while a lowering of the point Q will increase the angle of the gape, as in Fig. 288B, D.

The Suspensorium.—The suspensorium of long jaws is primitively inclined backward,

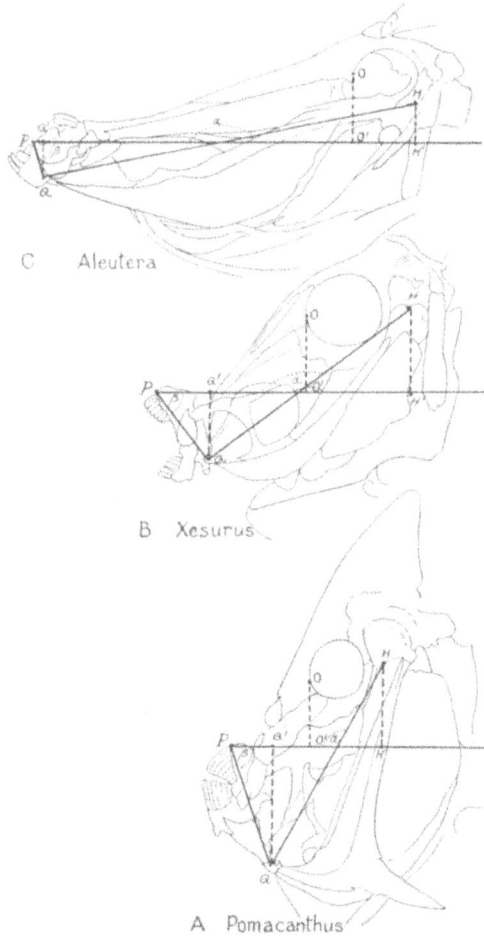

Fig. 290. Progressive forward swing of the suspensorial line (HQ) with protrusion of preorbital face and reduction of mouth in balistoids.

as in *Cheirolepis* (Fig. 286A), so that a good part of the jaws lies behind the orbiton (O'); but the teleosts appear to have been derived from small-mouthed forms (Fig. 287A) with a forwardly inclined suspensorium which, with some exceptions, retains its forward inclination even after the jaws become secondarily elongated, as in *Lepidosteus* (Fig. 287D) and the needle-gars (Fig. 289*A*). As to the exceptions, in the engraulids (Fig. 40), morays (Fig. 82*B*), gonostomids (Fig. 53), gastrostomoids (Fig. 56) and some others, the hyomandibular has become much inclined backward in correlation with the backward displacement of the quadrate-articular joint and the backward lengthening of the jaws.

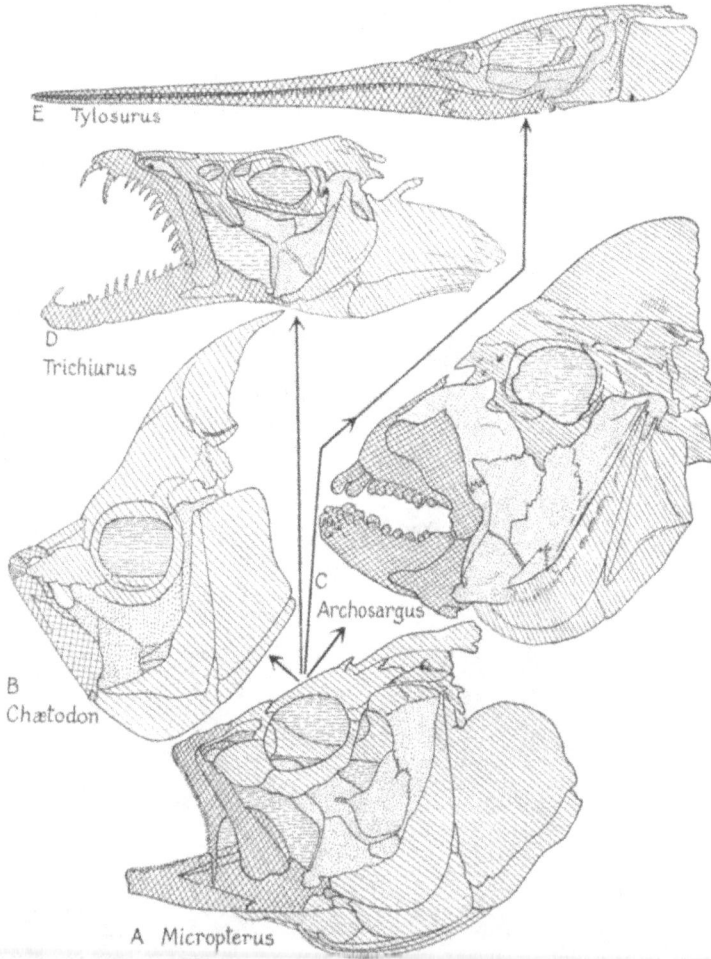

FIG. 291. Adaptive radiation of jaws and teeth in percomorphs.

The opposite case in which the suspensorium and the quadrate-articular joint are produced very far forward is exemplified in the balistoids (Fig. 290).

In the embryonic and larval stages of most teleosts the eyes are very large, the mouth small and the suspensorium is curved forward beneath the eye. This arrangement in earlier developmental stages may in many forms tend to be retained in the adult (see p. 431).

The Dentition.—The lining membrane of the stomadæum that covers the primary jaws and the branchial arches not only gives rise to the teeth in the premaxillæ, maxillæ and dentaries but also to those on the vomers, palatines, ento- and ecto-pterygoids, parasphenoid and coronoid or "splenial"; while from the same source arise the teeth on the pharyngeal surface of the hyoid and branchial arches.

In predaceous types as a rule the emphasis falls on the teeth of the outer jaws carried by the premaxillæ, maxillæ and dentaries, as already noted. But in the predaceous pikes, barracudas and other types the inner jaws and roof of the mouth also bristle with a *chevaux-de-frise* of long, more or less backwardly directed teeth, which obviously function to prevent the escape of the struggling and partly swallowed prey. Similar conditions are found in certain predaceous deep-sea forms such as *Omosudis* (Fig. 89). In such forms the primary or inner upper jaws (palatopterygoids) will naturally be sufficiently strong to support these large teeth. In the morays (Fig. 82*B*), on the other hand, the palatopterygoid tract degenerates, while the maxilla and vomer bear the principal teeth; the ceratohyals also bear recurved teeth and function like an inner pair of jaws.

The presence of stout teeth on the palate and inner sides of the jaw in *Osteoglossum* and *Lepidosteus* necessitates extra bracing of the palate (see pp. 164, 129).

Crushing teeth are developed on the massive premaxillæ and dentaries of the sparids (Fig. 123) and in *Anarrhichas* (Fig. 255) and bracing bones of the jaws are very strongly built. Figure 291 illustrates five of the principle types of jaws, namely, the normal (*A*), the predaceous (*D*), the small-mouthed or nibbling (*B*), the crushing (*C*) and the needle-gar (*E*) types. These have been evolved more than once independently among the lower and the higher fishes, as in the following examples:

	I Normal (with small teeth)	II Predaceous (usually with large sharp teeth)	III Small-mouthed, nibbling	IV Crushing (with molars or tritors)	V Needle-gar (with or without teeth)
Crossopterygii....................	*Osteolepis*	*Megalichthys*	———	*Mylacanthus*	———
Palæoniscoids....................	*Trissolepis*	*Cheirolepis*	*Mesolepis*	*Cheirodus* (tritors)	*Saurichthys*
Holostei.........................	*Amia*	*Protosphryæna*	*Mesodon*	*Lepidotus*	*Aspidorhynchus*
Isospondyli......................	*Salvinellus*	*Astronesthes*	*Chatoessus* (no teeth)	———	
Ostariophysi.....................	*Erythrinus*	*Hydrocyon*	*Distichodus*	———	———
Percomorphi.....................	*Micropterus*	*Trichiurus*	*Chætodon*	*Archosargus*	*Tylosurus*

In addition to these types the edentulous and protrusile types are noticed below.

In the extinct pycnodonts the teeth on the roof of the jaw became hemispherical and were arranged on a tapering cylindrical surface. They were opposed by similar teeth arranged in a U-shaped trough supported by the inner sides and floor of the mandible. In the Cretaceous and later Elopidæ, Albulidæ and related families, the roof of the mouth is thickly strewn with granular or small, more or less circular teeth, which oppose similar teeth on the tongue (glossohyal) and adjacent parts (Woodward, 1901, p. vii; Ridewood, 1904*a*, p. 50). In *Gonorhynchus* a circular patch of blunt teeth on the second basibranchial engages with teeth on the entopterygoids.

Pharyngeal teeth, on the fourth pharyngo-branchials in the upper, and on the fifth cerato-branchials, or the lower branchial elements, are present in many groups of fishes. In the carps and their allies the fifth ceratobranchials bear variously shaped processes which oppose a horny pad that rests on a bony projection from the basi-occipital. These pharyngeal teeth are drawn upward against the pad by powerful muscles that are attached to the lateral fossa of the cranium. In certain cyprinodonts (Microcyprini) the opposite toothed lower pharyngeals are united by a dentate suture and together form a triangular plate that opposed the teeth on the upper pharyngeals. In the skippers and flying-fishes the opposite lower pharyngeals are coalesced into a solid triangular block (whence the name Synentognathi). A more or less similar pharyngeal dental apparatus is developed in the pomacentrids, cichlids, chætodonts, embiotocids, labrids. In the scarids (Fig. 134) this pharyngeal mill becomes very elaborate, the elongate upper many-ridged plate sliding back and forth on the parasphenoid, while the lateral processes of the fused lower pharyngeals fit into grooves on the inner side of the cleithra.

Among the more remarkable types of teeth developed in the front of the mouth we may recall the incisiform front teeth of certain sparids (Fig. 123) and of *Gobiesox* (Fig. 249), the immensely long sabre-like lower tusks of *Chauliodus* (Fig. 55), the bristling front teeth of pomacentrids (Fig. 128), *Blennius* (Fig. 252) and *Chætodon* and the beak-like pincers, presumably developed from the fusion of many bristle-like teeth, in the scarids (Fig. 133), hoplegnathids (Fig. 124), siganids (Fig. 159), tetraodonts (Fig. 169), diodonts (Fig. 171).

The edentulous condition has been developed independently in various groups. In *Polyodon* (Fig. 17), for example, as well as in the whale-shark (*Rhineodon*), the manta (*Mobula*) and the basking-shark (*Cetorhinus*), the food consists of plankton or small fish, which are engulfed in the capacious mouth and swallowed whole. In all these cases the gill-slits are very large but the escape of the food is stopped by an extra-branchial sieve of one sort or another. At the same time the teeth are either more or less reduced (*Cetorhinus, Rhineodon, Mobula*) or completely absent (*Polyodon*). Somewhat similar adaptations for plankton feeding are seen in some of the Clupeidæ (*e.g., Brevoortia*). Teeth are also lost in some of the protrusile jaw types.

Protrusility.—The evolution of a strongly protrusile mouth has occurred quite independently in at least two widely removed orders, the Ostariophysi (p. 189 and Fig. 76) and the Acanthopterygii (p. 239 and Fig. 116). It has been noted by Delsman (1925) and by Regan (1924) that in the most extreme stage among certain wrasses even the quadrate, which is fixed by the surrounding bones in all preceding stages, has now acquired a flexible joint with the hyomandibular and can be swung far forward to permit extreme protrusion of the premaxillæ (Fig. 131). In *Phractolæmus* the toothless protractile mouth when at rest can be folded back on the top of the snout (Goodrich, 1909, p. 391).

The Branchial Arches.—As to the branchial arches themselves, obviously the most important factor affecting them is the size, number and character of the gills. In the primitive cephalaspid ostracoderms there were nine interbranchial septa (Stensiö, 1927, p. 161) or visceral arches, all bearing gills. In the existing elasmobranchs the normal number of gills is five and it may well be suspected that the six or seven gill-slits of the notidanoids and the six of *Pliotrema* represent a secondary increase in number. In all ganoids and primitive teleosts there are five gills but in many of the more specialized teleosts the number sinks to four and a half, four, three and one-half. In the morays and gastrostomoids there has been a reduction in the size of the gills and gill-arches, which is apparently compensated by the forced respiratory current set up by the syringe-like move-ments of the muscular opercular flap. In fishes with heavy pharyngeal dentition the supporting parts of the branchial apparatus are correspondingly strengthened or braced against such near-by bones as the cleithrum (in the scarids) or the base of the cranium (in the clupeoids).

Muscles of the Branchial Arches and Jaws.—In the sharks, as already noted, there is an elaborate system of muscles for flexing and extending the joints of the orobranchial appa-ratus, including superficial constrictors, which pass over the surface of the arches and the deep extensors and flexors (see Fig. 4). In the teleosts some of the dorsal muscles of the gill-arches spring from the side of the braincase, while the ventral ones form a complex system extending forward from the anterior border of the cleithrum and extending transversely across the floor of the throat between the hyoid arches. The branchial muscles, which have been beautifully figured by Allis in *Chlamydoselachus, Amia, Scomber,* are doubtless modified in many special ways in accordance with the various movements of the branchial arches, but about this subject little is known.

The jaw muscles, which are regarded by Vetter (1874; 1878, p. 542) as serially homol-ogous with the middle deep flexors of the branchial arches, extend fan-wise from the curved border of the preopercular to the mandible; one division (Ad 1) being connected with a tendon that is inserted on the medial surface of the maxilla. These muscles have been described in detail especially by Vetter (1874, 1878) and by Allis (1903, 1909). Their tendinous parts meet in a complex central tendon with several branches to the maxilla and mandible. The fan-like origin of *A1, A2,* from the preopercular and from the area in front of it probably insures smooth, continuous and efficient action as the mandible moves upward and as one part of the adductor after another comes into the phase of maximum extension. As noted above (p. 414) the mandible as a whole acts as a lever of the third class in which the point of application of the power lies between the resistance and the fulcrum. The coronoid process, either of the dentary or of the articular bone, acts as a lever to increase the power and decrease the range of movement of the teeth, especially of those on the front end of the mandible. Hence the coronoid process of fishes with strong-biting or pincers-like jaws is usually large or massive in proportion to the length of the mandible (Figs. 71, 167, 124, 133).

The Opercular Elements.—As to the history of the gill-covers, it is well known that these are foreshadowed in the opercular flaps of the chimæroids, and that all known earliest Crossopterygii and Actinopterygii were already in possession of a complete or nearly com-plete opercular system. The opercular, subopercular and branchiostegal bones arise as local dermal ossifications in a folded flap of skin that projects backward from the hyoid arch. The opercular is attached by a concave facet to the opercular pedicle of the hyo-mandibular, while the subopercular and branchiostegals are connected by ligaments with the lower segments of the hyoid arch. The interopercular according to Tate Regan (1929, p. 313) represents the lower end of the opercular, but Allis (1909, p. 69) suggests that it may be regarded as the branchiostegal ray of the interhyal, with which it is connected. It is also attached by a short ligament to the angular bone of the mandible, and by watching any large percoid fish in an aquarium one may see that in the respiratory movements the inter-opercular keeps the opercular series in step with the mandible, while the branchiostegals move up and down with the cerato- and epi-hyals.

The opercular bone is small in the primitive palæoniscids, which have a large sub-opercular (Fig. 12*A*), while in the sturgeons the large rounded "opercular" has no contact with the hyomandibular and appears to be an hypertrophied subopercular (see p. 119). In the morays (Fig. 82*B*) the upper two-thirds of the opercular series has been sacrificed in order to give room for the swelling movements of the branchial syringe. In the gonostomids (Fig. 53) the opercular system has become extremely narrow and in the gastrostomids it has disappeared, probably in connection with the great distensibility of the mouth and throat. In all other teleosts, so far as I know, the opercular retains its normal articulation with the pedicle of the hyomandibular. The opercular region is enlarged in *Ophiocephalus* (Fig. 145), *Osphronemus* (Fig. 147) and other fishes with enlarged respiratory chamber.

The opercular flap often extends posteriorly beyond the opercular bone and the curve of the posterior bevelled borders of the opercular and subopercular is always adjusted to the curve of the cleithrum so that there is a smooth fit, which is also insured by the flexibility of the border of the opercular flap and by the oblique position of the anterior surface of the cleithrum. The two projecting processes on the posterior border of the opercular (Figs. 114, 201) separate the spiracular region above from the vertically movable branchiostegal region below. The branchiostegal flaps act as pumps and valves for the rhythmic escape of the inspired water and coöperate with the movements of the breathing valves, which are folds of membrane on the fore part of the roof and floor of the mouth.

It can readily be seen that the stresses set up by the dilatator, adductor and levator operculi muscles (Fig. 285), together with the downward and forward pull of the interopercular, might well tend to cause a buckling of the smooth contour of the opercular, with consequent leaking of its valvular edge. This contingency is apparently eliminated in the typical Acanthopterygii by the development of two divergent tracts of folded trabeculæ (Fig. 202), radiating from the fulcrum or operculo-hyomandibular contact, respectively to the farthest points or projections on the upper posterior borders of the opercular. An oblique view of this region in a living fish shows that during dilatation of the opercular the principal radiating ridge, which is continued posteriorly into the main opercular spine, is raised just ahead of the flexible edge of skin beneath it. In the pediculate fishes (Figs. 267, 272, 279) these two divergent tracts of bony trabeculæ persist and stand out, even in very small specimens, while the main part of the bone has become tenuous and translucent.

The preopercular, although nominally belonging in the opercular series, lies between the field of the adductor muscles of the jaw (Fig. 285) and the opercular flap and has functional relations with these and other parts; it also supports the preopercular branch of the latero-sensory canal (Fig. 203). It is usually a crescentic or boomerang-like bone which follows the general curve of the suspensorium and fits between a vertical crest of the hyomandibular and the anterior border of the opercular fold. In the embryonic and larval stages (Fig. 204) the preopercular region is pressed close to the eye and doubtless its "circumorbital" position in the adult is partly determined by this early condition. At its lower end it fits behind the posterior crest on the quadrate; being concavo-convex as seen from the outer side, it often seems to play an important rôle in bracing and stiffening the quadrato-articular joint. The posterior border is sometimes serrated (as in various percoids) or provided with a number of projecting spikes, which in scorpænoids (Figs. 201, 212) have the appearance of protecting the enlarged lateral line organs.

From the almost invariable association of the preopercular with this latero-sensory canal, it seems at first that the course and position of the preopercular canal are the chief factors in determining the location of the preopercular itself. Indeed Ridewood (1904*a*, p. 68), following Cole and Johnstone (1902, p. 175), classifies the preopercular with the lacrymal, nasal, suborbital and supratemporal (pterotic, etc.) as bones "developed primarily around a portion of the lateral-line system," and therefore of a different nature from the other opercular bones. On the other hand, the preopercular fold seems to have essentially the same relation to the quadrate and metapterygoid that the opercular fold (including the inter- and sub-operculars as well as the opercular itself) has to the segments of the hyoid arch; moreover, it is only in such advanced groups as the scorpænoids that the preopercular seems to be so closely dependent upon the tunnel of the latero-sensory canal that perforates it. In the older ganoids the texture and general appearance of the preopercular is close to those of the postorbital and opercular series, which do not carry latero-sensory canals. Again, when the quadrate-articular joint is shifted forward, as in many small-mouthed fishes, the lower end of the preopercular follows it; the upper end also follows the general line of the suspensorium. Hence I conclude that the position of the suspensorium and the close appression of the preopercular to the eye in the embryo are the leading factors in the position of the preopercular and that the position and course of the preopercular-sensory canal are secondary factors. Finally the exact form of the concave anterior border of the preopercular is probably conditioned in part at least by the position of the three subdivisions ($A1$ $A2$ $A3$) of the adductor mandibulæ muscles, which are fastened to the border of the preopercular and the outer surface of the metapterygoid (see page 425 above).

From the posterior corner or elbow of the preopercular a single backwardly-directed spike sometimes attains great size, as in *Holocentrus* (Fig. 112), *Prionotus* (Fig. 218), *Dactylopterus* (Fig. 223), *Callionymus* (Fig. 242), *Gobiesox* (Fig. 249). Such spikes have the appearance of being useful either as defensive weapons (when moved by the wriggling of the body), or as "skids" upon which the body rests (*Dactylopterus* (Fig. 223)), or as accessory braces for a pelvic adhesive organ (*Gobiesox* (Fig. 250), *Callionymus* (Fig. 242)). Apparently such projecting spikes on the posterior border of the opercular have always originated from one of the less prominent spikes of more primitive acanthopts. These projections often start from near the center of growth of the preopercular and may have arisen in the first place as strengthening ribs or projections on that part of the bone which is subjected to the heaviest stresses (Figs. 114, 202).

In the pediculates the opercular apparatus as a whole is naturally much influenced by the great expansion of the pharynx (Figs. 265, 267, 272); here the opercular itself is reduced to a thin scale supported by two long diverging tracts representing the upper and lower spine of the opercular, while the subopercular, and still more the interopercular, are produced into trackers; the branchiostegals form very long, curved and delicate bands, which follow the swelling skin flap that forms the chief functional element in respiration.

In short, the bones of the opercular region teach us very clearly that they are only local precipitates in growing flexible membrane and that their boundaries are predetermined by the location of the movable creases in the membrane itself. At the same time it is evident that the presence of each bone of the series is due to hereditary factors. Thus the vast majority of fish inherit the normal elements of the opercular series, the only variability being in the number of branchiostegal rays (see pages 135 and 231 above).

Embryology of the Jaws, Hyoid Arch and Opercular Series.—W. K. Parker has figured the developmental stages of the skulls of the salmon, sturgeon, *Lepidosteus*, while Bigelow and Welsh (1925) have figured developmental stages of many other teleosts.   Larval stages and fry of many species of fishes were also studied by members of the New York Zoölogical Society's *Arcturus* expedition and in Dr. William Beebe's laboratory at Bermuda.   From such data it is easy to trace the ontogeny of the mouth parts, suspensorium and opercular region as well as of certain parts of the chondrocranium.

In the embryonic stages of many teleosts, the eye and the brain are accelerated and very large, so as to dominate the head.   The mouth becomes functional about the time that the yolk is used up.   At first (Fig. 292) the mouth is usually very small, suited for capturing

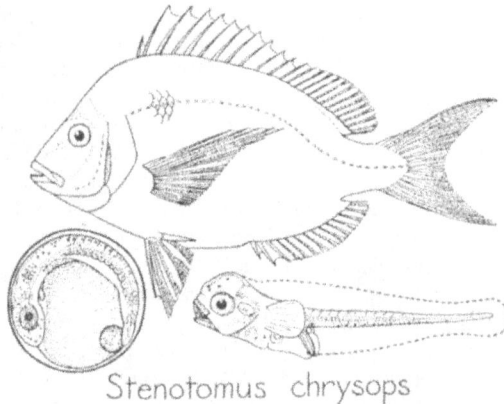

FIG. 292. *Stenotomus chrysops.* Egg, larva and adult. After Bigelow and Welsh.

only minute food such as diatoms or small copepods.   The snout is extremely short and the mouth upturned.   Consequently the suspensorium is produced forward beneath the very large eye.   The latter depresses the palatoquadrate bars, which thus appear falsely to be associated in origin with the eye (Parker, 1873, Pl. I).   The palatoquadrate arch of embryo vertebrates is indeed called the "subocular arch" by Kesteven (1925, pp. 42–66). The forward circumduction of the suspensorium brings the entire opercular fold into close proximity with the eye so as to form practically the posterior border of the eye.   Hence it is not surprising that even after the lengthening of the space between the opercular flap and the eye, the opercular series should retain a good deal of its "circumorbital" appearance. This is probably the explanation of the "circumorbital" arrangement in the oldest Protospondyli (p. 125) of the circum- and sub-orbitals, the cheek plates, preoperculars, opercular and branchiostegal folds, so that traces of the embryonic arrangement persist in the adult stages.   The proximity of the opercular flap to the circumorbital bones suggests how readily one of the latter could form an adhesion of its posterior border to the rim of the preopercular, such as we find in the scorpænoid fishes.   As the individual fish became older, the space between the opercular flap and the orbit would increase, perhaps to accommodate larger adductor mandibulæ muscles, while the contact between the nearest "suborbital" and the preopercular would be retained, thus necessitating the marked lengthening of the suborbital.

As development proceeds, in many teleosts the snout begins to lengthen as well as the

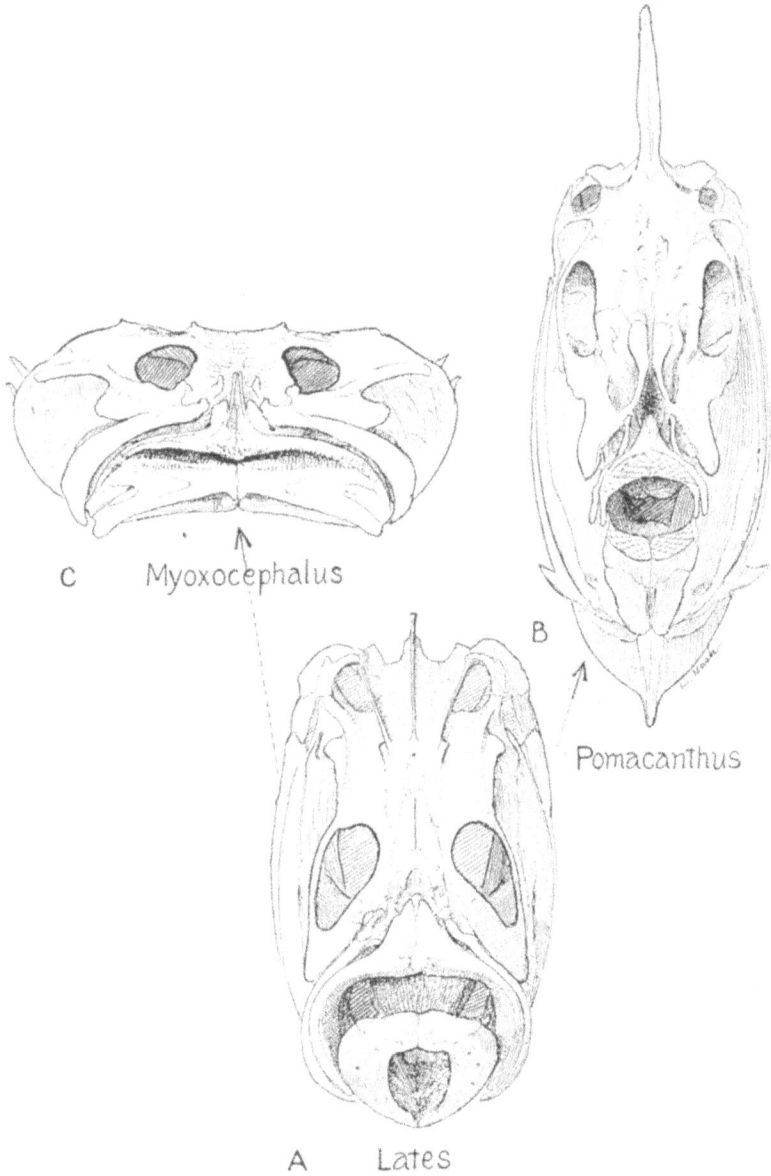

C Myoxocephalus

B

Pomacanthus

A Lates

FIG. 293. Divergence of primitive percoid (A) into very wide (C) and very high (B) types.

jaws, so that predaceous jaws and habits become more pronounced.   Meanwhile, the relative size of the orbits decrease rapidly, the jaw muscles increase so that, as noted above, the opercular fold is displaced backward.

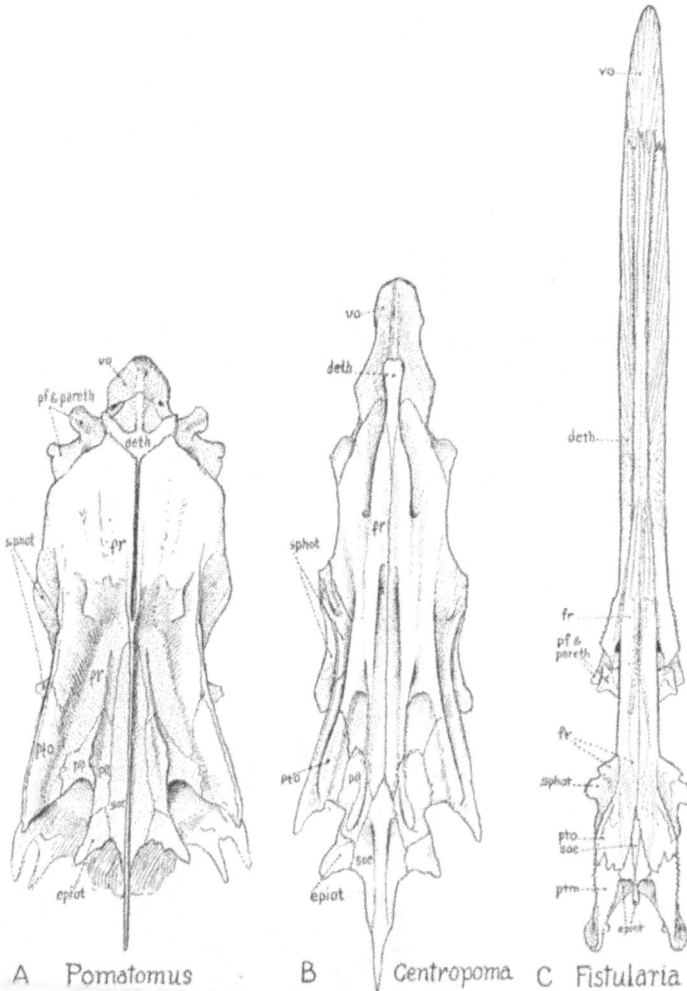

FIG. 294.   Differential lengthening of parts of the neurocranium.   A. Normal percoid.   B. Elongate percoid,   C. Excessive elongation of preorbital face.

At first (Fig. 292) the vertical diameter of the brain is greater than that of the future back and the myomeres are very small, especially in the vertical diameter.   In many teleosts, however, the myomeres grow rapidly in the vertical direction, adding an extra V and producing a high apex of the back. An extreme case of this tendency is shown in Fig. 297. At the same time the occiput deepens vertically but not as fast as the back, so that the skull roof slopes upward to the apex. In long-bodied teleosts, on the other hand, rapid multiplication of segments takes place in a fore and aft plane, while the occiput remains low (Fig. 297C). In either case forward growth of the snout naturally tends to lower the slope of the skull roof (Figs. 291E, 290C). In still other teleosts the transverse growth of the mouth (Fig. 293C) results in a wide, low head with a broad neurocranium.

We do not know exactly what conditions the dominance of one or another of these three more or less opposing tendencies toward length, height or width respectively, which have profound effects on all the bones of the neurocranium and branchiocranium (Fig. 294); but by analogy with mammals we may infer that these long, high or broad types have been produced through gradual changes in the partly hereditary qualities of the enzymes produced by the endocrine glands.

In most fishes, the food of the larvæ differs widely in size from that of the adult, as do the jaws and neurocranium. In small-mouthed fishes there may have been finally a tendency to retain the relatively small larval mouths in the adult (see p. 416 above).

The Form and Evolution of the Neurocranium

From the combined evidence of palæontology and taxonomic ichthyology there can be no substantial doubt that the several chief classes of true fishes (including the sharks and their allies, the crossopts and the actinopts) arose from animalivorous types. That is, all *typical* fishes are and were predaceous forms equipped to pursue and devour living prey. Some of the more specialized forms seek sedentary or slowly moving food and must therefore be equipped, for example, to break down the defensive armor of bivalves or crustaceans. Others by enlarging their mouths and developing gill-strainers browse on the limitless pastures of copepods and minute shrimps. But all such specialized forms have more conservative relatives that still pursue medium-sized prey in the way followed by their remote ancestors.

Consequently the neurocranium in any primitive type is adapted to support the organs of a fish that pursues animal prey in a fluid medium. It resists the various forces that impinge upon it in such a way as to afford both an immobile support and a protective cover for the cephalic sense organs, the brain and related parts. It also affords a fulcrum or anchor for the branchiocranium as well as for the backbone.

The neurocranium is a sort of passive cast or deposit in osseus tissue of spaces left between the more active parts of the head. Like the branchiocranium, it is a product of the growth of the vascular and connective tissue systems. Some of the factors affecting its form and evolution are considered below.

Mutual Adjustment of Head and Body

Stream-lining.—A modern student of piscine morphology, especially if he has studied fishes in their own habitat and has seen how adroitly they snatch and dispose of their chosen type of food, must admit that the various parts of any fish are in fact correlated with each other in such a way as to enable the fish to execute complicated manoeuvers with the greatest ease. This easy progress of the fish through the fluid medium is secured by the stream-lining of its body, that is, the body-form is presumably such that the fish at its highest normal speed will leave a flowing rather than a turbulent wake behind it.

My studies on the body-forms of fishes (1928), which were based primarily on measurements made on the *Arcturus* expedition of the New York Zoölogical Society, set forth the various ways in which the form of the cephalic and post-cephalic parts of the whole fish are correlated with each other and tend in *norma lateralis* to conform to a superposed kite-shaped frame of reference.

Stream-lining of the head, back and belly may be the resultant of growth along various "metabolic gradients" in the three principal axes of length, height and thickness. The kite-shaped figure results from the location of the peak of the vertical gradient somewhere in

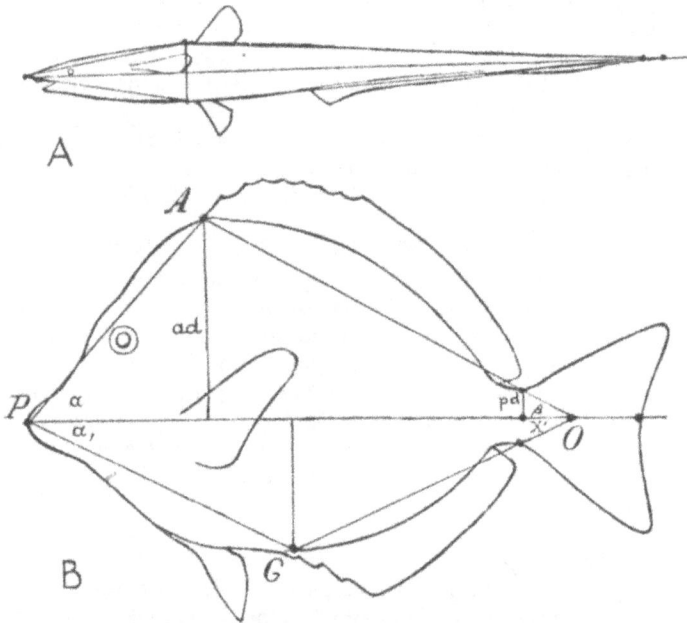

FIG. 295. The "entering angle" and the "run." A. Long-bodied fish (*Aldrovandia macrochir*). B. Deep-bodied fish (*Xesurus punctatus*). P, prosthion; A, apex; G, gasterion; O, opisthion.

the first third of the long diameter of the body. Hence as a rule the "entering angle" (which is determined at least in part by the shape of the head) is greater than that of the "run," or tapering part of the body (Fig. 295). The stream-lining of the body as a whole largely depends upon the contours of the surface bones and of the skin but the neurocranium and backbone must form an adequate support for the whole stream-lined shield.

Ceteris paribus, the detailed proportions of any given skull element will be in direct geometrical relations to tendencies that affect larger areas. Thus Fig. 295 shows that the slope of the cranial roof tends to vary directly with the height of the apex of the back; while Fig. 296 indicates that the slope of the cranial roof tends to vary inversely with the distance from the tip of the snout to the apex, projected on the horizontal.

Differential growth rates—Intrinsic and extrinsic.—In the individual development of anguilliform fishes the fastest growth increments have obviously been along the antero-

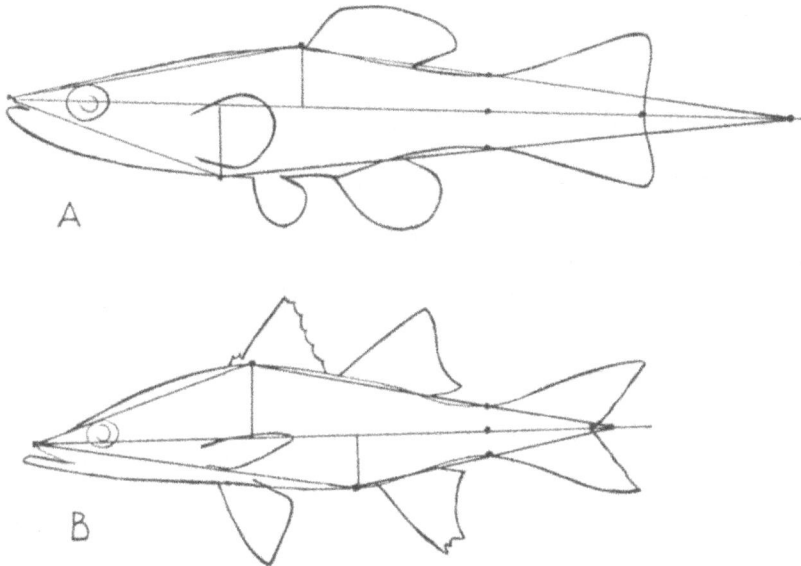

FIG. 296. Fishes with remote and low apex and sloping forehead. A. *Fundulus diaphanus*. B. *Oxylabrax undecimalis*.

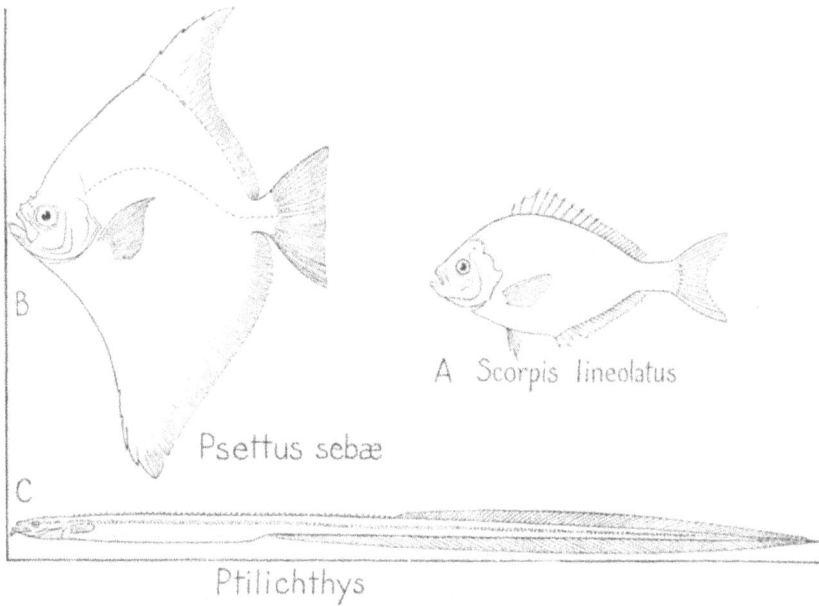

FIG. 297. Contrasting results of maximum growth along vertical and horizontal axes respectively. A. Primitive percoid (*Scorpis*). B. Extremely deep bodied form (*Psettus sebæ*). C. Extremely long-bodied form (*Ptilichthys*).

posterior axis, the slowest along the vertical axis, while in *Psettus* the reverse is the case (Fig. 297).   Moreover, in watching the development of any fish we see that the changes in form take place through differential growth rates affecting different regions at different periods of development.   Hence we arrive at once at the significant if obvious concept that the various parts of the skull share in general regional or extrinsic growth rates as well as having individual or intrinsic growth tendencies of their own.   For example, the circumorbital bones around the eye always fit into the space between the eye and the rim of the preopercular.   If the distance between the eye and the preopercular increases, several of the bones behind the eye may share the increased length, as in *Arapaima* (Fig. 59), or only one of them may do so, *e.g.* the suborbital stay of scorpænoids (Fig. 201).   Here we have a good example of the fact that in general each part has its own intrinsic growth rate, plus the extrinsic factor which it derives from other parts of the body.

Hence, to consider the neurocranium itself, it is now obvious that in a given type of fish the precise form and details of the neurocranium will depend upon the resultant of a multitude of regional (extrinsic) and local (intrinsic) influences; nevertheless, with the experience gained from a detailed study of a large number of different skull types it seems possible to distinguish a few of the more conspicuous evolutionary factors in any given skull.

FUNCTIONS OF THE FOUR MAIN PARTS OF THE NEUROCRANIUM

In a general way the neurocranium (Fig. 298) in the side view is typically more or less like a right triangle, its base being the vomer-parasphenoid-basioccipital, its hypothenuse

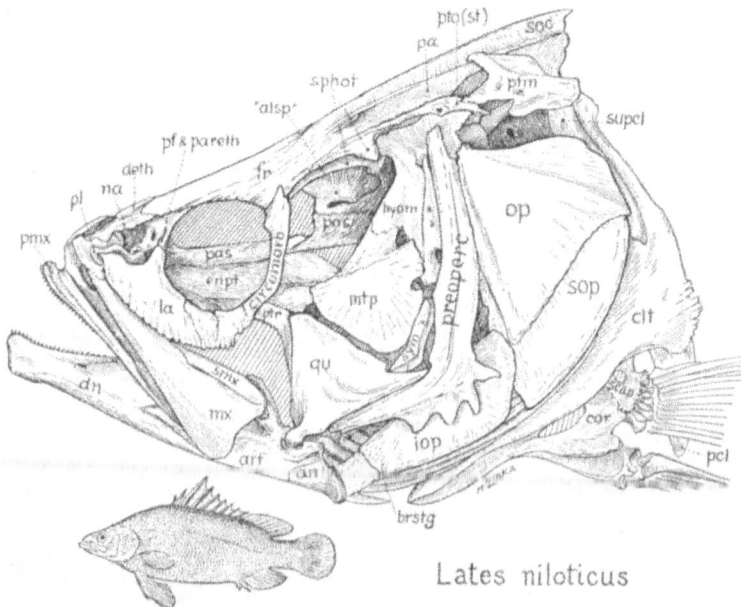

FIG. 298.   Syncranium of a typical percoid (*Lates niloticus*).

the roof of the ethmoid, frontal, parietal and supraoccipital bones. A circle formed by the orbit is superposed upon the triangle; it touches the hypothenuse above and extends below the base-line. The orbit divides the neurocranium into four regions with widely different functions. The first and most essential part is the *endocranial vault* or braincase proper behind the orbits, the second is the *interorbital bridge*, the third is the *ethmo-vomer block* and the fourth is the *keel-bone* or parasphenoid. In a general way the endocranial vault has eight great functions: (1) it must be so built as to resist the thrusts of the backbone in the rear and of the water in front; (2) it must resist the wrenching force of the great epaxial muscles of the back and of the anterior dorsal fin muscles; (3) it must afford a firm anchorage for the hyomandibular and thus resist all the wrenching, pulling and pushing forces that come from the jaws and their muscles and from the struggles of the living prey; (4) it must afford a support and anchorage for the shoulder-girdle and thus withstand a part of the strains occasioned by the various muscles that are attached to the cleithrum, coracoid-scapula, supracleithrum and posttemporal; (5) it must afford support to the often powerful muscles that operate the opercular flap and to others that operate the branchial apparatus; (6) it must at the same time protect from jarring the sensitive brain and cranial nerves that are lodged within it; (7) it must also protect from any disturbance the extremely delicate sensory-equilibrating apparatus, including the semicircular canals, the vestibule, the otolith, etc.; (8) it must afford protection to the eyeballs and provide a myodome for several of their muscles.

The second functional region, or *interorbital bridge*, sometimes aided by the interorbital septum and orbitosphenoid bone, must obviously be strong enough to resist the pressure of the water above and in front; it must brace the ethmo-vomerine block in front and transmit some of its thrusts to the endocranial vault behind. It must afford a channel for the olfactory nerves; and it must help to suspend the eyes and keep them in exact alignment.

The *ethmo-vomerine* block receives on its upper slope the thrusts of the water and transmits part of them to the interorbital bridge, part to the keel-bone below; the median and lateral ethmoids together give secure lodgment to the delicate olfactory capsules; the lateral ethmoids form the anterior margin of the orbits and receive the thrusts of the palatines, maxillæ, premaxillæ and vomers.

The *keel-bone*, or parasphenoid, ties the ethmo-vomerine block to the base of the braincase, forms the roof of the mouth and the floor of the interorbital septum, gives off dorsolateral wings or struts, which often form a secondary anterior wall of the braincase proper, affords attachment laterally to the elongate adductor arcus palatini muscle and forms the floor of the myodome in which certain of the eye-muscles are lodged. In embryonic and larval stages of typical teleosts the eyes are very large and the head very short. Consequently the ventral eye-muscles grow backward into the floor of the cranial vault, which yields to them and forms a funnel-like tunnel. As growth proceeds and the distances between this incipient myodome and the centers of the orbits increase, the eye-muscles retain their posterior connections and the myodome takes on its adult form.

To consider the *cranial vault* in somewhat greater detail, it seems evident that at least in typical fishes the final anchorage or base of the entire system is constituted by the rear portions of the basi- and ex-occipital bones, as well as the centrum, neural arches and anterior zygapophyses of the first vertebra. In the typical acanthopts the greatest concentration of bony tissue in this region is seen around the three occipital condyles (Fig. 299),

24

one median on the basioccipital and two lateral on the exoccipitals.   The articular surfaces
of these three condyles are inclined toward each other in such a way that forces coming, for
example, from a horizontal thrust of the backbone are divided into three streams which
spread out as they go forward and upward.   Conversely, thrusts coming from the front
are collected and concentrated into three streams, which find their exits in lines normal to
the three condyles.   Thus owing to the concentration of the bony tissue these three sets of
forces pass below and on either side of the medulla without in the least disturbing it.   In
the side view (Fig. 299) the half downwardly-facing exoccipital condyles and the half

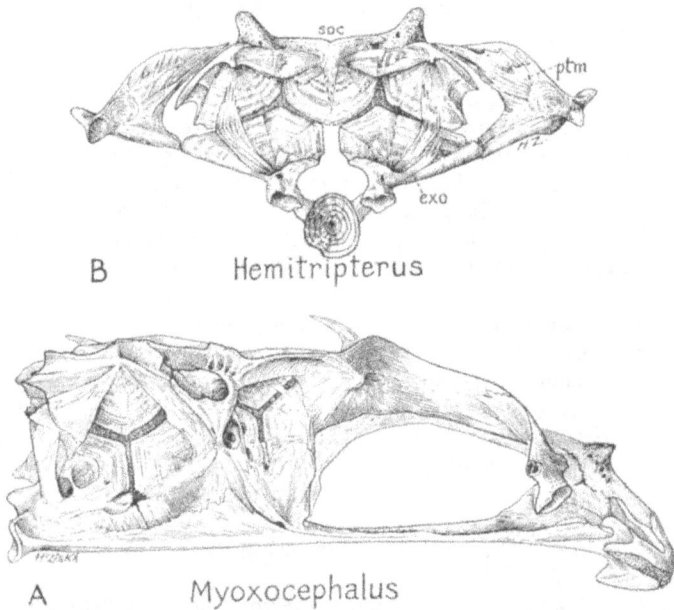

FIG. 299.   Neurocranium of scorpænoids.

upwardly-facing basioccipital condyle are inclined toward each other at an angle of about
ninety degrees; this arrangement makes it easy for the collateral ligaments to prevent dislocation of the neck and consequent strangulation of the spinal cord.   Moreover, in typical
fish the basioccipital condyle has a deep cotylus, which is filled in life with more or less
elastic intercentral tissue, and a raised rim, to which are attached the ligaments that tie it
to the first vertebra.   On the exoccipital condyles the articular surface is shallowly convex,
with raised rims.   Doubtless the ligaments and interarticular discs take up some of the
shocks, while permitting some movement of the head on the column in connection with the
alternate left and right movement of the head in swimming.
    The fact that the forces coming from the rear and impinging on the lateral condyles
are spread outward and upward is reflected in the construction of the occipital segment of
the skull, for from the arrangement of the main braces above the condyles it is seen that

each thrust-stream would be divided and distributed, one main division passing upward, inward and slightly forward toward the occipital crest, a second upward and forward to divide again into two branches, one leading to the epiotic process, the other to the posterior process of the pterotic. The forward thrusts on the basioccipital spread out radially on the basioccipital in a wide V but from the tip of the V a median tract with parallel sides runs forward to join the parasphenoid; apparently this median tract serves to resist tension rather than compression, and indeed such tension could readily be excited (through the intervention of the parasphenoid) by upward pressure on the under surface of the vomerine tooth patch.

The architecture of the roof of the cranial vault is no less challenging than that of the base. The interorbital bridge, including part of the frontals and alisphenoids, forms a beautiful arch, the thrusts of which are transmitted through spreading postorbital pillars, which rest on the proötics, the sphenotics and the pterotics. Between these diverging limbs at the right and left sides is a large median fissure, bounded below by a median keystone of an inverted arch formed by the basisphenoid. The median fissure affords a secure and immobile seat for the front part of the brain, including the olfactory stalks. Lateral to the base of the arch lies the trigemino-facialis chamber for the great ganglia of the trigeminus and facialis nerves, which are protected from being crushed by means of small but relatively strong pillars formed by the proötic bone.

The support for the hyomandibular, which has to withstand much of the forces that tend to wrench, push or pull the inner and outer jaws, is very appropriately located (Fig. 117) between the powerful postocular arch just described and the occipital buttress afforded by the lateral wing of the pterotic. The socket for the hyomandibular is always tripartite, with the sphenotic and proötic parts in front, the pterotic part in the rear. The proötic facet is supported by a pedicle that runs downward and forward between the facialis and the trigeminal ganglia, lateral to the trigemino-facialis chamber. The sphenotic facet is usually supported by the heavy postorbital process of the sphenotic, the pterotic facet is horizontally extended backward above the top of the opercular. This facet is supported by a very strong buttress on the exoccipital that runs obliquely downward and backward toward the exoccipital condyle, and sometimes by another smaller one that runs downward and forward.

The pterotic facet is often braced above by longitudinal ridges on the pterotic and parietal and in the tarpon by a stout transverse ridge running from the pterotic to the parietal. In the upper half of the occiput above the exoccipitals there are often deep fossæ bordered by ridges for the insertion of the trapezius muscles of the shoulder-girdle. In the tarpon (Fig. 32) and some other isospondyls these fossæ are very deep. The supraoccipital crest supports the anterior extensions of the muscles of the dorsal fins. In many scombroids (Figs. 183, 192) this crest extends forward to the snout and is supported by a great horizontal plate formed by the frontals.

The inner surface of the braincase (Figs. 32, 119) consists broadly of: (1) depressions for the various parts of the inner ear and of the brain; (2) elevated ridges and partitions between these parts. Many of the above-named ridges and buttresses on the outer side seem quite well adapted to resist tension as well as compression.

The entire cranial vault is divided by a system of *triradiate sutures* (Fig. 299). One triradiate suture separates the alisphenoid, sphenotic and proötic, another separates the exoccipital from the epiotic and pterotic and proötic.   These triradiate sutures seem to be the result of peripheral growth from three ossific centers, which have thus moved away from each other at nearly equal rates.

ADJUSTMENTS OF THE NEUROCRANIUM TO VARIOUS TYPES OF JAW, DIFFERENT POSITIONS OF MOUTH, ETC.

The neurocranium has solely a passive function in resisting stresses, but certain parts of the branchiocranium, especially the jaws, teeth and suspensorium, which are compound levers, assume an active function in modifying the neurocranium.   For example, in the sparids (Fig. 123) the front teeth become strong incisors, the tooth crowns on the side of the premaxilla become obtusely conical to massive and hemispherical, and the same is true of the teeth on the dentary; the fish is thus able to crush the shells of bivalves.   We have noted above (p. 249) the great enlargement of the maxilla and palatine to support these unusually heavy shocks; hence it is not surprising that we find equally great massiveness and strength in the construction of the cranial supports, namely the ethmo-vomer block, the interorbital bridge, the cranial vault and the keel-bone.

In large-mouthed, predaceous types the neurocranium must be variously strengthened in accordance with the particular type of specialization.   In the morays, for instance, the neurocranium (Fig. 82B) is long and narrow but very strongly built in order to support the huge backwardly-developed hyomandibular and swollen branchial syringe, whose dorsal muscles are fastened to it.   The ethmo-vomerine block and interorbital bridge are likewise stiffened in reaction to the stresses received through the enlarged vomerine teeth.   In the gulpers (Fig. 94) the neurocranium is remarkably wide and short because it has to be the fulcrum for the very thick neck and back muscles and to carry the great living scoop-net which is the mouth and pharynx.

Again, the position of the mouth affects the neurocranium.   In the sturgeon, for example (Fig. 20), the interorbital bridge and ethmo-vomer block are drawn downward and forward so that the lower surface of the ethmo-vomer block lies much below the level of the floor of the cranial vault.   Consequently the tactile ventral surface of the rostrum can reach the muddy base in search of food immediately in front of the small suctorial mouth.   This, however, does not explain why the neurocranium of the sturgeon has become secondarily cartilaginous.   A dorsal displacement of the mouth naturally affects profoundly the ethmo-vomer block and the anterior part of the keel-bone but is not often without marked indirect effects upon the cranial vault.   In *Pelor japonicus*, for example (Fig. 206), a strongly modified scorpænoid, the wide mouth is pushed upward and backward to such an extent that the ethmo-vomerine block is greatly shortened and widened; as the eyes also are turned upward and pushed backward, the cranial vault is considerably widened and shortened.

The opposite specialization is seen in the tube-fishes (Fig. 294C), in which the mouth, although equally upturned, has been drawn very far forward, greatly elongating the ethmo-vomer block.   Here also (Fig. 230C) the interorbital bridge has shared in the elongation but to a much less degree.   In the balistids also (Figs. 160–164) the ethmo-vomer block shares in the forward growth of the suspensorium and affords a firm support for the small but powerful nipping mouth.

The branchial apparatus as a whole is but loosely connected with the neurocranium through the intervention of the small rod-like interhyal, which fits in a socket furnished by the symplectic and the preopercular. But when the upper pharyngeal teeth become very powerful, as in the scarids (Fig. 134), they are furnished with a pedestal by the parasphenoid. In the Cyprinidæ, where the enlarged lower pharyngeals work against a horny pad, the latter is supported by an oval pedestal supplied by the basioccipital.

Responses to Cranial Diverticula of the Swim-bladder

The neurocranium is also moulded by certain prolongations of the swim-bladder, which put forth finger-like tips that, as it were, burrow their way into various places on the side of the cranial vault. This happens quite independently in the clupeoids (p. 147), mormyroids (p. 169) and, in a quite different way, in the Ostariophysi (p. 184).

Moulding Influence of the Eyes

The sizes and positions of the paired sense organs naturally influence profoundly the characters of the neurocranium. For example, when the eyes are greatly enlarged, as in Opisthoproctus (Fig. 43) and Periophthalmus (Fig. 229C), the interorbital bridge is constricted and the ethmo-vomer block is thrust forward. At the same time the circumorbitals tend to extend far backward over the cheeks as in the Semionotidæ (Fig. 22). It would seem also that the lunate arrangement of the preopercular, opercular and cleithral series is to a considerable extent conditioned both by the large size of the eyes and by the smallness of the mouth in larval stages (see p. 422) and that the basic teleost pattern may have been derived from a larval stage of an adult fish with small eyes and large predaceous jaws of the palæoniscoid type (see p. 112). Conversely, the reduction of the eyes, as in mormyrids (p. 170), naturally permits the interorbital bridge to widen and the parethmoid process to become reduced.

The size and position of the eyes naturally determine very largely the position and special characters of the interorbital bridge, or at least of the rear surface of the ethmo-vomer block, as well as the characters of the postorbital process of the sphenotic and of the entire postorbital arch or front wall of the cranial vault. When the large eyes are moved to the top of the skull the interorbital bridge may become very narrow, as in Astroscopus (Fig. 248), while certain of their muscles, forced to withdraw from the myodome on the base of the skull, fill a great space behind the orbits and become transformed into electric batteries; these in turn push back the front wall of the cranial vault. In Salmo one pair of the eye-muscles (the superior oblique) extending forward invade the ethmo-vomer block and doubtless influence the form and position of the mesethmoid septum in ways not yet sufficiently understood (Goodrich, 1909, p. 325). In the flat-fishes, where the eyes are twisted into new positions in accordance with the inclination of the body to one side or the other, the interorbital bridge is pushed in front of the migrating eye and a secondary brace is developed from the lateral ethmoid and the sphenotic (Figs. 224, 225).

In the acanthurids and still more in the balistids (Figs. 161–164) the eyes have moved upward and backward to such an extent that they actually override the hyomandibulars, pushing the suborbitals into contact with them, merging the interorbital bridge with the roof of the cranial vault and causing the pterotic process to point downward rather than backward. The opposite specialization is seen in Cyclothone (Fig. 54), in which the very small eyes are pushed far forward to the front end of the interorbital bridge. In the gulpers

(Fig. 94) the minute eyes and entire neurocranium have been pushed forward to near the tip of the snout, so that the neurocranium is remarkably short and wide.

Influence of the Auditory Capsules

The auditory capsule with its various subdivisions is naturally a potent element in moulding the plastic osseous tissue of the cranial vault. In many fish a "bulla" is formed by the proötic and basioccipital for the reception of the sagitta, or main otolith. The latter in turn varies considerably in size. In the deep-sea isospondyl *Opisthoproctus* (Fig. 43) it forms a very large, more or less circular disc. Similarly the semicircular canals and ampullæ require corresponding canals and swellings in the osseous cast surrounding them. As these structures therefore subtract from the area available for support in a region subject to severe stresses, the remaining skeletal parts are built up into more or less dense or protruding pillars, in accordance with the principles already mentioned.

Adjustments in the Ethmoid Region

The variations in the olfactory capsules in sharks play an important part in modifying the fore part of the skull, as we see in an extreme form in the hammer-head; but even in the teleosts, with their usually great reduction of the olfactory capsule, it must be protected by the nasals and by the lateral ethmoid, which is usually perforated by the olfactory nerve. Hence stability of the lateral ethmoids is necessary and so we find them securely tied in place by the surrounding parts. Moreover the ethmo-vomer block must ordinarily serve as the keystone for the articulation of the opposite palatines, maxillæ, premaxillæ; hence either the lateral ethmoids or the mesethmoids must be stoutly built and variously modified to suit each case. In the gurnards (*Prionotus*), for example, the mesethmoid forms a large median plate at the front end of the cranial shield; in *Callionymus* and *Drepane* the mesethmoid is displaced behind the enlarged lateral ethmoids (Starks 1926a). It seems not unlikely that further study of the ethmo-vomer block from a combined functional and phylogenetic viewpoint may permit further utilization and helpful interpretation of the accurate topographic data recorded by Starks.

Where the bony rostrum is greatly produced in widely different families, as in the swordfish, scombresocids, *Protosphyræna*, etc., it is natural to expect the variable composition of the rostrum and different ways of strengthening its junction with the interorbital bridge and palatopterygoid arch.

Effects of Special Accessory Organs (Trigger, Illicium, etc.)

It seems hardly necessary to state that various special organs such as the "trigger" of the balistids, the illicium of the ceratioids and the sucking-disc of the echeneids, have each brought about special modifications of the ethmo-vomer block, which are readily recognizable in every case so far noted and which largely account for the decidedly queer characters found in the ethmoid regions of these forms.

The development of backwardly directed spikes on top of the neurocranium as well as on the preoperculars, operculars and elsewhere is especially characteristic of the scorpænoids. The initiative in this instance perhaps comes from the integument rather than from the underlying bone, which responds to it by developing a concentration of bony tissue for the support of the spike.

Adjustments to Different Types of Skin

The varying characters of the basal layers of the skin have left their clear imprints on the surface of the neurocranium, but the functional meanings of the patterns of pits, ridges, etc., is far from clear and would well repay investigation. Contrast, for example, the dense, closely pitted surface of the cranial shield of some of the large catfishes (Fig. 77) with the cavernous, porous surface of the skull of the Sciænidæ (Fig. 125), in which the enlarged mucous-secreting lateral line organs sink into deep pits. The varying texture of the surface in the scorpænoids has been described above (pp. 341, 343) and its connection with the arrangements of the bony trabeculæ noted.

In *Mola* the skin has become very thick and forms a tough hide-like shield covering the whole surface of the skull and body. This tissue has invaded and interpenetrated the surface skull bones, burying them deep and profoundly modifying their texture (see p. 295).

Spinescence

Whenever spines are developed, as on the surface bones of the skull and pectoral arch (see p. 323) the phenomenon seems to involve the rapid proliferation of growing tips and the laying down of relatively inelastic continuous layers so as to produce internal pressure and crowding of layers. The location of these spinescent points frequently coincides with the growth centers of the bones themselves and is evidently controlled by the hereditary or genic pattern of the species.

Development of a Cranial Buckler

The development of a cranial shield has taken place quite independently in several groups, such as certain families of siluroids (Fig. 80), the gurnards (Figs. 218–220) and *Dactylopterus* (Fig. 223). In these cases some or many of the roofing bones become greatly enlarged; they may grow backward over the neck (*Dactylopterus*) or a separate nuchal shield may be developed (catfishes). In some of the gurnards the cranial shield becomes heavy and seems to serve as ballast. For sluggish or partly sedentary fishes the protective value of such a shield would seem to be high.

Adjustments for Balance and Flotation

Varying adjustments for balance and flotation have doubtless had considerable influence upon the skull in many and complex ways. In *Chætodipterus faber*, for example, the great supraoccipital crest (Fig. 152) and interorbital bridge are much swollen and very dense, so that they must measurably tend to depress the forepart of the fish and perhaps require correctional movements of the pectoral fins or counterbalancing in some other parts, as in the swollen epineural and interhæmal spines. Again, in the sheepshead *Archosargus* (Fig. 123), it would seem that the heavy dentition and jaws and the excessive amount of dense bony tissue in the skull would overweight the head, if the body itself were not so huge. The presence of a gas-filled air-bladder conditions an upward curve of the vertebral column and this probably affects many details of the occiput. Apparently in high-backed fishes the heightening of the back must tend to raise the center of gravity and thus increase the instability. This is possibly compensated by lowering the center of gravity of the jaws, throat and abdomen. Thus the position of the quadrate center (Q), the angle of the gape, the position of the eyes and the detailed manoeuvers in capturing the prey might all be influenced by a lowering of the center of gravity of the abdominal cavity or by a decrease in size of the swim-bladder.

In minute larval fishes the head is sometimes equipped with spikes, some of which occasionally attain great length, as in holocentrids. Here it would seem that the spikes and spicules contribute to the flotation of the organism perhaps within certain depth zones, besides being of distinct advantage in imposing relatively high lower limits on the size of hostile mouths capable of devouring the owner of the spikes. Doubtless various considerations of hydrostatic balance and stability must set limits to the relations of the weight of the head to that of the body, as well as to the distance of the head from the center of gravity of the body. Thus in fishes with either very long or large heads the junction of the skull with the backbone is braced in one way or another, as by coalescence of vertebræ or of neural arches.

WHY THE FISH SKULL IS DIVIDED INTO SEPARATE BONES

Some of the reasons why the fish skull is divided into separate bones may be as follows: (1) Since the skeleton is ultimately a product of the circulatory and connective tissue systems, the individual bones are mere inert deposits made by growing tissues and these tissues respond to the growth of the larger regions or organ-systems of which they are parts; (2) since throughout the members of any given taxonomic group there is a pretty constant number of skull bones, each with its characteristic topographic position and contacts, it follows that the number and position of osseous growth centers are predetermined by hereditary factors, just as are the number and position of the chief arteries and veins that supply these bony centers with their materials; (3) the boundaries between bones result either from bending or movement of one part on another, or from the meeting of peripheral zones growing from different ossific centers; (4) such growth centers, the loci of greatest concentration of growth and of bony material, seem often to be located at foci of greatest stress, while weak contact sutures are located along zones of least stress; (5) almost every bone has complex relations with surrounding parts, its own parts sharing the growth of functionally different regions; (6) the cranial vault, the interorbital bridge and septum, the ethmovomerine block and the keel-bone are, as it were, casts in osseous tissue of spaces left vacant between more dominant organs such as the olfactory capsules, the eyes, the brain, the roof of the mouth, the skin and the laterosensory canals; (7) similarly, the branchiocranium is a complex cast in osseous tissue of spaces left between such embryonic pockets or folds as the stomadæum, the gill-pouches, the opercular folds, etc.

The osseous cast which is the neurocranium originally included two layers named ectosteal and endosteal bones: the outer layer deposited on the under side of the lateral line pockets, folds, skin areas, etc., and the inner layer deposited on either side of the notochord and on or between deep-seated organs such as the brain, the cranial nerves and the main sense organs.

DISCUSSION: THE FISH SKULL AS A DOCUMENT OF EVOLUTION

DEVELOPMENT AND EVOLUTION

From all that precedes it is evident that a fish is a natural machine which inherits sufficient latent energy from its parents to begin its own career of capturing, storing and spending energy and of preparing the seed for coming generations. Energy is extracted from the environment chiefly by means of the gills and the digestive tract.

In most teleosts the energy-containing yolk is small and the young larvæ of 4 to 6 millimeters in length are already provided with small jaws, by means of which they can feed upon still smaller organisms. *These adaptive mechanisms of the larval stages tell us little if anything about the adult stages of remote ancestors except in cases of "pædogenesis."*

In the late fry stages the body becomes deeper, especially in deep-bodied species, and with it the occiput deepens vertically. Crests and ridges appear only in late fry and immature stages as the muscles increase in strength (see below, p. 445). Although the characteristic pattern of the adult skull is thus late in its appearance, it may nevertheless be regarded as the result of the interaction of hereditary forces with the normal factors of growth and environment.

For information as to the evolution of the adult skull of any particular type, we must therefore seek to understand both its adult functions and its developmental history; we must compare it with less specialized skulls of its own group and assemble the available palæontological evidence as to its derivation.

As to the functions of the skull in the adult, we recall that the branchiocranium served originally as a mechanism of jointed levers for operating the pumps and valves of the oxygen-intake system. Secondarily, certain of its arches function as compound levers for raking in food, or for breaking and crushing food, as well as for the support of the tunnel that leads to the assimilative system. The neurocranium is primarily a thrust-box, anchorage or fulcrum, fitted to receive and neutralize various thrusts and pulls from the body, from the resisting medium, and from the active parts of the branchiocranium. At the same time it groups and directs all the stresses and lines of force, above, below and around the central nervous system, which is perfectly suspended from it without shock or strangulation of the delicate nerve cables or brain parts.

Both neurocranium and branchiocranium may be conceived as a complex system of casts of osseous, cartilaginous and connective tissue material that has been excreted, so to speak, in the spaces left between the more dynamic tissues. In general the skeletal tissues are passive and plastic in relation to the nervous, muscular, nutritive and excretory tissues; but skeletogenous tissue sometimes appears to assert its own growth force, as when it produces thickenings and excrescences.

The skull is thus a living palimpsest of many writings and from it we may in many cases, after extended comparisons, decipher the story of its owner's way of life and even a good part of his ancestral lineage.

In our endeavor to decipher the habits of an individual fish from its skeleton we look first at the jaws and teeth, the position and direction of the mouth, the pharyngeal mill, etc., which tell us whether the fish was typically either predaceous or a plankton feeder, a "nibbler," a "pincers" fish, a crusher of bivalves and crustaceans, a mud-grubbing, sucking type, or an animated fish-trap. Checking of these inferences by examination of stomach contents in members of the same species would of course be desirable. Then we look at the general body-form for the marks of swift pelagic swimming, of the sudden rushes of a lurking robber, of the quick dodging by a deep-bodied fish of the coral reefs, etc.

As to how far one can read the ancient lineage of a fish from study of skeletal characters, it is well known that if the skeleton is well preserved there is no difficulty in detecting diagnostic characters, including proportional lengths of various parts by which to determine the systematic position of the specimen. This in turn would enable us to look up whatever general palæontological or taxonomic evidence there might be as to the steps by which the group under consideration acquired its peculiar characters. Unfortunately, perhaps ninety per cent or more of the energies of ichthyologists have been expended in determining the marks of new species rather than in endeavoring to discover the stages by which a given species has come by its present characteristics.

In connection with inquiries into the evolutionary history of any given type of fish skull, it is important to realize that the habitus (or totality of hereditary adaptations to a given way of life) in the ancestor becomes the phylogenetic "heritage" of its descendant (Gregory, 1913). For example, the predaceous habitus of the ancestral percoid becomes the phylogenetic heritage of such specialized forms as the beaked parrot wrasses, the nibbling balistids, the trap-mouthed anglers and many others. In other words, the earlier functions and structures of the predaceous habitus had to be modified progressively away from this relatively primitive condition; but traces of these earlier habitus characters may still be seen in many basic features of the branchiocranium and neurocranium of even the most specialized teleosts. In the light of these facts and principles we have already gained a probably fairly accurate historical concept as to the main steps (see pp. 85, 416) by which some remote prechordates that fed by ciliary ingestion were transformed into the central predaceous type of teleost fish, and as to the subsequent steps by which this fish in turn was changed into such diverse types as "nibblers," "pincers," "crushers," "tube-mouths," "scoop-nets" and so forth.

From all this emerges the generalization that even on the unlikely hypothesis that structural changes preceded new feeding habits and merely made possible the more thorough exploitation of possible food sources, there were in each phylogenetic series a large group of changes, affecting many parts of the skull and body that were correlated with each other, and that every evolutionary change in structure implied corresponding changes in food preferences, involving shifts in subtle and intricate correlations of sensory stimuli and motor response.

The Mechanism of Regulation

Accordingly we are led to inquire as to the physiological mechanism for regulating the size-relations of various parts of the body. The measurements made by systematists on thousands of species establish the fact that the ratios of head-length to body-length, of preorbital length to head-length, and many others, vary only within relatively narrow limits in adults of a given species. So that the species owes its characteristic contours and all its specific adult patterns to a specific mechanism for the regulation of differential growth rates.

From the fact that in certain cases, such as the hook-jawed salmon and the eel-pout, the ripening of the male gonads brings about marked proportional differences in certain parts, we have an indication that the interaction of the various endocrine glands plays a large part in the regulatory mechanism, as it is known to do in other classes of vertebrates. Moreover, in fishes we have many examples of extreme elongation of the body, as in the eels, and of extreme broadening, as in *Lophius;* these are at least analogous with Stockard's "linear" and "lateral" types of men and dogs; they are likewise probably due to contrasting activities of whatever agencies may correspond in function to the thyroid and pituitary extracts of mammals.

The studies of D'Arcy Thompson ("On Growth and Form," 1917), Julian Huxley (1931) and others have shown that in many animals the growth rate of a given part during a given period is a logarithmic function of the increase in mass of the entire organism, and that the growth of a localized excrescence again bears a logarithmic relation to the mass of the structure that bears it. The fact that the dependence of growing parts upon their sources of supplies may be expressed in tables like interest and compound interest, is enlightening as regards the processes of individual development and growth. But the chief problem of the student of the evolution of the fish skull is to discover how changes from one schedule of rates of interest to another have been induced during the millions of years which appear to have been necessary for the transformations from one feeding type to a radically different one.

In many, if not all, teleosts the changes in the proportions of the jaws and skull from the embryonic to the larval stages and from the later to the adult skull are very striking. Thus in the larval stages of many teleosts, as figured by Bigelow and Welsh (1925), the mouth and jaws are small and upturned, the muzzle very short, eyes very large, braincase large and swollen. Here as in later stages the bones of the skull simply share in the rates of growth that affect whole regions. As growth proceeds the jaws lengthen, the snout lengthens, the eyes are retarded, locomotor muscles overgrow the occiput, crests and ridges appear upon the cranial vault, etc.

These changes are doubtless brought about by the interaction of growth forces (lasting from embryonic and cleavage stages) with nutritive and other environmental influences. Thus Stockard (1931, pp. 108, 135) found that in young *Fundulus* cyclopean monsters could be produced by the reaction of the hereditary mechanism with several different chemical and physical agents, while Hubbs (1926) has noted many factors (including temperature, salt concentration, etc.) that induce changes in the proportions of parts in the growing fish, and which may under certain circumstances cause a departure from the adult norm.

From the consideration of hundreds of species of fishes of all orders, recent and fossil, it would seem that the factors that determine the *presence of the standard parts* of the fish skull have been on the whole amazingly constant for perhaps several hundred millions of years; but that the factors that determine *departures from normal proportions* in the adult are, in geologic time, more or less highly variable. Perhaps this might be because the embryological factors for the presence of mouth-parts, paired eyes, otic capsules, branchial arches, etc. bring the *Anlagen* of these parts into existence during the early embryonic stages, while their adult proportions are approached only in late larval stages. Consequently the parts once established are seldom crowded out, but are subjected during late embryonic, larval and immature changes to disturbances in growth rates that can hardly fail to affect their adult dimensions.

The Paradoxes of Adaptation

So far as I am aware, no subspecific difference in proportion from a parent type has ever yet been shown to have an adaptive or selective value. But this negative finding by no means offsets the positive evidence for the reality of adaptive changes. We must suppose either that the number of cases already investigated thoroughly is not yet sufficiently large, or that the time during which any given subspecies has been studied is not a sufficiently large fraction of a geologic epoch to reveal, for example, sensible differences in food preferences among related subspecies, or correlated differences in average size of jaw and size of food. Such correlations must be facts because the end results of some process of differential selection are already facts of record in the well established "adaptive radiation" of such groups as the percoids, the scorpænoids, the cyprinoids, characins, siluroids and many others in which the jaws are mechanically adapted for very diverse purposes.

Thus the first paradox of adaptation in fish skulls is that they have as yet to be demonstrated in their initial stages, as between related subspecies, but are perfectly patent as between related genera.

Another seeming paradox lies in the apparent conflict between "Wolf's Law" and the non-inheritance of individually "acquired" characters. The facts cited in this paper sufficiently prove that in the skull of fish, just as in the mammalian skeleton, bony trabeculæ, ridges, buttresses, etc., arise in response to specific stresses, such as those generated by the thrusts of one moving part upon another, in other words, that bones are usually strengthened in proportion to the loads they bear (which is essentially "Wolf's Law"). But, if we may judge from analogous cases, any special modifications which might be induced in the bony trabeculæ of an adult fish under laboratory conditions could not reappear in the subsequent generations unless the same stimulus were applied to each. On the other hand, we should expect from numerous analogies that a young sparid would develop perfect incisors and massive molar teeth, even though he might be deprived of the opportunity to use them.

Thus the second paradox of adaptation, long well known to palæontologists, is that adaptive changes take place as if they had arisen through the inherited effects of habit, use or disuse, without having really so arisen.

The student of the fish skull as a natural mechanism must be impressed again and again with the "purposive" or adaptive value of its individual parts. It is a fact of record that the complicated parts of a natural mechanism, like the upper and lower jaws, their teeth, their muscles and supporting elements, do evolve and have evolved with reference to, or in causal relations to, each other, so that natural mechanisms often bear a surprising but, as I do believe, spurious resemblance in function to the products of human design. Similarly, two individuals may assume a "purposive" relation to each other, as in courtship, mating, care of young, nest-building, and so forth. But it has been shown elsewhere (Gregory, 1924) that all vital processes are essentially anticipatory in character: the trap is made and set before it ever goes off; the teeth are made up before they come into use; the needs of the next generation are anticipated long before the needs are felt by the yet unborn. Since all vital processes are essentially anticipatory at least in part it is not surprising that the incipient "rectigradations" of Osborn, which are said to antedate their own usefulness, should also share this anticipatory character.

Thus the third paradox of adaptations is that they have the appearance of being purposive designs but are more probably the result of consistent action by an unconscious selective mechanism.

It has also been established by Spemann, Harrison and other modern embryologists that during development "organizers," probably of a chemico-physical nature, preside over certain regions of the developing embryo and have the power of directing and controlling the development of parts and of linking them up one with another in such a way as to produce viable individuals under a given set of environmental conditions. The problem of inducing changes in the adult skull doubtless involves the problem of inducing the "organizers" to relax their routine enough to permit one or more growth factors of skull form to be modified at an early stage of development. If we can locate the "organizer" or rather the organization that produces separate teeth on the premaxillæ, we may some day hope to modify it so that the teeth will coalesce into a beak. The operations of these "organizers" show strong analogies with the operations of intelligent localized minds but as these organizers in all probability do not have such minds, their behavior only indicates that "mental" operations are caused by chemico-physical means.

Thus the fourth paradox of purposive adaptations is that they are probably the unexpected by-products of blind chemico-physical organizers, forced by selective elimination to vary from their "purposive" standards.

In view of what I have seen of living fishes and their behavior, I would not be inclined to deny that each individual fish has a mind, or that the summation of the responses of individual fish minds has played an important part in adaptation and evolution. Just as the responses of the mammalian and human nervous systems have played a great part in the structural evolution of man, so have the gradually divergent responses of the fish nervous system had an equally profound effect in determining the evolution of the mouth, jaws, endocranium and all other parts in the adaptive radiation of fishes. In other words, the hereditarily determined likes and dislikes of the fish itself, together with its individual psychic experience, have been an important factor in the evolution of its racial adaptations.

Thus the fifth paradox of adaptation in the fish skull is that adaptation has been brought about in part through the selective influence of individual fish minds, preferring this kind of food and avoiding that, yet gradually changing their preferences, even through slow shifts in the characters of whatever wholly unconscious genes may determine hereditary types of reaction to different foods.

However, even though there is a progressive integration of responses to more and more remote contingencies, Ritter (1929) has well noted the shortsightedness or foolishness of certain responses even in such relatively intelligent creatures as birds. In other words, a generally beneficial reaction may, under changed conditions, become a dangerous or fatal weakness. Thus each response of the individual fish is made without regard to remote consequences and in the long run the guardian "minds" either of the fishes or of the organizers are not sufficiently wise to meet all the moves of their opponents and, in thousands of cases, even the species itself loses the game. And seeing how full the organic world has always been of waste, violence and stupidity, we can realize that natural mechanisms, including fishes, are at best only animated, imperfectly conscious automata more or less blindly struggling to capture food-energy and mates, according to their hereditary likes and dislikes. The conception of the individual fish contributing through its own reactions to the general evolution of the race has already been emphasized by MacFarlane (1923). The coincidence and integration of helpful and appropriate responses by generations of individual fish minds with the essentially anticipatory values of all physiological processes, may be the chief cause of the striking resemblance of the natural mechanism of the fish skull to machines designed by human minds.   I have already used similar considerations in support of the conclusion that human minds differ from fish minds only in complexity and that both are dependent upon the anticipatory values of the nervous system.   Doubtless a certain measure of hereditary variability in likes and dislikes has been prerequisite for progressive changes in any given line of fish skulls.

"Basic Patents" and Evolution

We have seen that in many lines of descent the excellent "basic patents" of the percomorph skull have often served as the starting-point for new and wholly unexpected devices, such as the nutcracker jaws of scarids or the folding scoop-net of *Stylophorus*. On the other hand, a new and successful invention often begins to lose its distinctive characteristics and may finally degenerate and disappear or become disguised so that we can only recognize its origin after prolonged study.   Thus the protractile apparatus of typical percoids loses its protractility in many scombroids and other groups, the long jaws of the primitive scombresocid become greatly shortened in the adult flying-fishes, the exquisitely designed illicium of the central pediculate degenerates in the sea-bats, and so forth.

Fish Skulls and Natural Selection

The foregoing studies afford some evidence that, considering the bony fishes as a whole, the mouth, jaws and teeth have varied in all directions according to the curve of probability, limited by certain fixed necessities of a flexible gateway to the digestive system.   Of all the conceivable modifications of the jaws and teeth, a surprisingly large number have proved to be useful at different times during the hundreds of millions of years of the geologic record. The student of deep-sea fishes also will readily agree that almost every conceivable combination and apparent misfit of grossly disproportionate development of certain parts has occurred, perhaps under the distorting stresses of great cold and darkness, and that a surprisingly large number have proved either useful or not fatal to their possessors.   But it should again be noted that within the hosts of the teleosts apparently random changes have never gone so far as to destroy such "basic patents" as the suspension of the jaws by the hyomandibular, or the functional integrity of the quadrate-mandibular joint, or the articulation of the hyomandibular with the pterotic, etc.   Nor is any record known to me of an individual fish that has been congenitally deficient in these features.   Such basic patents have thus been kept intact during the entire history of the ganoids and teleosts from the Lower Devonian to the present day,—a period of nearly four hundred million years.

It is significant that the researches of the physicists upon the ages of various rocks have multiplied by twenty-fold the earlier estimates of the length of time in which it was formerly thought that the observed evolution of fishes had taken place.   In other words, Nature, operating on small or large new hereditary tendencies, has had perhaps a hundred million years for her experiments, in order to change moderately sized jaws either into nibbling jaws or into "scoop-nets"; while for the entire transformation from the agnathous to the percoid type there may have been available as much as four hundred million years.

Such vast time periods imply equally enormous numbers of variably qualified individuals, conserving by heredity the advantageous characters of the past, subject in the long run to variable chemical and physical influences, especially during the sensitive periods of the division of the egg and the formation of the *Anlagen* of the adult skull.

The objection that Natural Selection does not originate individual changes and therefore does not account for phylogenetic differences between descendants and ancestors could be reconciled with available evidence of the evolution of the fish skull only if we quite arbitrarily limit Natural Selection to mean solely the selective results of competition of one fish with another. But if we use Natural Selection, as Darwin did, as a sort of personification of the vast complex of active forces and passive conditions which cumulatively result in hereditary differences between descendants and ancestors, then we can recognize that such cumulative actions, reactions and interactions have manifestly produced a wide diversity of effects upon the now scattered descendants of the primitive percomorph skull type. In some lines Selection has evidently penalized departure from the primitive skull patterns of various grades and has conserved for scores of millions of years the various skull types now exhibited by *Amia, Lepidosteus, Erythrinus, Megalops, Gonorhynchus, Alepisaurus, Percopsis, Beryx, Centrarchus, Serranus*, etc., each of which is a primitive or central type in its own ordinal group. In other cases "Natural Selection" has encouraged a high tendency to hereditary variability so that the skull patterns have lost the primitive percomorph type and assumed many new disguises to such a degree as to baffle the best efforts of several generations of investigators, who have sought in vain to determine the precise relationships of gobies, gobiesocids, echeneiids, symbranchoids, mastacembelids and others. In short the present study, which combines field observations on the varied functions of the fish skull, museum studies on the structure of recent and fossil fish skulls and the chief results of the literature of the subject, is submitted in evidence of the power of Natural Selection to produce wide secular differentiation among the descendants of a never entirely stable ancestral germ plasm.

SUMMARY: THE PROBLEM OF DIFFERENTIAL GROWTH AND EVOLUTION IN FISH SKULLS

The Rôle of Development

In the development of the typical teleost skull the eyes and brain become very large at an early stage of development, so that in the lateral aspect the tissues that give rise to the palatoquadrate bar, the suspensorium, the suborbitals, preopercular and opercular series are squeezed between the enlarged eye and the yolk. Hence at their first appearance these structures are arranged around the eye as a center. The mouth is at first extremely small, but by the time of the disappearance of the yolk it is provided with small jaws so that the young larva may catch minute organisms. As development proceeds the typical fish increases rapidly in size, its mouth and jaws grow larger and it feeds on larger and larger prey. Meanwhile the fore parts of the head are released from the restraining membranes of the egg stage and grow forward, while the body grows backward. Thus the preopercular and opercular series, which arose so close beneath the eye, move backward from it. But while the eye decreases in relative size and the preopercular and opercular series increase, the latter still retain more or less of their circumorbital arrangement, even in the adult stages of typical teleosts. Hence *the retention, to a greater or less degree, of embryonic and œtal characters in the adult stage explains a number of characters of certain adult teleost skulls.*

The Three Main Skull Types: Long, High, Broad

The late embryonic and larval stages of typical teleosts have swollen brains and very shallow bodies, so that the future occipital roof is higher than the back. In long-bodied teleosts, such as the eels, the number of vertebral segments in the adult is very large and there has been very little vertical growth. The neurocranium too remains vertically shallow and becomes long and narrow. In deep compressed teleosts, such as the porgy or other sparids, on the contrary, the number of segments remains small and as maturity approaches the apex of the back becomes very high, so that it far overtops the occiput. The latter meanwhile has been increasing in vertical depth only at lesser rates than that of the apex, so that the roof of the skull finally slopes sharply upward, while the throat has been extending downward so as to slope toward the rapidly descending gasterion. This great deepening of the body as a whole appears to be associated with the rapid vertical growth of the myomeres. In a third type of teleost, such as *Lophius*, with wide depressed body, growth in length and height is overshadowed by growth in the transverse planes, so that the mouth becomes very broad, the skull roof broad, flat and low. Thus in fishes no less than in crystals there are primary axes around which the characteristic body-form develops.

Correlation of Skull–form and Body–form

Obviously in these and hundreds of intermediate cases the skull tends to conform to the shape of the body of which it is a part; its rates of growth in the longitudinal, vertical or transverse planes are correlated to a greater or less extent with those of the body, so that the ratio of, for instance, head height (occipital crest to isthmus) to total body length will vary

450

only within relatively narrow limits in adults of a given species. The size and shape of every bone in the skull are therefore correlated to a greater or less extent with the general growth tendencies toward excessive length, height or thickness which pervade the body as a whole. In short, there is obviously some sort of regulating mechanism between the growth of any part and that of the body as a whole, such as has already been revealed in other organisms, for example, crabs, dogs and man. Moreover, there must evidently be a large hereditary factor in many of the diagnostic indices of adult head and body length, just as there is in the mammalian skull. But it is equally plain that many proportions change profoundly during individual development, due to the interaction of hereditary with environmental forces.

Now different rates of growth in either the longitudinal, vertical or transverse planes must apparently be dependent upon one or more of the following: (a) different rates of cell division, (b) a different orientation of the longest diameter of cells to the three primary axes of the dividing egg, (c) some combinations of these two categories, (d) a change in the direction of the axis of most rapid growth. Histological studies of embryonic and larval teleosts,

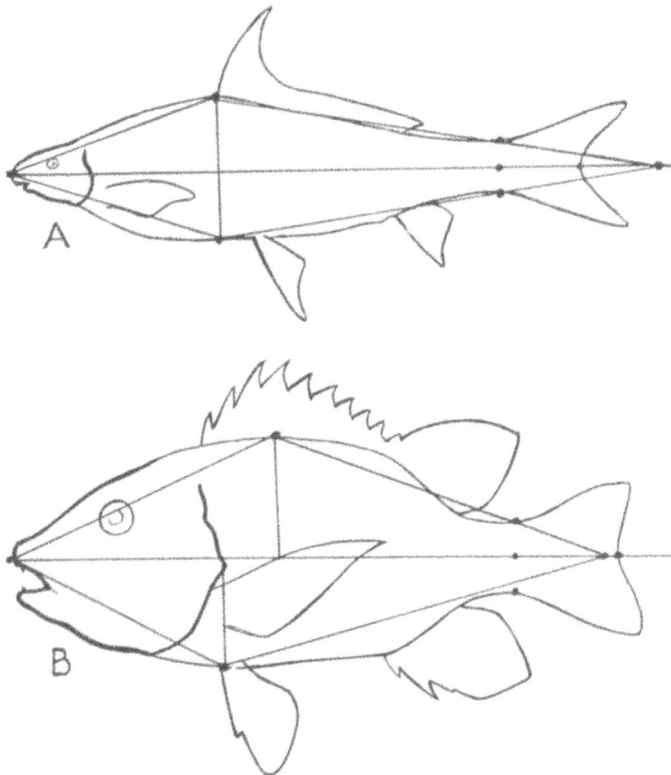

Fig. 300. Marked contrast in relative size of head to body. A. *Cycleptus.* B. *Hoplopagrus.*

25

checked if possible by experimental transplantation of parts, would possibly give clues to this problem.

In general, long bodies bear long skulls, short bodies have short skulls and wide bodies have wide skulls. But there are exceptions which are equally significant. In the peculiar

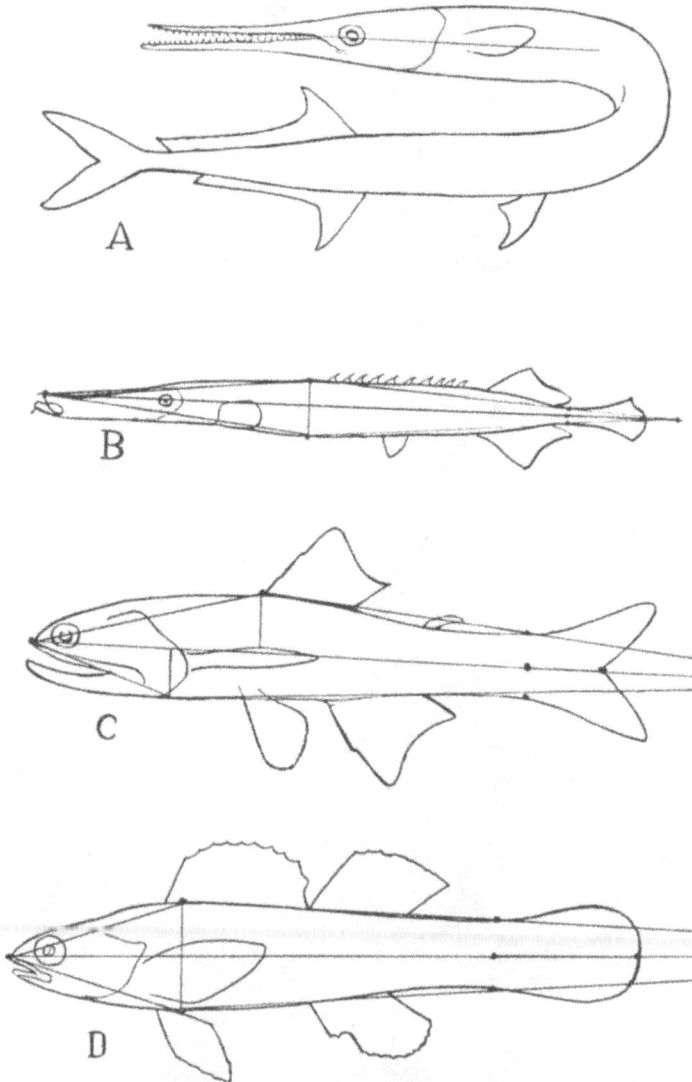

FIG. 301. A. Long jaw, long snout (*Tylosurus*). B. Short jaw, long snout (*Aulorhynchus*). C. Long jaw, short snout (*Lampanyctus*). D. Short jaw, short snout (*Copelandellus*).

cyprinoid genus *Cycleptus* (Fig. 300*A*) the head is remarkably small in proportion to the body, while in the percoid *Hoplopagrus guntheri* (Fig. 300*B*), on the contrary, the head is remarkably large. And there is the greatest diversity in the association of jaw length with snout length (Fig. 301). In short, every region of the head, while sharing more or less in the general tendencies of the body as a whole, seems to have its own intrinsic tendencies to attain specific characteristic proportions.

If different general tendencies toward length, height or thickness of skull and body be correlated in part with different amounts of enzymes received from the different ductless glands at different periods of growth, then each region or part of the skull must have its average specific rate of reacting to the enzymes that are distributed by the circulatory system. At any rate, each part seems to respond differently to a given enzyme, so that, for instance, the ripening of the gonads in the male hook-jawed salmon produces something that accelerates the growth of the tip of the lower jaw and surrounding parts without producing any equally conspicuous changes in other parts of the skull.

The Interaction of Growing Parts

Besides these general and regional responses perhaps to growth-stimulating enzymes, we must take into account the lively reactions of the growing parts to each other. Osseous tissue seems always to retreat from or grow around nerves, sense organs, brain, blood-vessels and glands and to strengthen itself against tensions transmitted by tendons, ligaments, muscles or connective tissue, and against compression or shearing stresses received from these or from other bones. For example, all the bones that come near the eyes have concave surfaces toward the eyes, while bones that carry the organs of the lateral line system develop pits and tunnels for them. On the other hand, bones that give attachments to muscles or tendons build themselves out into knobs or crests or develop ribs, trabeculæ or other forms of stiffening.

The principal muscle crests appear late in development and are dependent in part upon growth stimuli transmitted from surrounding parts. Nonetheless they are clearly determined in part also by the reaction of hereditary factors with the bodily environment. Moreover, the presence and size of certain crests may be just as specific as color spots or number characters. Every crest on the skull roof of a certain Cretaceous teleost figured by A. S. Woodward, for example, compares very closely with that of its modern descendant.

Evolution through Differential Growth Rates

Yet if we compare the skull of any highly specialized modern percoid with that of its primitive early Tertiary or Cretaceous ancestor, we shall see at once that there has been a progressive increase in certain parts and a relative decrease in others, so that, for example, in the case of the sheepshead (*Archosargus*) the jaws have become smaller but stouter, the lateral teeth have become molariform, and there have been many correlated adjustments for the support of the crushing dentition. Hence arises a paradox: for while adjustments have evidently taken place both in response to changing internal conditions and in response to changing reactions toward the environment, the experimental evidence weighs most strongly against any crude form of the Lamarckian hypothesis of the inheritance of "acquired characteristics." The immense extent of geologic time that has been required to overcome the inertia of ordinary heredity against new modifications also weighs against the Lamarck-ian hypothesis.

THE MECHANISM OF CORRELATION

In order to account for the correlation of adaptive changes in different parts of the body during the same period, *e.g.*, correlated changes in the upper and lower jaws and teeth, in the jaw muscles, supporting structures, digestive system, food preferences, etc., it would therefore seem necessary to infer that some sets of physiologically connected adult charac-ters are, at least to some extent, predetermined in connected parts of one genic system, and that shift in the survival value in a given direction (*i.e.*, toward shorter teeth and stronger jaw muscles) would encourage the survival of individuals in which the optimum combination of characters had been predetermined in one genic system.   If this be so, it may be possible some day to parallel the evolutionary history by artificial means, transforming the preda-tory mouth of a lutianid into the crushing mouth of a sparid.

In commenting on the foregoing section, Dr. G. K. Noble has kindly submitted the following note, under the title "What Molds the Skeleton?"

"Twitty (1929) has shown that if the eye of *A. tigrinum* is transplanted into the site of *A. maculatum* eye, the trabecula on this side would show a marked enlargement throughout most of its length.   Thus the eye controls the size of the trabecula.   Again, the auditory vesicle of Amphibia migrating from the overlying ectoderm induces the development of a cartilaginous capsule about itself even when transplanted to the region of the eye (Luther, 1925).   The cartilaginous nasal capsules of *Ambystoma* were shown by Burr (1916) to be dependent upon the nasal sacs for their conformation.   Hence the skull is not merely molded by the paired sense organs; parts of it are unable to develop unless the sense organs are present.

"Limb rudiments of chicks have been grown in vitro, and have produced rudimentary trochanters.   Further, there is abundant evidence that bone architecture may be assumed independent of function.   Perhaps the most recent paper on this subject is Benninghoff, A., 1930, Morph. Jahrb., 65; 11–45.   Of course function may later modify or completely reverse the type of architecture, but my point that architecture can be established without function seems to be definitely proved."

Thus experimental evidence confirms the inferences derived from the present study that the skeleton is molded by the soft parts, and that there are "intrinsic hereditary fac-tors" in each skeletal part.

In conclusion, my studies of the skulls of fishes suggest that the following factors or conditions have been conspicuous in moulding the skull into observed types:

1. Heredity, implying

    (a) individuality, manifested in differential growth of each skull bone;
    (b) more or less resistance to change,—general skull patterns often persisting through millions of years;
    (c) marked organizing ability of individual growth centers;
    (d) responsiveness, perhaps to hormones stimulating differential and successive growth along the three axes; regional and local responsiveness;

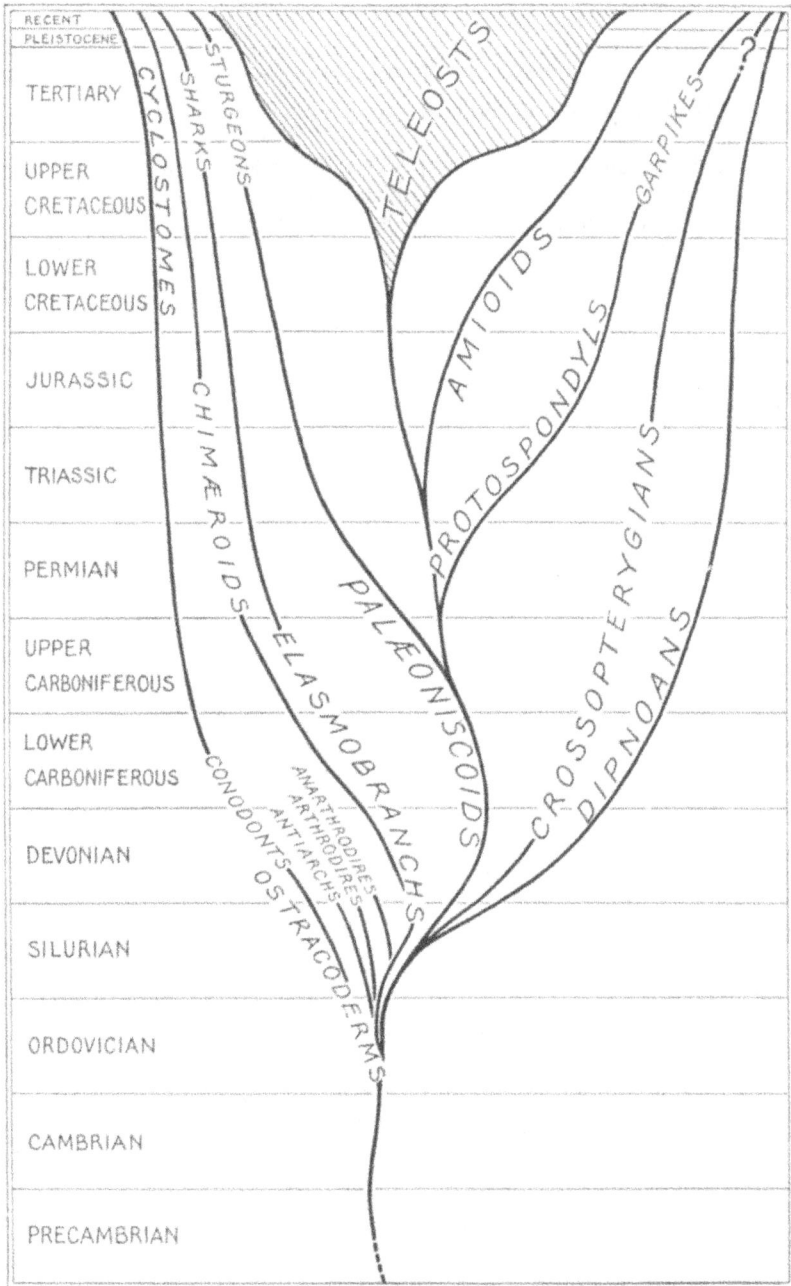

FIG. 302. Evolution of the Fishes in Geologic Time.

2. Changeableness, initiated perhaps by progressive breaking down of very complex molecules, but constantly checked or controlled by
   (a) viability (negative selection);
   (b) usefulness (positive selection);
3. Multiplicity of each species: the power to engender innumerable descendants endowed with hereditary advantages gained in the past;
4. Time—the four hundred-odd million years of fish history.

BIBLIOGRAPHY

ADAMS, L. A. 1908. Description of the Skull and Separate Cranial Bones of the Wolf-Eel (*Anarrhichthys ocellatus*). Kansas Univ. Sci. Bull., IV, No. 16 (Whole Series, XIV, No. 16), pp. 331–355, Pls. 25–36.

1919. A Memoir on the Phylogeny of the Jaw Muscles in Recent and Fossil Vertebrates. Ann. N. Y. Acad. Sci., XXVIII, pp. 51–166, 13 Pls.

ALLIS, EDWARD PHELPS, JR. 1895. The Cranial Muscles and Cranial and First Spinal Nerves in *Amia calva*. Journ. Morph., XI, pp. 485–491.

1897 *a*. The Cranial Muscles and Cranial and First Spinal Nerves in *Amia calva*. Journ. Morph., XII, Part II, pp. 487–808, Pls. 20–38.

1897 *b*. The Morphology of the Petrosal Bone and of the Sphenoidal Region of the Skull of *Amia calva*. Zool. Bull., I, 26 pp.

1898. On the Morphology of Certain of the Bones of the Cheek and Snout of *Amia calva*. Journ. Morph., XIV, pp. 425–466, Pl. 33.

1899. On Certain Homologies of the Squamosal, Intercalar, Exoccipitale and Extrascapular Bones of *Amia calva*. Anat. Anz., XVI, No. 3, 4, pp. 49–72.

1900. The Premaxillary and Maxillary Bones, and the Maxillary and Mandibular Breathing Valves of *Polypterus bichir*. Anat. Anz., XVIII, No. 11, 12, pp. 257–289, 3 figs.

1903. The Skull and the Cranial and First Spinal Muscles and Nerves in *Scomber scomber*. Journ. Morph., XVIII, No. 2, pp. 45–328, Pls. 3–12.

1905. The Laterosensory Canals and Related Bones in Fishes. Intern. Monats. Anat. Physiol., XXI, pp. 401–502.

1909. The Cranial Anatomy of the Mail-cheeked Fishes. Zoologica, Stuttgart, XXII, Heft 57, 219 pp., 8 Pls.

1913. The Homologies of the Ethmoidal Region of the Selachian Skull. Anat. Anz., XLIV, pp. 322–328.

1914. The Pituitary Fossa and Trigemino-facialis Chamber in Selachians. Anat. Anz., XLVI, No. 9, 10, pp. 225–253.

1915. The Homologies of the Hyomandibula of the Gnathostome Fishes. Journ. Morph., XXVI, No. 4, pp. 563–624.

1917. The Lips and the Nasal Apertures in the Gnathostome Fishes and Their Homologues in the Higher Vertebrates. Proc. Nat. Acad. Sci., III, pp. 73–78.

1918. On the Origin of the Hyomandibula of the Teleostomi. Anat. Rec., XV, No. 5, pp. 257–265.

1919*a*. The Lips and the Nasal Apertures in the Gnathostome Fishes. Journ. Morph., XXXII, No. 1, pp. 145–205, 4 Pls.

1919*b*. The Homologies of the Maxillary and Vomer Bones of *Polypterus*. Am. Journ. Anat., XXV, No. 4, pp. 349–394, 3 Pls.

1919*c*. On the Homologies of the Squamosal Bone of Fishes. Anat. Rec., XVII, No. 2, pp. 73–87.

1922. The Cranial Anatomy of *Polypterus*, with Special Reference to *Polypterus bichir*. Journ. Anat., LVI, Parts 3, 4, pp. 189–294, Pls. 3–24.

1923*a*. The Cranial Anatomy of *Chlamydoselachus anguineus*. Acta Zoologica, IV, pp. 123–221, 23 Pls.

1923*b*. Are the Polar and Trabecular Cartilages of Vertebrate Embryos the Pharyngeal Elements of the Mandibular and Premandibular Arches? Journ. Anat., LVIII, Part I, pp. 37–51.

1925*a*. On the Origin of the V-shaped Branchial Arch in the Teleostomi. Proc. Zool. Soc. Lond., Part 1, pp. 75–77.

1925*b*. In Further Explanation of My Theory of the Polar and Trabecular Cartilages. Journ. Anat., LIX, Part III, pp. 333–335.

1928. Concerning the Homologies of the Hyomandibula and Preoperculum. Journ. Anat., LXII, Part II, pp. 198–220.

ASSHETON, RICHARD. 1907. The Development of *Gymnarchus niloticus*. Budgett Memorial Volume. (Ed. J. Graham Kerr.) Art. XIV, pp. 293–422, Pls. 16–21, figs. 86–165. Cambridge Univ. Press.

AYERS, HOWARD. 1891. Concerning Vertebrate Cephalogenesis. Journ. Morph., IV, No. 2, pp. 221–245.

1892. Vertebrate Cephalogenesis. II. A Contribution to the Morphology of the Vertebrate Ear, with a Reconsideration of its Functions. Journ. Morphol., VI, Nos. 1, 2, pp. 1–360, 12 Pls., 26 figs.

1907. Vertebrate Cephalogenesis III. *Amphioxus* and *Bdellostoma*. Author's reprint. Pp. 1–40.

1919. Vertebrate Cephalogenesis. IV. Transformation of the Anterior End of the Head, Resulting in the Formation of the 'Nose.' Journ. Comp. Neurol., XXX, No. 4, pp. 323–342, 26 figs.

1921. Vertebrate Cephalogenesis. V. Origin of Jaw Apparatus and Trigeminus Complex—*Amphioxus*, *Ammocoetes*, *Bdellostoma*, *Callorhynchus*. Journ. Comp. Neurol., Pan-American Ed., XXXIII, No. 4, pp. 339–404, 36 figs.

1931. Vertebrate Cephalogenesis. VI. A. The Velum—Its Part in Head-Building—The Hyoid. The Velata. The Origin of the Vertebrate Head Skeleton. B. Myxinoid Characters Inherited by the Teleostomi. Journ. Morph., LII, No. 2, pp. 309–371, 22 figs.

BALFOUR, F. M. 1876. On the Development of Elasmobranch Fishes. Journ. Anat. and Physiol., X, pp. 377–410, Pls. 15, 16; pp. 517–570, Pls. 21–26; pp. 672–688, 1 Pl.

1877. The Development of Elasmobranch Fishes. Journ. Anat. and Physiol., XI, pp. 128–172, Pls. 5, 6; pp. 406–490, Pls. 15–19; pp. 674–706, Pls. 24, 25.

BEEBE, WILLIAM. 1929. *Haplophryne hudsonius*, a New Species; Description and Osteology. Zoologica, XII, No. 2, pp. 21–36, figs. 2–5.

1931. Notes on the Gill-finned Goby *Bathygobius soporator* (Cuvier and Valenciennes). Zoologica, XII, No. 5, pp. 55–66, 13 figs.

1932. Nineteen New Species and Four Post-Larval Deep-Sea Fish. Zoologica, XIII, No. 4, pp. 47–107, figs. 8–31. [Jaws, etc. of *Aceratias*.]

1932. Ontological Notes on *Remora remora*. Zoologica, XIII, No. 6, pp. 121–132, figs. 32–37.

DE BEER, G. R. 1924a. Studies on the Vertebrate Head. Part I. Fish. Quart. Journ. Microscopical Sci., LXVIII, Part 2, pp. 287–341.

1924b. The Prootic Somites of *Heterodontus* and of *Amia* and Contributions to the Study of the Development of the Head in *Heterodontus*. Quart. Journ. Microscopical Sci., LXVIII, Part 1, pp. 17–65.

1925. Contributions to the Development of the Skull in Sturgeons. Quart. Journ. Microscopical Sci., LXIX, No. 276. . pp. 671–687.

1926. Studies on the Vertebrate Head. II. The Orbito-temporal Region of the Skull. Quart. Journ. Microscopical Sci., LXX, pp. 263–370.

1927. The Early Development of the Chondrocranium of *Salmo fario*. Quart. Journ. Microscopical Sci., LXXI, Part II, pp. 259–312.

1928. Vertebrate Zoology: An Introduction to the Comparative Anatomy, Embryology and Evolution of Chordate Animals. 8vo., The Macmillan Company, New York, 505 pp., 185 figs.

BENHAM, W. B. AND DUNBAR, W. J. 1906. On the Skull of a Young Specimen of the Ribbon-fish *Regalecus*. Proc. Zool. Soc. London, I, pp. 544–556, Pls. 38, 39.

BIGELOW, HENRY B. AND WELSH, W. W. 1925. Fishes of the Gulf of Maine. Bull. U. S. Bureau of Fisheries, XL, Part I, pp. 1–567, 278 figs.

BLOCH, D. MARCUS GLIEFER. 1787. Naturgeschichte der auslandischen Fische mit vier und fünfzig Kupfertafeln nach Originalen. Zweiter Theil. Berlin, xii, 260 pp., Pls. 163–216.

BOULENGER, G. A. 1895. Catalogue of the Perciform Fishes in the British Museum. [Second edition.] Vol. I: Centrarchidæ, Percidæ and Serranidæ (part). 8vo, London, 395 pps., 15 Pls.

1898. A Revision of the Genera and Species of Fishes of the Family Mormyridæ. Proc. Zool. Soc. London, pp. 775–821.

1901a. Notes on the Classification of Teleostean Fishes. I. On the Trachinidæ and their Allies. Ann. Mag. Nat. Hist., Series 7, VIII, pp. 261–271.

1901b. Les Poissons du Bassin du Congo. Bruxelles, lxii, 532 pp., 25 Pls.

1902a. Notes on the Classification of Teleostean Fishes. II. On the Berycidæ. Ann. Mag. Nat. Hist., Series 7, IX, pp. 197–204.

1902b. Notes on the Classification of Teleostean Fishes. III. On the Systematic Position of the Genus *Lampris*, and on the Limits and Contents of the Suborder Catosteomi. Ann. Mag. Nat. Hist., Series 7, X, pp. 147–152.

1902c. Notes on the Classification of Teleostean Fishes. IV. On the Systematic Position of the Pleuronectidæ. Ann. Mag. Nat. Hist., Series 7, X, pp. 295–304.

1904, 1910. Fishes (Systematic Account of Teleostei). In The Cambridge Natural History (Edited by S. F. Harmer and A. E. Shipley), Chapt. VII, pp. 541–727, figs. 325–440. Also Edition of 1910.

BRAUER, A. 1908. Die Tiefseefische. I. Systematischer Teil. In Wiss. Ergebnisse der Deutschen Tiefsee-Expedition auf dem Dampfer "Valdivia," 1898–1899. (Ed. Carl Chun), XV, pp. 1–420, 16 Pls., 176 figs. II. Anatomischer Teil. *Idem*, XV, 2 Teil, pp. 1–266, Pls. 19–44, 11 figs. Jena.

BREDER, C. M. 1926. The Locomotion of Fishes. Zoologica, IV, No. 5, pp. 159–297, 83 figs.

1932. An Interesting Scorpion Fish. Bull. N. Y. Zool. Soc., XXXV, No. 1, pp. 30–32, 3 Pls.

BRIDGE, T. W. 1877. The Cranial Osteology of *Amia calva*. Journ. Anat. and Physiol., XI, pp. 605–622, Pl. 23.

1895. On Certain Features in the Skull of *Osteoglossum formosum*. Proc. Zool. Soc. London, pp. 302–310, Pl. 22.

1900. The Air-bladder and its Connection with the Auditory Organ in *Notopterus borneensis*. Journ. Linn. Soc., London, Zool., XXVII, pp. 503–540, Pls. 36, 37.

BRIDGE, T. W. AND HADDON, A. C. 1893. The Air-Bladder and Weberian Ossicles of the Siluroid Fishes. Contributions to the Anatomy of Fishes. II. Philos. Trans. Roy. Soc. London, CLXXXIV (1893), B, pp. 65–333, Pls. 11–19.

BROILI, FERDINAND. 1930. Ueber *Gemündina Stürtzi* Traquair. Abh. Bayerischen Akad. d. Wissen., Neue Folge VI, pp. 1–24, 4 Pls.

BROUGH, JAMES. 1931. On Fossil Fishes from the Karroo System, and Some General Considerations on the Bony Fishes of the Triassic Period. Proc. Zool. Soc. London, Part 1, pp. 235–296, 19 figs., 4 Pls.

BRYANT, W. L. 1919. On the Structure of *Eusthenopteron*. Bull. Buffalo Soc. Nat. Sci., XIII, No. 1, pp. 1–23, Pls. 1–18.

BUDGETT, JOHN SAMUEL. 1901. On Some Points in the Anatomy of *Polypterus*. Trans. Zool. Soc. London, XV, Part 7, 14 pp., 3 Pls., 7 figs.

1902. On the Structure of the Larval *Polypterus*. Trans. Zool. Soc. London, XVI, Part 7, 32 pp., 3 Pls.

CANTOR, THEODORE. 1849. Catalogue of Malayan Fishes. Journ. Asiatic Soc., Bengal, XVIII, pp. i–xii, 983–1443, Pls. 1–14.

CHILD, C. M. 1925. The Physiological Significance of the Cephalo-caudal Differential in Vertebrate Development. Anat. Rec., XXXI, No. 4, pp. 369–383.

CHUN, CARL. 1908. (Editor). Wissenschaftliche Ergebnisse der Deutschen Tiefsee-Expedition auf dem Dampfer "Valdivia" 1898–99. 15 Band. Jena.

CLELAND, JOHN. 1862. On the Anatomy of the Short Sun-Fish (*Orthragoriscus mola*). Nat. Hist. Review, pp. 170–185, Pls. 5, 6.

COCKERELL, T. D. A. 1916. The Scales of the Brotulid Fishes. Ann. Mag. Nat. Hist., (8), XVIII, pp. 317–325.

1924. Fossils in the Ondai Sair Formation, Mongolia. Bull. Amer. Mus. Nat. Hist., LI, Art. 6, pp. 129–144.

1925. The Affinities of the Fish *Lycoptera middendorffi*. Bull. Amer. Mus. Nat. Hist., LI, Art. 8, pp. 313–317.

COLE, F. J. AND JOHNSTONE, J. 1901. *Pleuronectes* (The Plaice). L. M. B. C. Memoirs, No. 8. Proc. and Trans. Liverpool Biol. Soc., XVI, 1901–1902, pp. 145–396, Pls. 1–9 (36 figs.), 1 Table.

COPE, E. D. 1884. Vertebrata of the Tertiary Formations of the West. Book I. Rept. U. S. Geol. Surv. of the Territories, III, pp. i–xxxiv, 1–1007, Pls. 1–75a [Fishes, pp. 1–100.]

CROOK, ALVA ROBINSON. 1892. Ueber einige fossile Knochenfische aus der mittleren Kreide von Kansas. Palæontographica, XXXIX, pp. 108–124, Pls. 15–18.

CUVIER, GEORGES. 1817. Sur le genre *Chironectes* Cuv. (*Antennarius* Commers.). Mém. Mus. d'Hist. Nat. Paris, III, pp. 418–435, Pls. 16–18.

1828–1849. (With Achille Valenciennes.) Histoire naturelle des poissons. Paris. 22 Vols., 650 Pls.

DARESTE, M. C. 1872. Études sur les types ostéologiques des poissons osseux. Comptes rendus de l'acad., LXXV, pp. 942–946; 1018–1021; 1086–1089; 1172–1175; 1253–1256.

DAY, ARTEMAS L. 1914. The Osseous System of *Ophiocephalus striatus* Bloch. Philippine Journ. Sci., IX, Section D, No. 1 pp. 19–55, Pls. 1–16, figs. 1–7.

DAY, FRANCIS. 1865. The Fishes of Malabar. London. 4to. xxxii, 293 pp., 20 Pls.

1878–88. The Fishes of India; being a Natural History of the Fishes Known to Inhabit the Seas and Fresh Waters of India, Burma, and Ceylon. Text and Atlas [198 Pls.]. London: Printed for the Author. 4to.

1889. Fauna of British India, including Ceylon and Burma. Vol. II. Fishes. London. 8vo. xiv, 509 pp., 177 figs.

DAY, HENRY. 1915. A Note on the Parasphenoid of a Palæoniscoid. Ann. and Mag. Nat. Hist., (8), XVI, pp. 421–434.

DEAN, BASHFORD. 1899. The "Devonian Lamprey" *Palæospondylus gunni*, Traquair. Mem. N. Y. Acad. Sci., II, Part I, pp. 1–32, Pl. 1.

1906. Chimæroid Fishes and Their Development. Publ. Carnegie Inst. Washington, No. 32. 195 pp., 144 figs.

1907. Notes on Acanthodian Sharks. Amer. Journ. Anat., VII, pp. 209–226, 36 figs.

1908. Accidental Resemblance among Animals. A Chapter in Un-natural History. Popular Science Monthly, LXXII, pp. 304–312, 10 figs.

1909. Studies on Fossil Fishes (Sharks, Chimæroids and Arthrodires). Mem. Amer. Mus. Nat. Hist., IX, pp. 211–287. 16 Pls., 65 figs.

DELAGE, YVES AND HÉROUARD, EDGARD. 1898. Traité de Zoologie Concrète. Tome VIII. Les Procordés. Paris. 8vo. Pp. i–vii, 1–379, 54 Pls., 275 figs.

DELSMAN, H.C. 1925. Fishes with Protrusile Mouths. Treubia, VI, Livr. 2, pp. 98–106.

DERSCHEID, J. M. 1923. Contributions à la Morphologie Cephalique des Vertébrés. A.—Structure de l'Organe olfactif chez les Poissons. Première Partie. Osteichthyes, Teleostei, Malacopterygii. Annales de la Société royale Zoologique de Belgique, LIV, pp. 79–162, 26 figs.

DIETZ, P. A. 1914. Beiträge zur Kenntnis der Kiefer- und Kiemenbogenmuskulatur der Teleostier. 1. Die Kiefer- und Kiemenbogenmuskeln der Acanthopterygier. Mitteil. a.d. Zool. Station zu Neapel, XXII, Nr. 4, pp. 99–162, 45 figs.

DOHRN, ANTON. 1876. Der Ursprung der Wirbelthiere und das Princip des Functionswechsels: Genealogische Skizzen. Wilhelm Engelmann. Leipzig. 8vo.

DOLLO, LOUIS. 1904. Poissons. Expédition Antarctique Belge. Résultats du Voyage du S. Y. Belgica en 1897–1898–1899. Rapports Scientifiques . . . Zoologie. Anvers. 4to. Pp. 1–240, Pls. 1–12.

1909. Les Téléostéens à Ventrales Abdominales Secondaires. Verhandl. d. K. K. zool.-bot. Gesellsch. in Wien, LIX, pp. 135–140.

DUNBAR, W. J. 1906. See Benham, W. B. and Dunbar, W. J.

EDGEWORTH, F. H. 1926. On the Hyomandibula of Selachii, Teleostomi and *Ceratodus*. Journ. Anat., LX, Part II, pp. 173–193.

EDINGER, TILLY. 1929. Ueber Knocherne Scleralringe. Zool. Jahrb., LI, pp. 163–226.

EDWARDS, LINDEN F. 1926. The Protractile Apparatus of the Mouth of the Catastomid Fishes. Anat. Rec., XXXIII, No. 4, pp. 257–270.

ELLIS, MAX MAPES. 1913. The Gymnotid Eels of Tropical America. Mem. Carnegie Mus., VI, No. 3, pp. 109–195, Pls. 15–23.

EMERY, CARLO. 1880. Le Specie del Genere *Fierasfer* nel Golfo di Napoli e Regioni Limitrofe. In "Fauna und Flora des Golfes von Neapel und der Angrenzenden Meeres-Abschnitte," No. 2, pp. 1–76, Pls. 1–9. Leipzig.

ERDL, M. P. 1847. Beschreibung des Skeletes von *Gymnarchus niloticus*, nebst Vergleichung mit Skeleten formverwandter Fische. Abh. k. Bayer. Akad. Wiss., math.-phys. Kl., V, pp. 209–252, Taf. 5.

FROST, G. ALLAN. 1925a. A Comparative Study of the Otoliths of the Neopterygian Fishes. [I. Isospondyli.] Ann. Mag. Nat. Hist., (9), XV, pp. 152–163, Pls. 11–13.

1925b. A Comparative Study of the Otoliths of the Neopterygian Fishes (continued). [II. A. Ostariophysi.] Ann. Mag. Nat. Hist., (9), XV, pp. 553–561, Pl. 29.

1925c. A Comparative Study of the Otoliths of the Neopterygian Fishes (continued). [II. B. Ostariophysi.] Ann. Mag. Nat. Hist. (9), XVI, pp. 433–446, Pls. 22, 23.

1925d. Eocene Fish Otoliths from the London District and the Isle of Wight. Ann. Mag. Nat. Hist., (9), XIV, pp. 160–164, Pl. 10.

1926a. Otoliths of Fishes from the Jurassic of Buckinghamshire and Dorset. Ann. Mag. Nat. Hist., (9), XVIII, pp. 81–85, Pl. 4.

1926b. A Comparative Study of the Otoliths of the Neopterygian Fishes (continued). [III. Apodes.] Ann. Mag. Nat. Hist., (9), XVII, pp. 99–104.

1926c. A Comparative Study of the Otoliths of the Neopterygian Fishes (continued). IV. Haplomi. V. Heteromi. VI. Iniomi. VII. Lyomeri. VIII. Hypostomides. IX. Salmopercæ. X. Synentognathi. XI. Microcyprini. XII. Solenichthyes. Ann. Mag. Nat. Hist., (9), XVIII, pp. 465–482, Pls. 20, 21.

1926d. A Comparative Study of the Otoliths of the Neopterygian Fishes (continued). [XIII. Anacanthini.] Ann. Mag. Nat. Hist., (9), XVIII, pp. 483–490, Pl. 22.

1927a. A Comparative Study of the Otoliths of the Neopterygian Fishes (continued). XIV. Allotriognathi. XV. Berycomorphi. XVI. Zeomorphi. Ann. Mag. Nat. Hist., (9), XIX, pp. 439–445, Pl. 8.

1927b. A Comparative Study of the Otoliths of the Neopterygian Fishes (continued). [XVIII. Percomorphi.] Ann. Mag. Nat. Hist., (9), XX, pp. 298–305, Pl. V.

1928a. A Comparative Study of the Otoliths of the Neopterygian Fishes (continued). [XVIII. Percomorphi (cont'd).] Ann. Mag. Nat. Hist., (10), I, pp. 451–456, Pl. XVII.

1928b. A Comparative Study of the Otoliths of the Neopterygian Fishes (continued). [XVIII. Percomorphi (cont'd).] Ann. Mag. Nat. Hist., (10), II, pp. 328–331, Pl. XII.

1929a. A Comparative Study of the Otoliths of the Neopterygian Fishes (continued). [XVIII. Percomorphi (concl'd). XVII. Symbranchii.] Ann. Mag. Nat. Hist., (10), IV, pp. 120–130, Pls. I, II.

1929b. A Comparative Study of the Otoliths of the Neopterygian Fishes (continued). [XIX. Scleroparei.] Ann. Mag. Nat. Hist., (10), IV, pp. 257–264, Pl. III.

1930a. A Comparative Study of the Otoliths of the Neopterygian Fishes (continued.) [XX. Heterosomata.] Ann. Mag. Nat. Hist., (10), V, pp. 231–239, Pl. IX.

1930b. A Comparative Study of the Otoliths of the Neopterygian Fishes (concluded). [XXI. Discocephali. XXII. Plectognathi. XXIII. Malacicthyes. XXIV. Xenopterygii. XXV. Haplodoci. XXVI. Pediculati. XXVII. Opisthomi.] Ann. Mag. Nat. Hist., (10), V, pp. 621–627, Pl. XXIII.

Fürbringer, Max. 1875. Untersuchungen zur vergleichenden Anatomie der Muskulatur des Kopfskelets der Cyclostomen. Jena. Zeitschr. Naturw., IX, pp. 1–129.

Garman, Samuel. 1899. Reports on an Exploration off the West Coasts of Mexico, Central and South America, and off the Galapagos Islands, in Charge of Alexander Agassiz, by the U. S. Fish Commission Steamer "Albatross," during 1891. XXVI. The Fishes. Mem. Mus. Comp. Zool. Harvard Coll., XXIV, pp. 1–431, Pls. 1–97. [2 vols., Text and Atlas.]

1913. The Plagiostomia (Sharks, Skates and Rays). Mem. Mus. Comp. Zool. Harvard Coll., XXXVI, pp. 1–528, 77 Pls. [Text and Atlas.]

Garstang, Walter. 1931. The Phyletic Classification of the Teleostei. Proc. Leeds Philos. and Lit. Soc., Scient. Sect., II. Part 5, pp. 240–260, 1 phyletic chart.

Gaskell, W. H. 1898–1906. On the Origin of Vertebrates, Deduced from the Study of Ammocoetes. Journ. Anat. Physiol., London.

1898a. 1. The Origin of the Brain. Journ. Anat. Physiol., London, 1898, XXXII, pp. 513–533, 4 figs.

1898b. 2. The Origin of the Vertebrate Cranio-facial Skeleton. Idem, pp. 553–581. 1 Pl., 3 figs.

1899a. 3. On the Origin of the Branchial Segmentation. Idem, 1899, XXXIII, pp. 154–188. 1 Pl., 6 figs.

1899b. 4. The Thyroid or Opercular Segment; the Meaning of the Facial Nerve. Idem, pp. 638–671. 1 Pl., 15 figs.

1900a. 5. On the Origin of the Pro-otic Segmentation; the Meaning of the Trigeminal and Eye-muscle Nerves. Idem, 1900, XXXIV, pp. 465–513. 11 figs.

1900b. 6. The Old Mouth and the Olfactory Organ; the Meaning of the First Nerve. Idem, pp. 514–537.

1900c. 7. On the Evidence of Prosomatic Appendages in Ammocoetes, as Given by the Course and Distribution of the Trigeminal Nerve. Idem, pp. 537–561. 5 figs.

1900d. 8. The Palæontological Evidence: Ammocoetes a Cephalaspid. Idem, pp. 562–587. 3 Pls., 8 figs.

1901. 9. On the Origin of the Optic Apparatus; the Meaning of the Optic Nerves. Idem, 1901, XXXV, pp. 224–267. 12 figs.

1902. 10. On the Origin of the Auditory Organ; the Meaning of the Eighth Cranial Nerve; together with a Consideration of the Origin of the Cranial Nerves as a Whole, in Accordance with the Principles Laid Down in Part I. Idem, 1902, XXXVI, pp. 164–208. 13 figs.

1903. 11. The Origin of the Vertebrate Body Cavity and Excretory Organs; the Meaning of the Somites of the Trunk and of the Ductless Glands. Idem, 1903, XXXVII, pp. 168–219. 6 figs.

1905. 12. The Principles of Embryology. Idem, 1905, XXXIX, pp. 371–401.

1906. 13. The Origin of the Notochord and Alimentary Canal. Idem, 1906, XL, pp. 305–317. 5 figs.

Gaupp, E. 1898. Zur Entwicklungsgeschichte des Eidechsenschädels. Berichte d. Naturforsch. Gesell. z. Freiburg i. Br., X, pp. 302–316. [Homologies of skull bones.]

1900. Alte Probleme und neuere Arbeiten über den Wirbeltierschädel. Ergebnisse d. Anat. u. Entwickelungsgesch., X, (1901), 2 Teil, pp. 847–1001, 5 figs.

1904. Das Hyobranchialskelet der Wirbeltiere. Ergebnisse d. Anat. u. Entwickelungsgesch., XIV, (1905), No. 9, pp. 808–1037, 46 figs.

1905. Neue Deutungen auf dem Gebiete der Lehre vom Säugetierschädel. Anat. Anz., XXVII, No. 12, 13, pp. 273–310, 9 figs. [Septomaxilla.]

1911. Beiträge zur Kenntnis des Unterkiefers der Wirbeltiere. I. Der Processus anterior (Folii) des Hammers der Säuger und das Goniale der Nichtsäuger. Anat. Anz., XXXIX, No. 4, 5, pp. 97–135, 16 figs. [Goniale = prearticular, p. 125; homologies of jaw elements in teleosts, pp. 126, 127.]

1912. Die Reichertsche Theorie (Hammer-, Amboss- und Kieferfrage). Archiv f. Anat. u. Entwickelungsgesch., Jahr. 1912 (1913), Suppl.-Bd., ix–xiii, 1–416, 149 figs. ø

Gegenbaur, Carl. 1872. Untersuchungen zur Vergleichenden Anatomie der Wirbelthiere. Heft 3. Das Kopfskelet der Selachier: ein Beitrag zur Erkenntniss der Genese des Kopfskeletes der Wirbelthiere. x, 316 pp., 22 Pls. Leipzig.

1878. Ueber das Kopfskelet von *Alepocephalus rostratus* Risso. Morphol. Jahrb., IV; Fünfzigjährigen Jubelfeier, Carl Theodor Ernst von Siebold, pp. 1–42, Pls. I, II.

1898–1901. Vergleichende Anatomie der Wirbelthiere mit Berücksichtigung der Wirbellosen. I, II. 974 figs. Leipzig.

GILBERT, CHARLES HENRY. 1905. The Deep-Sea Fishes of the Hawaïian Islands. In "The Aquatic Resources of the Hawaiian Islands" by David Starr Jordan and Barton Warren Evermann. Bull. U. S. Fish Commission, XXIII, 1903 (1905), Part 2, Section 2, pp. 575–713. 45 Pls., 44 figs.

GILL, E. LEONARD. 1923a. The Permian Fishes of the Genus *Acentrophorus*. Proc. Zool. Soc. London, pp. 19–40. 16 figs.

1923b. An Undescribed Fish from the Coal Measures of Lancashire. Ann. Mag. Nat. Hist., (9), XI, pp. 465–472.

GILL, T. N. 1893. Families and Subfamilies of Fishes. Mem. Nat. Acad. Sci., VI, pp. 127–138.

GISLÉN, TORSTEN. 1930. Affinities between the Echinodermata, Enteropneusta, and Chordonia. Zool. Bidrag f. Uppsala, Band XII, pp. 199–304, 46 figs.

GMELIN, CARL CHRISTIAN. 1829. Gemeinnützige Systematische Naturgeschichte der Fische. 2. Auflage. 28 Parts, 164 figs. Mannheim.

GOODE, GEORGE BROWN AND BEAN, TARLETON H. 1895. Oceanic Ichthyology. A Treatise on the Deep-sea and Pelagic Fishes of the World. Based chiefly upon the Collections Made by the Steamers Blake, Albatross, and Fish Hawk in the Northwestern Atlantic. Text and Atlas (417 figs.). Smithsonian Inst., U. S. Nat. Mus., Special Bull., pp. i–xxiii, 1–26.

GOODRICH, E. S. 1908a. On the Systematic Position of *Polypterus*. Rept. Brit. Assoc. Adv. Sci., 77th Meeting, 1907 (1908), pp. 545, 546.

1908b. On the Scales of Fish, Living and Extinct, and Their Importance in Classification. Proc. Zool. Soc. London, 1907 (1908), pp. 751–774. 4 Pls., 9 figs.

1909. A Treatise on Zoology (Ed. Sir Ray Lankester). Part IX. Vertebrata Craniata (First Fascicle: Cyclostomes and Fishes). London. 8vo.

1919. Restorations of the Head of *Osteolepis*. Journ. Linnean Soc. London, Zool., XXIV, pp. 181–188, figs. 1–5.

1928. *Polypterus* a Palæoniscid? Palæobiologica, I Bd., pp. 87–92.

1930. Studies on the Structure and Development of Vertebrates. xxx, 837 pp., 754 figs.

GRABER, VITUS. 1886. Die äussern mechanischen Werkzeuge der Tiere. I. Wirbeltiere. Leipzig. 224 pp., 171 figs. [Mechanism of tongue, p. 114; carp skull, pp. 79–83, figs. 44, 45.]

GREGORY, WILLIAM K. 1904. The Relations of the Anterior Visceral Arches to the Chondrocranium. Biol. Bull., VII, No. 1, pp. 55–69, 1 Pl.

1913. Locomotive Adaptations in Fishes Illustrating "Habitus" and "Heritage." Ann. N. Y. Acad. Sci., Feb. 10, pp. 266–268.

1915. Present Status of the Problem of the Origin of the Tetrapoda with Special Reference to the Skull and Paired Limbs. Ann. N. Y. Acad. Sci., XXVI, pp. 317–383, Pl. IV.

1917. Second Report of the Committee on the Nomenclature of the Cranial Elements in the Permian Tetrapoda. Bull. Geol. Soc. Amer., XXVIII, pp. 973–986.

1920. Studies in Comparative Myology and Osteology. IV. A Review of the Evolution of the Lacrymal Bone of Vertebrates with Special Reference to that of Mammals. Bull. Amer. Mus. Nat. Hist., XLII, Art. 2, pp. 95–263, Pl. XVII.

1923. A Jurassic Fish Fauna from Western Cuba, with an Arrangement of the Families of Holostean Ganoid Fishes. Bull. Amer. Mus. Nat. Hist., XLVIII, Art. 8, pp. 223–242, 6 figs.

1924. On Design in Nature. The Yale Review, January, 1924. 12 pp.

1927. The Palæomorphology of the Human Head: Ten Structural Stages from Fish to Man. Part I. The Skull in *Norma Lateralis*. Quart. Rev. Biol., II, No. 2, pp. 267–279, 5 figs.

1928a. Studies on the Body-Forms of Fishes. Zoologica (New York), VIII, No. 6, pp. 325–421, figs. 117–155.

1928b. The Body-Forms of Fishes and Their Inscribed Rectilinear Lines. Palæobiologica, I. Bd., pp. 93–100, Taf. VIII.

1929. The Palæomorphology of the Human Head: Ten Structural Stages from Fish to Man. Part II. The Skull in *Norma Basalis*. Quart. Rev. Biol., IV, No. 2, pp. 233–247, 3 figs.

1932. Some Strange Teleost Skulls and Their Derivation from Normal Forms. Copeia, July 1, 1932, No. 2, pp. 53–60, 3 figs.

GRENHOLM, ÅKE. 1923. Studien über die Flossenmuskulatur der Teleostier. Uppsala. ix, 296 pp., 168 figs.

GUDGER, E. W. 1925. The Crucifix in the Catfish Skull. Natural History, XXV, No. 4, pp. 371–380.

1926. A Study of the Smallest Shark-suckers (Echeneididæ) on Record, with Special Reference to Metamorphosis. Amer. Mus. Novitates, No. 234, 26 pp., 6 figs.

GUNTHER, A. 1880. Report on the Shore Fishes Procured during the Voyage of H. M. S. Challenger in the Years 1873–1876. In "Zoology of the Voyage of H. M. S. 'Challenger,'" I, Part 6, 82 pp., 32 Pls. London, 4to.

HARRISON, ROSS G. 1907. Experiments in Transplanting Limbs and Their Bearing upon the Problems of the Development of Nerves. Journ. Exper. Zool., IV, No. 2, pp. 239–281.

1915. Experiments on the Development of the Limbs in Amphibia. Proc. Nat. Acad. Sci., I, pp. 539–544, 3 figs.

1929. Correlation in the Development and Growth of the Eye Studied by Means of Heteroplastic Transplantation. Archiv. f. Entwicklungs-mechanik d. Organismen. 120 Band. Festschrift f. Hans Spemann. Fünfter Teil, pp. 1–55, 47 figs.

HAY, O. P. 1902. Bibliography and Catalogue of the Fossil Vertebrata of North America. Bull. U. S. Geol. Surv., No. 179, iii, 868 pp. [Classification of Fishes, pp. 252–255.]

1903a. On Certain Genera and Species of North American Cretaceous Actinopterous Fishes. Bull. Amer. Mus. Nat. Hist., XIX, Art. 1, pp. 1–95, Pls. I–V, figs. 1–72.

1903b. On a Collection of Upper Cretaceous Fishes from Mount Lebanon, Syria, with Descriptions of Four New Genera and Nineteen New Species. Bull. Amer. Mus. Nat. Hist., XIX, Art. X, pp. 395–452.

HEINTZ, ANATOL. 1931. A New Reconstruction of *Dinichthys*. Amer. Mus. Novitates, No. 457, 5 pp., 3 figs.

HERRE, ALBERT W. AND MONTALBAN, HERACLIO R. 1928. The Philippine Siganids. Philippine Jour. Sci., XXXV, pp. 151–185, Pls. I–VI.

HERRICK, C. JUDSON. 1899. The Cranial and First Spinal Nerves of *Menidia;* a Contribution upon the Nerve Components of the Bony Fishes. Journ. Comp. Neurology, IX, pp. 153–455, Pls. XIV–XX.

HILDEBRAND, SAMUEL F. AND CABLE, LOUELLA E. 1930. Development and Life History of Fourteen Teleostean Fishes at Beaufort, N. C. U. S. Dept. Commerce, Bull. Bureau of Fisheries, XLVI, Doc. No. 1093, pp. 383–488, figs. 1–101.

HOLMQUIST, OTTO. 1910. Der Musculus Protractor Hyoidei (Geniohyoideus Auctt.) und der Senkungsmechanismus des Unterkiefers bei den Knochenfischen. Lunds Universitets Arsskrift. N.F. Afd. 2, Bd. VI, Nr. 6. Kongl. Fysiografiska Sällskapets Handlingar, N.F. Bd. XXI, Nr. 6, pp. 1–25, figs. 1, 2.

HUBBS, CARL L. 1919. A Comparative Study of the Bones Forming the Opercular Series of Fishes. Journ. Morphol., XXXIII, No. 1, pp. 61–71.

1926. The Structural Consequences of Modifications of the Developmental Rate in Fishes, Considered in Reference to Certain Problems of Evolution. Amer. Naturalist, LX, No. 666, pp. 57–81.

HUSSAKOF, LOUIS. 1906. Studies on the Arthrodira. Mem. Amer. Mus. Nat. Hist., IX, Part 3, pp. 105–154, Pls. XII, XIII.

1911. The Permian Fishes of North America. Carn. Inst. Washington, Publ. No. 146, pp. 153–178, Pls. XXVI–XXXII, figs. 53–56.

HUXLEY, JULIAN S. 1931. Notes on Differential Growth. Amer. Nat., LXV, July–Aug., pp. 289–315, 7 figs., 8 tables.

HYRTL, JOSEPH. 1856. Anatomische Mittheilungen über *Mormyrus* und *Gymnarchus.* Denkschr. d. Kaiserlichen Akad. d. Wissensch. Mat.-Naturv. Kl., XII, pp. 1–23, Taf. 1–6.

JAEKEL, O. 1927. Der Kopf der Wirbeltiere. Ergebnisse der Anat. u. Entwick., XXVII, pp. 815–974.

JENKINS, J. TRAVIS. 1925. The Fishes of the British Isles, both Fresh Water and Salt. London. Frederick Warne and Co. vii, 376 pp., 278 figs.

JENNINGS, H. S. 1930. The Biological Basis of Human Nature. W. W. Norton and Co. 384 pp. 8vo.

JORDAN, DAVID STARR. 1905. A Guide to the Study of Fishes. Henry Holt and Co. 2 Vols. 427 figs. 8vo.

1923. A Classification of Fishes, including Families and Genera as Far as Known. Stanford Univ. Publ., Univ. Ser. Biol. Sci., III, No. 2, pp. i–x, 79–243.

JORDAN, DAVID STARR AND EVERMANN, BARTON WARREN. 1896. The Fishes of North and Middle America. A Descriptive Catalogue of the Species of Fish-like Vertebrates Found in the Waters of North America North of the Isthmus of Panama. Bull. U. S. Nat. Mus., XLVII, Parts 1–4, 3313 pp., 392 Pls.

1905. The Aquatic Resources of the Hawaiian Islands. Part I.—The Shore Fishes. Bull. U. S. Fish Comm., XXIII, 1903 (1905), Part 1, pp. 1–574, 138 Pls., 229 figs.

JORDAN, DAVID STARR AND HUBBS, CARL LEAVITT. 1919. Studies in Ichthyology: A Monographic Review of the Family of Atherinidæ or Silversides. Leland Stanford Junior Univ. Publ., Univ. Series, pp. 1–87, Pls. 1–12.

JUNGERSEN, H. F. E. 1908. Ichthyotomical Contributions. I. The Structure of the Genera *Amphisile* and *Centriscus.* Danske Vidensk. Skrift. Naturv., (7), VI, pp. 41–109. 2 Pls., 33 figs.

1910. Ichthyotomical Contributions. II. The Structure of the Aulostomidæ, Syngnathidæ and Solenostomidæ. Danske Vidensk. Skrift. Naturv., 1910–1911, (7), VIII, pp. 268–364, 7 Pls., 1 fig.

KESTEVEN, H. LEIGHTON. 1922. A New Interpretation of the Bones in the Palate and Upper Jaw of Fishes. Journ. Anat., LVI, Parts 3, 4, pp. 307–324.

1923. Are the Polar and Trabecular Cartilages of Vertebrate Embryos the Pharyngeal Elements of the Mandibular and Premandibular Arches? Journ. Anat., LVIII, Part I, pp. 37–51.

1925a. Contributions to the Cranial Osteology of the Fishes. No. 1. *Tandanus tandanus* Mitchell. Rec. Australian Mus., XIV, No. 4, pp. 271–288.

1925b. A Third Contribution on the Homologies of the Parasphenoid, Ectopterygoid and Pterygoid Bones and of the Metapterygoid. Journ. and Proc. Roy. Soc. N. S. Wales, LIX, pp. 41–107, 59 figs.

1926a. Contributions to the Cranial Osteology of the Fishes. No. II. The Maxillæ in the Eels and the Identification of These Bones in the Fishes Generally. Rec. Australian Mus., XV, No. 1, pp. 132–140, figs. 1–8.

1926b. Contributions to the Cranial Osteology of the Fishes. No. III. The Teleostome Skull: An Attempt to Provide an Ichthyocraniological Nomenclature. Rec. Australian Mus., XV, No. 3, pp. 201–208.

1926c. Contributions to the Cranial Osteology of the Fishes. No. IV. Some Scleropareian Skulls. Rec. Australian Mus., XV, No. 3, pp. 208–232.

1926d. Contributions to the Cranial Osteology of the Fishes. No. V. A Discussion of the Maxillo-ethmoid Articulation in the Skulls of Bony Fishes. Rec. Australian Mus., XV, No. 3, pp. 233–236.

1928. Contributions to the Cranial Osteology of the Fishes. No. VI. Some Percomorph Skulls. Rec. Australian Mus., XVI, No. 7, pp. 316–345.

KIAER, JOHAN. 1924. The Downtonian Fauna of Norway. I. Anaspida. Vidensk. Skrift. I. Mat.-Naturv. Kl., No. 6, pp. 1–139, Pls. 1–14.

KINDRED, J. E. 1919. The Skull of *Amiurus.* Illinois Biol. Monogr., V, No. 1, 120 pp., 8 Pls.

1921. The Chondrocranium of *Syngnathus fuscus.* Journ. Morph., XXXV, No. 2, pp. 425–456, 14 figs.

1924. An Intermediate Stage in the Development of the Skull of *Syngnathus fuscus.* Amer. Journ. Anat., XXXIII, No. 3, pp. 421–447, 3 Pls.

KINGSBURY, B. F. 1926a. Branchiomerism and the Theory of Head Segmentation. Journ. Morph., XLII, No. 1, pp. 83–109.

1926b. On the So-called Law of Anteroposterior Development. Anat. Rec., XXXIII, No. 2, pp. 73–87.

KISHINOUYE, KAMAKICHI. 1923. Contributions to the Comparative Study of the So-called Scombroid Fishes. Journ. Coll. Agriculture, Imp. Univ. Tokyo, CIII, No. 3, pp. 293–475, Pls. 13–34.

Lanshina, T. N. 1932. Correlations in the Maxillary and Branchial Structures of Carnivorous and Plankton-feeding Fishes. [Text in Russian with French summary.] Bull. Acad. Sci. U. S. S. R., Cl. Sci. Phys.-Math., 1928 (3), pp. 253–272, 8 figs.

Longman, Heber A. 1932. A New Cretaceous Fish. Mem. Queensland Mus., X, Part 2, pp. 89–98, Pls. 10, 11, figs. 1–3.

Loomis, Frederic B. 1900. Die Anatomie und die Verwandtschaft der Ganoid- und Knochen-Fische aus der Kreide-Formation von Kansas. Palæontographica, XLVI, pp. 213–283, Pls. 19–27.

Lubosch, W. 1923. Der Kieferapparat der Scariden und die Frage der Streptognathie. Anat. Anz., LVII, pp. 10–29.

Lütken, Christian Frederik. 1864. Om en ved Sevedöi Begyndelsen 1862 opdreven "Kæmpe-Klumpfisk" (Mola nasus Raf.). Vidensk. Meddel. Naturhist. Foren. Kjobenhavn, pp. 379, 380.

———. 1871. Oneirodes eschrichtii Ltk., en ny gronlandsk Tudsefisk. Overs. Dansk. Vid. Selsk. Kjobenhavn, pp. 56–74, Pl. 2.

———. 1880a. Spolia Atlantica. Bidrag til kundskab om formforandringer hos fiske under deres vaext og udvikling, saerligt hos nogle af Atlanterhavets Hojsofiske. Dansk. Vidensk. Selsk. Skrift. Kjobenhavn, (5), XII, pp. 409–613, 5 Pls., 25 figs.

———. 1880b. Contributions pour servir à l'histoire de deux genres de poissons de la famille des baudroies, Himantolophus et Ceratias, habitant les grandes profondeurs des mers arctiques. Arch. Zool. Expér. Gén., 1879 (1880), VIII, pp. 23–26.

———. 1886. Fortsatte bidrag til kundskab om de arktiske dybhavstudsefiske, særligt, slægten Himantolophus. Dansk. Vidensk. Selsk. Skrift. Kjobenhavn, 1886–1888, pp. 323–334.

Luther, A. F. 1909a. Beiträge zur Kenntniss von Muskulatur und Skelett des Kopfes des Haies Stegostoma tigrinum Gm. und der Holocephalen, mit einem Anhang über die Nasenrinne. Acta Soc. Sci. Fenn., XXXVII, No. 6, pp. 1–60, 36 figs.

———. 1909b. Untersuchungen über die vom N. trigeminus innervierte Musculatur der Selachier (Haie und Rochen) unter Berücksichtigung ihrer Beziehungen zu benachbarten Organen. Acta Soc. Sci. Fenn., XXXVI, No. 3, pp. 1–176, 5 Pls., 23 figs.

Lydekker, Richard. 1903. (Editor.) The New Natural History, V, Section X. Fishes. Pp. 314–584. New York.

MacFarlane, J. M. 1923. The Evolution and Distribution of Fishes. Macmillan Co. New York. 8vo.

Mangold, O. 1931. Das Determinationsproblem. Dritter Teil. Das Wirbeltierauge in der Entwicklung und Regeneration. Ergebnisse d. Biologie, VII, pp. 193–403, 50 figs.

Mayhew, Roy L. 1924. The Skull of Lepidosteus platostomus. Journ. Morph., XXXVIII, No. 3, pp. 315–346, Pls. I–IV.

McCulloch, Allan R. 1921. Check-list of the Fish and Fish-like Animals of New South Wales. Australian Zoologist, II, pp. 24–68, Pls. IV–XXIV.

———. 1922. Check-list of the Fish and Fish-like Animals of New South Wales. Australian Zoologist, II, pp. 86–130, Pls. XXV–XLIII.

Meek, Seth E. and Hildebrand, Samuel F. 1923. The Marine Fishes of Panama. Part I. Field Mus. Nat. Hist., Publ. No. 215, Zool. Ser., XV, v–xi, 1–330 pp., 24 Pls.

———. 1925. The Marine Fishes of Panama. Part II. Field Mus. Nat. Hist., Publ. 226, Zool. Ser., XV, xv–xix, 331–707 pp., Pls. XXV–LXXI.

———. 1928. The Marine Fishes of Panama. Part III. Field Mus. Nat. Hist., Publ. 249, Zool. Ser., XV, xxv–xxx, 709–1045 pp., Pls. LXXII–CII.

Moodie, Roy L. 1922. The Influence of the Lateral-line System on the Peripheral Osseous Elements of Fishes and Amphibia. Journ. Comp. Neurology, XXXIV, No. 2, pp. 319–335.

Müller, Johannes. 1846. Über den Bau und die Grenzen der Ganoiden und über das natürliche System der Fische. Abhandl. d. Königl. Akad. d. Wissensch. zu Berlin, 1844 (1846). Pp. 117–216, Pls. I–VI.

———. 1848. Fossile Fische. In "Reise in den äussersten Norden und Osten Siberiens während . . ." by A. T. von Middendorff. St. Petersburg, 1847–75. 4to.

Müller, Johannes and Troschel, Franz Hermann. 1845. Horæ Ichthyologicæ. Beschreibung und Abbildung neuer Fische; die Familie Characinen. Nos. 1, 2, 40 pp., 11 Pls. Berlin. 4to.

Murray, Sir John and Hjort, Dr. Johan. 1912. The Depths of the Ocean. Macmillan and Co., Ltd. London. 8vo.

Naef, A. 1926. Notizen zur Morphologie und Stammesgeschichte der Wirbeltiere. 13. Die systematisch-morphologischen Vorstufen der Tetrapodenhand. Zool. Jahrb., XLVIII, pp. 405–456, figs. A–J.

———. 1931. Phylogenie der Tiere. Handb. d. Vererbungswiss., III, pp. 1–200, figs. 1–77.

Neal, H. V. 1898. The Problem of the Vertebrate Head. Journ. Comp. Neurology, VIII, No. 3, pp. 153–161, 2 figs.

Nichols, John Treadwell, and Murphy, Robert Cushman. 1916. Long Island Fauna. IV. The Sharks. Brooklyn Mus. Sci. Bull., III, No. 1, pp. 1–34, 19 figs.

Nichols, John Treadwell, and Griscom, Ludlow. 1917. Fresh-water Fishes of the Congo Basin Obtained by the American Museum Congo Expedition, 1909–1915. Bull. Amer. Mus. Nat. Hist., XXXVII, Art. 25, pp. 653–756, Pls. LXIV–LXXXIII, figs. 1–31, 3 maps.

Nichols, J. T. and Breder, C. M. 1927. The Marine Fishes of New York and Southern New England. Zoologica, IX, No. 1, 192 pp., 263 figs.

———. 1928. An Annotated List of the Synentognathi with Remarks on their Development and Relationships. Zoologica, VIII, No. 7, Dept. Tropical Research Contrib. No. 283, pp. 423–448.

Norman, J. R. 1926. The Development of the Chondrocranium of the Eel (Anguilla vulgaris). Philos. Trans. Roy. Soc. Lond., (B), CCXIV, pp. 369–464, 56 figs.

———. 1929. The Teleostean Fishes of the Family Chiasmodontidæ. Ann. Mag. Nat. Hist., (10), III, pp. 529–544.

———. 1931. A History of Fishes. London. Ernest Benn Limited. 8vo.

Nusbaum-Hilarowicz, Joseph. 1923. Études d'anatomie comparée sur les Poissons provenant des Campagnes scientifiques de S. A. S. le Prince de Monaco (Deuxième partie). Résultats des Campagnes Scientifiques Accomplis sur son Yacht par Albert I, Prince Souverain de Monaco, LXV, pp. 1–100, Pls. I–XII.

Pander, C. H. 1858. Die Ctenodipterinen des devonischen Systems. St. Petersburg. Pp. i–viii, 1–65, 4to.

Parker, T. J. 1886. Studies in New Zealand Ichthyology. I. On the Skeleton of Regalecus argenteus. Trans. Zool. Soc. Lond., XII, pp. 5–33. 5 Pls.

PARKER, W. K. 1873a. On the Structure and Development of the Skull of the Salmon (*Salmo salar*, L.). Bakerian Lecture.
  Philos. Trans. Roy. Soc. Lond., CLXIII, Part 1, pp. 95–145, Pls. I–VIII.
  1873b. On the Development of the Face in the Sturgeon (*Acipenser sturio*). Monthly Micros. Journ., IX, pp. 254–257,
  Pl. 20.
  1881. On the Structure and Development of the Skull in Sturgeons (*Acipenser ruthenus* and *A. sturio*). Proc. Roy. Soc.
  Lond., XXXII, pp. 142–145.—Philos. Trans. Roy. Soc. Lond., 1883, CLXXIII, pp. 139–185, Pls. XII–XVIII.
  1882. On the Development of the Skull in *Lepidosteus osseus*. Proc. Roy. Soc. Lond., XXXIII, pp. 107–112.—Philos.
  Trans. Roy. Soc. Lond., 1883, CLXXIII, pp. 443–492, Pls. XXX–XXXVIII.
  1884. On the Skeleton of the Marsipobranch Fishes.—Part II. *Petromyzon*. Philos. Trans. Roy. Soc. Lond., CLXXIV,
  pp. 373–409; 411–457, Pls. VIII–XVII.
PARR, ALBERT E. 1927a. The Stomiatoid Fishes of the Suborder Gymnophotodermi (Astronesthidæ, Melanostomiatidæ,
  Idiacanthidæ) with a Complete Review of the Species. Bull. Bingham Oceanographic Collection, III, Art. 2, 123 pp., 62 figs.
  1927b. Ceratioidea. Bull. Bingham Oceanographic Collection, III, Art. 1, 34 pp., 13 figs.
  1927c. On the Functions and Morphology of the Postclavicular Apparatus in *Spheroides* and *Chilomycterus*. Zoologica, IX,
  No. 5, pp. 245–269, figs. 285–293.
  1929. A Contribution to the Osteology and Classification of the Orders Iniomi and Xenoberyces with Description of a New
  Genus and Species of the Family Scopelarchidæ from the Western Coast of Mexico; and Some Notes on the Visceral Anatomy,
  of Rondeletia. Occasional Papers of the Bingham Oceanographic Collection, Peabody Mus. Nat. Hist., Yale Univ., No. 2,
  45 pp., 19 figs.
  1930a. On the Probable Identity, Life-history and Anatomy of the Free-living and Attached Males of the Ceratioid Fishes.
  Copeia, No. 4, Dec. 31, pp. 129–135.
  1930b. A Note on the Classification of the Stomiatoid Fishes. Copeia, No. 4, Dec. 31, p. 136.
  1930c. On the Osteology and Classification of the Pediculate Fishes of the Genera *Aceratias*, *Rhynchoceratias*, *Haplophryne*,
  Lævoceratias, *Allector* and *Lipactis;* with Taxonomic and Osteological Description of *Rhynchoceratias longipinnis*, New
  Species, and a Special Discussion of the Rostral Structures of the Aceratiidæ. Occasional Papers of the Bingham Oceano-
  graphic Collection, Peabody Mus. Nat. Hist., Yale Univ., No. 3, 23 pp., 6 figs.
  1932. On a Deep-sea Devilfish from New England Waters and the Peculiar Life and Looks of Its Kind. Bull. Boston Soc.
  Nat. Hist., No. 63, April, 1932, pp. 3–16, figs. 1–4.
PATTEN, WILLIAM. 1912. The Evolution of the Vertebrates and Their Kin. Philadelphia. 8vo. xvi, 486 pp., 309 figs.
PEHRSON, TORSTEN. 1922. Some Points in the Cranial Development of Teleostomian Fishes. Acta Zool., III, pp. 1–63.
POLLARD, H. B. 1895. The Suspension of the Jaws in Fish. Anat. Anz., X, pp. 17–25, 5 figs.
REGAN, C. TATE. 1902. On the Classification of the Fishes of the Suborder Plectognathi; with Notes Descriptive of New
  Species from Specimens in the British Museum Collection. Proc. Zool. Soc. London, II, pp. 284–303, Pls. XXIV, XXV,
  figs. 56–59.
  1903. On the Systematic Position and Classification of the Gadoid or Anacanthine Fishes. Ann. Mag. Nat. Hist., (7), XI,
  pp. 459–466.
  1904. The Phylogeny of the Teleostomi. Ann. Mag. Nat. Hist., (7), XIII, pp. 329–349, Pl. VII.
  1905. A Revision of the Fishes of the South American Cichlid Genera *Acara*, *Nannacara*, *Acaropsis* and *Astronotus*. Ann.
  Mag. Nat. Hist., (7), XV, No. 88, pp. 329–347.
  1906a. A Classification of the Selachian Fishes. Proc. Zool. Soc. London, 1906, pp. 722–758, figs. 115–124.
  1906b. Pisces. In Biologia Centrali-Americana (1906–1908), xxxii, 203 pp., Pls. I–XXVI, 2 maps.
  1907a. On the Anatomy, Classification and Systematic Position of the Teleostean Fishes of the Suborder Allotriognathi.
  Proc. Zool. Soc. London, II, pp. 634–643, figs. 166–171.
  1907b. Descriptions of the Teleostean Fish *Velifer hypselopterus* and of a New Species of the Genus *Velifer*. Proc. Zool.
  Soc., 1907, Oct. 8, pp. 633, 634.
  1909a. The Classification of Teleostean Fishes. Ann. Mag. Nat. Hist., (8), III, pp. 75–86.
  1909b. On the Anatomy and Classification of the Scombroid Fishes. Ann. Mag. Nat. Hist., (8), III, pp. 66–75, 1 fig.
  1910a. Notes on the Classification of the Teleostean Fishes. Proc. Seventh International Zool. Congress, Boston Meeting,
  August 19–24, 1907 (1910). 16 pp.
  1910b. The Origin of the Chimæroid Fishes. Proc. Seventh International Congress, Boston Meeting, August 19–24, 1907
  (1910). 2 pp.
  1911a. The Anatomy and Classification of the Teleostean Fishes of the Order Iniomi. Ann. Mag. Nat. Hist., (8), VII, pp.
  120–133, 7 figs.
  1911b. The Anatomy and Classification of the Teleostean Fishes of the Order Salmopercæ. Ann. Mag. Nat. Hist., (8), VII,
  pp. 294–296, 1 fig.
  1911c. The Osteology and Classification of the Gobioid Fishes. Ann. Mag. Nat. Hist., (8), VIII, pp. 729–733, 2 figs.
  1911d. The Osteology and Classification of the Teleostean Fishes of the Order Microcyprini. Ann. Mag. Nat. Hist., (8),
  VII, pp. 320–327, Pl. VIII.
  1911e. The Anatomy and Classification of the Teleostean Fishes of the Orders Berycomorphi and Xenoberyces. Ann. Mag.
  Nat. Hist., (8), VII, pp. 1–9, Pl. I, 2 figs.
  1911f. The Classification of the Teleostean Fishes of the Order Ostariophysi.—Siluroidea. Ann. Mag. Nat. Hist., (8),
  VIII, pp. 553–577, 3 figs.
  1911g. On the Cirrhitiform Percoids. Ann. Mag. Nat. Hist., (8), VII, pp. 259–262.
  1911h. The Classification of the Teleostean Fishes of the Order Synentognathi. Ann. Mag. Nat. Hist., (8), VII, pp. 327–334,
  Pl. IX, 1 fig.

1912a. The Osteology and Classification of the Teleostean Fishes of the Order Apodes. Ann. Mag. Nat. Hist., (8), X, pp. 377–387.

1912b. The Anatomy and Classification of the Teleostean Fishes of the Order Lyomeri. Ann. Mag. Nat. Hist., (8), X, pp. 347–349.

1912c. The Anatomy and Classification of the Symbranchoid Eels. Ann. Mag. Nat. Hist., (8), IX, pp. 387–390, Pl. IX.

1912d. The Osteology of the Teleostean Fishes of the Order Opisthomi. Ann. Mag. Nat. Hist., (8), IX, pp. 217–219.

1912e. The Classification of the Blennioid Fishes. Ann. and Mag. Nat. Hist., (8), X, pp. 265–280, 4 figs.

1912f. The Classification of the Teleostean Fishes of the Order Pediculati. Ann. Mag. Nat. Hist., (8), IX, pp. 277–289.

1912g. The Anatomy and Classification of the Teleostean Fishes of the Order Discocephali. Ann. Mag. Nat. Hist., (8), X, pp. 634–637, 2 figs.

1913a. Classification of the Percoid Fishes. Ann. Mag. Nat. Hist., (8), XII, pp. 111–145.

1913b. The Osteology and Classification of the Teleostean Fishes of the Order Scleroparei. Ann. Mag. Nat. Hist., (8), XI, pp. 169–184, 5 figs.

1914. The Antarctic Fishes of the Scottish National Antarctic Expedition. Trans. Roy. Soc. Edinburgh, XLIX, Part II (No. 2), pp. 229–292, Pls. I–XI.

1916. The British Fishes of the Subfamily Clupeinæ and Related Species in Other Seas. Ann. Mag. Nat. Hist., (8), XVIII, pp. 1–19, Pls. I–III.

1923a. The Skeleton of Lepidosteus with Remarks on the Origin and Evolution of the Lower Neopterygian Fishes. Proc. Zool. Soc. London, June, pp. 445–461, 8 figs.

1923b. The Classification of the Stomiatoid Fishes. Ann. Mag. Nat. Hist., (9), XI, pp. 612–614.

1924. The Morphology of a Rare Oceanic Fish, Stylophorus chordatus Shaw; Based on Specimens Collected in the Atlantic by the "Dana" Expeditions, 1920–1922. Proc. Roy. Soc. London, (B), XCVI pp. 193–207, 12 figs.

1925a. The Fishes of the genus Gigantura, A. Brauer; Based on Specimens Collected in the Atlantic by the "Dana" Expeditions, 1920–1922. Ann. Mag. Nat. Hist., (9), XV, pp. 53–59, Pl. VII.

1925b. New Ceratoid Fishes from the North Atlantic, the Caribbean Sea and the Gulf of Panama Collected by the "Dana." Ann. Mag. Nat. Hist., (9), XV, pp. 561–567.

1926. The Pediculate Fishes of the Suborder Ceratioidea. In "The Danish 'Dana' Expeditions, 1920–1922, in the North Atlantic and Gulf of Panama." Oceanogr. Repts., 1926, No. 2, 45 pps., 13 pls., 27 figs.

1929. Fishes. Encycl. Brit. (14), IX, pp. 305–328.

REGAN, C. TATE AND TREWAVAS, E. 1929. The Fishes of the Families Astronesthidæ and Chauliodontidæ. The Danish "Dana" Expeditions, 1920–1922. Oceanogr. Repts., No. 5, 39 pp., 7 pls., 25 figs.

1930. The Fishes of the Families Stomiatidæ and Malacosteidæ. The Danish "Dana" Expeditions, 1920–1922. Oceanogr. Repts., No. 6, 143 pp., 14 pls., 138 figs.

REINHARDT, JOHAN THEODOR. 1852. Om svömmeblæren hos familien Gymnotini. Arch. f. Naturgesch., XX Jahrg., pp. 169–184.

REIS, OTTO M. 1896. Ueber Acanthodes bronni Agassiz. Morphol. Arbeiten, Jena, VI, pp. 143–220, 2 pls., 3 figs.

RENDAHL, HJALMAR. Pegasiden-Studien. Arkiv f. Zoologi, K. Svenska Vetenskaps akad., Bd. 21 A, No. 27, 56 pp., 15 figs.

RICHARDSON, (SIR) JOHN. 1844. Fishes. In "The Zoology of the Voyage of H. M. S. 'Erebus and Terror,' under the command of Capt. Sir J. C. Ross . . . during . . . 1839–43." Ed. J. Richardson and J.Æ. Gray. II. 39 pls. London. 4to.

RIDEWOOD, W. G. 1904a. On the Cranial Osteology of the Fishes of the Families Elopidæ and Albulidæ, with Remarks on the Morphology of the Skull in the Lower Teleostean Fishes Generally. Proc. Zool. Soc. London, II, pp. 35–81.

1904b. On the Cranial Osteology of the Clupeoid Fishes. Proc. Zool. Soc. London, II, pp. 448–493, figs. 118–143.

1904c. On the Cranial Osteology of the Fishes of the Families Mormyridæ, Notopteridæ and Hyodontidæ. Journ. Linnean Soc. London, Zoology, XXIX (1903–1906), pp. 188–217, Pls. XXII–XXV.

1905a. On the Cranial Osteology of the Fishes of the Families Osteoglossidæ, Pantodontidæ, and Phractolæmidæ. Journ. Linnean Soc. London, Zoology, XXIX (1903–1906), pp. 252–282, Pls. XXX–XXXII.

1905b. On the Skull of Gonorhynchus Greyi. Ann. Mag. Nat. Hist., (7), XV, pp. 361, 372, Pl. XVI.

1913. Notes on the South American Freshwater Flying-fish, Gastropelecus and the Common Flying-fish, Exocoetus. Ann. Mag. Nat. Hist., (8), XII, pp. 544–548, Pl. XVI.

RITTER, WILLIAM E. 1929. The Nutritional Activities of the California Woodpecker (Balanosphyra formiscivora). Quart. Rev. Biol., IV, No. 4, pp. 455–483.

1930. Is Man a Rational Animal? Human Biol., II, No. 4, pp. 457–472.

ROULE, LOUIS. 1926. Les Poissons et le Monde Vivant des Eaux. Études Ichthyologiques. Tome Premier. Les Formes et les Attitudes. Paris. Librairie Delagrave. 355 pp., 16 Pls., 50 figs.

1929. Présentation d'un Squelette de Lampris luna. Bull. Mus. Hist. Nat. Paris, (2), I, 133 pp.

RUCKES, HERBERT. 1929a. Studies in Chelonian Osteology. Part I. Truss and Arch Analogies in Chelonian Pelves. Ann. N. Y. Acad. Sci., XXXI, pp. 31–80, Pls. IV–VII.

1929b. Studies in Chelonian Osteology. Part II. The Morphological Relationships between the Girdles, Ribs and Carapace. Ann. N. Y. Acad. Sci., XXXI, pp. 81–120.

RUGE, GEORG. 1897. Ueber das Peripherische Gebiet des Nervus Facialis bei Wirbelthieren. Festschr. zum Siebenzigsten Geburtstage von Carl Gegenbaur. Leipzig. 4to.

SAGEMEHL, M. 1884. Beiträge zur vergleichenden Anatomie der Fische. I. Das Cranium von Amia calva L. Morphol. Jahrb., IX, pp. 177–228, Pl. X.

1885. Beiträge zur vergleichenden Anatomie der Fische. III. Das Cranium der Characiniden nebst allgemeinen Bemerkungen über die mit einem Weber'schen Apparat versehenen Physostomen-familien. Morphol. Jahrb., X, 119 pp., 2 pls.

1887.   Die accessorischen Branchialorgane von *Citharinus*. Morphol. Jahrb., XII, pp. 307–324, Pl. XVIII.
1891.   Beiträge zur vergleichenden Anatomie der Fische. IV. Das Cranium der Cyprinoiden. Morphol. Jahrb., XVII, pp. 489–595, Pls. XXXVIII, XXXIX.
SCHMALHAUSEN, J. J. 1923a.   Der Suspensorialapparat der Fische und das Problem der Gehörknöchelchen. Anat. Anz., LVI, Nr. 1–24, pp. 534–543.
1923b.   Über die Autostylie der Dipnoi und der Tetrapoda. Anat. Anz., LVI, Nr. 1–24, pp. 543–550.
SEWERTZOFF, A. N. 1899.   Die Entwickelung des Selachierschädels: Ein Beitrag zur Theorie der Korrelativen Entwickelung. Festschrift zum Siebenzigsten Geburtstag von Carl von Kuppfer, pp. 281–320, Pls. 29–31, 4 figs.
1923a.   The Place of the Cartilaginous Ganoids in the System and the Evolution of the Osteichthyes. Journ. Morph., XXXVIII, No. 1, pp. 105–145, 4 figs.
1923b.   Die Morphologie des Visceralapparates der Elasmobranchier. (Vorläufige Mitteilung.) Anat. Anz., LVI, Nr. 1–24, pp. 389–410.
1925.   Beiträge zu einer Theorie des Knochenschädels der Wirbeltiere. Anat. Anz., LX, (1926), pp. 427–443, 15 figs.
1927.   Études sur l'evolution des Vertébrés inférieurs. Structure primitive de l'appareil viscéral des Elasmobranches. Pubbl. d. Stazione Zool. d. Napoli, VIII, Fasc. 3–4, pp. 475–554, Pls. XXIX–XXXI.
1928.   The Head Skeleton and Muscles of *Acipenser ruthenus*. Acta Zool., IX, pp. 193–319, 9 pls.
1931.   Morphologische Gesetzmässigkeiten der Evolution. Gustav Fischer. Jena. xv, 372 pp., 21 diagr., 131 figs. Gr. 8vo.
SIEBOLD, PHILIPP FRANZ VON. 1842.   Fauna Japonica, sive descriptio animalium quae in itinere per Japoniam suscepto annis 1823–30 collegit, notis, observationibus et adumbrationibus illustravit P. F. de Siebold. Conjunctis studiis C. J. Temminck, et H. Schlegel pro vertebratis atque W. de Haan pro invertebratis elaborata. 6 vols. in 4. Lugduni Batavorum, 1833–50 Pisces, p. 323. 160 pls.
SIEBOLD, CARL THEODOR ERNST VON, AND STANNIUS, F. H. 1854–1856.   Handbuch der Zootomie. Theil II. Die Wirbelthiere, von H. Stannius. Berlin. 8vo.
SPEMANN, H. 1927.   Neue Arbeiten über Organisatoren in der tierischen Entwicklung. Die Naturw. Wochenschr. f. d. Fortschr. d. Reinen u. d. Angewandten Naturw., XV, pp. 946–951.
STANNIUS, HERMANN. 1854–1856.   Handlb. d. Anatomie d. Wirbelthiere. In "Handbuch der Zootomie" (with Siebold, v.), Zweiter Theil: Die Wirbelthiere. Erster Buch: Die Fische. 279 pp. Berlin. 8vo.
STARKS, E. C. 1898a.   The Osteological Characters of the Genus *Sebastolobus*. Proc. California Acad. Sci., (3), I, No. 11, pp. 361–370, 3 pls.
1898b.   The Osteology and Relationships of the Family Zeidæ. Proc. U. S. Nat. Mus., XXI, pp. 469–476, Pls. XXXIII–XXXVIII.
1899a.   The Osteology and Relationship of the Percoidean Fish, *Dinolestes lewini*. Proc. U. S. Nat. Mus., XXII, pp. 113–120, Pls. VIII–XI.
1899b.   The Osteological Characters of the Fishes of the Suborder Percesoces. Proc. U. S. Nat. Mus., XXII, pp. 1–10, Pls. I–III.
1901.   Synonymy of the Fish Skeleton. Proc. Washington Acad. Sci., III, pp. 507–539, Pls. LXIII–LXV, figs. 45, 46.
1902a.   The Shoulder-girdle and Characteristic Osteology of the Hemibranchiate Fishes. Proc. U. S. Mus., XXV, pp. 619–634, figs. 1–6.
1902b.   The Relationship and Osteology of the Caproid Fishes or Antigoniidæ. Proc. U. S. Nat. Mus., XXV, pp. 565–572, figs. 1, 2.
1904a.   A Synopsis of Characters of Some Fishes Belonging to the Order Haplomi. Biol. Bull., VII, No. 5, pp. 254–262.
1904b.   The Osteology of Some Berycoid Fishes. Proc. U. S. Nat. Mus., XXVII, pp. 601–619, figs. 1–10.
1904c.   The Osteology of *Dallia pectoralis*. Zool. Jahrb., XXI, Heft 3, pp. 249–262, 2 figs.
1905.   The Osteology of *Caularchus mæandricus* (Girard). Biol. Bull., IX, No. 5, pp. 292–303, 2 figs. [Cling-fishes.]
1907.   On the Relationship of the Fishes of the Family Siganidæ. Biol. Bull., XIII, No. 4, pp. 211–218, 1 fig.
1908a.   The Characters of *Atelaxia*, a New Suborder of Fishes. Bull. Mus. Comp. Zool., LII, No. 2, pp. 1–22, 5 pls. [*Stylephorus*.]
1908b.   On a Communication between the Air-bladder and the Ear in Certain Spiny-rayed Fishes. Science, N.S., XXVIII, No. 732, pp. 613, 614.
1908c.   On the Orbitosphenoid in Some Fishes. Science, N.S., XXVIII, No. 717, pp. 413–415.
1909.   The Scombroid Fishes. Science, N.S., XXX, No. 773, pp. 572–574.
1910.   The Osteology and Mutual Relationships of the Fishes Belonging to the Family Scombridæ. Journ. Morph., XXI, No. 1, pp. 77–99, Pls. I–III.
1911a.   Osteology of Certain Scombroid Fishes. Leland Stanford Junior Univ. Publ., Univ. Ser., No. 5, pp. 1–49, 2 pls.
1911b.   A Possible Line of Descent of the Gobioid Fishes. Science, N.S., XXXIII, pp. 747–748.
1916.   The Sesamoid Articular, a Bone in the Mandible of Fishes. Leland Stanford Junior Univ. Publ., pp. 1–40, 15 figs.
1923.   The Osteology and Relationships of the Uranoscopoid Fishes. Stanford Univ. Publ., Biol. Sci., III, No. 3, pp. 261–290, Pls. I–V.
1926a.   Bones of the Ethmoid Region of the Fish Skull. Stanford Univ. Publ., Biol. Sci., IV, No. 3, pp. 139–388, 58 figs.
1926b.   Factors of Fish Classification. Amer. Naturalist, LX, Jan.–Feb., pp. 82–94.
1930.   The Primary Shoulder Girdle of the Bony Fishes. Stanford Univ. Publ., Biol. Sci., VI, No. 2, pp. 149–239, 38 figs.
STEENSTRUP, JAPETUS AND LÜTKEN, CHR. 1898–1901.   Spolia Atlantica. Bidrag til Kundskab om Klumpeller Maanefiskene (Molidæ). Det Kongelige Danske Vidensk. Selsk. Skrift., Niende Bind, 102 pp., 4 pls.

STENSIÖ, ERIK A: SON. 1921. Triassic Fishes from Spitzbergen. Part I. Vienna. 4to. xxviii, 307 pps., 87 figs., 35 pls.
    1922. Notes on Certain Crossopterygians. Proc. Zool. Soc. London, Dec., pp. 1241–1271, Pl. I.
    1925a. Triassic Fishes from Spitzbergen. Part II. Kungl. Svenska Vetenskaps-akad. Handl., (3), II, No. 1, 261 pp., 58 figs., 34 pls.
    1925b. On the Head of the Macropetalichthyids, with Certain Remarks on the Head of the Other Arthrodires. Field Mus. Nat. Hist., Publ. 232, Geol. Ser., IV, No. 4, pp. 87–197, Pls. XIX–XXXI.
    1927. The Downtonian and Devonian Vertebrates of Spitsbergen. Part I. Family Cephalaspidæ. A. Text. B. Plates. Skrifter om Svalbard og Nordishavet. Resultater av de Norske Statsunderstottede Spitsbergen-ekspeditioner. No. 12. Det Norske Videnskaps-akad. i Oslo. xii, 391 pp., 103 figs., 112 pls.
STETSON, H. C. 1931. Studies on the Morphology of the Heterostraci. Journ. Geol., XXXIX, No. 2, pp. 141–154.
STOCKARD, CHARLES R. 1923. The Significance of Modifications in Body Structure. Harvey Society Lectures, 1921–1922. J. B. Lippincott Co., Philadelphia. [Experiments on double-monsters, etc., in fishes.]
    1931. The Physical Basis of Personality. New York. 8vo. 320 pp., 73 figs.
SWINNERTON, H. H. 1902. A Contribution to the Morphology of the Teleostean Head Skeleton Based upon a Study of the Developing Skull of the Three-spined Stickleback (Gasterosteus aculeatus). Quart. Journ. Micros. Sci., XLV, pp. 503–593. 4 pls., 5 figs.
    1903. Osteology of Cromeria nilotica and Galaxias attenuatus. Zool. Jahrb. Abt. f. Anat. u. Ont., XVIII, Heft 1, pp. 58–70. 15 figs.
    1905. A Contribution to the Morphology and Development of the Pectoral Skeleton of Teleosteans. Quart. Journ. Micros. Sci., XLIX, pp. 363–385, 1 pl., 3 figs.
TAKAHASI, NISUKE. 1925. On the Homology of the Cranial Muscles of the Cypriniform Fishes. Journ. Morph. Physiol., XL, No. 1, 109 pp., 3 pls., 20 figs.
THILO, OTTO. 1920. Das Maulspitzen der Fische: Das Entstehen und Vergehen seiner Mechanik. Biologisches Zentralblatt, XL, pp. 216–238.
THOMPSON, D'ARCY W. 1917. On Growth and Form. Univ. Press, Cambridge (Eng.). 8vo. [Body-form of Orthagoriscus mola, p. 751.]
THYNG, F. W. 1906. Squamosal Bone in Tetrapodous Vertebrata. Proc. Boston Soc. Nat. Hist., XXXII, 1904–1906, pp. 387–425, Pls. XXXIX–XLII.
TRAQUAIR, R. H. 1875. On the Structure and Systematic Position of the Genus Cheirolepis. Ann. Mag. Nat. Hist., (4), XV, No. 88, pp. 237–249.
    1877. The Ganoid Fishes of the British Carboniferous Formations. Part I. No. 1. Palæoniscidæ. Monogr. Palæontogr. Soc., 60 pp., 8 pls.
    1879. On the Structure and Affinities of the Platysomidæ. Trans. Roy. Soc. Edinburgh, XXIX, pp. 343–391, Pls. III–VI.
    1901. The Ganoid Fishes of the British Carboniferous Formations. Part I. No. 2. Palæoniscidæ. Monogr. Palæontogr. Soc., pp. 61–87, Pls. VIII–XVIII.
    1907. The Ganoid Fishes of the British Carboniferous Formations. Part I, No. 3. Palæoniscidæ. Monogr. Palæontogr. Soc., LXI, pp. 87–106, Pls. XIX–XXIII.
    1909. The Ganoid Fishes of the British Carboniferous Formations. Part I, No. 4. Palæoniscidæ. Monogr. Palæontogr. Soc., LXIII, pp. 107–122, Pls. XXIV–XXX.
    1911. The Ganoid Fishes of the British Carboniferous Formations. Part I, No. 5. Palæoniscidæ. Monogr. Palæontogr. Soc., pp. 123–158, Pls. XXXI–XXXV.
    1912. The Ganoid Fishes of the British Carboniferous Formations. Part I, No. 6. Palæoniscidæ. Monogr. Palæontogr. Soc., pp. 159–180, Pls. XXVI–XL.
TROTTER, E. S. 1926. Brotulid Fishes. Zoologica (New York), VIII, No. 3, pp. 107–125, figs. 29–33, Pl. C.
TURNER, C. L. 1921. Food of the Common Ohio Darters. Ohio Journ. Sci., XXII (1921–1922), No. 2, pp. 41–62.
TWITTY, VICTOR C. AND SCHWIND, JOSEPH L. 1931. The Growth of Eyes and Limbs Transplanted Heteroplastically between Two Species of Amblystoma. Journ. Exper. Zool., LIX, No. 1, pp. 61–86.
UHLMANN, EDUARD. 1921. Studien zur Kenntnis des Schädels von Cyclopterus lumpus L. 1 Teil: Morphogenese des Schädels. 2 Teil: Entstehung des Schädelknochen. Jena. Zeitschr. f. Naturw., LVII, pp. 275–370, Pls. XVI, XVII, 62 figs.
VAILLANT, LÉON. 1905. Le Genre Alabes de Cuvier. Comptes Rend., CXL, pp. 1713–1715. [On the affinities of Alabes with blennioids rather than with symbranchioid eels.]
VETTER, BENJAMIN. 1874. Untersuchungen zur vergleichenden Anatomie der Kiemen- und Kiefermusculatur der Fische. I. Jena. Zeitschr. f. Naturw., VIII, Neue Folge, I, pp. 405–458, Pls. 14, 15.
    1878. Untersuchungen zur vergleichenden Anatomie der Kiemen- und Kiefermusculatur der Fische. II. Jena. Zeitschr. f. Naturw., XII, Neue Folge, V, pp. 431–550.
VROLIK, A. J. 1873. Studien über die Verknöcherung und die Knochen des Schädels der Teleostei. Niederländisches Archiv f. Zool., I, Heft 3, pp. 219–318, Pls. 18–22.
WAITE, EDGAR R. 1902. Skeleton of Luvarus imperialis, Rafinesque (A Fish New to the Western Pacific Fauna). Rec. Australian Mus., IV, No. 7, pp. 292–297, Pls. 45, 46.
    1906. Descriptions of and Notes on Some Australian and Tasmanian Fishes. Rec. Australian Mus., VI, 1905–1907, pp. 194–210, Pls. 34–36.
WALTER, EMIL. 1913. Unsere Süsswasserfische. Eine Übersicht über die heimische fisch fauna nach norwiegend biologischen und fischereiwirtschaftlichen Gesichtspunkten. Leipsig. 52 pp., 50 pls. [Good figures of freshwater fishes.]
WATSON, D. M. S. 1921. On the Coelacanth Fish. Ann. Mag. Nat. Hist., (9), VIII, pp. 320–337, 5 figs.
    1925. The Structure of Certain Palæoniscids and the Relationships of that Group with Other Bony Fish. Proc. Zool. Soc. London, Part 3, No. 54, pp. 815–870.

1926. The Evolution and Origin of the Amphibia. Croonian Lecture. Philos. Trans. Roy. Soc. London, (B), CCXIV,
        pp. 189–257, 39 figs.
1928. On Some Points in the Structure of Palæoniscid and Allied Fish. Proc. Zool. Soc. London, Part 1, No. 4, pp. 49–70,
        15 figs.
WATSON, D. M. S. AND DAY, HENRY. 1916. Notes on Some Palæozoic Fishes. Manchester Memoirs, LX, No. 2, 52 pp.,
        3 Pls.
WATSON, D. M. S. AND GILL, E. L. 1923. The Structure of Certain Palæzoic Dipnoi. Journ. Linnaean Soc. London.—Zool.,
        XXXV, pp. 163–216, 34 figs.
WEBER, MAX AND DE BEAUFORT, L. F. 1922. The Fishes of the Indo-Australian Archipelago. IV. Heteromi, Solenichthyes,
        Synentognathi, Percesoces, Labyrinthici, Microcyprini. Leiden. 8vo.
WHITE, E. GRACE. 1918. The Origin of the Electric Organs in *Astroscopus guttatus*. Carn. Inst. Washington, Publ. No. 252,
        pp. 139–172.
WILLEY, ARTHUR. 1894. *Amphioxus* and the Ancestry of the Vertebrates. Columbia Univ. Biol. Ser., II. Macmillan and
        Co. New York. 8vo.
WILLIAMS, STEPHEN R. 1902. Changes Accompanying the Migration of the Eye and Observations on the Tractus opticus
        and Tectum opticum in *Pseudopleuronectes americanus*. Bull. Mus. Comp. Zool. at Harvard Coll., XL, No. 1, pp. 1–57,
        5 Pls., figs. A–F.
WILLISTON, SAMUEL WENDALL. 1914. Water Reptiles of the Past and Present. Univ. Chicago Press. vi, 251 pp., 131 figs.
        8vo. [Progressive reduction of skull elements in vertebrates.]
WOODWARD, A. S. 1891. Catalogue of the Fossil Fishes in the British Museum (Natural History). Part II. Containing the
        Elasmobranchii (Acanthodii), Holocephali, Ichthyodorulites, Ostracodermi, Dipnoi, and Teleostomi (Crossopterygii and
        Chondrostean Actinopterygii). Printed by Order of the Trustees. xliv, 567 pp., Pls. 1–16.
1895. Catalogue of the Fossil Fishes in the British Museum (Natural History). Part III. Containing the Actinopterygian
        Teleostomi of the Orders Chondrostei (concluded), Protospondyli, Ætheospondyli and Isospondyli (in part). Printed by
        Order of the Trustees. xl, 534 pp., 18 pls.
1896. On Some Extinct Fishes of the Teleostean Family Gonorhynchidæ. Proc. Zool. Soc. London, pp. 500–504, Pl. XVIII.
1897. A Contribution to the Osteology of the Mesozoic Amioid Fishes, *Caturus* and *Osteorachis*. Ann. Mag. Nat. Hist.,
        (6), XIX, No. 30, pp. 292–297, Pls. VIII–XI.
1898. The Antiquity of the Deep-sea Fish-Fauna. Natural Science, XII, pp. 257–260.
1901. Catalogue of the Fossil Fishes in the British Museum (Natural History). Part IV. Containing the Actinopterygian
        Teleostomi of the Suborders Isospondyli (in part), Ostariophysi, Apodes, Percesoces, Hemibranchii, Acanthopterygii and
        Anacanthini. Printed by Order of the Trustees. xxxviii, 636 pp., 19 pls.
1902. The Fossil Fishes of the English Chalk. Part I. Palæontographical Soc., LVI, 56 pp., 13 pls.
1903. The Fossil Fishes of the English Chalk. Part II. Palæontographical Soc., LVII, pp. 57–96, Pls. XIV–XX.
1907. The Fossil Fishes of the English Chalk. Part III. Palæontographical Soc., LXI, pp. 97–128, Pls. XXI–XXVI.
1908. The Fossil Fishes of the English Chalk. Part IV. Palæontographical Soc., LXII, pp. 129–152, Pls. XXVII–XXXII.
1909. The Fossil Fishes of the English Chalk. Part V. Palæontographical Soc., LXIII, pp. 153–184, Pls. XXXIII–
        XXXVIII.
1911. The Fossil Fishes of the English Chalk. Part VI. Palæontographical Soc., LXIV, pp. 185–224, Pls. XXXIX–XLVI.
1912. The Fossil Fishes of the English Chalk. Part VII. Palæontographical Soc., LXV, pp. 225–264, Pls. XLVII–LIV.
1916. The Fossil Fishes of the English Wealden and Purbeck Formations. Part I. Palæontographical Soc., 1915 (1916),
        LXIX, pp. 1–48, Pls. I–X.
1917. The Fossil Fishes of the English Wealden and Purbeck Formations. Part II. Palæontographical Soc., 1916 (1917),
        LXX, pp. 49–104, Pls. XI–XX.
1918. The Fossil Fishes of the English Wealden and Purbeck Formations. Part III. Palæontographical Soc., 1917 (1918),
        LXXI, pp. i–xiii, 105–148, Pls. XXI–XXVI.
1920. The Dentition of *Climaxodus*. Quart. Journ. Geol. Soc., LXXV, Part 1, pp. 1–6, Pl. I.
1928. President's Address. Dorset Natural History and Antiquarian Field Club. Pp. 1–14. Friary Press, Dorchester.
        1928.
WOSKOBOJNIKOFF, M. M. 1914. Études sur la Branchiomérie des Vertébrés. Parts I, II, III (en russe). Journ. d. Soc. des
        natur. russes de Kieff. Kieff.
YOUNG, R. T. 1923. Origin of the Notochord in Chordates. Anat. Rec., XXV, No. 5, pp. 289–290.
ZUGMAYER, E. 1911. Poissons Provenant des Campagnes du Yacht Princesse Alice (1901, 1910). Résult. Campagn. Scient.,
        Monaco, XXXV. 159 pp., 6 pls., 48 figs.

INDEX

www.ingramcontent.com/pod-product-compliance
Lightning Source LLC
Chambersburg PA
CBHW081340190326
41458CB00018B/6057